THE SURVIVAL OF VEGETATIVE MICROBES

IE SURVIVAL OF VEGETATIVE MICROBES: twenty-sixth symposium of the Society for General Microbiology held at Lady Mitchell Hall, the University of Cambridge, April 1976, ed. by T. R. G. Gray and J. R. Postgate. Cambridge, 1976. 432p ill bibl index (Symposia — Society for General Microbiology, 26) 75-31399. 29.50. ISBN 0-521-21094-1. C.I.P.

CHOICE *NOV. '76*

Biology

The aim of this symposium volume is to explore the fundamental processes of death and survival when physical and chemical stresses are put upon microorganisms. This book is the *first* definitive edition on the survival of vegetative bacteria. It is composed of 14 extensive reviews written by invited specialists in their various fields of expertise. The authors have surveyed fundamental aspects of the effect of stress upon viability, including nutrient deprivation, freezing, desiccation, osmotic stress, ultraviolet and ionizing radiations, heat, hydrostatic pressure, and chemical stress. These reviews have not been previously published and are, on the whole, well written. For those libraries subscribing to this series, purchase will be automatic. For others, if they maintain a book section on bacteriology or microbiology, this volume is highly recommended.

312

Other Publications of the
*Society for General Microbiology**

THE JOURNAL OF GENERAL MICROBIOLOGY
THE JOURNAL OF GENERAL VIROLOGY

SYMPOSIA

1 THE NATURE OF THE BACTERIAL SURFACE
2 THE NATURE OF VIRUS MULTIPLICATION
3 ADAPTATION IN MICRO-ORGANISMS
4 AUTOTROPHIC MICRO-ORGANISMS
5 MECHANISMS OF MICROBIAL PATHOGENICITY
6 BACTERIAL ANATOMY
7 MICROBIAL ECOLOGY
8 THE STRATEGY OF CHEMOTHERAPY
9 VIRUS GROWTH AND VARIATION
10 MICROBIAL GENETICS
11 MICROBIAL REACTION TO ENVIRONMENT
12 MICROBIAL CLASSIFICATION
13 SYMBIOTIC ASSOCIATIONS
14 MICROBIAL BEHAVIOUR, 'IN VIVO' AND 'IN VITRO'
15 FUNCTION AND STRUCTURE IN MICRO-ORGANISMS
16 BIOCHEMICAL STUDIES OF ANTIMICROBIAL DRUGS
17 AIRBORNE MICROBES
18 THE MOLECULAR BIOLOGY OF VIRUSES
19 MICROBIAL GROWTH
20 ORGANIZATION AND CONTROL IN PROKARYOTIC AND EUKARYOTIC CELLS
21 MICROBES AND BIOLOGICAL PRODUCTIVITY
22 MICROBIAL PATHOGENICITY IN MAN AND ANIMALS
23 MICROBIAL DIFFERENTIATION
24 EVOLUTION IN THE MICROBIAL WORLD
25 CONTROL PROCESSES IN VIRUS MULTIPLICATION

* Published by the Cambridge University Press, except for the first Symposium, which was published by Blackwell's Scientific Publications Limited.

THE SURVIVAL OF VEGETATIVE MICROBES

EDITED BY

T. R. G. GRAY AND J. R. POSTGATE

TWENTY-SIXTH SYMPOSIUM OF THE SOCIETY FOR GENERAL MICROBIOLOGY HELD AT LADY MITCHELL HALL THE UNIVERSITY OF CAMBRIDGE APRIL 1976

Published for the Society for General Microbiology

CAMBRIDGE UNIVERSITY PRESS

CAMBRIDGE

LONDON · NEW YORK · MELBOURNE

Published by the Syndics of the Cambridge University Press
The Pitt Building, Trumpington Street, Cambridge CB2 1RP
Bentley House, 200 Euston Road, London NW1 2DB
32 East 57th Street, New York, NY 10022, USA
296 Beaconsfield Parade, Middle Park, Melbourne 3206, Australia

Library of Congress Cataloguing in Publication Data

Main entry under title:
The Survival of vegetative microbes.
(Symposia – Society for General Microbiology; 26)
Includes Index.
1. Micro-organisms – Physiology – Congresses.
2. Stress (Physiology) – Congresses. 3. Adaption (Physiology) – Congresses.
I. Gray, Timothy R. G. II. Postgate, John Raymond.
III. Society for General Microbiology.
IV. Series: Society for General Microbiology. Symposium; 26.
QR1. S6233 no. 26 [QR97. A1] 576'.08s

ISBN 0 521 21094 1 [589.9'05'22] 75–31399

First published 1976

Printed in Great Britain
at the
University Printing House, Cambridge
(Euan Phillips, University Printer)

CONTRIBUTORS

BRIDGES, B. A., MRC Cell Mutation Unit, University of Sussex, Falmar, Brighton BN1 9QG, Sussex.

CALCOTT, P. H., Department of Microbiology, Macdonald Campus, McGill University, Montreal, Quebec, Canada.

COX, C. S., Microbiological Research Establishment, Porton Down, Salisbury SP4 0JG, Wiltshire.

DAWES, E. A., Department of Biochemistry, University of Hull, Hull HU6 7RX, Humberside.

GRAY, T. R. G., Hartley Botanical Laboratories, University of Liverpool, PO Box 147, Liverpool L69 3BX, Merseyside.

HUGO, W. B., Department of Pharmacy, University of Nottingham, Nottingham NG7 2RD, Nottinghamshire.

KRINSKY, N. I., Department of Biochemistry and Pharmacology, Tufts University School of Medicine, 136 Harrison Avenue, Boston, Massachusetts 02111, USA.

MACLEOD, R. A., Department of Microbiology, Macdonald Campus, McGill University, Montreal, Quebec, Canada.

MORITA, R. Y., Department of Microbiology and School of Oceanography, Oregon State University, Corvallis, Oregon 97331, USA.

POSTGATE, J. R., ARC Unit of Nitrogen Fixation, The Chemical Laboratory, University of Sussex, Brighton BN1 9QJ, Sussex.

ROSE, A. H., Zymology Laboratory School of Biological Sciences, University of Bath, Claverton Down, Bath BA2 7AY, Avon.

SMITH, H., Department of Microbiology, University of Birmingham, PO Box 363, Birmingham B15 2TT, West Midlands.

STACEY, K. A., Biological Laboratory, The University of Kent, Canterbury CT2 7NJ, Kent.

STRANGE, R. E., Microbiological Research Establishment, Porton Down, Salisbury SP4 0JG, Wiltshire.

THURSTON, C. F., Department of Microbiology, Atkins Building, Queen Elizabeth College, Campden Hill, London W8 7AH.

TRINCI, A. P. J., Department of Microbiology, Atkins Building, Queen Elizabeth College, Campden Hill, London W8 7AH.

WELKER, N. E., Department of Biochemistry and Molecular Biology, Northwestern University, Evanston, Illinois 60201, USA.

CONTENTS

EDITORS' PREFACE

A scientific visitor from another planet, studying the state of Earth's science, might well conclude that a taboo exists among microbiologists regarding the death of bacteria. Pick almost any standard microbiology textbook and you will find discussed growth, multiplication, variation, transformation of the chemical environment and all the other varied activities which make this group of creatures so fascinating. But you will seek in vain for a discussion of their death; if it is mentioned at all, it will be almost furtively – 'the phase of decline' – and probably dismissively, as if there is nothing more to be said except that they die. Only in discussions of sterilization and chemotherapy will the death of microbes become respectable – and then from the anthropomorphic viewpoint of killing the beasties outright, rather than the microbomorphic view of studying their survival.

In this Symposium the Society has taken a step towards redressing this imbalance for, at a more serious level, it must be obvious that we need, as microbiologists, to know as much about the fundamentals of the death of microbes as we do about their life. Death is the final, irreversible response to a variety of stresses, natural or imposed by man, which microbes encounter in the biosphere; in every aspect of agriculture and industry, medicine and research – even in our own interiors – the viability of the microbial population involved is of great, sometimes critical, importance. So we have assembled specialists, whose work may only have bordered on the subject of the death of microbes, and asked them to survey fundamental aspects of their specialization which bear particularly on the mechanisms and character of the survival of microbes. We are aware that a complementary symposium, organized by the Society for Applied Bacteriology last July, covering more practical aspects of microbial survival, is to be published, so we restricted our programme to more academic matters. But the coverage of even this restricted scope is incomplete: we are conscious that pH as a distinct stress has been inadequately discussed; survival of viruses is scarcely mentioned (we were advised, and surprised to discover, that very little is known of this seemingly very important topic); exotic stresses of interest in the space programme have been neglected, so have complex stresses such as predation; the emphasis is regrettably heavy on bacteria, which is wrong for our Society of *general* microbiologists; resistant forms – cysts and spores – have been almost completely excluded from the discussion. Availability and willingness of contributors determined some of these omissions, but the decision to exclude resistant forms was a deliberate one, since these have been treated in considerable detail in other meetings and could provide a symposium topic on their own. It is the survival of the vegetative forms which have been most neglected by microbiologists over the years.

Accepting these omissions, we are confident that contributors have covered a very substantial part of this subject and, in doing so, have illustrated how badly this area of microbiology needed bringing together. Perhaps this symposium is a step in the direction of breaking the curious reluctance of microbiologists to discuss the bacterial way of death.

Hartley Botanical Laboratories, University of Liverpool, T. R. G. GRAY
P.O. Box 147, *Liverpool L*69 3*BX*
ARC Unit of Nitrogen Fixation J. R. POSTGATE
*University of Sussex, Brighton, BN*1 9*QJ*

DEATH IN MACROBES AND MICROBES

J. R. POSTGATE
University of Sussex, Brighton BN1 9QJ, Sussex

DEATH IN HIGHER ORGANISMS

The term 'death' presents the same semantic problems as the term 'life': neither is capable of explicit definition and the scientist is obliged, in order to avoid becoming entangled in word play, to achieve some degree of agreement on the usage of these terms. At a coarse level, the death of a complex, differentiated macro-organism is normally a self-evident event, but even the layman, in these days of organ transplants and tissue cultures, recognizes that the death of a complex macrobe is a very different event from the death of its component cells. In the absence of a catastrophe, the death of a complete macrobe precedes the death of most of its component cells and is the result of a loss of effective interaction between these cells; the component cells themselves die later, if for a similar reason. Cells can be cultured from higher organisms and, in those circumstances, their behaviour has some analogy to populations of microbes in laboratory culture. Such analogies must not be taken too far, however, because differentiation in higher organisms seems to carry the consequence that normal somatic cells are 'committed' to death: they are capable of only a limited number of divisions unless they become neoplastic (Hayflick, 1965) or, in plants, lose their ability to re-differentiate (see Street, 1975). Indeed, in some insects such as *Drosophila*, the somatic cells of the animal's body undergo their final division as adulthood is reached, and do not divide again.

As a generalization, one can say that macro-organisms and their component cells are both intrinsically mortal: death is a normal, inevitable part of their biology. Like most generalizations, this statement is faulty when applied in detail. Creatures from lower phyla, both plant and animal, have powers of regeneration which, as far as we know, permit something approaching immortality if they are regularly chopped into pieces and allowed to regenerate. The organs or other differentiated portions of more complex organisms can survive the death of the whole creature and, since techniques for transplantation are now well developed, fragments of one individual can, even after his death, become living, functioning parts of another. This situation is familiar and useful to gardeners, but disconcerting when applied to men – the question of the instant of death then becomes a question of legal rather than scientific definition. For present purposes the message is that the term death has an entirely different sense in such contexts from that which will be useful to this Symposium.

DEATH IN DIFFERENTIATED MICROBES

Death has been held to be a consequence of the separation of the 'germ plasm' from the somatic plasm or 'soma' (Weissman, 1885) the differentiated macro-organism being viewed as a sort of transient temporal appendage of the germ plasm, which is in a sense 'immortal'. While this proposition is philosophically interesting, it is not of profound scientific value because the separation of germ plasm and soma need not be intercellular. Among unicellular microbes situations exist in which ageing and death are an intrinsic part of the life cycle of each individual cell (see Trinci & Thurston, this volume). An example is the budding yeasts, where a scar distinguishes the mother from the daughter cell and a restriction seems to exist which prevents budding at a scarred area. Thus multiple scarring indicates increasing age; a commitment to death of the individual occurs because organisms eventually become completely scarred and therefore incapable of further multiplication (Liebelova, Beran & Striebelova, 1964).

In such eukaryotic systems, as in filamentous moulds and algae, ageing and death of individuals or of parts of filamentous individuals, can be observed, at least in principle. Studies can be made of the physiology of ageing and death in such organisms and those, apart from their intrinsic scientific value, can be held to bear analogy to comparable studies in higher organisms.

DEATH IN BUDDING BACTERIA

It is pertinent to discuss whether ageing and a consequent commitment to death can be expected in prokaryotes at all. Caulobacteria (see Dow & Whittenbury, 1975) have a two-phase life cycle represented by swarmer cells which develop into a non-flagellate stalked form which is sessile and buds off new swarmer cells, adding a length to its stalk each time (Fig. 1). The swarmer cells are juvenile forms, siblings derived from a stalked mother, and the mothers can be distinguished in age according to the number of crossbands in their stalk (Staley & Jordan, 1973). That in Fig. 1 is four divisions old; Staley & Jordan

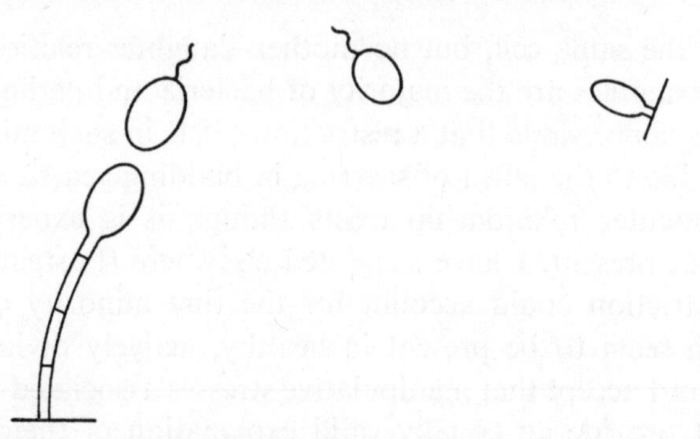

Fig. 1. Sketch of the life cycle of *Caulobacter*. The mother is depicted as four divisions old. The three swarmer offspring increase in age from left to right, the oldest having become sessile and formed a stalk.

illustrated one 18 divisions old. Clearly possibilities for senescence leading to a 'natural' death exist here, but at present there is no evidence that the mothers are other than immortal: there is no known restriction on the number of daughters a mother may produce nor on the number of crossbands in her filament (Professor R. Whittenbury, personal communication). Similar considerations apply to *Rhodopseudomonas palustris* in which mother and daughter cells are distinguishable on a morphological basis but there seems to be no restriction on the multiplication of the older forms. In contrast, *Rhodomicrobium vannielli* appears to be subject to such a restriction (Dow & Whittenbury, 1975). In its branching phase it buds off successively no more than four offspring (a number equal to the number of spores a branched form can produce). Thus a parent with four bud scars is senescent and, presumably, committed to death. It seems, therefore, that the budding bacteria provide transitional forms between those microbes, such as budding yeasts, which show senescence and death on the macrobiological pattern, and the fissile microbes which, as discussed below, probably do not. It is relevant to note, however, that the facts that organisms are capable of differentiation, or that they can show distinctions in age (observable as parent–offspring relationships) do not necessarily carry with them a commitment to senescence and death. In this context, more information on the mortality of elderly budding microbes is obviously needed.

DEATH IN FISSILE MICRO-ORGANISMS

Consider the sort of fissile, haploid, vegetative microbes with which this Symposium is mainly concerned. Here the germ plasm and soma are within the same cell, but no mother–daughter relationship can be detected; examples are the majority of bacteria and perhaps the fission yeasts. It is conceivable that a restriction exists in such microbes which is comparable to the effect of scarring in budding yeasts, so that a real mother–daughter relationship exists though it is experimentally inaccessible at present. I have suggested elsewhere (Postgate, 1967) that such a restriction could account for the tiny minority of non-viable cells which seem to be present in healthy, actively dividing bacterial cultures, but I accept that manipulative stresses associated with viability assessment provide an equally valid explanation of their appearance. At present the limited evidence available indicates that fissile microbes give rise to progeny of equal youth when they divide. The classical experiments of Meselson & Stahl (1958), who followed the segregation of [^{15}N]DNA among progeny of labelled *Escherichia coli*, established the semi-conservative character of DNA replication. It follows that each daughter cell has equivalent amounts of 'young' and 'old' DNA, so there is no reason to suspect other than random distribution among daughter cells of any old versus newly synthesized material. The evidence for localized growth points on the cell envelopes of *E. coli* and possibly other bacteria (see Donachie, Jones & Teather, 1973) implies that some portions of the envelope may be very much older than the individual carrying them, but in principle it seems that the progeny of binary fission do not differ in age even at the molecular level.

To summarize: fissile, vegetative microbes do not grow old and die, they vanish. The parent is replaced by two equally young individuals. No equivalent can be observed of the commitment to death found in whole macrobes, in their component cells and in certain budding microbes, so no 'natural' process of ageing and death can be studied.

Yet such microbes do die if cultures are left to age, or if other circumstances arrest multiplication. The basic difference between fissile microbes and organisms with more complex mechanisms of multiplication is that, in the former, death only results from some environmental stress. The investigator may often be only partly aware, or completely unaware, of what stress may be, and considerable research effort may be necessary to analyse the stresses imposed on moribund bacterial populations and to decide which ones are critical. As an example of this situation, I shall take our studies in starvation (Postgate & Hunter, 1962*a*; see also Dawes's contribution to this volume), but first it is necessary to define more clearly the terms 'death' and 'viability' as applied to fissile unicellular microbes.

TERMINOLOGY

At present one must accept that the death of a microbe can only be discovered retrospectively: a population is exposed to a recovery medium, incubated, and those individuals which do not divide to form progeny are taken to be dead. Whether they were dead at the time they were exposed to the recovery medium or whether they died on it is a second question, which can sometimes be answered by subsidiary experiments (as in 'substrate-accelerated death' or 'metabolic injury', q.v.); the important point is that there exist at present no short cuts which would permit assessment of the moment of death: vital staining, optical effects, leakage of indicator substances and so on are not of general validity (see Postgate, 1967, 1969). The ratio of the number of dead organisms, so described, to the total number of organisms is the viability. The term 'viability' applies to populations, not to individuals (except in an all or none sense: an individual is either viable or non-viable). The term 'viability' is sometimes used in another sense: an individual clone or population is sometimes described as more viable than another in the sense that it is more resistant to stress, more metabolically active or capable of more rapid or extensive multiplication. Though this usage is legitimate in ordinary English, I recommend that it be avoided* in the context of this Symposium: 'vigour' or 'vitality' are possible appropriate homonyms.

This description of death as applied to fissile microbes is an operational one. It has the obvious defect that it involves an element of opinion, albeit informed opinion: the investigator chooses a recovery medium on the basis of the best available information and, of course, he may be wrong. Subsequent experiments may show a choice of recovery medium to be misguided – this Symposium will include examples – but in principle it is impossible to avoid an element of microbiologists' intuition contributing to the choice of recovery environment. In addition, it is usual to assess viability after a standard time, yet one can never be completely confident that a few 'dead' organisms would not have multiplied had incubation been extended for a few more hours or days. In practice, this objection is often trivial because conditions can be devised in which the proportion of such slow-starting organisms, or of filaments capable of fragmentation into viable progeny, is

* The term 'survival', the maintenance of viability in adverse circumstances, need not suffer from this linguistic restriction. Survival of individual microbes has a clear meaning, distinct from (and usually incompatible with) survival of a clone or population. In many areas of microbiology only clonal survival is normally susceptible to experimentation, as the contribution of Harry Smith illustrates.

negligibly small; but the question can be important, for example, in slide culture (Postgate, Crumpton & Hunter, 1961; Strange & Shon, 1964), where the number of doubling times for which a slide chamber can be incubated is limited because colonies may overgrow bacterial corpses.

The description of death used here also begs a question concerning a state which vegetative bacteria can assume that has some analogy to the senescent state of eukaryotes. As a result of certain stresses, bacteria can lose the ability to multiply but remain biologically completely functional as individuals. An example familiar to me arose with populations of *Klebsiella aerogenes* which, after starvation for 24 hours, reached a viability of 20 per cent yet over 80 per cent of the population had an intact osmotic barrier, with a complete amino acid pool, and were able to respond to a mild osmotic stress (Postgate & Hunter, 1962*a*). A proportion of the 'dead' organisms after freezing and thawing often shows a similar biological integrity (Postgate & Hunter, 1963*b*). Are such organisms any more dead than, say, a woman past menopause? For most practical purposes they must be regarded as dead, but the existence of such moribund populations raises interesting questions about the nature (and possible reversibility) of their pseudosenescent state; it also emphasizes, if emphasis were necessary, that death, like life, can be described and qualified, but not defined.

VIABILITY AND STRESS

Accepting the pragmatic view that (*a*) conventional viability assessment truly assesses the proportion of dead organisms and (*b*) that all bacterial death is 'unnatural' in the sense of being a response to stress, the question arises what stresses can cause death. The tautologous answer is, of course, any stress which prevents multiplication, because this protracts the existence of each individual organism and sooner or later something nasty will happen to each one – even a lyophilized bacterial population in a dark refrigerator will die over the years, though the death rate must be very slow indeed. Microbes normally suffer a variety of mild stresses, several of which are specifically dealt with in this volume, and it is usually the case that one stress presupposes, or entrains, several others. In our studies of starvation (Postgate & Hunter, 1962*a*), we set about investigating the survival of *K. aerogenes* populations starved in buffer at their growth pH and temperature. Our strategy at the outset was carefully, if somewhat naively, planned. To ensure reproducibility in the behaviour of our populations, we used a glycerol-limited chemostat culture which had run for 150 or so doublings, the intention being to 'get over' any mutations or phenotypic variations which transfer to continuous culture might have induced. We obtained evidence that such changes had indeed occurred, and that they continued, though on a much diminished scale, throughout the four years for which the master chemostat was run. We also standardized our handling procedure most carefully. The survival patterns of our organisms were, happily, very reproducible (though a long-term trend to a faster death rate occurred over the four years) but it emerged clearly that transfer to warm buffer followed by incubation involved at least five stresses. These were:

(1) The overt stress of ageing a bacterial population in the non-nutrient environment.

(2) A stress arising from Mg^{2+} deprivation, expressed as a decreased death rate consequent on adding Mg^{2+} to the buffer.

(3) A stress due to Cu^{2+} present as a hitherto unsuspected contaminant of the laboratory glass-distilled water. Cu^{2+} was present at about 2×10^{-7} M; this was one-fiftieth of the minimum concentration required to inhibit growth of a culture yet it accelerated death markedly. (In practice the effect was overcome by adding a small amount of a chelating agent, which neutralized the toxic effect of the Cu^{2+}, so this stress did not contribute to the published data.)

(4) A stress which arose from centrifuging and resuspending the bacteria. To avoid carry-over of salts, the bacteria were suspended in distilled water before exposure to the starvation environment; this mild osmotic stress committed them to a faster death rate than if buffer had been used for the final re-suspension, though their viability at the time starvation was initiated was unchanged.

(5) A pH stress – arising from the discovery that the optimal pH for survival in non-nutrient conditions (about 6) was not that used for growth (about 7).

The message here is that, since the death of such populations involves stress, death in ostensibly mild conditions may be a consequence of a variety of stresses, some of which may not be immediately obvious even to the experienced investigator. The proposition that manipulation will pre-dispose a population to accelerated death in otherwise mild stress conditions is fundamental to studies of microbial survival and even extends to other aspects of microbial behaviour. An impressive example is the demonstration by Quesnel (1963) that the 'shock' of transfer of *E. coli* from an overnight culture to a warm membrane supplied with nutrient induced a wide distribution of individual doubling times which damped out in the second generation, though it was in this generation that the highest proportion of non-viable progeny was observed.

THE DEATH CURVE

The conventional microbiology or bacteriology textbook will often divide the history of a laboratory bacterial culture into a number of phases: the lag phase (when growth of each cell occurs without multiplication), the phase of logarithmic multiplication, the stationary phase (in which the viable population remains unchanged but multiplication occurs) and the phase of decline or death (when the viable population decreases). Though the more positive aspects of this sequence – mobilization of resources in the lag phase followed by multiplication* – have been subject to considerable research and discussion, even whole textbooks, the phase of decline has been curiously neglected. Perhaps there is something melancholy about a preoccupation with death, even that of bacteria. Loss of viability has been dismissed as due to nutrient deprivation, occupancy of all available biological space, accumulation of toxic products, adverse pH and so on, often with rather little evidence of an experimental nature. Death is often held to be a logarithmic process, but again this is probably true of only a minority of the initial population: inspection of many published curves illustrating the 'logarithmic death' revealed to the author that survivals were almost always plotted logarithmically, the first point being taken when more than 90 per cent of the population had died. While legitimate in their contexts, such plots probably give an illusory impression of the physiology of death of that first 90 per cent, which is, after all the majority of the population. In our starvation studies, a technique of viability assessment was used which was accurate to ± 3 per cent and, though the 'tails' of our death curves were often logarithmic, the first 80 per cent of the populations showed survival curves of a marvellous variety of shapes – linear, sigmoid, probit, concave – depending on the nutritional status of the population. It is not unreasonable to expect that similar survival curves would occur in the phase of decline of batch cultures which had entered the stationary phase because of exhaustion of a single nutrient, though in many instances

* I avoid the term 'growth' most carefully, since microbiologists use it in two senses which readily become confusing in this context. Growth of a microbial population (i.e. increase in number of individuals) is preceded by growth (i.e. increase in mass) of the individuals comprising that population, but the two growth processes are quite distinct and in fact alternate.

the survival pattern would then be complicated by cryptic growth (see below). Though it is reasonable to suppose that death from a single, isolated stress might occur logarithmically, it is safest to assume that the classical 'phase of decline' might take almost any form depending on the multiplicity of stresses to which the population is responding.

Hinshelwood (1951, 1957; see also Dean & Hinshelwood, 1966) was one of the few scientists to study and speculate on the death of bacterial cultures. He pointed out that, while a logarithmic decay curve would arise if death were a single hit process resulting from a chance event unaffected by the cell's previous history, logarithmic curves did not necessarily imply that single hit processes were operating. He regarded lysis and other degradative processes in bacteria as the consequence of unchecked reactions which were normally balanced by biosyntheses, and he deduced equations, analogous to Volterra's which describe fluctuations of interacting microbial populations, which had periodic solutions, implying that the balance of lytic and synthetic processes fluctuated. In the phase of decline, he considered that a conjunction of such fluctuations, themselves random, could form a lethal event, and he showed that the distribution of such events throughout a population would be logarithmic. Hinshelwood's treatment left open the analysis of stresses used here. A crucial stage in his argument is illustrated in this passage from Hinshelwood (1957) (my italics): '*When growth is interrupted* and the cells enter a non-proliferating phase the nicely balanced co-ordination which characterized the phase of exponential increase is upset'. The opening phrase of that sentence implies that his treatment is valid for whatever reason growth was interrupted. Yet if one considers four common stresses which might interrupt growth of a culture (anoxia, carbon-energy source deprivation, trace metal deprivation or pH stress) one sees that, as far as we know, they bring about death by very different processes and the survival curves, where they are known, have very different forms. Hinshelwood's speculations are a valuable caution against too rigid acceptance of the logarithmic death curve and its simple interpretation, but their applicability in specific instances depends on the detailed physiology of the microbe's response to whatever stress is dominant, and this is rarely known.

RELEVANCE TO DESIGN OF EXPERIMENTS

Accepting that death usually results from a variety of simultaneous stresses, or a conjunction of several stresses, what is one actually studying when one assesses viability of a microbial population after starvation, freezing and thawing, irradiation and so on? When the stress is a strong one – heating, disinfection or prolonged γ-irradiation – one can be confident that one is destroying almost the whole biological apparatus – proteins, lipids, nucleic acids – whose integrity is essential to life. With the milder stresses, however, the molecular architecture of the cell is receiving a much milder buffeting and co-operative effects may occur in which the overt stress is supplemented by unintentional stresses, the latter sometimes becoming dominant. The stresses involved in starvation provide an example in which much of the information obtained proved to refer primarily to magnesium deprivation. Nevertheless, it was possible to obtain a great deal of self-consistent data on the behaviour of such populations, and on the effect of nutritional status and of environmental conditions (pH, aeration, temperature etc.) on survival. It was also possible to obtain descriptive information on the physiological processes leading to death by starvation in the strain of *K. aerogenes* used: for example, degradation of RNA seemed to entrain a critical event leading to death, and carbohydrate reserves seemed to postpone that event. Such matters are discussed more exhaustively in Dawes's chapter. In addition one could plan, on the basis of such data, experiments in which continuous cultures were run so slowly that the death rate made a serious contribution to their dynamics: steady states were successfully obtained in which a constant proportion of the population was dead (Postgate & Hunter, 1962*a*). Such studies were developed further by Tempest, Herbert & Phipps (1967) and probably have ecological relevance (Postgate, 1973).

Most biological scientists are familiar with the situation thus exemplified: that the phenomenon one is ostensibly studying proves to be compounded of a number of other phenomena of which one was only partly aware if at all. It is a happy circumstance for the advance of science, if at times an irritating one; a special case of the general proposition that research opens up more questions than it answers. Stresses are not the only unwanted factors which can influence the study of survival; examples will appear elsewhere in this Symposium, but a few factors, reasonably obscure until they were understood, are discussed briefly below as possible traps for the unwary scientist.

The population effect

This phenomenon, so named by Harrison (1960), can be readily observed when a population of bacteria is starved. Dense populations are killed to a lesser extent than sparse populations. This is not simply that there are fewer microbes to kill, so the whole population dies sooner; the *specific* death rate of dense populations (below a certain maximum) is slower than that of sparse ones. Population effects of a comparable kind can be observed in freeze-thawed (Postgate & Hunter, unpublished), heat-stressed (Strange & Shon, 1964) and cold-shocked (Strange & Dark, 1962) *K. aerogenes*. There is evidence that a population effect can be observed in moribund chemostat populations: Postgate & Hunter (1963*a*) showed that the steady state viability of slow-growing Mg^{2+}-limited *K. aerogenes* was significantly greater the denser the population. In some circumstances cryptic growth (see below) can simulate a population effect, but in freezing damage or starvation (Postgate & Hunter, 1963*a*), cryptic growth can be excluded. Population effects of this kind generally imply threshold phenomena: either that substances leak from dead organisms which, above a threshold concentration, can protect surviving neighbours, or that toxic materials are present which a population can neutralize above a threshold density. In special circumstances a more precise explanation may be possible. Brown & Howitt (1969) related the population effect in oxygen-killing of *E. coli* to carbon dioxide deprivation; Strange & Dark (1962) and Strange & Shon (1964) demonstrated leakage of protective materials in cold shock and heat shock respectively. For the design of survival experiments it is obvious that, in studying any stress that could be subject to a population effect, populations of roughly similar densities should be compared. In addition, interpretation of experiments must take into account whatever stress (or alleviation of stress) causes the population effect in the special circumstances being studied.

Cryptic growth

This phenomenon, so named by Ryan (1959) and later given names such as 'cell turnover', 'regrowth' or 'cannibalism' (see Postgate, 1967), occurs because dying bacteria leak all sorts of metabolizable small molecules which can not only protect neighbours from stress but actually permit their multiplication. Postgate & Hunter (1962*a*) observed that death by starvation of 50 individuals of their glycerol-limited *K. aerogenes* was necessary to permit doubling of one survivor

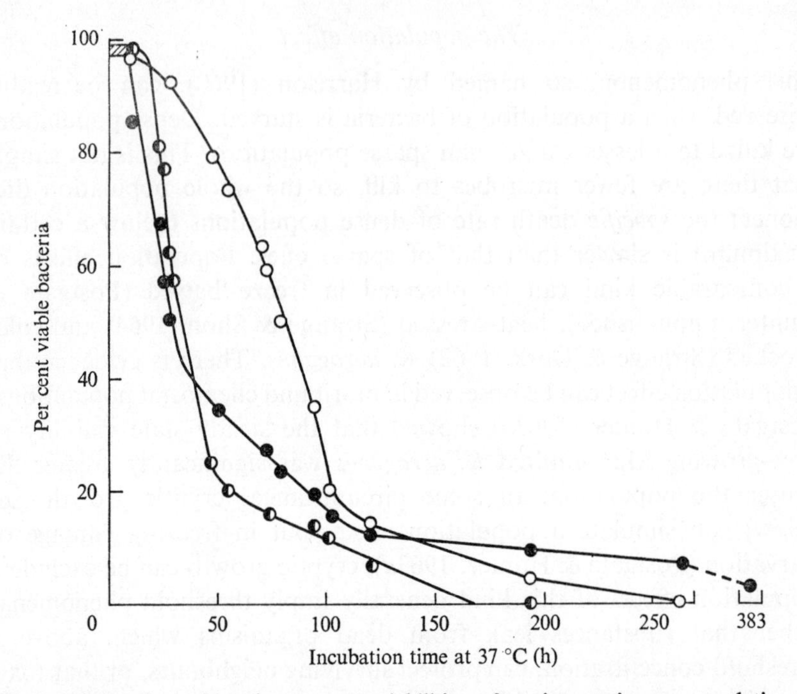

Fig. 2. Effect of nutritional status on viabilities of ageing stationary population of *K. aerogenes* (after Postgate & Hunter, 1963*a*). Flasks shaken at 37 °C in air in defined minimal media formulated to give glycerol-limited (◐), NH_4^+-limited (○) and Mg^{2+}-limited (●) stationary populations. Viability determinations by slide culture.

in non-nutrient buffer. A factor of this order of magnitude was unimportant in their experiments; it is also rare: NH_4^+-limited and Mg^{2+}-limited populations can show cryptic growth factors in the region of 3 to 5. If a population is allowed to age in a nutrient medium limited by a component, such as Mg^{2+}, which can be re-cycled, cryptic growth ratios considerably less than one can be obtained as generation after generation of organisms multiplies, dies and passes on its Mg^{2+} to survivors; Fig. 2 illustrates the survival characters of flask cultures of *K. aerogenes* in various defined media: the Mg^{2+}-limited population died most rapidly but achieved what amounted to a steady state of low viability based on persistent cryptic growth. I discussed implications of cryptic growth elsewhere (Postgate, 1967) because these ramify beyond the study of survival and have a bearing on the study of such topics as maintenance coefficients; for present purposes the message is that, in survival studies in which cryptic growth is possible, it should be expected to be the rule rather than the exception. Experimental stratagems are available to assess or avoid it; if it is not taken into account, the population at the end of stresses such as irradiation, starvation, chilling or mild heating, may not consist of the same individuals as it contained at the start. Physiological data and viabilities will then be effectively uninterpretable.

Substrate-accelerated death

This is a phenomenon whereby growth substrates can, in special circumstances, accelerate the death of populations in a non-nutrient environment. Essentially the conditions are these. The traumatic substrate is most usually either the carbon source whose exhaustion limited growth in the medium from which the bacterial population was taken, or a metabolite which lies on the degradation pathway of that substrate. For death to be accelerated, a traumatic substrate must also be present in both the non-nutrient environment and the recovery medium. Substrate-accelerated death is discussed further in the contribution of Dawes; for present purposes it provides an example of a situation in which a recovery medium which may at first sight seem most suitable – that enriched with the growth-limiting substrate – proves to be the worst possible for the assessment of the viability of a certain class of starved populations. It follows that populations exposed to any stress which involves leakage (freeze-thawing, starvation, irradiation, drying and dehydration) are likely to encounter a traumatic substrate and might then give an aberrant viability on a casually chosen complete medium. As a different example, taken from personal experience, growth substrates leached from a commercial brand of membrane filter were able to accelerate markedly the death of *K. aerogenes* starved in warm buffer. Experimental stratagems such as careful choice of recovery media, use of cyclic AMP or of Mg^{2+}, are available to avoid substrate-accelerated death; the message here is that experimental design must take the existence of substrate-accelerated death into account.

Hypersensitivity

The contribution of MacLeod & Calcott discusses metabolic injury in which the survivors of freezing and thawing show enhanced sensitivity to the nature of the recovery medium. Death of some of a freeze-thawed population during the lag phase in the recovery media accounts in part for metabolic injury; increased sensitivity to toxic agents such as copper may also play a part. A comparable phenomenon is the observation of Klein & Wu (1974) that starvation makes organisms hypersensitive to a mild heat stress, a point against the use of ‘pour-plates' to count populations in samples of natural origin. For present purposes two general points of importance emerge. Firstly, a degree of fragility analogous to metabolic injury can result from stresses other then freeze-thawing, such as a mild pH stress (Roth & Keenan, 1971), exposure to sub-lethal amounts of disinfectants (Jacobs & Harris, 1961; Harris & Winfield, 1963) or lyophilization (see Strange & Cox, this volume). Secondly, the effect of the recovery medium is not predictable: in the case of the 'metabolic injury' which results from phenol damage, minimal media can give greater viabilities than rich media.

The message here is that survival of an overt stress will probably render microbes hypersensitive to mild stresses to which they would not normally react at all.

Cold shock and dilution shock

It would be improper for me to pre-empt too much of the contributions of MacLeod & Calcott or of Rose, but it is worth emphasizing that abrupt exposure to cold media or buffers causes cold shock, a stress which, in exponentially growing coliforms, can be lethal and which could add itself to any other stress which might be being studied. One cannot, for example, study freeze-thaw damage without exposing organisms to conditions which, in certain circumstances, would cause cold shock. Similarly, distilled water or 'isotonic' saline, sometimes used as diluents for viable counts, can cause osmotic stresses (Mager *et al.*, 1956) or strip Mg^{2+} from the cell envelope (Strange & Shon, 1964; Sato *et al.*, 1971) leading to a 'dilution shock' which can be lethal. I mentioned earlier an osmotic stress resulting from exposure to distilled water which, without killing any individuals, committed the whole population to accelerated death from starvation. Expedients to avoid these traps are self-evident once they are recognized; their recognition can be less easy.

Storage and reversal of stress

In the study of microbial survival of stress, it is necessary, each time the viability of the population is assessed, to reverse that stress, at least so far as a sample of the population is concerned. The manner in which these operations are done can sometimes be critically important: the studies of Record, Taylor & Miller (1962) on lyophilization, a complex stress involving freezing, desiccation and rehydration, showed that populations could be recovered from the dried state with almost 100 per cent viability if reconstituted in an osmotically strong environment and dialysed slowly to a more normal environment: the most traumatic event in lyophilization was the osmotic shock resulting from causal reconstitution; much of the *mystique* of lyophilization media lies in their ability to protect organisms from osmotic shock during recovery. In freeze–thawing damage, as MacLeod & Calcott point out in this volume, rates of freezing and thawing can be crucially important. In ultraviolet-damage, as Bridges's chapter discusses, exposure to visible light can reactivate seemingly non-viable cells.

In many circumstances, a period of storage occurs between the application of an overt stress, such as freezing or drying, and its reversal by thawing or rehydration. It is important to recognize that storage may permit longer term, sometimes unsuspected, stresses. An impressive example occurred with bacteria suspended in 10 per cent erythritol which survived freezing in liquid nitrogen, if this was followed closely by thawing, without loss of viability. If, however, the freezing and thawing steps were separated by a few minutes' storage at 15 °C (a typical deep freeze temperature), a 95 per cent loss of viability took place (Postgate & Hunter, 1962*b*). The reasons for this phenomenon are still obscure – I suspect devitrification of the frozen erythritol solution – but other protective solutes also gave divergent store lives, though none so dramatic as erythritol. So storage can involve unsuspected stresses, though after lyophilization, those organisms which survive the initial stress seem to last for remarkably long periods. The chapter of Strange & Cox deals with situations involving elaborate combinations of overt and storage stresses. The message here, then, is that all sorts of normal laboratory operations – illumination, cold storage, suspension in buffer – are capable of influencing survival drastically and alertness to these experimental hazards is essential.

CONCLUSION

The reader could be forgiven for concluding that research on the survival of vegetative microbes is operationally impossible: that there are too many invisible stresses and traps for meaningful experiments to be designed. Indeed, it is true that the perfect experiment on survival of a known stress is difficult to plan, because the milder the overt stress, the greater seems to be the contribution of ancillary and even unsuspected side-stresses. Can one prescribe an ideal condition for the study of survival of vegetative microbes in response to specified mild stresses? Moribund continuous cultures with a steady state viability of about 80 per cent might seem to be the best research material for meaningful work of this kind, for one can be confident that the survivors are not drastically stressed, in that they are multiplying. Perhaps students of survival should always set up such chemostats and, when they are in the steady state, apply to the population such additional stress (temperature, radiation, pH, osmotic pressure, chemical stress) as may interest them and observe its effects. Yet we know already that the nutritional status of such populations has dramatic effects on the steady state viability. So how many types of nutritional status should the operator examine? And what bearing would the information from so elaborate and invariant a system have on our knowledge of microbial behaviour and its normal (or abnormal) ecological niche? Sometimes it seems that the more one approaches a perfect experiment, the less flexible and informative does it become. There is something to be said for cocking a snook at perfectionism with an inverted adage: if an experiment is worth doing at all, it is worth doing badly. If I insult fellow contributors by suggesting that their experiments, like mine, have sometimes been less than perfect in design, they will perhaps forgive me, because the information content of this volume gives clear evidence of the effectiveness of doing the best you can when you can.

I am grateful to Professor R. Whittenbury for constructive discussion of the physiology and mortality of budding bacteria.

REFERENCES

Brown, O. R. & Howitt, H. F. (1969). Growth inhibition and death of *Escherichia coli* from CO_2 deprivation. *Microbios*, **1**, 241–246.

Dean, A. C. R. & Hinshelwood, C. N. (1966). *Growth, Function and Regulation in Bacterial Cells*. Clarendon Press: Oxford.

Donachie, W. D., Jones, N. C. & Teather, R. (1973). The bacterial cell cycle. *Symposium of the Society for General Microbiology*, **23**, 9–44. London.

Dow, C. S. & Whittenbury, R. (1975). Budding bacteria as models for the study of morphogenesis and differentiation: *Rhodomicrobium vannielli* as one example. *Bacteriological Reviews*, (in press).

Harris, N. D. & Winfield, M. (1963). A lethal effect on damaged *Escherichia coli* associated with the counting techniques. *Nature, London*, **200**, 606–607.

Harrison, A. P. (1960). The response of *Bacterium lactis aerogenes* when held at growth temperatures in the absence of nutriment: an analysis of survival curves. *Proceedings of the Royal Society of London, Series B*, **152**, 418–428.

Hayflick, L. (1965). The limited *in vitro* lifetime of human diploid cell strains. *Experimental Cell Research*, **37**, 614–636.

Hinshelwood, C. N. (1951). Decline and death of bacterial populations. *Nature, London*, **167**, 666–669.

Hinshelwood, C. N. (1957). Ageing in bacteria. In *The Biology of Ageing*, ed. W. B. Yapp & G. H. Bourne. *Symposia of the Institute of Biology No.* **6**, pp. 1–7. London: Institute of Biology.

Jacobs, S. E. & Harris, N. D. (1961). The effect of modifications in the counting medium on the viability and growth of bacteria damaged by phenols. *Journal of Applied Bacteriology*, **24**, 172–181.

Klein, D. A. & Wu, S. (1974). Stress: a factor to be considered in heterotrophic microorganisms in enumeration from aquatic environments. *Applied Microbiology*, **27**, 429–431.

Liebelova, J., Beran, K. & Striebelova, E. (1964). Fractionation of a population of *Saccharomyces cerevisiae* yeasts by centrifugation in a dextrose gradient. *Folia Microbiologica*, **9**, 205–213.

Mager, J., Kuczynski, M., Schatzberg, G. & Avi-dor, Y. (1956). Turbidity changes in bacterial suspensions in relation to osmotic pressure. *Journal of General Microbiology*, **14**, 69–75.

Meselson, M. & Stahl, F. W. (1958). The replication of DNA in Escherichia coli. *Proceedings of the National Academy of Sciences, USA*, **44**, 671–682.

Postgate, J. R. (1967). Viability measurements and the survival of microbes under minimum stress. *Advances in Microbial Physiology*, **1**, 2–23.

Postgate, J. R. (1969). Viable counts and viability. In *Methods in Microbiology*, ed. J. R. Norris & D. W. Ribbons, **1**, 611–628. London: Academic Press.

Postgate, J. R. (1973). The viability of very slow-growing populations: a model for the natural eco-system. *Bulletin from the Ecological Research Committee* (Stockholm), **17**, 287–292.

Postgate, J. R., Crumpton, J. E. & Hunter, J. R. (1961). The measurement of bacterial viability by slide culture. *Journal of General Microbiology*, **24**, 15–24.

Postgate, J. R. & Hunter, J. R. (1962*a*). The survival of starved bacteria. *Journal of General Microbiology*, **29**, 233–267; Corrigenda: (1964) *ibid.*, **34**, 473.

Postgate, J. R. & Hunter, J. R. (1962*b*). On the survival of frozen bacteria. *Journal of General Microbiology*, **26**, 367–378.

Postgate, J. R. & Hunter, J. R. (1963*a*). The survival of starved bacteria. *Journal of Applied Bacteriology*, **26**, 295–306.

Postgate, J. R. & Hunter, J. R. (1963*b*). Metabolic injury in frozen bacteria. *Journal of Applied Bacteriology*, **26**, 405–414.

Quesnel, L. B. (1963). A genealogical study of clonal development in *Escherichia coli. Journal of Applied Bacteriology*, **26**, 127–151.

Record, B. R., Taylor, E. & Miller, D. S. (1962). The survival of *Escherichia coli* on drying and rehydration. *Journal of General Microbiology*, **28**, 585–598.

Roth, L. H. & Keenan, D. (1971). Acid injury of *Escherichia coli. Canadian Journal of Microbiology*, **17**, 1005–1008.

Ryan, F. J. (1959). Bacterial mutation in a stationary phase and the question of cell turnover. *Journal of General Microbiology*, **21**, 530–549.

Sato, T., Suzuki, Y., Isaki, K. & Takahashi, H. (1971). Some features of saline-sensitive phenomenon in *Escherichia coli* – protection and recovery by magnesium ion. *Journal of General and Applied Microbiology*, **17**, 371–382.

Staley, J. T. & Jordan T. L. (1973). Crossbands of *Caulobacter crescentus* stalks serve as indication of cell age. *Nature, London*, **246**, 155–156.

Strange, R. E. & Dark, F. A. (1962), Effect of chilling on *Aerobacter aerogenes* in aqueous suspension. *Journal of General Microbiology*, **29**, 719–730.

Strange, R. E. & Shon, M. (1964). Effects of thermal stress on viability and ribonucleic acid of *Aerobacter aerogenes* in aqueous suspension. *Journal of General Microbiology*, **34**, 99–114.

STREET, H. (1975). Plant cell cultures: present and projected applications for studies in genetics. In *Genetic Manipulations with plant material*, ed. L. Ledoux, pp. 231–244. New York: Plenum Publishing Company.

TEMPEST, D. W., HERBERT, D. & PHIPPS, P. J. (1967). Studies on the growth of *Aerobacter aerogenes* at low dilution rates in a chemostat. *Microbial physiology and continuous culture, Proceedings of the Third International Symposium*, ed. E. O. Powell, C. G. T. Evans, R. E. Strange & D. W. Tempest, pp. 240–254. London: HMSO.

WEISSMAN, A. (1885). *Die Kontinuität des Keimplasmas.* Jena.

ENDOGENOUS METABOLISM AND THE SURVIVAL OF STARVED PROKARYOTES

E. A. DAWES
Department of Biochemistry, University of Hull,
Hull HU6 7RX, Humberside

INTRODUCTION

It is a fact of microbial life that the majority of bacteria can survive, sometimes for considerable periods, in the absence of nutrients. Aqueous suspensions of aerobic and facultative organisms display measurable respiratory quotients indicating the oxidation of cellular materials, presumably to provide energy and possibly to permit the resynthesis from endogenous carbon sources of essential cellular constituents which might have undergone degradation during starvation. The endogenous metabolism of a bacterial cell comprises the total metabolic reactions which occur when it is deprived of compounds or elements which may serve specifically as exogenous substrates (Dawes & Ribbons, 1962, 1964). The existence of such metabolism does not necessarily imply the existence of specialized reserves of carbon or energy within the cell and indeed, protein and RNA may be degraded by organisms whether or not they are endowed with reserves. The importance of the growth environment in determining the chemical composition of a bacterium has been admirably discussed by Herbert (1961).

The survival characteristics of starved bacteria depend upon the organism and such factors as the growth phase from which they are taken, growth rate, nutritional status, population density, biological history and the nature of the starvation environment. Attention here will be confined to conditions of minimum stress with organisms which have been subjected to the mildest possible treatment in setting up the starvation system. Such experiments usually involve either (1) suspension in buffer–saline solutions at the growth temperature; (2) continued incubation in the culture medium after growth has ceased due to exhaustion of one essential nutrient; or (3) growth at such a slow rate in continuous culture that the spontaneous death rate of the population contributes to the overall dynamics of the steady state system. Patently, these conditions of starvation are not equivalent.

The experimental problems associated with the assessment of viability need emphasis. The methodology of viable counting has been critically surveyed (Postgate & Hunter, 1962; Postgate, 1967, 1969) and the importance of proper controls to prevent the occurrence of cryptic growth in starving suspensions, which has bedevilled much published work, has also been discussed by Postgate in this volume. Measurement of endogenous metabolism is usually achieved by a combination of manometry with analyses of cellular composition (sometimes including enzymic activities) and of the suspending fluid, while isotopic labelling techniques afford information concerning turnover of macromolecular components. When no gas exchange occurs, micro-calorimetry (Forrest, 1972) can be a valuable adjunct.

Most of the earlier work involved aerobic and facultative organisms, but recently attention has been directed to anaerobes which, despite difficulties of handling for survival studies, offer some advantages since they secure their energy by substrate-level phosphorylation reactions which are generally well charted.

THE NATURE OF THE PROBLEM

Some of the pertinent questions that have been posed by research in this area are as follows. Does survival bear any direct or indirect relationship to endogenous metabolism? Is endogenous metabolism wholly catabolic or do available precursors and energy liberated in the process support some degree of macromolecular synthesis? If so, are cellular components whose loss is particularly likely to result in death selectively re-formed from dispensable materials? Does the possession of specialized reserve materials exert a sparing action on the degradation of protein and RNA, and does it confer longevity? Can death be attributed to the loss of any specific cellular constituent, or to the energetic state of the cell? As we shall see, answers can be offered to some but not all of these questions. However, it is worth noting that endogenous metabolism may be affected significantly by precisely those environmental factors which profoundly influence survival, as comparison of the observations of Postgate & Hunter (1962) and Ribbons & Dawes (1963) reveals. The existence of so many possible external influences on endogenous metabolism and survival, indeed, should caution the investigator against making generalized statements concerning these parameters.

Dawes & Ribbons (1962) proposed that endogenous metabolism fulfils certain functions in the absence of exogenous substrates. These are (1) to serve as a source of energy; (2) to provide carbon substrates for resynthesis of degraded cellular constituents; and (3) to perform special functions, such as furnishing a source of reducing power in some chemolithotrophic and phototrophic bacteria, or a source of phosphorus or sulphur in organisms that store polyphosphate or sulphur granules. The first of these functions will be considered in the following section.

ENERGETICS OF SURVIVAL

Energy of maintenance

Cellular processes, whether mechanical or chemical, require energy for their performance and unless a supply of energy is available these essential processes, which include osmotic regulation, maintenance of intracellular pH, turnover of macromolecules, and motility, will cease and the cell will die. In starvation, endogenous substrates must furnish the energy required for these activities. This 'energy of maintenance' is defined as the energy consumed for purposes other than the production of new cell material; clearly it will be required under both growth and starvation conditions, which has led to two approaches for its determination. McGrew & Mallette (1962, 1965) and Mallette (1963) added, intermittently, small quantities of an energy source (glucose) to starved *Escherichia coli* to ascertain the energy threshold necessary to sustain viability without permitting growth. Viability was prolonged by appropriate additions without an increase in turbidity, although the experimental design did not eliminate the possibility of cryptic growth and was also criticized on other counts by Postgate (1967). Marr, Nilson & Clark (1963) used carbon-limited continuous cultures of *E. coli* to show that the steady state concentration of bacteria decreased with decreasing specific growth rate as a greater proportion of the carbon source was diverted from growth and metabolized to meet the maintenance requirement. They computed that about half of the maintenance energy was required for resynthesis of protein and RNA during turnover.

Pirt (1965) reported maintenance energy requirements of about 0.076 to 0.094 g glucose g^{-1} (dry wt) h^{-1} for the aerobic metabolism of *E. coli*, *Klebsiella aerogenes* and *Klebsiella* (*Aerobacter*) *cloacae*; anaerobic growth of *K. cloacae* gave a much higher maintenance requirement of 0.473 g glucose g^{-1} (dry wt) h^{-1}, reflecting the lower energy yield from glucose under these conditions. Likewise, a marked increase in the maintenance requirement of carbon-limited, nitrogen-fixing populations of *Azotobacter chroococcum* in comparison with comparable ammonia-utilizing organisms was noted by Dalton & Postgate (1969). Stouthamer & Bettenhaussen (1973) have proposed that, in growth, much of the maintenance energy is required for stabilization of ionic composition and intracellular pH.

Evidence from several sources now suggests that rapid metabolism of endogenous substrates, which produces energy at a rate greatly in excess of that needed for maintenance, accelerates the death of starved bacteria, whereas prolonged viability is associated with a low rate of endogenous metabolism more nearly matched to the provision of maintenance energy requirements (Thomas & Batt, 1969*c*). Halvorson (1962) found, significantly, that the degradation of proteins and nucleic acids in starved bacteria was suppressed by inhibitors of energy-yielding reactions.

The adenylate energy charge in relation to survival

The adenylate energy charge (reviewed by Atkinson, 1971) indicates the energetic state of a living cell, a value of unity meaning that all the adenine nucleotide is present as ATP. Atkinson has suggested that the energy charge regulates the pathways that produce and utilize high energy compounds, a value of 0.85 representing the equilibrium point for the balance of these reactions.

Deposition of reserve materials in bacteria occurs when growth is restricted in the presence of an excess of the carbon and/or energy source, a situation which should lead to a favourable intracellular energetic state. Further, both glycogen and polyphosphate synthesis involve the direct participation of ATP whereas poly-β-hydroxybutyrate formation does not. Chapman, Fall & Atkinson (1971) were the first to relate energy charge to survival; their results implied that growth of *E. coli* could occur only at energy charges of 0.8 or higher, that viability was maintained at values between 0.8 and 0.5 and that death occurred when the energy charge fell below 0.5. They did not determine the bacterial glycogen content so that it was not possible to correlate energy charge and survival with the presence of this storage compound. Subsequently Dietzler, Lais & Leckie (1974) found that the energy charge increased from 0.74 to 0.87 when exogenous nitrogen was exhausted in cultures of *E. coli* containing excess glucose and that, concomitantly, the rate of glycogen synthesis increased 3.3-fold; however, they, in turn, did not measure viabilities or study the effect of prolonged starvation.

The relationship between energy and survival of the strict anaerobe *Peptococcus prévotii* was investigated by Montague & Dawes (1974).

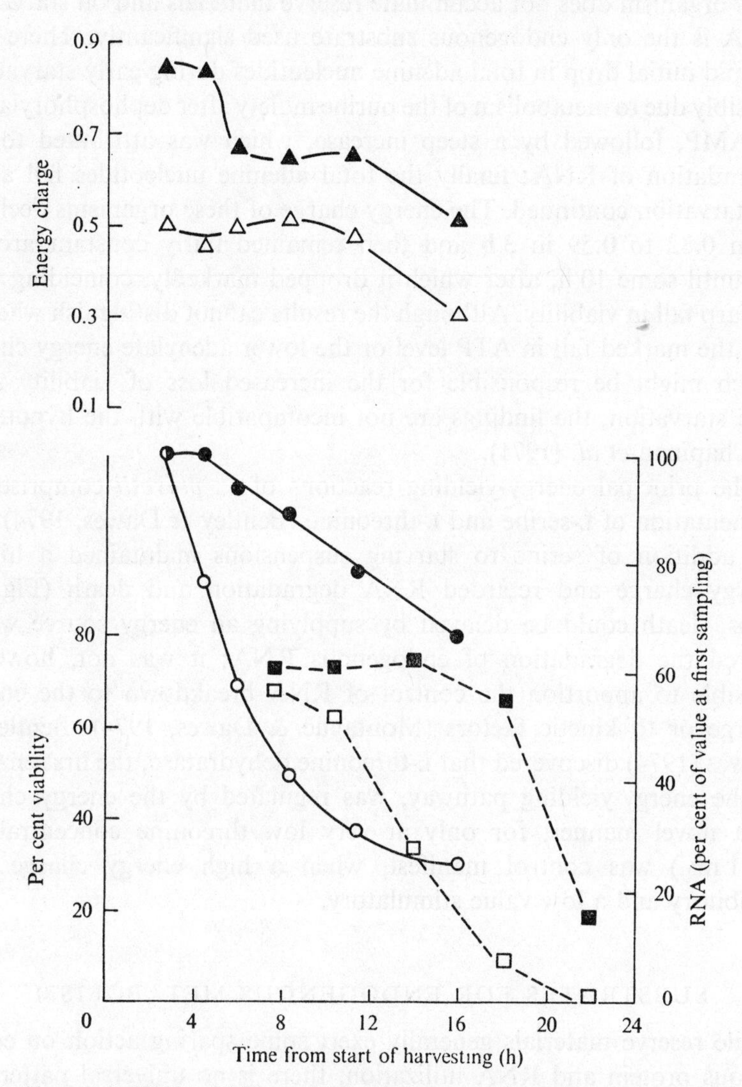

Fig. 1. Changes in RNA, viability and adenylate energy charge in starving *Peptococcus prévotii* supplemented with serine. Bacteria were harvested and half were washed with 67 mM phosphate buffer, pH 6.8, and half with similar buffer containing L-serine (20 mM). The washed organisms were suspended in fresh quantities of the wash buffers and the suspensions placed in separate starvation flasks, serine-supplemented at 0.67 and control bacteria at 0.72 mg (dry wt) ml^{-1}. RNA (○, ●), viabilities (□, ■) and adenylate energy charges (△, ▲) were determined. Open symbols: control bacteria; closed symbols: serine-supplemented bacteria. The RNA contents of both suspensions were 12.9 per cent of the dry bacterial weight at the time of first sampling (Montague & Dawes, 1974).

This organism does not accumulate reserve materials and on starvation RNA is the only endogenous substrate used significantly. There was a rapid initial drop in total adenine nucleotides during early starvation, possibly due to metabolism of the purine moiety after dephosphorylation of AMP, followed by a steep increase, which was attributed to the degradation of RNA; finally the total adenine nucleotides fell again as starvation continued. The energy charge of these organisms declined from 0.82 to 0.59 in 3 h and then remained fairly constant around 0.5 until some 10 h, after which it dropped markedly, coinciding with a sharp fall in viability. Although the results cannot distinguish whether it is the marked fall in ATP level or the lower adenylate energy charge which might be responsible for the increased loss of viability after 10 h starvation, the findings are not incompatible with the hypothesis of Chapman *et al.* (1971).

The principal energy-yielding reactions of *P. prévotii* comprise the fermentation of L-serine and L-threonine (Bentley & Dawes, 1974) and the addition of serine to starving suspensions maintained a higher energy charge and retarded RNA degradation and death (Fig. 1). Thus, death could be delayed by supplying an energy source which spared the degradation of endogenous RNA; it was not, however, possible to apportion the control of RNA breakdown to the energy charge or to kinetic factors (Montague & Dawes, 1974). Bentley & Dawes (1974) discovered that L-threonine dehydratase, the first enzyme in the energy-yielding pathway, was regulated by the energy charge in a novel manner, for only at very low threonine concentrations (< 1 mM) was control manifest, when a high energy charge was inhibitory and a low value stimulatory.

SUBSTRATES FOR ENDOGENOUS METABOLISM

While reserve materials generally exert some sparing action on endogenous protein and RNA utilization, there is no universal pattern of behaviour and sequential or simultaneous degradation of cellular materials can occur according to the organism.

The free amino acid pool

The free amino acid pool in some organisms can serve as an endogenous substrate. Thus about one-half of the free amino acid pool of the aerobic micrococcus *Sarcina lutea* was oxidized with ammonia release during early (5 h) starvation, and the endogenous Q_{O_2} fell to a negligible value; only certain amino acids were metabolized, with glutamate being both the principal pool constituent and endogenous substrate (Dawes & Holms, 1958). Viability was maintained during pool depletion and even after 40 h there were 90 per cent survivors. Glucose-grown *S. lutea* accumulated a polyglucan compound which, on subsequent starvation, was metabolized simultaneously with the pool amino acids and RNA, and led to accelerated death (Burleigh & Dawes, 1967).

There is little evidence for DNA degradation during bacterial starvation (see, however, MacKelvie, Campbell & Gronlund, 1968*b*), although increases in DNA content have been recorded during short-term starvation of exponential phase *E. coli* (Brdar, Kos & Drakulić, 1965).

Protein turnover and degradation

The turnover of protein requires energy and enables a bacterium to adjust its existing enzyme levels in response to a changing environment without the need for appreciable growth. Recent work, reviewed by Pine (1972), has indicated relationships between proteolysis and RNA metabolism, and Goldberg (1971) obtained evidence that aminoacyl-tRNA represses proteolysis, regulation of which has been attributed more specifically to the *RC* gene which couples the synthesis of protein to RNA synthesis (Sussman & Gilvarg, 1969). Pine (1972) has observed that the minimal proteolytic difference between starvation and growth is that proteolysis is more generalized and uniform in starvation but is not necessarily greater, owing to the continual regeneration of highly labile proteins in growing cells. The rate of protein turnover in growth of *E. coli* was 2.5 to 3 per cent h^{-1} at generation times of 50 to 170 min (Pine, 1970) and proteolytic rates of 6 to 7 per cent h^{-1} were attained in diauxic lag (Willetts, 1965). The loss of catalytic protein during starvation is discussed separately and general protein degradation in relation to survival is considered for individual organisms.

The role of RNA

Maaløe & Kjeldgaard (1966) proposed the hypothesis of constant efficiency of ribosome function, namely that, in balanced growth, each ribosome is functional and participates in protein synthesis at a rate independent of the growth rate. This hypothesis implies that under these conditions the RNA content of a bacterial cell is directly proportional to its growth rate, and these workers observed such proportionality for the ribosomal RNA at doubling times (t_D) about 0.6 h^{-1}. However, subsequent chemostat work by Koch and his colleagues (reviewed by Koch, 1971) revealed that glucose-limited *E. coli*, grown at much lower growth rates, contained six-fold the RNA that would be necessary for sustaining those growth rates based on the RNA content of faster-growing bacteria, and in the form of functional but unused synthetic machinery. Surprisingly, even phosphate-limited *E. coli* at slow growth rates diverted a considerable portion of the phosphate to synthesize five- to seven-fold the amount of RNA believed necessary to support these growth rates (Alton & Koch, 1974). Koch (1971) offered an ecological explanation for this unusual observation, suggesting that the behaviour benefits organisms subjected to a feast-and-famine existence in the human gut.

Clearly then, a starving bacterium, irrespective of its previous rate of growth, possesses a complement of ribosomes in excess of its immediate needs, which correlates with the general finding that the RNA content can fall to very low values without affecting viability. Conversely, glucose-grown *Azotobacter agilis* starved for 72 h retained all their RNA while losing 80 per cent of their viability (Sobek, Charba & Foust, 1966). Thus, no general relationship between the extent of RNA degradation and survival can be adduced.

The degradation of RNA yields ribose and purine and pyrimidine bases which, according to the organism, may or may not be used as endogenous substrates. Release of bases into the suspending fluid during starvation is a common observation, and such degradation also releases RNA-associated magnesium.

Effect of magnesium on survival

Magnesium usually plays an important role in prolonging the viability of bacteria during starvation. It is the most common metal co-factor of enzymes (Vallee, 1960), is essential for stabilizing bacterial ribosomes (Dagley & Sykes, 1957), counters carbon substrate-accelerated death (Strange & Hunter, 1967) and has a stabilizing effect on the permeability control mechanisms of bacteria which, in the absence of magnesium, are susceptible to cold shock. Magnesium is bound in ribosomal structures, and Tempest & Dicks (1967) demonstrated a stoichiometric relationship between intracellular magnesium and RNA in *Aerobacter* (*Klebsiella*) *aerogenes*.

Generally, starvation of bacteria in phosphate or saline in the absence of magnesium leads to a more rapid loss of viability and greater degradation of RNA than in comparable organisms starved in its presence, as found with *K. aerogenes* (Tempest & Strange, 1966), *Streptococcus lactis* (Thomas & Batt, 1968, 1969*a*) and *Zymomonas anaerobia* (Dawes & Large, 1970).

In contrast, while magnesium suppressed the degradation of RNA in *Sarcina lutea* it did not prolong survival (Burleigh & Dawes, 1967), although, as Thomas & Batt (1969*a*) pointed out, it is possible that the density of the suspensions used may have ensured that sufficient magnesium was released in the degradation of RNA in the control organisms to mask the effect of the added magnesium.

ENDOGENOUS METABOLISM AND SURVIVAL AT SLOW GROWTH RATES

Postgate & Hunter (1962) and Tempest, Herbert & Phipps (1967) studied the effect of decreasing growth rate on *K. aerogenes* in a chemostat and found that the 'steady state' viability of glycerol-limited cultures declined markedly below dilution rates of 0.1 h^{-1}. The doubling time attained a maximum value of about 80 h, i.e. a minimum specific growth rate of 0.009 h^{-1}. Ammonium-limited cultures, when the energy supply was always in excess, behaved similarly although the viabilities were higher than in the glycerol-limited culture at corresponding dilution rates and the maximum doubling time was about 100 h. Decreasing the temperature from 37 to 25 °C improved culture viability, indicating temperature dependence of the minimum growth rate. The endogenous Q_{O_2} values of glycerol-limited organisms were lower than those of ammonium-limited bacteria, e.g. Q_{O_2} values for dilution rates between 0.06 and 0.004 h^{-1} were 5 to 2 and 16 to 12 respectively but there was no significant correlation with viability.

The existence of a minimum growth rate, which clearly was not limited by the rate of supply of energy (c.f. Schulze & Lipe, 1964), was thus established and Tempest *et al.* (1967) suggested that it was a function of the inability of the bacteria to synthesize ribosomes at a rate less than some finite value.

SURVIVAL AND ESSENTIAL ENZYMIC ACTIVITY

Given that a starving bacterium must meet a certain minimum energy requirement and maintain a minimal concentration of essential cellular component(s) to remain viable, depletion, loss or inactivation of one or more enzymes in the energy-producing or macromolecule-synthesizing systems could lead to death of the organism. Postgate & Hunter (1962) found that glycerol oxidation in *K. aerogenes* declined in parallel with viability, as did glutamate and glucose oxidation in *S. lutea* (Burleigh & Dawes, 1967).

Peptococcus prévotii secures its energy principally by the fermentation of serine and threonine, involving threonine dehydratase, thioclastic enzyme, phosphotransacetylase and acetate kinase. During anaerobic starvation, viability fell to zero after 33 h and, of the aforementioned enzymes, only threonine dehydratase activity decreased (by 35 per cent), which was insufficient to explain the observed loss of ability to generate ATP from serine (Bentley & Dawes, 1974).

Constitutive enzymes were stable whereas inducible enzymes lost activity during 14 days starvation of *Arthrobacter crystallopoietes* which maintained full viability (Boylen & Ensign, 1970*b*). The inducible β-galactosidase of *E. coli* was likewise preferentially degraded during carbon and nitrogen starvation for 48 h, with only 10 per cent fall in viability (Strange, 1966).

Thus, in those cases where a decline in viability is associated with the loss of an enzymic activity which could conceivably be concerned with death, there is still no convincing evidence that such an enzymic defect is the responsible factor, and the parallelism could be fortuitous.

Recent investigations on the superoxide dismutases of bacteria (Fridovich, 1974) revealed that these enzymes of defence against the superoxide radical are, as might be anticipated, confined to aerobes and aerotolerant anaerobes. The dismutases of facultative organisms are inducible and they give protection against hyperbaric oxygen stress (Gregory & Fridovich, 1973*a*, *b*). It is thus feasible that if the superoxide dismutases were labile during starvation their loss could be a significant factor in hastening death, a possibility that awaits investigation.

RESERVE MATERIALS AND SURVIVAL

Many micro-organisms accumulate materials which can serve as reserves of carbon, phosphate and/or energy and which, in starvation, may be degraded to provide the cell with carbon for resynthesis of essential, degraded compounds and with energy for maintenance of viability (Dawes & Senior, 1973). Such materials embrace polyglucans ('glycogen'), lipids (including poly-β-hydroxybutyrate) and polyphosphate. Some organisms accumulate more than one type of reserve material and, in these cases, the environmental conditions and the regulatory mechanisms involved determine the proportion of the different compounds synthesized. Wilkinson (1959) has defined the criteria that are needed to establish the energy-storage function of a compound.

It must be emphasized that in by no means all instances has possession of a reserve material been correlated with an increased capacity for survival during starvation; attention here will be restricted to those investigations which have sought to relate survival to the possession of such reserves.

Glycogen and glycogen-like reserves

Many bacteria, including aerobic, facultative and anaerobic organisms, accumulate α-polyglucans similar to glycogen when growth is limited by the nitrogen, sulphur or phosphorus source, or by low pH, in the presence of an excess of the carbon and energy source.

Dawes & Ribbons (1965) found that glycogen was the preferentially utilized endogenous substrate in *E. coli*, which, however, degraded this reserve rapidly in 2 to 3 h when starved in phosphate buffer. They demonstrated isotopically that the effect of glycogen in suppressing the release of ammonia from starved cells (Ribbons & Dawes, 1963) was not due to a complete suppression of protein metabolism; turnover of protein occurred while the bacteria contained glycogen and thus the carbohydrate apparently supplied carbon and energy to permit reincorporation of ammonia released by protein breakdown. Wilkinson & Munro (1967) have suggested that glycogen reserves could be used by an organism as a method of detoxifying ammonia under certain conditions.

Glycogen-rich *E. coli* survived better than glycogen-poor organisms (Dawes & Ribbons, 1963) and a thorough investigation by Strange (1968) of the relation between glycogen and survival, in which comparisons were made between nitrogen- and carbon-limited *E. coli* at different growth rates, led to the conclusion that only bacteria with relatively large amounts of carbohydrate have superior survival properties. The addition of Mg^{2+} (0.05 to 1.0 mM) to carbon-limited organisms decreased the death rate but had little effect on nitrogen-limited cells with large glycogen reserves. The importance of bacterial magnesium content in survival was emphasized, especially since nitrogen-limited *K. aerogenes* contain more magnesium than carbon-limited organisms grown at the same rate (Tempest & Strange, 1966) and also survive better (Strange, Dark & Ness, 1961). *Klebsiella aerogenes* differs from *E. coli* in degrading its glycogen more slowly and in releasing ammonia during glycogen utilization, albeit at a slower rate than glycogen-free cells.

Facultatively anaerobic streptococci can display widely different endogenous metabolism, e.g. *Streptococcus salivarius* synthesizes an intracellular polyglucan (glycogen) to the extent of 50 per cent of the

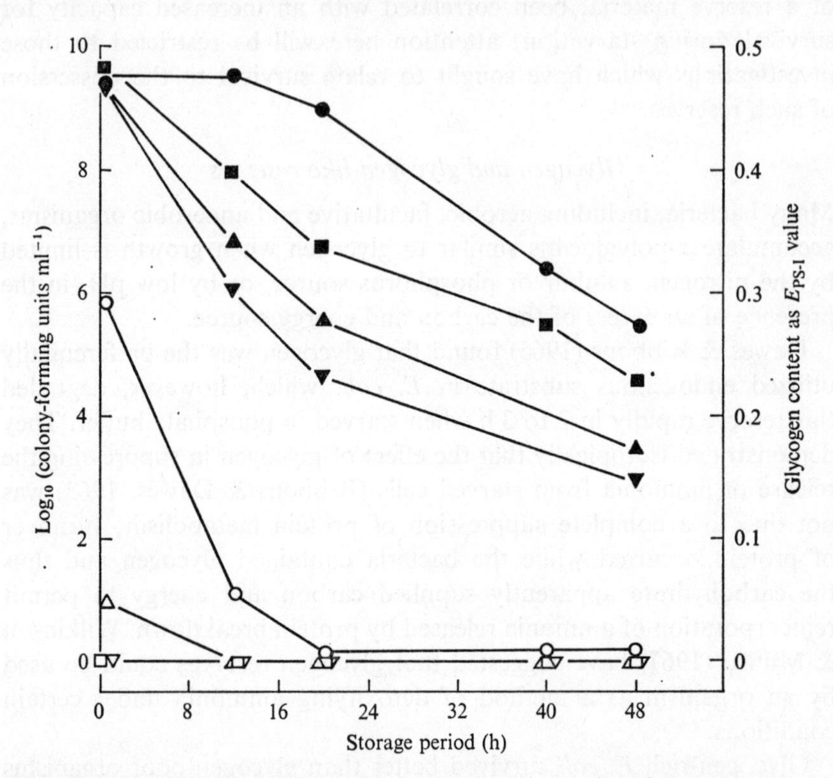

Fig. 2. Survival of glycogen-positive and glycogen-negative strains of *Streptococcus mitis* when starved in phosphate buffer, pH 6.5, at 37 °C. Glycogen-positive (●) and negative (■) strains for 6 h in 1 per cent glucose trypticase broth to obtain exponential phase cells. Glycogen-positive (▲) and negative (▼) strains grown in medium containing 0.1 per cent glucose for 16 h (stationary-phase cells). The corresponding open symbols record the glycogen content as the extinction at 565 nm of the polysaccharide-iodine complex (E_{PS-I}). The authors state that for suspensions containing 10^9 cells ml^{-1} an E_{PS-I} value of 0.4 is equivalent to a glycogen content of 50 per cent of the bacterial dry weight (van Houte & Jansen, 1970).

cell dry mass (Hamilton, 1968) while *S. lactis* does not accumulate any reserve material (Thomas & Batt, 1969*a*). Hamilton (1968) and his colleagues carried out some notable work on *S. salivarius* glycogen, but the endogenous metabolism of glycogen was not related to viability. However, the survival characteristics of *S. mitis*, which also accumulates glycogen (up to 37 per cent of its dry weight (Gibbons & Kapsimalis, 1963; Builder & Walker, 1970)), have been studied by van Houte & Jansen (1970). They showed that glycogen-rich organisms in saline phosphate survived well during the period when the glycogen was being degraded, while bacteria with little or no glycogen died rapidly (Fig. 2). The viability of glycogen-rich cells decreased by 40 per cent during a 16 h storage period, with a concomitant decrease in the amount of lactic acid produced. The survival of glycogen-free cells over the same storage period was only 0.01 per cent.

Apart from the reported synthesis of glycogen during exponential growth (Gibbons & Kapsimalis, 1963), this reserve in *S. mitis* obeys the criteria formulated for such materials (Wilkinson, 1959; Dawes & Senior, 1973). However, the oral environment of *S. mitis* may have selected an organism which synthesizes glycogen whenever exogenous glucose is in excess. As the presence of oral carbohydrate is intermittent, an organism capable of storing reserve glycogen under such conditions would be selected since subsequent degradation of the reserve during starvation would maintain viability.

Some species of *Clostridium* accumulate large, intracellular granules of 'granulose', an α-polyglucan which resembles amylopectin (Whyte & Strasdine, 1972) or may be less branched (Laishley, Brown & Otto, 1974). Robson, Robson & Morris (1974) found that granulose synthesis in *Clostridium pasteurianum* was not subject to the fine allosteric control encountered in polyglucan synthesis in other bacteria although its synthesis and degradation responded to the adenylate energy charge (Robson & Morris, 1974). Exogenous glucose completely inhibited granulose mobilization, in contrast to *C. botulinum* E which fermented reserve carbohydrate under these conditions (Strasdine, 1972). However, comparison of wild type and granulose-negative mutants revealed no simple relationship between possession of the carbohydrate and capacity for survival (R. Robson & J. G. Morris, unpublished).

Polyphosphates

The available evidence suggests that in the majority of bacteria which accumulate polyphosphate this material functions as a reserve of phosphorus rather than of energy (Harold, 1966; Dawes & Senior 1973). Its accumulation occurs usually under conditions of nutrient imbalance and an antagonistic relationship between polyphosphate and nucleic acid metabolism offers a reasonable explanation for its accumulation or degradation. Attempts to demonstrate a biological advantage for survival of wild-type *K. aerogenes* over a polyphosphate-less mutant were inconclusive (Harold, 1966), although possibly the capacity to synthesize polyphosphate does confer a small selective advantage which is not readily detected but which is sufficient to ensure the preservation of the enzymes of polyphosphate metabolism and to account for their retention during the course of evolution.

Poly-β-hydroxybutyrate

Poly-β-hydroxybutyrate (PHB) is accumulated by a wide variety of Gram-positive and Gram-negative aerobic and photosynthetic species and by lithotrophs as well as many organotrophs. The role of the polymer as a carbon- and energy-reserve material and the regulation of its biosynthesis and degradation were reviewed by Dawes & Senior (1973) and here attention will be confined to its function during starvation and in survival. A clear relationship between survival and the PHB content of the obligate halophile *Micrococcus halodenitrificans* was established by Sierra & Gibbons (1962). Polymer-rich (50 per cent PHB) bacteria retained 100 per cent viability for 100 h whereas polymer-poor (10 per cent PHB) organisms died rapidly and there were less than 10 per cent survivors after 30 h starvation in phosphate saline.

In the chemolithotrophic *Hydrogenomonas*, PHB serves as both a carbon and energy source and supports protein synthesis in the presence of a source of nitrogen (Schlegel, Gottschalk & Von Bartha, 1961). Hippe (1967) found PHB was the preferentially-utilized endogenous substrate in *Hydrogenomonas eutropha* H16 and its presence inhibited the net degradation of nitrogenous cellular components and the release of ammonia. The addition of uncoupling agents increased both the Q_{O_2} and the rate of PHB degradation some five-fold, suggesting a high degree of coupling between oxidative phosphorylation and endogenous respiration in *Hydrogenomonas*. If the polymer content was high, viability remained at maximum levels over longer periods of starvation (70 per cent survival at 70 h), although it was apparent that cells with a low polymer content, and after their PHB was exhausted, could maintain viability at the expense of cell protein.

Sphaerotilus discophorus, a filamentous, sheathed bacterium, accumulates PHB, and polymer-rich cells survive better than cells with little or no PHB when they are starved in phosphate buffer (Stokes & Parson, 1968). An interesting observation was that exponential-phase bacteria survived better than those from the stationary phase, in contrast to the more usual pattern of resistance to stress (Postgate & Hunter, 1962).

The azotobacters can accumulate large amounts of PHB but they display a low endogenous respiration rate, which might be considered a factor conducive to prolonged cell survival. Sobek *et al.* (1966) found the viability of glucose-grown *Azotobacter agilis*, containing 18 per cent PHB, did not decrease until over half of the polymer had been degraded, whereas PHB-poor organisms (4 per cent of the dry weight) displayed a rapid and immediate decline in viability on starvation. However, succinate-grown bacteria, with a consistently lower PHB content than glucose-grown cells, degraded the polymer at a slower rate and survived longer, suggesting a more efficient polymer utilization.

Stockdale (1967) made similar observations with carbon-starved *Azotobacter insigne*: PHB served as the immediate, major endogenous substrate, although other cell constituents were oxidized simultaneously and, after PHB exhaustion, a high Q_{O_2} was still manifest. The initial polymer content alone did not determine the survival pattern of the cells since stationary phase cells from glucose-limited media, and containing little PHB, survived better initially than cells grown in the presence of excess glucose, although subsequently they died at a faster rate.

Thus, under certain circumstances, bacteria with a lower content of PHB can survive longer than those with a higher one, either because they are not subjected to some additional adverse factor (since they may be better able to utilize alternative components, e.g. protein, as endogenous substrates) or because they can utilize their reserve material more efficiently.

It has been suggested that, in azotobacter species, PHB plays an important additional role in assisting survival under conditions of adverse oxygen concentration by serving as a redox regulator (Senior *et al.*, 1972). Oxygen concentrations in excess of about 20 per cent of air saturation are inhibitory to the nitrogenase system of azotobacters and this effect may be countered by the organism increasing its oxidative activity, thereby lowering the environmental partial pressure of oxygen to a tolerable value, a process termed respiratory protection (Dalton & Postgate, 1969). The possession of PHB may thus bestow an additional advantage by permitting an organism to increase its oxidative activity in the absence of an exogenous substrate, thereby securing respiratory protection and enhancing survival.

Inter-relationship of reserve materials

One of the few studies to examine the inter-relationship of different reserve materials was carried out by Zevenhuizen & Ebbink (1974) with *Pseudomonas* V-19, which they isolated from activated sludge. This bacterium can accumulate glycogen, PHB and lipid in amounts depending upon the growth medium, e.g. in either casein yeast extract (CYE) or ammonium salts medium glucose-grown organisms accumulated much more PHB than did glycerol-grown ones. In this organism PHB appeared to be a mobile reserve which was synthesized and utilized more rapidly than glycogen and neutral lipid; carbon from PHB was used for glycogen and lipid synthesis. On the addition of a nitrogen source (NH_4^+) to starved bacteria the carbon reserves declined, although the total dry weight did not significantly alter, indicating that some net cellular synthesis occurred at the expense of the reserves. Bacteria from the CYE medium were reported to remain 100 per cent viable for 21 days but after prolonged incubation, when their endogenous reserves were virtually spent, they began to die and were 50 per cent viable after 63 days. Those from the simple medium to which NH_4^+ was added during starvation, first increased in viable count but then displayed substrate-accelerated death and died at a faster rate than the control cells, which remained nearly 100 per cent viable for 28 days. Although the design of these experiments did not exclude the possibility of cryptic growth occurring over the long periods of incubation employed, there is no question that *Pseudomonas* V-19 degrades its endogenous reserves at a much slower rate than do organisms which die more rapidly. Zevenhuizen & Ebbink (1974) make the interesting point that bacteria which degrade their glycogen rapidly, e.g. *E. coli* and *K. aerogenes*, have glycogens structurally similar to animal glycogen with average chain lengths ($\overline{CL}$) of 12 to 15 hexoses, whereas the glycogen of *Pseudomonas* V-19, with $\overline{CL} = 8$, resembles that of *Arthrobacter* and *Mycobacterium* ($\overline{CL} = 7$ to 9) with a high degree of branching, comparable to phosphorylase limit dextrins, which are not susceptible to phosphorylase action and which must first be debranched by an α-1,6-glucosidase. It is suggested that in these organisms glycogen breakdown is limited by the low activity of the debranching enzyme. *Arthrobacter* (Zevenhuizen, 1966; Mulder & Zevenhuizen, 1967; Boylen & Ensign, 1970*a*, *b*; Gray, this volume) like *Pseudomonas* V-19, displays long-term survival on starvation with a low rate of glycogen utilization.

SUBSTRATE-ACCELERATED DEATH

'Substrate-accelerated death' is the name given to an interesting phenomenon first described in detail by Postgate & Hunter (1964). When a substrate, which by its exhaustion had limited the growth of a population of certain Gram-negative bacteria, was added to a starving suspension of the organisms in non-nutrient buffer, it greatly accelerated their death. The effect occurred with carbon-, ammonium- or phosphate-limited populations of *K. aerogenes* but not with those limited by magnesium ions or sulphate. Although their results with ammonium-

and phosphate-limited cells could not be verified by Strange & Dark (1965) with, putatively, the same strain, this was accounted for by changed properties of the original strain (Strange & Hunter, 1966) while apparent phosphate-limited death was attributed to K^+ toxicity, since KH_2PO_4 accelerated death in this buffer.

Survivors of substrate-accelerated death exhibited long division lags in the recovery medium, reminiscent of the recovery of organisms which have been subjected to repression of enzyme synthesis (Postgate & Hunter, 1964). Since 3′,5′-cyclic adenosine monophosphate (cAMP) alleviates catabolite repression in *E. coli* (De Crombrugghe *et al.*, 1969), Calcott & Postgate (1972) examined its effect on glycerol- and lactose-accelerated death of *K. aerogenes*. In both cases addition of cAMP to the starvation medium protected the populations, although it did not decrease the long lags exhibited by survivors of these stresses. Calcott, Montague & Postgate (1972) found a correlation between the intracellular pool level of cAMP and the death rate of chemostat-grown, lactose-limited *K. aerogenes* starved in saline phosphate. Substrates accelerating death caused the intracellular cAMP to fall by 90 per cent, compared with a 60 per cent drop in controls and in the presence of neutral substrates, e.g. glycerol. cAMP was released from the bacteria during starvation and its concentration in the suspending medium more than doubled when lactose was present.

Three possible factors could govern the cAMP levels of *K. aerogenes* undergoing lactose-accelerated death, namely the activity of adenyl cyclase which converts ATP to cAMP; the activity of the phosphodiesterase which degrades it to AMP; or the release of cAMP into the medium (Pastan & Perlman, 1970). Theophylline, an inhibitor of some microbial phosphodiesterases, significantly protected the population from lactose-accelerated death (Calcott *et al.*, 1972). Calcott & Postgate (1972) found that the overt stress, that of exposing the susceptible population to the growth-limiting substrate, did not actually kill the bacteria and that the substrate (or metabolite thereof) must be present in the recovery medium for death to occur. Calcott *et al.* (1972) observed that organisms starved for 4 h in the presence of lactose lost 90 per cent of their cAMP and appeared nearly 100 per cent dead on lactose recovery medium, yet were 80 per cent viable on glycerol medium. Further, bacteria starved for only 30 min likewise lost 90 per cent of their cAMP pool and were completely viable on lactose recovery medium, indicating that low internal cAMP concentrations render the organism gradually susceptible to substrate-accelerated death but recovery from this predisposition occurs in the absence of a traumatic substrate.

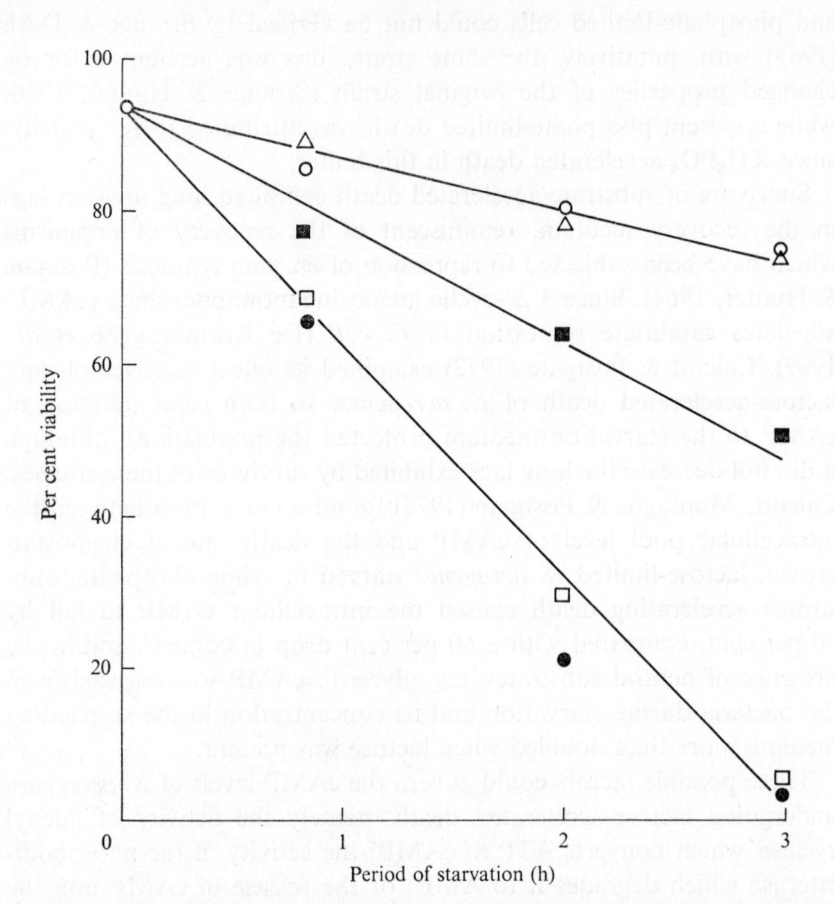

Fig. 3. Effect of cAMP in the recovery medium on lactose-accelerated death of *Klebsiella aerogenes*. Washed, lactose-limited bacteria were starved at 20 μg (dry wt) ml^{-1} saline phosphate with or without 5 mM lactose at 37 °C. Viabilities were determined by slide culture on a lactose-based recovery medium with or without cAMP: starved without lactose (○); starved with lactose, recovered on unsupplemented lactose medium (●); starved with lactose, recovered on lactose medium with (□) 1 mM, (■) 3 mM or (△) 5 mM cAMP (Calcott & Postgate, 1974).

Only inducers of the *lac* operon could mimic lactose-accelerated death of lactose-limited populations (Calcott & Postgate, 1972), again indicating similarity to acute enzyme repression. Comparisons were therefore made with *K. aerogenes* possessing hyper, intermediate and repressed levels of β-galactosidase; their sensitivity to lactose-accelerated death was directly related to their levels of this enzyme (Calcott & Postgate, 1974). cAMP added to the lactose-based recovery medium alleviated lactose-accelerated death completely at 5 mM, partially at 3 mM, and not at all at 1 mM (Fig. 3). Thus the bacteria need not necessarily die when transferred to a recovery medium containing lactose.

Although the precise mechanism of substrate-accelerated death of *K. aerogenes* is still uncertain, control of the intracellular cAMP concentration is obviously involved. For an effective cellular economy, Calcott & Postgate (1974) envisaged immediate control occurring at the phosphodiesterase and/or release level with long-term control at the cyclase level; ATP, the substrate for adenyl cyclase, is known to increase markedly in substrate-accelerated death (Strange, 1967). However, at the present time there is a paucity of information concerning the regulation of the cyclase, phosphodiesterase and release of cAMP from bacterial cells (Rickenberg, 1974). Calcott & Postgate (1974) suggested that the traumatic substrate might depress adenyl cyclase synthesis or function through a more central control mechanism linked to the system regulating the use of that substrate, in this case the *lac* operon.

It would be instructive to know more about the role of Mg^{2+} in protecting against substrate-accelerated death, particularly in relation to bacteria that are hyper-induced for β-galactosidase. Further, adenyl cyclase in *E. coli* is a Mg^{2+}-dependent enzyme and addition of Mg^{2+} decreased the elevated ATP concentration which occurs in substrate-accelerated death (Strange, 1967). Finally, these details have all been established with *K. aerogenes* and it would be useful to know whether they apply equally to other Gram-negative bacteria.

SOIL BACTERIA

The autochthonous soil bacteria, which maintain themselves over long periods in nutrient-poor soils, offer the best known examples of resistance to starvation. *Arthrobacter crystallopoietes* possesses champion survival abilities and also displays two distinct vegetative morphological forms according to the environmental conditions. In a nutritionally poor environment a coccoid form is manifest whereas in a rich medium the organism grows as rods (Ensign & Wolfe, 1964; see Gray, this volume). Despite the tempting inference that the coccoid form would be expected to be the more resistant species, Boylen & Ensign (1970*a*) found both forms were highly resistant to starvation, e.g. after 30 days, rods and spheres were 100 per cent viable, and after 60 days starvation, more than 65 per cent viable.

Despite the remarkable similarity of the behaviour of rods and spheres, Boylen & Ensign (1970*b*) found their biochemical patterns

Table 1. *Effect of starvation on the composition and viability of the two morphologically distinct forms of* Arthrobacter crystallopoietes

Bacteria (0.5 mg dry wt ml^{-1}) were starved by shaking in 0.03 M-phosphate buffer at 30 °C

(*a*) Cellular composition at time of harvest (per cent of dry weight)

	Rods	Spheres
Carbohydrate	10	40
Protein	41	30
RNA	21	15
DNA	2	2

(*b*) Effect of starvation on cellular composition and viability

Time of starvation (days)	Degradation of cellular materials (per cent of initial content)							
	Rods				Spheres			
	Carbohydrate	Protein	RNA	Viability	Carbohydrate	Protein	RNA	Viability
1	30	15	45	100	30	0	< 10	100
7			75	100			18	100
30	60	40	85	100	60	20	32	100

Based on data recorded by Boylen & Ensign (1970*b*).

on starvation differed significantly, as Table 1 reveals. The marked differences in chemical composition and of the rates of degradation of protein and RNA are apparent. The sparing effect of the carbohydrate of spherical organisms on the utilization of protein and RNA is evident yet both types of cell remained completely viable during the 30-day starvation period.

Electron microscopy of sections of *A. crystallopoietes* during prolonged starvation showed that the remarkable capacity for survival was not due to an altered morphology and that general structural integrity was maintained by both rods and spheres over an eight week period (Boylen & Pate, 1973). Such cytological changes as occurred could be correlated with the previous chemical analyses.

Another autochthonous soil organism which displays long-term survival characteristics is the actinomycete *Nocardia corallina*. Robertson & Batt (1973) observed a slow decline in viability with 50 per cent survivors after 20 days starvation in phosphate–Mg^{2+} buffer. Changes in cellular composition were studied over shorter periods. Carbohydrate content fell from 24 per cent to 6 per cent of the initial dry weight in 27 h and accounted for the rapid initial decrease in dry weight observed. Ammonia was released only after cellular carbohydrate was depleted; after 176 h the total ammonia released correlated with the decline in protein content from 35 per cent to 27 per cent of the initial dry weight. Microbial fatty acids fell from 14 per cent to 10 per cent during the same period. In keeping with the long doubling time of 5 h, the RNA content of *N. corallina* was only 4 per cent of the dry weight at the start of starvation and its decline to 1 per cent after 11 days was associated with the release of u.v.-absorbing materials; at this time cells were still about 80 per cent viable.

Thus, while the pattern of utilization of endogenous substrates resembles that of many other bacteria, the resistance to starvation is undoubtedly greater, even though the possibility of cryptic growth was not excluded. Robertson & Batt (1973) observed that the rate of endogenous metabolism of *N. corallina* is some five-fold lower than that of *E. coli*, which displays a similar overall biochemical behaviour but whose 50 per cent survival time is about five times shorter than that of *N. corallina*, and they suggest that loss of viability may, as with *A. crystallopoietes*, be related to the rate at which cellular constituents are metabolized rather than to the absolute level of a particular constituent at the onset of starvation. They further suggest that autochthonous micro-organisms might possibly have a mechanism for decreasing the rate of endogenous metabolism which enhances their chance of survival during conditions of starvation.

BACTERIA WITHOUT RESERVE MATERIALS

Streptococcus faecalis

Streptococcus faecalis does not synthesize carbohydrate or lipid reserves. In starved anaerobic suspension it exhibits no detectable respiration, releases amino acids but not ammonia, loses no lipid, and the pH does not change, yet heat evolution occurs (Forrest & Walker, 1963). Walker & Forrest (1964) measured glycolytic activity, as a parameter indicating the degree of organization of the cells, during starvation in phosphate buffer and found that endogenous metabolism correlated with the maintenance of glycolytic activity and also with a constant ATP pool concentration during several hours of starvation. Organisms harvested from an energy-limited medium displayed neither a constant ATP pool nor constant glycolytic activity (Forrest & Walker, 1965) and they showed no endogenous activity at a level detectable by microcalorimetry. It was suggested that when growth of *S. faecalis* ceases in the presence of excess energy source some unidentified reserve material(s) is synthesized. There are curious features, however, since only amino acids appear during anaerobic endogenous metabolism (in contrast to the report of ammonia release under aerobic conditions

(Gronlund & Campbell, 1961)) and the large amount of heat evolved and the production of ATP cannot be accounted for by hydrolysis of peptide bonds.

The most favourable environment for *S. faecalis* was one of low redox potential and high cell density (Forrest & Walker, 1963). In a less favourable situation more heat was produced indicating that more work was required for maintenance of the organisms. Unfortunately, in none of their work did Forrest & Walker measure the viability of their organisms.

Zymomonas mobilis *and* Z. anaerobia

These aerotolerant anaerobes ferment glucose via the Entner-Doudoroff pathway with a theoretical energy yield of 1 mole ATP/mole glucose fermented. However, molar growth yield studies with *Z. mobilis* indicated rather lower yields than expected (Belaich & Senez, 1965; Dawes, Ribbons & Rees, 1966) and values for *Z. anaerobia* were even lower (McGill & Dawes, 1971), suggesting that uncoupled growth occurred. Neither organism synthesized reserve materials and on anaerobic starvation the only component to undergo significant degradation was RNA, which in *Z. anaerobia* declined from 22 per cent to 3 per cent of the dry weight in 138 h (Dawes & Large, 1970). The RNA loss could be equated with the appearance of E_{260}-absorbing material in the medium. Degradation began immediately and continued in linear fashion; it was suppressed by the addition of magnesium. *Zymomonas mobilis* displayed marked differences in the rate of RNA breakdown; in 24 h the RNA content fell from 22 per cent to 11 per cent of the dry weight and then it declined much more slowly. There was a correspondingly rapid release of E_{260}-absorbing material in the first 24 h, followed by a slow re-utilization during the succeeding 50 h, an effect never observed with *Z. anaerobia*. Magnesium ions again suppressed RNA degradation and the release of nucleotides.

The ability to ferment glucose remained unimpaired after 143 h starvation of *Z. anaerobia* in the absence of magnesium ions, but starvation for 15 days seriously impaired this property. Under these conditions the bacteria died rapidly within 24 h and it is clear that the ability to produce ATP in response to glucose is not related to viability. Magnesium ions prolonged survival but had no significant effect on the ATP content of the organisms which decreased exponentially, being most marked during the initial 6 h. Although there appeared to be some correlation of ATP content with viability, other adenine nucleotides were not measured and so the overall energetic state of the organisms during starvation is not known.

Streptococcus lactis

The endogenous metabolism and survival of *S. lactis*, economically important in the dairy industry as a cheese starter, has been extensively studied by Batt and his colleagues (Thomas & Batt, 1968; 1969*a*, *b*, *c*; Thomas *et al.*, 1969). Reserve materials were not accumulated and on starvation RNA was the only component to undergo substantial breakdown yet, in contrast to most other bacteria, *S. lactis* did not utilize the ribose and bases from the degraded RNA. Similarly, the small amount of protein lost appeared to involve only hydrolytic reactions with release of undegraded products to the suspending fluid; depletion of the free amino acid pool was likewise accounted for by cellular release to the environment. Prolonged starvation resulted in substantial membrane phospholipid breakdown but the DNA was not depleted (Thomas & Batt, 1969*a*). These observations led to the conclusion that endogenous metabolism could not be yielding appreciable energy for survival, especially since *S. lactis* is a typical lactic acid organism with its energy-yielding reactions restricted to glycolysis and arginine degradation (significantly this amino acid was not detected in the free amino acid pool). This belief was strengthened by the finding that an exogenous energy source was essential for resynthesis of protein and RNA in starving *S. lactis* (Thomas & Batt, 1969*b*).

The survival of *S. lactis* in Na–K phosphate buffer at the growth temperature (30 °C) was optimal at pH 7.0 and enhanced by the presence of Mg^{2+}; lower temperatures also increased survival, as did the use of dense populations, possibly because of the protection afforded by Mg^{2+} which was released into the environment (Thomas & Batt, 1968). Both glucose and arginine accelerated the death rate while the addition of Mg^{2+} decreased this effect with glucose and extended survival times with arginine.

These observations were correlated with the respective rates of metabolism of the energy sources (supplied at 10 mM), e.g. glucose yielded ATP at 7.5-fold the rate with arginine. However, continuous feeding of glucose over a range of feed rates to starving cultures containing Mg^{2+} revealed that when the rate approximated to that calculated to yield ATP at the rate characteristic of ATP generation from arginine metabolism, the survival behaviour in the presence of glucose resembled that with the amino acid. Further decrease of feed rate extended survival of most of the population to four days and corresponded to a maintenance energy requirement of 0.045 to 0.090 g glucose (g dry wt)$^{-1}$ h^{-1}, a requirement approximating to that for aerobic bacteria (Pirt, 1965). Although glycolysis yields substantially less energy than aerobic metabolism, half the maintenance requirement of *E. coli* may be for resynthesis (Marr *et al.*, 1963) and since Thomas & Batt (1969*b*) found that little resynthesis of RNA and protein occurs during starvation of *S. lactis*, it seems probable that the maintenance requirement of *S. lactis* is less than that for *E. coli*. This extensive work did not, however, encompass measurements of cellular adenine nucleotide concentrations and therefore it is not known how the adenylate energy charge behaved during starvation; Thomas *et al.* (1969) did comment that if the organism contained constitutive kinases for nucleotide metabolism, the ATP pool could be maintained.

Starved bacteria were extremely sensitive to pH but provision of an exogenous energy source (arginine) enhanced the survival of *S. lactis* at adverse pH values (Thomas & Batt, 1969*c*) thus furnishing the first experimental support for the proposal (Dawes & Ribbons, 1962, 1964) that internal pH control is one of the functions of endogenous metabolism.

Electron microscopy of sections of *S. lactis* taken at intervals during starvation revealed stabilization of ribosomes by addition of Mg^{2+} plus amino acids (Thomas *et al.*, 1969). Death of the organism in phosphate without Mg^{2+} was accompanied by the disappearance of mesosomes and a marked dislocation of nuclear material. In the presence of Mg^{2+} lysis occurred principally between 30 and 40 h, after which time most of the organisms had lysed and the viability was < 1 per cent. Both the membrane and cell wall were ruptured in lysed organisms and the walls subsequently fragmented, suggesting that wall autolysis might be responsible for loss of viability.

Pseudomonas aeruginosa

The endogenous metabolism of *Pseudomonas aeruginosa*, a bacterium which does not accumulate any reserve material (MacKelvie, Campbell & Gronlund, 1968*a*), has been the subject of detailed investigations by Campbell and his collaborators. The starved organism released NH_4^+ and its protein and ribosomal RNA were degraded (Gronlund & Campbell, 1963, 1965). Exogenous substrates suppressed endogenous RNA oxidation to an extent which correlated with the amount of oxygen consumption, and hence the energy produced by their oxidation. Endogenous protein oxidation was decreased by some exogenous substrates and increased by others, a difference attributed to the requirement for NH_4^+ imposed upon the cells for assimilation of the individual exogenous substrate (Gronlund & Campbell, 1966) since Duncan &

Campbell (1962) had found that oxidative assimilation of carbon substrates by *P. aeruginosa* is obligatorily linked to the re-incorporation of NH_4^+ released by endogenous metabolism. Polynucleotide phosphorylase, firmly bound to 50 S and 30 S ribosomes, permitted degradation of RNA to nucleoside phosphates and the presence of adenylate kinase would enable energy to be secured. Interconversion of nucleotides and their breakdown to free bases, inorganic phosphate and ribose also occurred, the sugar gaining entry to the pentose phosphate cycle.

The survival of *P. aeruginosa* in relation to its cellular composition was studied over a period of 48 h starvation in Tris buffer, pH 7.3, with cells harvested from carbon- and nitrogen-limited defined glucose media and from complex tryptone–yeast extract–glucose (complete) medium (MacKelvie *et al.*, 1968*b*). The viability of both carbon- and nitrogen-limited bacteria had declined by some 25 per cent after 48 h but no precautions were taken to eliminate cryptic growth; in contrast, the viability of cells from complete medium increased by 57 per cent in 12 h and then remained constant. The concentration of 50 S ribosomes was twice as great as in the 'complete' cells as in the carbon- and nitrogen-limited ones but 60 to 75 per cent of these particles were in each case degraded during the initial 12 h and the rate of degradation during this period was proportional to the 50 S-ribosomal concentration. Chemical fractionation revealed a net loss of 16 per cent protein from all three types of cell and an increase in DNA of 10 per cent with nitrogen-limited and 'complete' cells while the carbon-limited organisms lost 22 per cent of their DNA.

MacKelvie *et al.* (1968*b*) found that after 48 h starvation both carbon- and nitrogen-limited organisms retained the ability to re-incorporate ammonia and u.v.-absorbing material when either glucose or 2-oxoglutarate was added. Within 7 h of the addition of either carbon source, the concentration of 50 S ribosomes had increased; the effect on viability was not measured. As glucose-grown cells do not possess the permease for 2-oxoglutarate, clearly the organisms retained the capacity for induced protein synthesis after 48 h starvation, an asset valuable for scavenger bacteria which persist in environments, such as water, where the nutritional state often fluctuates between famine and the provision of bizarre carbon and nitrogen sources. Survival will depend on the ability of the organism to synthesize the necessary permeases and induced systems to feed these substrates into the amphibolic pathways of metabolism.

Bdellovibrio bacteriovorus

The bdellovibrios are unique among known bacteria in their ability to attack, penetrate and multiply in the intraperiplasmic space of other bacteria (Hespell *et al.*, 1973). They are obligate aerobes which, shortly after attacking their prey and before initiating growth *per se*, render the respiratory apparatus of the victim non-functional (Rittenberg & Shilo, 1970). Consequently the bdellovibrios must generate their own energy during intraperiplasmic growth and cannot be parasitic upon the host organism. Hespell *et al.* (1973) concluded that amino acids, derived from the breakdown of the protein of the prey, serve as a major energy source during such growth of *B. bacteriovorus*. These bacteria, whose natural environments include soil, sewage and polluted water, have a very high rate of endogenous metabolism (six- to seven-fold that of *E. coli* on an equivalent protein basis) and its possible relationship to viability during starvation led Hespell, Thomashow & Rittenberg (1974) to a detailed study of the process.

Bdellovibrio bacteriovorus did not contain glycogen-like compounds or poly-β-hydroxybutyrate when grown intraperiplasmically in *E. coli*. On starvation, the high rate of endogenous respiration remained constant for 6 h and then declined rapidly; the R.Q. remained essentially constant at 0.88 throughout. Ultraviolet-absorbing compounds, ammonia, amino acids and orcinol-positive material (probably not free ribose or RNA) were released. Analysis of the cellular composition during starvation showed that 50 per cent of the cell carbon was lost within 4 h although the cells were still 95 per cent viable. During this period, decreases of 36 per cent of the initial RNA and 14 per cent of the protein contents occurred. By 10 h the viability had declined to 50 per cent, and 68 per cent of the RNA and 27 per cent of the protein had been lost, while the DNA remained constant, indicating that no lysis had occurred. Exogenous glutamate, or glutamate plus a balanced amino acid mixture, exerted a marked sparing action on the degradation of these endogenous components.

The large decrease in RNA content during starvation was correlated with a substantial reduction in the amounts of 70 S, 50 S and 30 S ribosomal particles, the effect being marked after only 3 h. Concurrent degradation of RNA and protein occurred in contrast to the sequential degradation of RNA followed by protein recorded for *E. coli* (Dawes & Ribbons, 1965; Jacobson & Gillespie, 1968). It is noteworthy that the transition from a high to a low rate of endogenous respiration coincided with the onset of the phase of rapid viability loss and, further, that as the R.Q. was unaltered during this transition, the chemical nature of the endogenous substrate did not change.

Hespell *et al.* (1974) have considered two possible reasons for the high endogenous respiration and concluded that, since protein turnover of *B. bacteriovorus* is of approximately the same magnitude as that of *E. coli*, energy expenditure to sustain rapid motility is the most likely explanation. They speculate that if a chemotactic attraction exists between the bdellovibrio and its substrate organisms, then rapid motility of the former would allow it to hunt efficiently and overtake a more slowly moving prey. These workers also evaluated susceptibility to death by starvation in relation to the chance of the organism finding its prey within 10 h, in order to have a 50 per cent chance of survival. Clearly various factors are important but, in general, the usual environments from which bdellovibrios have been isolated should provide a sufficient density of substrate organisms to enable the bdellovibrio to survive the period between being released from one substrate organism and finding the next; survival in unpolluted, fresh- and marine waters would be much less likely.

PHOTOSYNTHETIC BACTERIA

Light is the only acceptable exogenous energy source for many blue-green bacteria and consequently, in darkness, these organisms meet their energy requirements for survival by endogenous metabolism of energy-reserve compounds. Doolittle & Singer (1974) obtained a mutant strain (704) of the obligately photoautotrophic bacterium *Anacystis nidulans* to study its dark endogenous metabolism and to answer the specific questions of what metabolic pathway(s) operates and whether functions other than the minimal energy requirements for survival are furnished by it. This organism photoassimilates carbon dioxide via the reductive pentose phosphate cycle and accumulates glycogen which, in the dark, is degraded by the oxidative pentose phosphate cycle.

The mutant, which was defective in 6-phosphogluconate dehydrogenase activity, did not consume oxygen and died rapidly in the dark on solid media (< 1 per cent viable in 30 min compared with some 90 per cent survival of the wild type after 20 h). Partial phenotypic revertants (R1, R5), which survived dark incubation, were defective in both glucose-6-phosphate and 6-phosphogluconate dehydrogenases; it thus appeared that accumulated 6-phosphogluconate was the lethal agent, possibly by inhibiting glucose-6-phosphate isomerase, rather than the inability of the organisms to oxidize hexose phosphate via the pentose phosphate cycle in the dark. It was concluded that maintenance energy for survival of the revertants must be derived from some other pathway which did not involve oxygen uptake, possibly glycolysis, although Pelroy, Rippka & Stanier (1972) reported that *A. nidulans* has only low levels of phosphofructokinase. However, the authors did not measure the glycogen content of mutant cells during dark incubation which would have yielded direct information on this point.

If glycogen cannot be mobilized by mutants in the dark, RNA degradation might furnish the necessary maintenance energy and also account for the delayed photosynthetic oxygen evolution observed when the revertants were re-illuminated. This situation would also offer an explanation for the greatly decreased incorporation of [^{3}H]uracil recorded during dark incubation of R1 and R5 as compared with the wild type.

CONCLUSIONS AND OUTLOOK

Despite considerable effort, it has not proved possible to associate the death of vegetative bacteria with the loss of any specific cellular constituent or with the destruction of osmotic integrity. The pattern of depletion of cellular components during starvation is a characteristic of the individual organism and its growth environment, and marked differences are apparent between various species. However, although the DNA content of bacteria does not alter significantly during starvation and the accompanying loss of viability, the possibility exists that its structure may have altered in some way that predisposes organisms to die (Burleigh & Dawes, 1967; Postgate, 1967). Recently, for example, it has been found that in ageing mice the DNA displays an increasing proportion of single strands (Chetsanga *et al.*, 1975). At the time of writing there are no reports of similar studies with starved bacteria and such investigations might well be rewarding.

It is possible, too, that death is the result of non-specific changes of conformation in other individual macromolecules such as enzymes, even though those enzymes so far studied afford no unequivocal support for the idea. Lethal stresses may be physico-chemical resultants of the composition, ionic strength and osmotic pressure of the environment, and cleavage of a few covalent bonds within such molecules could contribute to the ultimate weakening of cell structure and organization.

Of the current evidence, the most promising explanation for bacterial survival has been derived from studies with organisms which display long term survival capacity. Their rates of metabolism of endogenous substrates are low, possibly generating ATP at a rate compatible with the maintenance requirements of the cell. Thus energy reserve polymers appear to enhance survival only if they are degraded at comparatively slow rates; if faster metabolism occurs the energy runs to waste and the bacteria die more rapidly.

Evolution appears to have selected bacteria with patterns of endogenous metabolism related to survival prospects in their normal environment. Thus, organisms subjected to a regular, recurring feast-and-famine existence, such as in the mouth or intestines, have a more rapid metabolism than organisms in the soil, where the rate of nutrient supply is low and uncertainty attends the next period of plenty. The enzymic constitution of the latter group presumably facilitates a low rate of metabolism but whether by regulation or altered enzyme characteristics is currently unknown, while the structure of glycogen reserves also appears related to the survival behaviour (Zevenhuizen & Ebbink, 1974). The findings with extremely starvation-resistant soil bacteria therefore highlight the aptness of Harrison & Lawrence's (1963) citation of Aesop's fable of the tortoise and the hare.

REFERENCES

ALTON, T. H. & KOCH, A. L. (1974). Unused protein synthetic capacity of *Escherichia coli* grown in phosphate-limited chemostats. *Journal of Molecular Biology*, **86**, 1–9.

ATKINSON, D. E. (1971). Adenine nucleotides as stoichiometric coupling agents in metabolism and as regulatory modifiers: the adenylate energy charge. In *Metabolic Pathways*, ed. H. J. Vogel, **5**, pp. 1–21. New York: Academic Press.

BELAICH, J-P. & SENEZ, J. C. (1965). Growth yields of *Zymomonas mobilis*. *Journal of Bacteriology*, **89**, 1195–1200.

BENTLEY, C. M. & DAWES, E. A. (1974). The energy-yielding reactions of *Peptococcus prévotii*, their behaviour on starvation and the role and regulation of threonine dehydratase. *Archives of Microbiology*, **100**, 363–387.

BOYLEN, C. W. & ENSIGN, J. C. (1970*a*). Long-term starvation survival of rod and spherical cells of *Arthrobacter crystallopoietes*. *Journal of Bacteriology*, **103**, 569–577.

BOYLEN, C. W. & ENSIGN, J. C. (1970*b*). Intracellular substrates for endogenous metabolism during long-term starvation of rod and spherical cells of *Arthrobacter crystallopoietes*. *Journal of Bacteriology*, **103**, 578–587.

BOYLEN, C. W. & PATE, J. L. (1973). Fine structure of *Arthrobacter crystallopoietes* during long-term starvation of rod and spherical stage cells. *Canadian Journal of Microbiology*, **19**, 1–5.

BRDAR, B., KOS, E. & DRAKULIĆ, M. (1965). Metabolism of nucleic acids and protein in starving bacteria. *Nature, London*, **208**, 303–304.

Builder, J. E. & Walker, G. J. (1970). Metabolism of the reserve polysaccharide of *Streptococcus mitis*. Properties of glycogen synthetase. *Carbohydrate Research*, **14**, 35–51.

Burleigh, I. G. & Dawes, E. A. (1967). Studies on the endogenous metabolism and senescence of starved *Sarcina lutea*. *Biochemical Journal*, **102**, 236–250.

Calcott, P. H., Montague, W. & Postgate, J. R. (1972). The levels of cyclic AMP during substrate-accelerated death. *Journal of General Microbiology*, **73**, 197–200.

Calcott, P. H. & Postgate, J. R. (1972). On substrate-accelerated death in *Klebsiella aerogenes*. *Journal of General Microbiology*, **70**, 115–122.

Calcott, P. H. & Postgate, J. R. (1974). The effects of β-galactosidase activity and cyclic AMP on lactose-accelerated death. *Journal of General Microbiology*, **85**, 85–90.

Chapman, A. G., Fall, L. & Atkinson, D. E. (1971). Adenylate energy charge in *Escherichia coli* during growth and starvation. *Journal of Bacteriology*, **108**, 1072–1086.

Chetsanga, C. J., Boyd, V., Peterson, L. & Rushlow, K. (1975). Single-stranded regions in DNA of old mice. *Nature, London*, **253**, 130–131.

Dagley, S. & Sykes, J. (1957). Effect of starvation upon the constitution of bacteria. *Nature, London*, **179**, 1249–1250.

Dalton, H. & Postgate, J. R. (1969). Growth and physiology of *Azotobacter chroococcum* in continuous culture. *Journal of General Microbiology*, **56** 307–319.

Dawes, E. A. & Holms, W. H. (1958). Metabolism of *Sarcina lutea*. III. Endogenous metabolism. *Biochimica et Biophysica Acta*, **30**, 278–293.

Dawes, E. A. & Large, P. J. (1970). Effect of starvation on the viability and cellular constituents of *Zymomonas anaerobia* and *Zymomonas mobilis*. *Journal of General Microbiology*, **60**, 31–42.

Dawes, E. A. & Ribbons, D. W. (1962). The endogenous metabolism of micro-organisms. *Annual Review of Microbiology*, **16**, 241–264.

Dawes, E. A. & Ribbons, D. W. (1963). Endogenous metabolism and survival of *Escherichia coli* in aqueous suspensions. *Journal of Applied Bacteriology*, **26**, vi.

Dawes, E. A. & Ribbons, D. W. (1964). Some aspects of the endogenous metabolism of bacteria. *Bacteriological Reviews*, **28**, 126–149.

Dawes, E. A. & Ribbons, D. W. (1965). Studies on the endogenous metabolism of *Escherichia coli*. *Biochemical Journal*, **95**, 332–343.

Dawes, E. A., Ribbons, D. W. & Rees, D. A. (1966). Sucrose utilization by *Zymomonas mobilis*: formation of a levan. *Biochemical Journal*, **98**, 804–812.

Dawes, E. A. & Senior, P. J. (1973). The role and regulation of energy reserve polymers in micro-organisms. *Advances in Microbial Physiology*, **10**, 135–266.

De Crombrugghe, B., Perlman, R. L., Varmus, H. E. & Pastan, I. (1969). Regulation of inducible enzyme synthesis in *Escherichia coli* by cyclic adenosine 3′,5′-monophosphate. *Journal of Biological Chemistry*, **244**, 5828–5835.

Dietzler, D. N., Lais, C. J. & Leckie, M. P. (1974). Simultaneous increases of the adenylate energy charge and the rate of glycogen synthesis in nitrogen-starved *Escherichia coli* W 4597(K). *Archives of Biochemistry and Biophysics*, **160**, 14–25.

Doolittle, W. F. & Singer, R. A. (1974). Mutational analysis of dark endogenous metabolism in the blue-green bacterium *Anacystis nidulans*. *Journal of Bacteriology*, **119**, 677–683.

Duncan, M. G. & Campbell, J. J. R. (1962). Oxidative assimilation of glucose by *Pseudomonas aeruginosa*. *Journal of Bacteriology*, **84**, 784–792.

ENSIGN, J. C. & WOLFE, R. S. (1964). Nutritional control of morphogenesis in *Arthrobacter crystallopoietes*. *Journal of Bacteriology*, **87**, 924–932.

FORREST, W. W. (1972). Microcalorimetry. In *Methods in Microbiology*, **6B**, ed. J. R. Norris & D. W. Ribbons, pp. 285–318. London: Academic Press.

FORREST, W. W. & WALKER, D. J. (1963). Calorimetric measurements of energy of maintenance of *Streptococcus faecalis*. *Biochemical and Biophysical Research Communications*, **13**, 217–222.

FORREST, W. W. & WALKER, D. J. (1965). Synthesis of reserve materials for endogenous metabolism in *Streptococcus faecalis*. *Journal of Bacteriology*, **89**, 1448–1452.

FRIDOVICH, I. (1974). Superoxide dismutases. *Advances in Enzymology*, **41**, 35–97.

GIBBONS, R. J. & KAPSIMALIS, B. (1963). Synthesis of intracellular iodophilic polysaccharide by *Streptococcus mitis*. *Archives of Oral Biology*, **8**, 319–329.

GOLDBERG, ALFRED L. (1971). A role of aminoacyl-(transfer) RNA in the regulation of protein breakdown in *Escherichia coli*. *Proceedings of the National Academy of Sciences, USA*, **68**, 362–366.

GREGORY, E. M. & FRIDOVICH, I. (1973*a*). Induction of superoxide dismutase by molecular oxygen. *Journal of Bacteriology*, **114**, 543–548.

GREGORY, E. M. & FRIDOVICH, I. (1973*b*). Oxygen toxicity and the superoxide dismutase. *Journal of Bacteriology*, **114**, 1193–1197.

GRONLUND, A. F. & CAMPBELL, J. J. R. (1961). Nitrogenous compounds as substrates for endogenous metabolism in micro-organisms. *Journal of Bacteriology*, **81**, 721–724.

GRONLUND, A. F. & CAMPBELL, J. J. R. (1963). Nitrogenous substrates of endogenous respiration in *Pseudomonas aeruginosa*. *Journal of Bacteriology*, **86**, 58–66.

GRONLUND, A. F. & CAMPBELL, J. J. R. (1965). Enzymatic degradation of ribosomes during endogenous metabolism of *Pseudomonas aeruginosa*. *Journal of Bacteriology*, **90**, 1–7.

GRONLUND, A. F. & CAMPBELL, J. J. R. (1966). Influence of exogenous substrates on the endogenous respiration of *Pseudomonas aeruginosa*. *Journal of Bacteriology*, **91**, 1577–1581.

HALVORSON, H. O. (1962). The function and control of intracellular protein turnover in micro-organisms. In *Amino Acid Pools*, ed. J. T. Holden, pp. 646–654. Amsterdam: Elsevier.

HAMILTON, I. R. (1968). Synthesis and degradation of intracellular polyglucose in *Streptococcus salivarius*. *Canadian Journal of Microbiology*, **14**, 65–77.

HAROLD, F. M. (1966). Inorganic polyphosphates in biology: structure, metabolism and function. *Bacteriological Reviews*, **30**, 772–794.

HARRISON, A. P., JR & LAWRENCE, F. R. (1963). Phenotypic, genotypic and chemical changes in starving populations of *Aerobacter aerogenes*. *Journal of Bacteriology*, **85**, 742–750.

HERBERT, D. (1961). The chemical composition of micro-organisms as a function of their environment. *Symposia of the Society for General Microbiology*, **11**, 391–416.

HESPELL, R. B., ROSSON, R. A., THOMASHOW, M. F. & RITTENBERG, S. C. (1973). Respiration of *Bdellovibrio bacteriovorus* Strain 109J and its energy substrates for intraperiplasmic growth. *Journal of Bacteriology*, **113**, 1280–1288.

HESPELL, R. B., THOMASHOW, M. F. & RITTENBERG, S. C. (1974). Changes in cell composition and viability of *Bdellovibrio bacteriovorus* during starvation. *Archives for Microbiology*, **97**, 313–327.

HIPPE, H. (1967). Abbau und Wiederverwertung von Poly-β-hydroxy-buttersäure durch Hydrogenomonas H16. *Archiv für Mikrobiologie*, **56**, 248–277.

Jacobson, A. & Gillespie, D. (1968). Metabolic events occurring during recovery from prolonged glucose starvation in *Escherichia coli*. *Journal of Bacteriology*, **95**, 1030–1039.

Koch, A. L. (1971). The adaptive responses of *Escherichia coli* to a feast and famine existence. *Advances in Microbial Physiology*, **6**, 147–217.

Laishley, E. J., Brown, R. G. & Otto, M. C. (1974). Characteristics of a reserve α-glucan from *Clostridium pasteurianum*. *Canadian Journal of Microbiology*, **20**, 559–562.

Maaløe, O. & Kjeldgaard, N. O. (1966). In *Control of Macromolecular Synthesis*, pp. 109–114. New York: W. A. Benjamin, Inc.

MacKelvie, R. M., Campbell, J. J. R. & Gronlund, A. F. (1968*a*). Absence of storage products in cultures of *Pseudomonas aeruginosa* grown with excess carbon or nitrogen. *Canadian Journal of Microbiology*, **14**, 627–631.

MacKelvie, R. M., Campbell, J. J. R. & Gronlund, A. F. (1968*b*). Survival and intracellular changes of *Pseudomonas aeruginosa* during prolonged starvation. *Canadian Journal of Microbiology*, **14**, 639–645.

Mallette, M. F. (1963). Validity of the concept of energy of maintenance. *Annals of the New York Academy of Sciences*, **102** (3), 521–535.

Marr, A. G., Nilson, E. H. & Clark, D. J. (1963). The maintenance requirement of *Escherichia coli*. *Annals of the New York Academy of Sciences*, **102**, 536–548.

McGill, D. J. & Dawes, E. A. (1971). Glucose and fructose metabolism in *Zymomonas anaerobia*. *Biochemical Journal*, **125**, 1059–1068.

McGrew, S. B. & Mallette, M. F. (1962). Energy of maintenance in *Escherichia coli*. *Journal of Bacteriology*, **83**, 844–850.

McGrew, S. B. & Mallette, M. F. (1965). Maintenance of *Escherichia coli* and the assimilation of glucose. *Nature, London*, **208**, 1096–1097.

Montague, M. D. & Dawes, E. A. (1974). The survival of *Peptococcus prévotii* in relation to the adenylate energy charge. *Journal of General Microbiology*, **80**, 291–299.

Mulder, E. G. & Zevenhuizen, L. P. T. M. (1967). Coryneform bacteria of the *Arthrobacter* type and their reserve material. *Archiv für Mikrobiologie*, **59**, 345–354.

Pastan, I. & Perlman, R. (1970). Cyclic adenosine monophosphate in bacteria. *Science, Washington*, **169**, 339–344.

Pelroy, R. A., Rippka, R. & Stanier, R. Y. (1972). Metabolism of glucose by unicellular blue-green algae. *Archives of Microbiology*, **87**, 303–322.

Pine, M. J. (1970). Steady-state measurement of the turnover of amino acid in the cellular proteins of growing *Escherichia coli*: existence of two kinetically distinct reactions. *Journal of Bacteriology*, **103**, 207–215.

Pine, M. J. (1972). Turnover of intracellular proteins. *Annual Review of Microbiology*, **26**, 103–126.

Pirt, S. J. (1965). The maintenance energy of bacteria in growing cultures. *Proceedings of the Royal Society of London*, Series B, **163**, 224–231.

Postgate, J. R. (1967). Viability measurements and the survival of microbes under minimum stress. *Advances in Microbial Physiology*, **1**, 1–23.

Postgate, J. R. (1969). Viable counts and viability. In *Methods in Microbiology*, **1**, ed. J. R. Norris & D. W. Ribbons, pp. 611–628. London: Academic Press.

Postgate, J. R. & Hunter, J. R. (1962). The survival of starved bacteria. *Journal of General Microbiology*, **29**, 233–263.

Postgate, J. R. & Hunter, J. R. (1964). Accelerated death of *Aerobacter aerogenes* starved in the presence of growth-limiting substrates. *Journal of General Microbiology*, **34**, 459–473.

Ribbons, D. W. & Dawes, E. A. (1963). Environmental and growth conditions affecting the endogenous metabolism of bacteria. *Annals of the New York Academy of Sciences*, **102**, 564–586.

Rickenberg, H. V. (1974). Cyclic AMP in prokaryotes. *Annual Review of Microbiology*, **28**, 353–369.

Rittenberg, S. C. & Shilo, M. (1970). Early host damage in the infection cycle of *Bdellovibrio bacteriovorus*. *Journal of Bacteriology*, **102**, 149–160.

Robertson, J. G. & Batt, R. D. (1973). Survival of *Nocardia corallina* and degradation of constituents during starvation. *Journal of General Microbiology*, **78**, 109–117.

Robson, R. L. & Morris, J. G. (1974). Mobilization of granulose in *Clostridium pasteurianum*. Purification and properties of granulose phosphorylase. *Biochemical Journal*, **144**, 513–517.

Robson, R. L., Robson, R. M. & Morris, J. G. (1974). The biosynthesis of granulose by *Clostridium pasteurianum*. *Biochemical Journal*, **144**, 503–511.

Schlegel, H. G., Gottschalk, G. & Von Bartha, R. (1961). Formation and utilization of poly-β-hydroxybutyric acid by Knallgas bacteria (*Hydrogenomonas*). *Nature, London*, **191**, 463–465.

Schulze, K. L. & Lipe, R. S. (1964). Relationship between substrate concentration, growth rate, and respiration rate of *Escherichia coli* in continuous culture. *Archiv für Mikrobiologie*, **48**, 1–20.

Senior, P. J., Beech, G. A., Ritchie, G. A. F. & Dawes, E. A. (1972). The role of oxygen limitation in the formation of poly-β-hydroxybutyrate during batch and continuous culture of *Azotobacter beijerinckii*. *Biochemical Journal*, **128**, 1193–1201.

Sierra, G. & Gibbons, N. E. (1962). Role and oxidation pathway of poly-β-hydroxybutyric acid in *Micrococcus halodenitrificans*. *Canadian Journal of Microbiology*, **8**, 255–269.

Sobek, J. M., Charba, J. F. & Foust, W. N. (1966). Endogenous metabolism of *Azotobacter agilis*. *Journal of Bacteriology*, **92**, 687–695.

Stockdale, H. (1967). A comparative survey of poly-β-hydroxybutyrate in the Azotobacteriaceae with special reference to the endogenous metabolism and survival of *Azotobacter insigne*. Ph.D. Thesis, University of Hull.

Stokes, J. L. & Parson, W. L. (1968). Role of poly-β-hydroxybutyrate in survival of *Sphaerotilus discophorus* during starvation. *Canadian Journal of Microbiology*, **14**, 785–789.

Stouthamer, A. H. & Bettenhaussen, C. (1973). Utilization of energy for growth and maintenance in continuous and batch cultures of micro-organisms. A reevaluation of the method for the determination of ATP production by measuring molar growth yields. *Biochimica et Biophysica Acta*, **301**, 53–70.

Strange, R. E. (1966). Stability of β-Galactosidase in starved *Escherichia coli*. *Nature, London*, **209**, 428–429.

Strange, R. E. (1967). In *Microbial Physiology and Continuous Culture*, ed. E. O. Powell, C. G. T. Evans, R. E. Strange & D. W. Tempest, p. 122. London: HMSO.

Strange, R. E. (1968). Bacterial glycogen and survival. *Nature, London*, **220**, 606–607.

Strange, R. E. & Dark, F. A. (1965). 'Substrate-accelerated death' of *Aerobacter aerogenes*. *Journal of General Microbiology*, **39**, 215–228.

Strange, R. E., Dark, F. A. & Ness, A. G. (1961). The survival of stationary phase *Aerobacter aerogenes* stored in aqueous suspension. *Journal of General Microbiology*, **25**, 61–76.

STRANGE, R. E. & HUNTER, J. R. (1966). 'Substrate-accelerated death' of nitrogen-limited bacteria. *Journal of General Microbiology*, **44**, 255–262.

STRANGE, R. E. & HUNTER, J. R. (1967). Effect of magnesium on the survival of bacteria in aqueous suspension. In *Microbial Physiology and Continuous Culture*, ed. E. O. Powell, C. G. T. Evans, R. E. Strange & D. W. Tempest, pp. 102–121. London: HMSO.

STRASDINE, G. A. (1972). The role of intracellular glucan in endogenous fermentation and spore maturation in *Clostridium botulinum* type E. *Canadian Journal of Microbiology*, **18**, 211–217.

SUSSMAN, A. J. & GILVARG, C. (1969). Protein turnover in amino-acid-starved strains of *Escherichia coli* K-12 differing in their ribonucleic acid control. *Journal of Biological Chemistry*, **244**, 6304–6306.

TEMPEST, D. W. & DICKS, J. W. (1967). Inter-relationships between potassium, magnesium, phosphorus and ribonucleic acid in the growth of *Aerobacter aerogenes* in a chemostat. In *Microbial Physiology and Continuous Culture*, ed. E. O. Powell, C. G. T. Evans, R. E. Strange & D. W. Tempest, pp. 140–153. London: HMSO.

TEMPEST, D. W., HERBERT, D. & PHIPPS, P. J. (1967). Studies on the growth of *Aerobacter aerogenes* at low dilution rates in a chemostat. In *Microbial Physiology and Continuous Culture*, ed. E. O. Powell, C. G. T. Evans, R. E. Strange & D. W. Tempest, pp. 240–253. London: HMSO.

TEMPEST, D. W. & STRANGE, R. E. (1966). Variation in content and distribution of magnesium and its influence on survival, in *Aerobacter aerogenes* grown in a chemostat. *Journal of General Microbiology*, **44**, 273–279.

THOMAS, T. D. & BATT, R. D. (1968). Survival of *Streptococcus lactis* in starvation conditions. *Journal of General Microbiology*, **50**, 367–382.

THOMAS, T. D. & BATT, R. D. (1969*a*). Degradation of cell constituents by starved *Streptococcus lactis* in relation to survival. *Journal of General Microbiology*, **58**, 347–362.

THOMAS, T. D. & BATT, R. D. (1969*b*). Synthesis of protein and ribonucleic acid by starved *Streptococcus lactis* in relation to survival. *Journal of General Microbiology*, **58**, 363–369.

THOMAS, T. D. & BATT, R. D. (1969*c*). Metabolism of exogenous arginine and glucose by starved *Streptococcus lactis* in relation to survival. *Journal of General Microbiology*, **58**, 371–380.

THOMAS, T. D., LYTTLETON, P., WILLIAMSON, K. I. & BATT, R. D. (1969). Changes in permeability and ultrastructure of starved *Streptococcus lactis* in relation to survival. *Journal of General Microbiology*, **58**, 381–390.

VALLEE, B. L. (1960). Metal and enzyme interactions: correlation of composition, function and structure. In *The Enzymes*, **3**, ed. P. D. Boyer, H. Lardy & K. Myrback, pp. 334–344. New York: Academic Press.

VAN HOUTE, J. & JANSEN, H. M. (1970). Role of glycogen in survival of *Streptococcus mitis*. *Journal of Bacteriology*, **101**, 1083–1085.

WALKER, D. J. & FORREST, W. W. (1964). Anaerobic endogenous metabolism in *Streptococcus faecalis*. *Journal of Bacteriology*, **87**, 256–262.

WHYTE, J. N. C. & STRASDINE, G. A. (1972). An intracellular α-D-glucan in *Clostridium botulinum* E. *Carbohydrate Research*, **25**, 435–443.

WILKINSON, J. F. (1959). The problem of energy-storage compounds in bacteria. *Experimental Cell Research Supplement*, **7**, 111–130.

WILKINSON, J. F. & MUNRO, A. L. S. (1967). The influence of growth limiting conditions on the synthesis of possible carbon and energy storage polymers in *Bacillus megaterium*. In *Microbial Physiology and Continuous Culture*, ed. E. O. Powell, C. G. T. Evans, R. E. Strange & D. W. Tempest, pp. 173–184. London: HMSO.

WILLETTS, N. S. (1965). Protein degradation during diauxic growth of *Escherichia coli*. *Biochemical and Biophysical Research Communications*, **20**, 692–696.

ZEVENHUIZEN, L. P. T. M. (1966). Formation and function of the glycogen-like polysaccharide of *Arthrobacter*. *Antonie van Leeuwenhoek Journal of Microbiology and Serology*, **32**, 356–372.

ZEVENHUIZEN, L. P. T. M. & EBBINK, A. G. (1974). Interrelations between glycogen, poly-β-hydroxybutyric acid and lipids during accumulation and subsequent utilization in a *Pseudomonas*. *Antonie van Leeuwenhoek Journal of Microbiology and Serology*, **40**, 103–120.

TRANSITION TO THE NON-GROWING STATE IN EUKARYOTIC MICRO-ORGANISMS

A. P. J. TRINCI AND C. F. THURSTON
Department of Microbiology, Queen Elizabeth College, Campden Hill, London W8 7AH

INTRODUCTION

Growth of a microbial population may decelerate and eventually cease either because the cells age and become senescent, or because some environmental condition becomes unfavourable for growth, such as nutrient exhaustion or changes in pH or temperature. Processes which can be included within even fairly loose definitions of ageing have been observed in very few eukaryotic micro-organisms, and where age dependent changes are observed, these processes are ill defined. Consequently ageing and senescence will only be considered briefly in our article, the larger part being devoted to organisms which are not growing because they are starved rather than aged. Although several detailed studies have been made of the morphology, endogenous metabolism and survival of stationary phase cultures of bacteria, few comparable studies have been made with eukaryotic micro-organisms. Further, interpretation of the studies which have been made is often equivocal because it is not known which environmental factor caused the cultures to enter the stationary phase.

Growth on solid media inevitably results in the formation of colonies which are heterogeneous with respect to age (the middle is older than the periphery) and this heterogeneity is reflected in their morphology, physiology and biochemistry (Yanagita & Kogané, 1962). Some fungi even form colonies (pellets) when grown in submerged culture, but under these conditions most micro-organisms form populations which are superficially homogeneous. Transition to a non-growing state in a population growing as a colony and in a homogeneous population are quite different processes.

TRANSITION TO THE NON-GROWING STATE DUE TO AGEING

Ageing is an endogenous process which proceeds more or less independently of the exogenous conditions. Mammalian diploid cell strains appear to age in culture, since only a limited number of divisions

(40 to 60) occur before they become senescent and die (Hayflick & Moorehead, 1961). Ageing of human diploid fibroblasts in culture may be related to ageing of the whole organism (Holliday, 1972) since the number of divisions during culture of cells is reduced as the age of the individual from which they were isolated is increased (Martin, Sprague & Epstein, 1970). Fibroblasts isolated from patients with Werner's syndrome (a condition of premature ageing) divide fewer times than cultured fibroblasts from normal individuals of a similar age (Epstein *et al.*, 1966).

There is good evidence that at least some eukaryotic micro-organisms age in culture in a way which is independent of exogenous conditions. The life span of *Paramecium* varies from about 20 to about 200 vegetative generations and ageing of a clone is usually accompanied by increases in the duration of the generation time and the period of DNA synthesis (the S period in the cell cycle), but not the G_1 period (Smith-Sonneborn & Klass, 1974). Vegetative cells of an aged clone of *Paramecium* are rejuvenated when they undergo autogamy or conjugation (Smith-Sonneborn, Klass & Cotton, 1974).

Saccharomyces cerevisiae cells also have a finite life span. The number of times a particular cell has divided, and hence its age, may be determined from the number of bud scars on its surface, since budding never occurs on the site of a previous scar (Bartnicki-Garcia & McMurrough, 1971). Although a yeast cell apparently has sufficient surface area to form up to about 100 bud scars, they never produce more than 20 to 50 buds (Cook, 1963). Ageing in *S. cerevisiae*, unlike that in *Paramecium*, appears to be a sudden process rather than a gradual degeneration (Cook, 1963). However, variation in the sporulation ability of *S. cerevisiae* appears to be age dependent (Sando *et al.*, 1973). Like *Paramecium*, yeast cells may be rejuvenated by conjugation as the ascus mother cell wall is discarded when the ascospores are released.

Some coenocytic moulds, e.g. *Podospora anserina* and *Aspergillus glaucus*, undergo an ageing process, known as clonal senescence or vegetative death, which is exhibited as a fairly sudden drop in colony radial growth rate, hyphal distortion and the eventual cessation of colony expansion (Holliday, 1969). Ageing in these moulds is claimed to have a cytoplasmic rather than a nuclear basis and may be consistent with Orgel's (1963) hypothesis that degeneration can be caused by failure to maintain the accuracy of protein synthesis (Holliday, 1969; Lewis & Holliday, 1970). Orgel suggested that during protein synthesis there are random errors in amino acid sequences of proteins and when the errors are in proteins which are themselves involved in transcription and translation they will increase the rate at which further defective proteins accumulate. Eventually the whole machinery of protein synthesis becomes defective and the organism dies.

TRANSITION TO THE NON-GROWING STATE WITHOUT AGEING

The behaviour of stationary phase organisms will depend upon the nature of the growth-limiting substrate or the inhibitory condition (Pirt, 1975). Growth of microbial populations often decelerates and stops, because of the exhaustion of an essential nutrient in the culture medium. The media may be designed so that a particular nutrient is the first to become exhausted, but unfortunately the limiting nutrient is not known for many common culture media. During the deceleration phase, which normally precedes cessation of growth and complete nutrient exhaustion, the specific growth rate of the organism will be substrate limited (Pirt, 1975). Microbial populations are often experimentally starved of one or more nutrients by harvesting them from growing cultures and resuspending the cells, after appropriate washing procedures, in a deficient medium.

The behaviour of a microbial population starved of an exogenous supply of a particular nutrient will vary depending on the amount of endogenous reserve of the nutrient and whether growth is possible in its absence. Microbial populations starved of an exogenous carbon and energy source lose viability and autolyse once their endogenous reserves are depleted, but nitrogen-starved cultures may increase in biomass for some considerable period after the onset of starvation and lose viability less rapidly, although their growth will be 'unbalanced' (Campbell, 1957).

The rate of growth of a microbial population may decelerate and growth may stop even before the advent of nutrient exhaustion. This may be associated with changes in the pH of the medium and/or the accumulation of products of secondary metabolism.

In many species transition to the non-growing state is associated with differentiation resulting in the formation of structures which can withstand unfavourable conditions (Ashworth & Smith, 1973). The ecological advantage in such a developmental sequence is obvious.

AUTOLYSIS OF MOULD CULTURES

Autolysing cultures of *Penicillium chrysogenum* (Trinci & Righelato, 1970) and *Aspergillus nidulans* (Bainbridge *et al.*, 1971) were cytologically heterogeneous throughout four to five day periods of carbon starvation in that 'normal' hyphae were observed alongside autolysing hyphae. The cytological heterogeneity of the whole culture was reflected in individual hyphae since compartments (the coenocytic cell between adjacent septa) which were clearly autolysing were observed next to others which appeared to be 'normal'. The cytological disorganization of a particular compartment proceeded uniformly rather than by a sequential breakdown of its organelles by autophagy; the individuality of each hyphal compartment was maintained as the pores in the septa were plugged. The cytological heterogeneity of the culture suggested that cryptic growth occurred during the starvation period, and that some hyphal compartments were maintained (Pirt, 1965) or even grew using substances released during the autolysis of other compartments. The observed increase in intrahyphal hyphae during the starvation period may indicate that cryptic growth took this particular form. In the later stages of autolysis, the hyphae contained only an accumulation of disorganized membranes and discontinuities appeared in the hyphal walls and septa. The degeneration process eventually reached a stage where the hyphae were devoid of visible contents. Mycelial disruption (caused partly by the impeller) resulted in a four-fold increase in the number of hyphal fragments in cultures of *A. nidulans* which had been carbon-starved for about seven days. About 60 per cent of the mycelial fragments were viable after four days starvation. Since each mycelial fragment consisted of a large number of hyphal compartments, the significance of such viability measurements is doubtful.

THE METABOLISM OF STARVING ORGANISMS

Storage materials and starvation

Changes in cell composition of algae which have been nitrogen-starved have been reviewed by Fogg (1959) and Syrett (1962). Nitrogen-starved algae continue to make both polysaccharides and lipids, but whereas increase in polysaccharide content is of limited duration, lipid synthesis continues for long periods (weeks). Some algae such as *Monodus subterraneus* are unable to store polysaccharide at all and hence during nitrogen-starvation accumulate only lipid (Fogg, 1959).

Algae such as *Chlorella* which accumulate starch probably do so as a means of storing a source of respiratory substrate (Cramer & Myers, 1949). In light–dark synchronized *Chlorella fusca,** starch made during the light period is consumed during the dark period (Atkinson, John & Gunning, 1974). It has been shown that both starch and sucrose contribute to the metabolism of starving *C. pyrenoidosa*. In cells starved of both an energy source and a nitrogen source (nitrogen-free medium, in the dark), starch is steadily depleted but sucrose is not. Addition of ammonium to the medium does not change the rate of starch utilization, but sucrose is consumed as well (Kanazawa *et al*., 1972). It appears that starch is used as substrate for endogenous respiration whereas sucrose supplies the carbon skeletons for synthesis of new nitrogen-containing molecules such as amino acids. Dual carbohydrate stores also occur in fungi, e.g. glycogen and trehalose in *S. cerevisiae*. In carbon- and nitrogen-starved cells, trehalose is conserved, but it is consumed within one hour if an assimilable nitrogen source is supplied (Sols, Gancedo & Delafuente, 1971).

Both carbohydrate and lipid accumulate during phosphate-starvation of *Chlorella* and *Ankistrodesmus* (Kuhl, 1974). Lipid accumulation has also been been observed during phosphate starvation in *S. cerevisiae* (Suomalainen & Oura, 1971).

It is not known whether any nitrogen-containing compound produced during growth can be considered as a reserve of this nutrient during nitrogen-starvation. The nitrogen content of both fungi and algae may decrease during nitrogen-starvation. In *Chlorella*, *Scenedesmus* and *Monodus* the nitrogen content falls from about 10 per cent of dry weight to about two per cent (Fogg, 1959; Syrett, 1962). Mycelium-nitrogen of *Penicillium griseofulvum* is reduced by half after 24 hours nitrogen-starvation under anaerobic conditions, although loss of nitrogen is not observed under aerobic conditions (Morton, Dickerson & England, 1960), contrasting with data for carbon-starved *P. chrysogenum* described later. Although amounts of protein, nucleic acids and acid-soluble nitrogen compounds may be reduced during nitrogen-starvation, it is not known whether any of these are degraded to allow continued synthesis of other nitrogen-containing compounds. During nitrogen-starvation, chlorophyll content of algae falls faster than total nitrogen content (Fogg, 1959; Syrett, 1962; Yentsch, 1962), but the

* The *Chlorella* strain used by Atkinson *et al.* (1974) is *C. fusca* var. *vacuolata* Cambridge culture collection number 211/8p. This strain was originally called *C. vulgaris*, Pearsall's strain, as in Syrett (1958) and Morris & Syrett (1965). It was reclassified as *C. pyrenoidosa* by Syrett (1966) as in John, Thurston & Syrett (1970). The present name follows the classification of Fott & Novakova (1969) and is used throughout.

fate of nitrogen made available by chlorophyll breakdown has not been studied.

The behaviour of carbon-starved cultures of *P. chrysogenum* (Trinci & Righelato, 1970) and *Aspergillus nidulans* (Bainbridge *et al.*, 1971) has been studied after stopping the nutrient supply to glucose-limited chemostat cultures. The mycelia were almost immediately starved of a carbon source in the presence of an excess of all other nutrients. Autolysis began without an appreciable lag and was characterized by decreases in dry weight and protein, RNA and DNA contents of the culture. The total carbohydrate content of mycelia remained constant for about 48 hours in *P. chrysogenum* and 24 hours in *A. nidulans*, but subsequently decreased. Similar decreases have been observed in the DNA, RNA and protein content of starved cultures of *Saccharomyces carlsbergensis* during autolysis at 45 °C (Hough & Maddox, 1970), and in DNA and RNA contents of stationary phase cultures of *Protomyces inundatus* (Valadon, Myers & Manners, 1962).

Energy metabolism

In carbon-starved fungi and algae (the latter in the dark) a low level of respiration is found (Gibbs, 1962). This endogenous respiration has been shown to be fuelled by consumption of stored carbohydrate (see above). The rate of respiration in starved cells is not determined by the activity of the respiratory enzyme system as addition of exogenous substrate strongly stimulated respiration (Gibbs, 1962). Although endogenous respiration rate falls at the beginning of starvation, there may be little subsequent change in rate for many days. For instance, respiration rate in nitrogen-starved *M. subterraneus* is essentially constant from the second to the seventh day of starvation, after which measurements were not recorded (Fogg, 1959). In *Prototheca zopfii*, endogenous respiration is essentially constant for 30 days, but has declined to less than five per cent of its initial value after 45 days of starvation (Lloyd, 1974). Interestingly, substrate stimulated respiration, in the presence of 10 mM acetate in *P. zopfii* starved for 45 days, is about 25 per cent of the rate measured at the onset of starvation and is at least as fast as the initial endogenous respiration rate. Substrate-stimulated respiration rate in carbon-starved *S. cerevisiae* has been investigated by Miura & Yanagita (1972). After 20 days starvation, the rate had dropped by about two-thirds. Loss of respiratory activity occurred in two steps, the second of which began after 15 days and coincided with loss of activity of succinate–cytochrome *c* reductase and cytochrome oxidase.

Bainbridge *et al.* (1971) showed that there was a decline in the activity of the hexose-monophosphate pathway in carbon-starved cultures of *A. nidulans*. It has not been established if a eukaryotic cell can maintain its vitality during starvation conditions by degrading some of its functional components, e.g. ribosomes. It is possible that the maintenance of viability in carbon-starved cultures is dependent upon the presence of endogenous carbon-containing reserves and/or carbon-containing substances released from cells in the population which are lysing. Fungi, unlike bacteria (Dawes & Ribbons, 1964), may be able to use wall polymers as substrates for endogenous metabolism. Zonneveld (1972) claimed that the α-1,3-glucan of the walls of *A. nidulans*' hyphae represented the organism's main reserve polymer. During carbon-starvation there was an increase in α-1,3-glucanase activity, concomitant with a decrease in the α-1,3-glucan content of the hyphal walls. It is, however, not known if a particular hyphal compartment can use its own wall polymers as substrates for endogenous metabolism. It is possible that some hyphae degrade the walls of other hyphae or utilize the products of autolysis of the walls of other hyphae. It is perhaps significant that Mahadevan & Mahadkar (1970) have shown that *Neurospora crassa* walls contain bound enzymes (proteases and glucanases) which can degrade some of the wall polymers. During carbon-starvation of cultures of *Schizophyllum commune* the R glucan of the hyphal walls is degraded, enabling pileus formation to occur. The chitin component of hyphal walls is probably not degraded during autolysis (Lahoz & Ibeas, 1968).

In algae, the cellular content of ATP under starvation conditions is remarkably similar to values for growing organisms (Syrett, 1958; Holm-Hansen, 1970; Löppert & Broda, 1973), although prolonged nitrogen- or phosphorus-starvation (10 to 30 days) decreases cellular ATP to less than half the amount found in growing algae (Holm-Hansen, 1970). Such reductions in ATP appear to be no greater than the reduction of other nitrogen-containing cellular components. Although the amount of ATP per cell changes very little at the onset of starvation in *Chlorella fusca*, the rate of ATP formation is greatly reduced (Löppert & Broda, 1973).

The onset of starvation provokes an increase in the efficiency of respiratory energy conservation in some organisms. It has been shown for *Acanthamoeba castellani* that respiration becomes rotenone sensitive in stationary phase cultures and ADP:O ratio values approaching three are obtained (Evans, 1973). It is inferred that site 1 oxidative phosphorylation is absent in growing organisms, where ADP:O ratios

Table 1. *Maintenance coefficients for glucose and oxygen*

Organism	Maintenance coefficient for glucose (g g^{-1} (dry wt) h^{-1})	Source
*Saccharomyces cerevisiae**	0.036	Watson, 1970
Penicillium chrysogenum	0.022	Righelato *et al.*, 1968
Aspergillus nidulans†	0.029	Carter *et al.*, 1971
A. niger	0.018	Ng, Smith & McIntosh, 1973
Mucor hiemalis	0.04	Lynch & Harper, 1974
	Maintenance coefficient for oxygen (mmol g^{-1} (dry wt) h^{-1})	
A. nidulans	0.55	Carter *et al.*, 1971
P. chrysogenum	0.74	Righelato *et al.*, 1968

* Under anaerobic conditions.
† Corrected for dry weight due to melanin.

less than two are observed, but present in non-growing organisms. Similar changes in rotenone sensitivity have been obtained with *Torulopsis utilis* (Katz, 1971; Katz, Kilpatrick & Chance, 1971).

Maintenance energy

A microbial population will only survive and remain viable if it can generate sufficient energy from endogenous or exogenous sources to maintain its non-growth-associated functions such as osmotic regulation, pH control and the turnover of macromolecules. The energy used to maintain these functions is called the energy of maintenance. The values which have been obtained with fungi for the maintenance coefficients for glucose and oxygen are listed in Table 1. The cytological heterogeneity of 'maintained' cultures of *Penicillium chrysogenum* (Righelato *et al.*, 1968) and *Aspergillus nidulans* (Bainbridge *et al.*, 1971) suggest that it is difficult, if not impossible, to achieve stable maintenance states under experimental conditions.

Protein metabolism

During growth the rate of protein synthesis greatly exceeds the rate of protein degradation. In growing *S. cerevisiae* where the rate of protein synthesis (for doubling time of two hours) is about 35 per cent h^{-1}, the rate of protein degradation is less than one per cent of the rate of synthesis (Halvorson, 1958*a*). Such a difference in rate may be exceptional but even in organisms which grow more slowly the rate of protein degradation is much less than the rate of synthesis during growth. *C. fusca* cultures with a net rate of protein synthesis of 8.7 per cent h^{-1} (mass doubling in eight hours) degrade protein at a rate of 1.7 per cent h^{-1} (L. Richards & C. F. Thurston, unpublished data).

In non-growing organisms there is no increase in total protein, but protein synthesis does not stop. In starved *S. cerevisiae* (Halvorson, 1958*b*), protein turnover is 0.7 per cent h^{-1}; that is, the rate of protein synthesis is 50 times less than during growth, but in addition the rate of protein degradation is increased. Our measurements with nitrogen-starved *C. fusca* show a similar pattern, although when growth ceases the decrease in rate of protein synthesis is less, as the rate of protein degradation is always much greater than in yeast. Protein concentration is constant during the first 24 hours of nitrogen-starvation of this organism (L. Richards & C. F. Thurston, unpublished data). A more extreme change in protein metabolism occurs when a *Penicillium chrysogenum* culture is carbon-starved (Trinci & Righelato, 1970). Protein concentration falls exponentially with an initial rate of 5.7 per cent h^{-1}. This presumably reflects a large increase in protein degradation rate at the onset of starvation. The absolute rates of protein degradation during growth and during carbon-starvation were not measured. Similar results were obtained with *A. nidulans* (Bainbridge *et al.*, 1971). Where organisms differentiate in response to starved conditions, very high values of protein turnover are found, as for instance in *Bacillus subtilis* (Mandelstam & Waites, 1968), *Dictyostelium discoideum* (Wright & Anderson, 1960) and *Blastocladiella emersonii* (Lodi & Sonneborn, 1974), but discussion of these is outside the scope of this review.

It may be supposed that non-growing organisms degrade protein at high rates, *vis à vis* growing organisms, to provide precursors for continued protein synthesis, which is consistent with the rapid depletion of free amino acid pools observed when *S. cerevisiae* cultures are starved of nitrogen (Halvorson, Fry & Schwemmin, 1955). This in turn poses the question: why do starving organisms require ongoing synthesis? The following may be contributory reasons, albeit far short of a full explanation. First, some proteins may be inherently unstable and require constant degradation and resynthesis. This together with replacement of 'damaged' protein has frequently been proposed as a rationale for protein degradation which may be applied both to growing and non-growing organisms (Siekevitz, 1972; Goldberg *et al.*, 1974). Second, as Mandelstam (1960) has suggested, turnover allows non-growing organisms to change their protein composition, so that

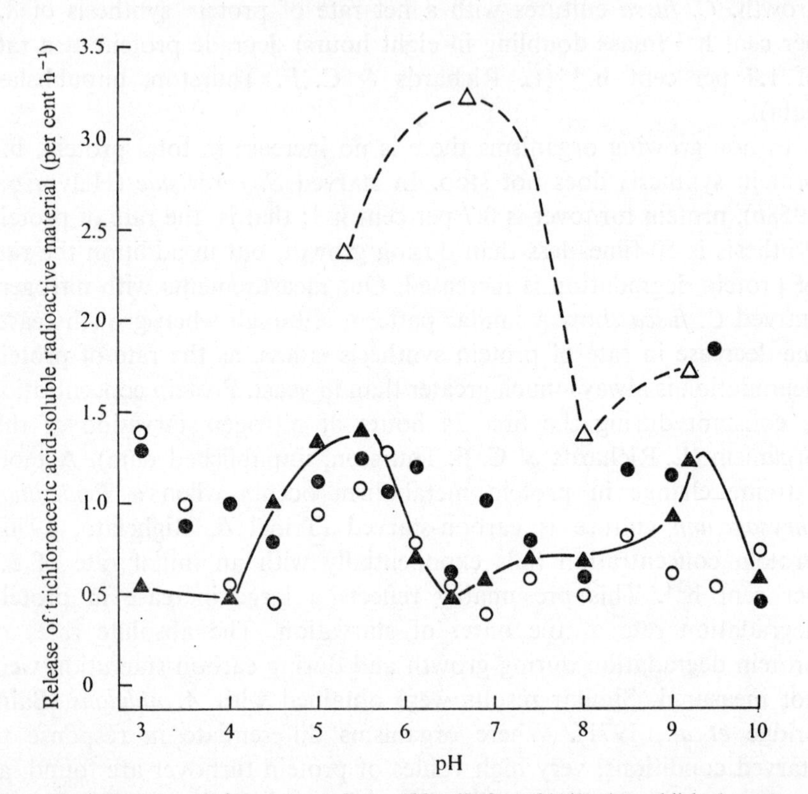

Fig. 1. Proteolysis in cell-free extracts from *Chlorella fusca* 211/8p (unpublished results of I. D. Faulkner & C. F. Thurston). Extracts of $10^5 g$ supernatant, in which protein was radioactively labelled with ^{35}S, were prepared as described by Thurston *et al.*, 1973. Each assay (total vol 600 μl) contained 1 mg ^{35}S-labelled protein (2.0–2.1×10^6 d min^{-1}). Increase in trichloroacetic acid-soluble radioactivity after 1 h incubation at 30 °C was measured. ▲, ^{35}S-labelled extract protein from growing organisms; ○, ^{35}S-labelled extract protein from growing organisms + 1 mg unlabelled extract protein from growing organisms; ●, ^{35}S-labelled extract protein from growing organisms + 1 mg unlabelled extract protein from nitrogen-starved organisms; △, ^{35}S-labelled extract protein from nitrogen-starved organisms.

The addition of unlabelled extract from nitrogen-starved organisms to ^{35}S-labelled extract from growing organisms does not greatly increase the rate of degradation of the radioactive protein. Hence, extracts from nitrogen-starved organisms do not contain greater amounts of proteases, sufficient to account for the much more rapid degradation of radioactive protein observed in extracts where proteins made during nitrogen-starvation are labelled. It is inferred that nitrogen-starved organisms contain a greater proportion of protein, which is susceptible to degradation by intracellular proteases, by comparison with growing organisms. Assays were performed at a range of pH values as the pH optima of *C. fusca* intracellular proteases are unknown. Substantial proteolytic activity was found in extracts from growing cells at pH 5.5, pH 8.5 and pH 9.5. In nitrogen-starved organisms, activity with nitrogen-starved cell protein as substrate is greatest at pH 6.7.

a change in the environment can be exploited, where growth would be possible if additional proteins could be made (typically permeases or inducible enzymes). Third, although many organisms do not show major morphological changes during starvation, changes in protein composition of the cell are not precluded. It may be that starvation provokes biochemical differentiation even where the vegetative state is apparently conserved. Fourth, it is possible that where organisms do not respond to starvation by differentiating into a resting stage (such as a spore), survival depends on the machinery of protein synthesis being kept in operation.

The response of *C. fusca* to nitrogen-starvation certainly includes changes in protein composition. Increased rate of protein degradation results from a larger proportion of the protein in starved organisms being susceptible to degradation, rather than from elevated levels of intracellular proteases (Fig. 1). A comparable situation is found in bacteria (Pine, 1972). Increases in intracellular proteolysis have been demonstrated in carbon-starved *P. chrysogenum* (Trinci & Righelato, 1970), *A. nidulans* (Bainbridge *et al.*, 1971), and in several other filamentous fungi during carbon- or nitrogen-starvation (Miller, Sullivan & Shepherd, 1974; Morton *et al.*, 1960). However, it has not been shown whether these changes are due to increased levels of intracellular proteases, increased susceptibility of cell protein to degradation, or both. This ambiguity arises where protein degradation is measured *in vivo* or where endogenous protein is used as substrate for cell-free extract measurements. Increases in levels of intracellular proteases when growth stops have been observed in *P. cyaneo-fulvum* (Hill & Martin, 1966). In contrast, an intracellular neutral protease of *P. chrysogenum* loses activity in the stationary phase (Brunner, Rohr & Brunner, 1971).

Specific mechanisms of protein degradation and inactivation

Present knowledge of the mechanism of protein degradation as a component of intracellular protein turnover is extremely fragmentary. Data concerning the metabolism of starved eukaryotic micro-organisms are sparse but there is evidence that some very specific mechanisms of protein degradation and inactivation operate during starvation.

Under some conditions some cell proteins can be rapidly degraded (Thurston, 1972) and in the following instances degradation or inactivation are partially a response to starvation. Morris & Syrett (1965) showed that *C. fusca* loses nitrate reductase activity and isocitrate lyase activity when carbon- and nitrogen-starved. Loss of isocitrate lyase activity correlated with loss of enzyme protein (John *et al.*, 1970) and as isocitrate lyase is not rapidly turned over during growth on acetate, it is inferred that the enzyme changes at the onset of starvation from a form which is resistant to degradation to a form which is susceptible (John *et al.*, 1970). The alternative explanation that starvation greatly increases levels of intracellular proteases, is implausible as measurements of proteolytic activity *in vitro* show in Fig. 1. Furthermore, measurements *in vivo* show that in growing cells, the turnover rate of isocitrate lyase is less than that of four other abundant proteins, although during starvation it is far more rapidly degraded than these proteins (John *et al.*, 1970). How the susceptibility to degradation of isocitrate lyase is controlled *in vivo* is not known, but as the enzyme is strongly protected by its substrate when digested by papain *in vitro*, changes in susceptibility to a proteolytic enzyme *in vivo* are possible (Thurston, John & Syrett, 1973).

It is not known whether disappearance of nitrate reductase activity is due to degradation of enzyme protein or some other inactivation process in *C. fusca*. Assimilatory nitrate reductases in eukaryotic micro-organisms undergo inactivation in a variety of circumstances. In *Neurospora crassa*, nitrate reductase activity also disappears when organisms are nitrogen-starved (Subramanian & Sorger, 1972) but *Chlamydomonas reinhardii* loses this activity during carbon-starvation and not in response to nitrogen-starvation (Thacker & Syrett, 1972). Nitrate reductase inactivation in response to the addition of ammonium ions rather than starvation, has been observed in a different *Chlorella fusca* strain (Losada *et al.*, 1970), *C. vulgaris* (Solomonson, Jetschmann & Vennesland, 1973), *Chlamydomonas reinhardii* (Herrera *et al.*, 1972), *A. nidulans* (Cove, 1966) and the basidiomycete *Ustilago maydis* (Lewis & Fincham, 1970). In experiments performed by Losada's group rapid reactivation was observed *in vivo* and *in vitro*, which makes degradation at best a secondary component of loss of enzyme activity (Losada *et al.*, 1970; Maldonado *et al.*, 1973).

Why nitrate reductases are readily inactivated by starvation or the presence of ammonium ions is unexplained, but the degradation of the *Chlorella fusca* isocitrate lyase can be rationalized as follows. This enzyme is unusually abundant when fully induced, constituting about seven per cent of total soluble protein in *Chlorella fusca* 211/8p but in the absence of acetate it is not required for metabolism. Its subsequent degradation is most rapid when organisms are nitrogen-starved in the presence of glucose (John *et al.*, 1970). Hence, a redundant protein is degraded to release amino acids which may be used for the synthesis of other proteins required for the survival of the organism when nitrogen is at a premium.

Specificity in protein degradation of a different kind has been shown in *S. cerevisiae* by Holzer and his co-workers (Katsunuma *et al.*, 1972; Hasilik, Müller & Holzer, 1974). Three proteases have been identified in cell-free extracts from *S. cerevisiae*, two of which rapidly degrade tryptophan synthase *in vitro* but do not degrade alcohol dehydrogenase, hexokinase or glucose-6-phosphate dehydrogenase. One of these two proteases degrades threonine dehydratase and not aspartate transaminase, whilst the reverse is true for the other. It has been proposed that this degradation system of two proteases, with strong preference for pyridoxal phosphate-containing enzymes, allows starved *S. cerevisiae* to degrade biosynthetic enzymes which are not required for endogenous metabolism (Katsunuma *et al.*, 1972). This hypothesis has not been confirmed as tryptophan synthase is largely conserved in stationary-phase *S. cerevisiae* cultures. Furthermore, the inactivating proteases are found in the vacuole of *S. cerevisiae* while protease inhibitors and tryptophan synthase are present in the cytoplasm (Hasilik *et al.*, 1974; Matern, Betz & Holzer, 1974).

Neurospora crassa has a similar system to *S. cerevisiae* and rapidly loses tryptophan synthase activity in stationary phase. A protease has been isolated from *N. crassa* cell-free extracts which degrades tryptophan synthase *in vitro*, and an inhibitor of the protease is also present (Yu, Kula & Tsai, 1973). Unfortunately the activity of tryptophan synthase has not been followed in either organism in a nitrogen-free medium containing a good carbon source, the conditions for the maximum rate of degradation of isocitrate lyase in *C. fusca*.

VARIATIONS IN RESPONSE TO NUTRIENT EXHAUSTION

Colonies on solidified media and in liquid media (pellets)

Although for convenience fungal colonies are sometimes divided into four morphological zones (Yanagita & Kogané, 1962), there is a continuous differentiation from the periphery of a colony to its centre. This morphological differentiation reflects a transition in growth rate, from hyphae at the periphery growing at the organism's maximum specific growth rate (Trinci, 1971) to those at the centre which may be autolysing (Gillie, 1968). This transition in growth rate is paralleled by differences in the specific rate of uptake of [$^{32}PO_4$] (Yanagita & Kogané, 1963) (Table 2). In *Rhizoctonia solani* colonies, the proportion of

Table 2. *Uptake of [^{32}P]orthophosphate by colonies of* Aspergillus niger (*adapted from Yanagita & Kogané, 1963*)

Region of the colony	Distance from the margin of the colony (mm)	Specific rate of [^{32}P]orthophosphate uptake (ct min^{-1} mg^{-6} (dry wt)) expressed as a percentage of the rate in the peripheral growth zone
Peripheral growth zone	0–1.5	100
Productive zone	1.5–3.5	31
Fruiting zone	3.5–8.0	6
Aged zone	8.0–11.5	3

ribosomes present in hyphae as polysomes decreases with distance from the margin of the colony, as does the ability of the mycelium to synthesize new proteins (Ricciardi, Holloman & Gottlieb, 1974). The decline in growth rate with distance from the margin of the colony (Gillie, 1968) is most probably due to changes in the composition of the medium; as colonization proceeds, changes will occur in the colonized medium which inhibit growth, such as altered pH, reduction in oxygen tension, accumulation of products of secondary metabolism and nutrient exhaustion. The morphological changes which occur in ascomycete hyphae during transition to the non-growing state within a colony include increases in cytoplasmic vacuolation (Park & Robinson, 1967), septal plugging (Trinci & Collinge, 1973*a*), increased hyphal wall thickness (Trinci & Collinge, 1975) and increased frequency of intrahyphal hyphae (Trinci & Collinge, 1973*b*); asexual and/or sexual reproductive structures may be formed at the centre of the colony. The increase in cytoplasmic vacuolation occurs concurrently with an increase in the concentration of solids in the cytoplasm (Park & Robinson, 1967). A factor isolated from the medium of stationary phase cultures of *Fusarium oxysporum* which causes vacuolation when applied to the tips of fungal hyphae (Park & Robinson, 1966) has been identified as bikaverin (Cornforth *et al.*, 1971), and this has been implicated in the normal 'ageing' process which proceeds behind the margin of a fungal colony (Park & Robinson, 1966, 1967). Bu'lock *et al.* (1974) showed that bikaverin is mainly produced during the deceleration phase of batch cultures of *Gibberella fujikuroi* and it is probably a secondary metabolite of a number of fungi.

Gottlieb & Van Etten (1966) studied the change in the macromolecular composition of hyphae with distance from the margin of colonies of *Rhizoctonia solani* and *Sclerotium bataticola*. In *R. solani* the total and soluble carbohydrate content of the mycelium increased with distance from the colony margin, whilst the total lipid content remained constant. The reverse situation was observed in *S. bataticola*. It was concluded that the main storage reserve in *R. solani* was carbohydrate whilst in *S. bataticola* it was lipid. The DNA and soluble amino-nitrogen content of the mycelia of both species decreased with distance from the margin of the colony. The observed changes in the macromolecular composition of mycelia during transition to a non-growing state within colonies presumably occur as a result of 'unbalanced' growth (Campbell, 1957) and autolysis.

Transition to the non-growing state almost certainly occurs more rapidly in fungal pellets in submerged culture than in fungal colonies growing on solid media. In pellets, hyphae in all but the peripheral region will be subjected to anaerobic or near-anaerobic conditions (Pirt, 1966; Trinci, 1970). This is indicated by the observation that the specific respiration rate of pellets of *A. niger* decreases with pellet diameter (Yano, Kodama & Yamada, 1961; Kobayashi, Vandedem & Moo-Young, 1973). Autolysis at the centre of fungal pellets is often complete so that a hollow develops (Plate 1) which increases in diameter with time (Camici, Sermonti & Chain, 1952). Eventually the hollow centre may occupy almost the entire volume of the pellet.

Autophagy and starvation

Golgi are present in many eukaryotic micro-organisms, but absent from others, such as *Tetrahymena pyriformis* (Nilsson, 1970), most moulds (Bracker, 1967) and perhaps some yeasts (Matile, Moor & Robinow, 1969). In organisms which do not possess Golgi, the endoplasmic reticulum has additional comparable functions. Golgi and the endoplasmic reticulum form the principal components of the cytoplasmic vacuolar system or 'vacuome' (DeDuve, 1969) of the cell. This is a dynamic system made up of numerous vesicular components interconnected by permanent or transient channels. One role of the vacuome of higher animal cells is the autophagic degradation of redundant cell components such as mitochondria. In this process primary lysosomes, containing acid hydrolases and other degradative enzymes (DeDuve, 1969), separate off from the Golgi (or from the endoplasmic reticulum in organisms lacking Golgi) and fuse with autophagic vacuoles containing cell components. The substances released from the digestion of the contents of the autophagic vacuole,

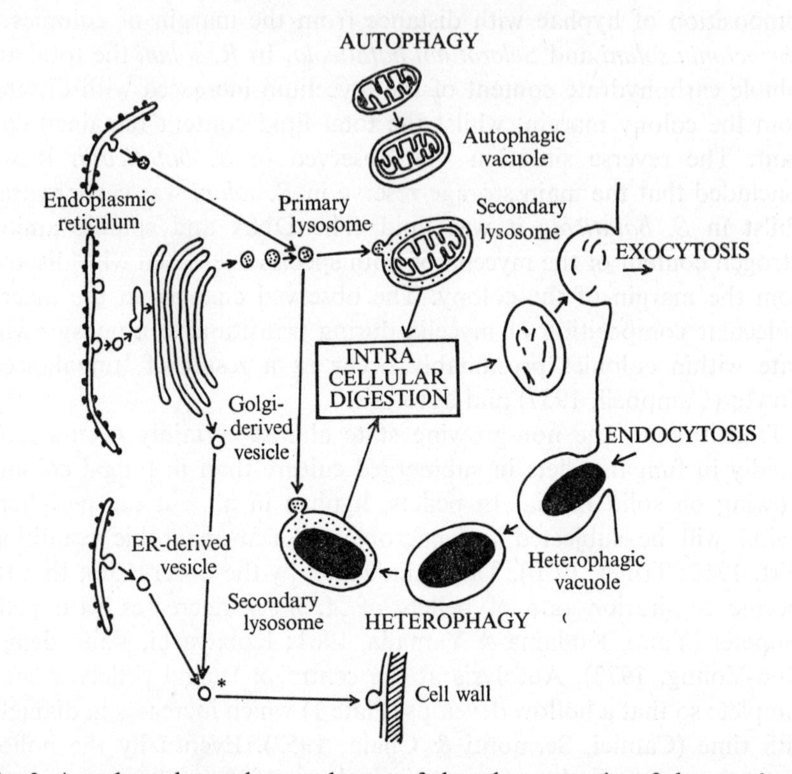

Fig. 2. Autophagy, heterophagy and some of the other properties of the cytoplasmic vacuolar system in eukaryotic micro-organisms. *Vesicles derived from the endoplasmic reticulum or Golgi may contain cellulase, e.g. in *Achlya* (Nolan & Bal, 1974), proteases, e.g. in *Neurospora* (Matile, 1969), acid phosphatase and acid protease, e.g. in *Microsporum*, (Page & Stock, 1972), cell wall precursors, e.g. in *Pythium* (Grove, Bracker & Morré, 1970), scales and flagellar hairs, e.g. in *Heteromastix* (Manton, Rayns, Ettl & Parke, 1965), β-1,3-glucanases, mannan and protein, e.g. in *Saccharomyces* (Cortat, Matile & Wiemken, 1972).

like those released as a result of heterophagy, may be used for anabolic or catabolic processes. Fusions between primary lysosomes and autophagic vacuoles may be determined by the properties of the membranes involved. Clearly some regulatory mechanism is essential. Autophagy, heterography and some of the other properties of the vacuome of eukaryotic micro-organisms are summarized in Fig. 2.

Brandes *et al.* (1964) suggested that autophagy plays a role in the survival of *Euglena gracilis* under conditions of carbon-starvation. Structures which appear to be autophagic vacuoles have been observed in fungi during differentiation (Bracker, 1966; Thornton, 1968; Hemmes & Hohl, 1973), and in stationary phase cultures of *E. gracilis* (Brandes *et al.*, 1964) and *T. pyriformis* (Nilsson, 1970). Vesicles containing acid phosphatase and acid protease activities have been implicated in spore germination in *Microsporum gypseum* (Page & Stock, 1972) and vesicles have been isolated from *N. crassa* which contain two acid proteases but no other lytic enzymes (Matile, 1969). *Achlya bisexualis* may have a lysosomal system (Horgen & O'Day, 1975). Matile *et al.* (1969) suggest that the large central vacuole of *S. cerevisiae* plays a role in turnover and represents the lysosomal system of this organism. Certainly the yeast vacuole contains high activities of some hydrolytic enzymes including proteases (but not acid phosphatase), as mentioned in an earlier section. However, autophagic vacuoles were not observed during the autolysis of carbon-starved cultures of *P. chrysogenum* (Trinci & Righelato, 1970) and *A. nidulans* (Bainbridge *et al.*, 1971). There is little evidence that eukaryotic micro-organisms possess a lysosomal system identical to that observed in the cells of higher animals. Even where such organelles are found, their significance as part of the mechanism of survival in non-growing organisms is uncertain. There is no firm evidence that eukaryotic micro-organisms can prolong their survival under starvation conditions by the autophagic degradation of functional components of the cell.

Cell division

Many organisms are arrested at particular stages in the cell division cycle when growth stops as a result of nutrient exhaustion; for instance *T. pyriformis* cells accumulate in the G_1 phase (Mowat, Pearlman & Engberg, 1974). Consequently starved organisms may grow synchronously when re-inoculated into complete medium (Mitchison, 1971). *Saccharomyces cerevisiae* cultures contain two size classes of cells in the stationary phase, indicating two stages of the cell cycle which can be blocked by carbon-starvation (Golombek & Wintersberger, 1974). In low-phosphate medium, but where growth stops as a result of glucose exhaustion, *Schizosaccharomyces pombe* undergoes an anomalous division at the onset of stationary phase, resulting in lysis of half the progeny (Johnson, 1967).

Stationary-phase cultures of algae and protozoa may contain unusually large cells (Gomez, Harris & Walne, 1974), the behaviour of which varies from organism to organism. In *E. gracilis* giant cells are reported to be 'quiescent and short-lived' (Gomez *et al.*, 1974) but a rather different situation occurs in the Emerson strain of *Chlorella vulgaris*. This organism will divide only in the light and hence, in darkened cultures containing glucose, giant cells are produced. When synchronous cultures are darkened shortly after cell separation, virtually all organisms become giant cells but if returned to the light, these divide producing up to 32 normal-size autospores (Griffiths, 1970). The mechanisms which block the cell cycle in these examples are unknown.

Viability

Miura & Yanagita (1972) have shown that carbon-starved *S. cerevisiae* cultures retained maximum viability during 14 days, after which viability declined exponentially, with a half-life of about nine days. We have not found comparable quantitative data in the literature for starved algal cultures, perhaps because it is difficult to produce colonies of algae originating from single cells on solidified media. However, some algal cultures probably retain viability for long periods under starvation conditions. We have already referred to the decline of endogenous respiration rate in *Prototheca zopfii* during 40 days of incubation under carbon- and nitrogen-starved conditions. This was accompanied by 'little loss of cell viability' (Lloyd, 1974). Ten days incubation under non-growing conditions (extreme light-limitation) did not increase the lag phase for *Chlorella* suspensions transferred back to growing conditions (von Witsch, 1948). Similarly, after 10 days of phosphate-starvation *Dunaliella tertiolecta* and *Monochrysis lutheri* cultures grew with negligible lag when phosphate was added to the medium (Holm-Hansen, 1970).

If long retention of viability is common in starving populations, two explanations are possible. The starvation response could have evolved so as to maximize conservation of vitality in the majority of organisms. Alternatively, the rapid death and destruction (autolysis) of a large proportion of the population could result in the release of products of autolysis which would allow cryptic growth of the surviving organisms (see Postgate, this volume). These alternatives are not mutually exclusive but may be regarded as extreme cases where the majority of species would probably show some intermediate response. The metabolism of starving algae and fungi, as presently understood, give few indications of the importance of cryptic growth in eukaryotes. Where starved organisms maintain uniform vitality, their metabolism represents the essential core of functions required for survival. Such functions may be reflected in a different way as the 'structural-genetic' components of cell function which can be distinguished kinetically from 'synthetic' functions (Williams, 1971).

CONCLUSIONS

There is evidence that a few eukaryotic micro-organisms stop growing because of ageing processes but it appears that changes in cultural conditions are more usually responsible and perhaps of wider significance. At present we cannot give a comprehensive, coherent description of how eukaryotic micro-organisms respond to starvation. There is a considerable variety of responses, but it is not clear to what extent these variations reflect the diversity of experimental approaches rather than inherent differences between species. It is not known whether survival of populations depends frequently upon cryptic growth rather than retention of cell vitality. Conversely, it may be important that populations of vegetative organisms which die rapidly under starvation conditions have not been reported frequently. Nevertheless, sufficient is known to justify the assertion that adaptation for survival of vegetative microbes is a complex phenomenon as is illustrated by the specificity of some of the mechanisms of protein degradation which have been described.

It is less clear what advantages are gained by organisms which do not differentiate into spores or other structures, more obviously adapted to survive starvation conditions. One advantage might be the ability to start growing again more quickly than a spore when conditions allowing growth are re-established. It is also possible that non-growing vegetative organisms can grow when nutrients are available at concentrations insufficient to trigger germination of spores.

The authors are grateful to Dr G. E. Mathison for reading the manuscript and for his helpful suggestions.

REFERENCES

Ashworth, J. M. & Smith, J. E. (1973). *Microbial Differentiation. Symposium of the Society for General Microbiology*, **23**.

Atkinson, A. W., John, P. C. L. & Gunning, B. E. S. (1974). The growth and division of the single mitochondrion and other organelles during the cell cycle of *Chlorella*, studied by quantitative stereology and three-dimensional reconstruction. *Protoplasma*, **81**, 77–109.

Bainbridge, B. W., Bull, A. T., Pirt, S. J., Rowley, B. & Trinci, A. P. J. (1971). Biochemical and structural changes in non-growing maintained and autolysing cultures of *Aspergillus nidulans*. *Transactions of the British Mycological Society*, **56**, 371–385.

Bartnicki-Garcia, S. & McMurrough, I. (1971). Biochemistry of morphogenesis in yeasts. In *The Yeasts* **2**, ed. A. H. Rose & J. S. Harrison, pp. 441–491. New York and London: Academic Press.

BRACKER, C. E. (1966). Ultrastructural aspects of sporangiospore formation in *Gilbertella persicaria*. In *The Fungus Spore*, ed. M. F. Madelin, pp. 39–60. London: Butterworth Science Publications.

BRACKER, C. E. (1967). Ultrastructure of fungi. *Annual Review of Phytopathology*, **5**, 343–374.

BRANDES, D., BUETOW, D. E., BERTINI, F. & MALKOFF, D. B. (1964). Roles of lysosomes in cellular lytic processes. 1. Effect of carbon starvation in *Euglena gracilis*. *Experimental and Molecular Pathology Supplement*, **3**, 583–609.

BRUNNER, R., ROHR, M. & BRUNNER, H. (1971). Evidence of an intracellular proteinase of *Penicillium chrysogenum*. *Enzymologia*, **40**, 209–216.

BU'LOCK, J. D., DETROY, R. W., HOŠŤÁLEK, Z. & MUNIM-AL-SHAKARCHI, A. (1974). Regulation of secondary biosynthesis in *Gibberella fujikuroi*. *Transactions of the British Mycological Society*, **62**, 377–389.

CAMICI, L., SERMONTI, G. & CHAIN, E. B. (1952). Observations on *Penicillium chrysogenum* in submerged culture. I. Mycelial growth and autolysis. *Bulletin of the World Health Organization*, **6**, 265–275.

CAMPBELL, A. (1957). Synchronization of cell division. *Bacteriological Reviews*, **21**, 263–272.

CARTER, B. L. A., BULL, A. T., PIRT, S. J. & ROWLEY, B. I. (1971). Relationship between energy substrate utilization and specific growth rate in *Aspergillus nidulans*. *Journal of Bacteriology*, **108**, 309–313.

COOK, A. H. (1963). Protein synthesis in yeast. *Pure and Applied Chemistry*, **7**, 621–637.

CORNFORTH, J. W., RYBACK, G., ROBINSON, P. M. & PARK, D. (1971). Isolation and characterization of a fungal vacuolation factor (bikaverin). *Journal of the Chemical Society C*, 2786–2788.

CORTAT, M., MATILE, P. & WIEMKEN, A. (1972). Isolation of glucanase-containing vesicles from budding yeast. *Archiv für Mikrobiologie*, **82**, 189–205.

COVE, D. J. (1966). The induction and repression of nitrate reductase in the fungus *Aspergillus nidulans*. *Biochimica et Biophysica Acta*, **113**, 51–56.

CRAMER, M. & MYERS, J. (1949). Effects of starvation on the metabolism of *Chlorella*. *Plant Physiology*, **24**, 255–264.

DAWES, E. A. & RIBBONS, D. W. (1964). Some aspects of the endogenous metabolism of bacteria. *Bacteriological Reviews*, **28**, 126–149.

DEDUVE, C. (1969). The lysosome in retrospect. In *Lysosomes in Biology and Pathology*, ed. J. T. Dingle & H. B. Fell, pp. 3–42. Amsterdam and London: North-Holland Publishing Company.

EPSTEIN, C. J., MARTIN, G. M., SCHULTZ, A. L. & MOTULSKY, A. G. (1966). Werner's Syndrome: a review of its symptomatology, natural history, pathologic features, genetics and relationship to the natural ageing process. *Medicine, Baltimore*, **45**, 177–190.

EVANS, D. A. (1973). Growth phase and number of phosphorylation sites in the mitochondrial electron transport chain of *Acanthamoeba castellani*. *Journal of Protozoology*, **20**, 336–338.

FOGG, G. E. (1959). Nitrogen nutrition and metabolic patterns in algae. *Symposium of the Society for Experimental Biology*, **13**, 106–125.

FOTT, B. & NOVAKOVA, M. (1969). A monograph of the genus *Chlorella*. The freshwater species. In *Studies of Phycology*, ed. B. Fott, pp. 10–74. Prague: Academia.

GIBBS, M. (1962). Fermentation. In *Physiology and Biochemistry of Algae*, ed. R. A. Lewin, pp. 91–98. New York and London: Academic Press.

GILLIE, O. J. (1968). Observations on the tube method of measuring growth rate in *Neurospora crassa*. *Journal of General Microbiology*, **51**, 185–194.

Goldberg, A. L., Howell, E. M., Li, J. B., Martel, S. B. & Prouty, W. F. (1974). Physiological significance of protein degradation in animal and bacterial cells. *Federation Proceedings*, **33**, 1112–1120.

Golombek, J. & Wintersberger, E. (1974). Glucose uptake in the cell cycle of *Saccharomyces cerevisiae*. *Experimental Cell Research*, **86**, 199–202.

Gomez, M. P., Harris, J. B. & Walne, P. L. (1974). Studies of *Euglena gracilis* in ageing culture. I. Light microscopy and cytochemistry. *British Phycological Journal*, **9**, 163–174.

Gottlieb, D. & Van Etten, J. L. (1966). Changes in fungi with age. I. Chemical composition of *Rhizoctonia solani* and *Sclerotium bataticola*. *Journal of Bacteriology*, **91**, 161–168.

Griffiths, D. J. (1970). The growth of synchronous cultures of the Emerson strain of *Chlorella vulgaris* under heterotrophic conditions. *Archiv für Mikrobiologie*, **71**, 60–66.

Grove, S. N., Bracker, C. E. & Morré, D. J. (1970). An ultrastructural basis for hyphae tip growth in *Pythium ultimum*. *American Journal of Botany*, **57**, 245–266.

Halvorson, H. O. (1958*a*). Studies on protein and nucleic acid turnover in growing cultures of yeast. *Biochimica et Biophysica Acta*, **27**, 267–276.

Halvorson, H. O. (1958*b*). Intracellular protein and nucleic acid turnover in resting yeast cells. *Biochimica et Biophysica Acta*, **27**, 255–266.

Halvorson, H. O., Fry, W. & Schwemmin, D. (1955). A study of the properties of the free amino acid pool and enzyme synthesis in yeast. *Journal of General Physiology*, **38**, 549–573.

Hasilik, A., Müller, H. & Holzer, H. (1974). Compartmentation of the tryptophan-synthase proteolysing system in *Saccharomyces cerevisiae*. *European Journal of Biochemistry*, **48**, 111–117.

Hayflick, L. & Moorhead, P. S. (1961). The serial cultivation of human diploid cell strains. *Experimental Cell Research*, **25**, 585–621.

Hemmes, D. E. & Hohl, H. R. (1973). Mitosis and nuclear degeneration: simultaneous events during secondary sporangia formation in *Phytophthora palmivora*. *Canadian Journal of Botany*, **51**, 1673–1675.

Herrera, J., Paneque, A., Maldonado, J. M., Barea, J. L. & Losada, M. (1972). Regulation by ammonia of nitrate reductase synthesis and activity in *Chlamydomonas reinhardi*. *Biochemical and Biophysical Research Communications*, **48**, 996–1003.

Hill, P. & Martin, S. M. (1966). Cellular protolytic enzymes of *Penicillium cyaneo-fulvum*. *Canadian Journal of Microbiology*, **12**, 243–248.

Holliday, R. (1969). Errors in protein synthesis and clonal senescence in fungi. *Nature, London*, **221**, 1224–1228.

Holliday, R. (1972). Ageing of human fibroblasts in culture: studies on enzymes and mutation. *Humangenetik*, **16**, 83–86.

Holm-Hansen, O. (1970). ATP levels in algal cells as influenced by environmental conditions. *Plant and Cell Physiology*, **11**, 689–700.

Horgen, P. A. & O'Day, D. H. (1975). The developmental patterns of lysosomal enzyme activities during Ca^{++} induced sporangium formation in *Achlya bisexualis*. II. α-mannosidase. *Archives of Microbiology*, **102**, 9–12.

Hough, J. S. & Maddox, I. S. (1970). Yeast autolysis. *Process Biochemistry*, **5**, 50–52.

John, P. C. L., Thurston, C. F. & Syrett, P. J. (1970). Disappearance of isocitrate lyase enzyme from cells of *Chlorella pyrenoidosa*. *Biochemical Journal*, **119**, 913–919.

JOHNSON, B. F. (1967). Growth of the fission yeast, *Schizosaccharomyces pombe*, with late, eccentric, lytic fission in an unbalanced medium. *Journal of Bacteriology*, **94**, 192–195.

KANAZAWA, T., KANAZAWA, K., KIRK, M. R. & BASSHAM, J. A. (1972). Regulatory effects of ammonia on carbon metabolism in *Chlorella pyrenoidosa* during photosynthesis and respiration. *Biochimica et Biophysica Acta*, **256**, 656–669.

KATSUNUMA, T., SCHOTT, E., ELSASSER, S. & HOLZER, H. (1972). Purification and properties of tryptophan-synthetase- inactivating enzymes from yeast. *European Journal of Biochemistry*, **27**, 520–526.

KATZ, R. (1971). Growth phase and rotenone sensitivity in *Torulopsis utilis*: difference between exponential and stationary phase. *FEBS Letters*, **12**, 153–156.

KATZ, R., KILPATRICK, L. & CHANCE, B. (1971). Aquisition and loss of rotenone sensitivity in *Torulopsis*. *European Journal of Biochemistry*, **21**, 301–307.

KOBAYASHI, T., VANDEDEM, G. & MOO-YOUNG, M. (1973). Oxygen transfer into mycelial pellets. *Biotechnology and Bioengineering*, **15**, 27–45.

KUHL, A. (1974). Phosphorus. In *Algal Physiology and Biochemistry*, ed. W. D. P. Stewart, pp. 636–654. Oxford: Blackwells.

LAHOZ, R. & IBEAS, J. G. (1968). The autolysis of *Aspergillus flavus*. *Journal of General Microbiology*, **53**, 101–108.

LEWIS, C. M. & FINCHAM, J. R. S. (1970). Regulation of nitrate reductase in the basidiomycete *Ustilago maydis*. *Journal of Bacteriology*, **103**, 55–61.

LEWIS, C. M. & HOLLIDAY, R. (1970). Mistranslation and ageing in *Neurospora*. *Nature, London*, **228**, 877–880.

LLOYD, D. (1974). Dark respiration. In *Algal Physiology and Biochemistry*, ed. W. D. P. Stewart, pp. 505–529. Oxford: Blackwells.

LODI, W. R. & SONNEBORN, D. R. (1974). Protein degradation and protease activity during the life cycle of *Blastocladiella emersonii*. *Journal of Bacteriology*, **117**, 1035–1042.

LÖPPERT, H. G. & BRODA, E. (1973). ATP-Gehalt und ATP-Umsatz von *Chlorella* in Abhängigkeit von Belichtung und Belüftung. *Zeitschrift für Allgemeine Mikrobiologie*, **13**, 499–506.

LOSADA, M., PANEQUE, A., APARICIO, P. J., VEGA, J. M., CÁRDENAS, J. & HERRERA, J. (1970). Inactivation and repression by ammonium of the nitrate reducing system in *Chlorella*. *Biochemical and Biophysical Research Communications*, **38**, 1009–1015.

LYNCH, J. M. & HARPER, S. H. T. (1974). Fungal growth rate and the formation of ethylene in soil. *Journal of General Microbiology*, **85**, 91–96.

MAHADEVAN, P. R. & MAHADKAR, U. R. (1970). Role of enzymes in growth and morphology of *Neurospora crassa*: cell-wall-bound enzymes and their possible role in branching. *Journal of Bacteriology*, **101**, 941–947.

MALDONADO, J. M., HERRERA, J., PANEQUE, A. & LOSADA, M. (1973). Reversible inactivation by NADH and ADP of *Chlorella fusca* nitrate reductase. *Biochemical and Biophysical Research Communications*, **51**, 27–33.

MANDELSTAM, J. (1960). The intracellular turnover of protein and nucleic acids and its role in biochemical differentiation. *Bacteriological Reviews*, **24**, 289–308.

MANDELSTAM, J. & WAITES, W. M. (1968). Sporulation in *Bacillus subtilis*. The role of exoprotease. *Biochemical Journal*, **109**, 793–801.

MANTON, I., RAYNS, D. G., ETTL, H. & PARKE, M. (1965). Further observations on green flagellates with scaly flagella. The genus *Heteromastix* Korshikov. *Journal of the Marine Biology Association of the United Kingdom*, **45**, 241–255.

MARTIN, G. M., SPRAGUE, C. A. & EPSTEIN, C. J. (1970). Replicative life-span of cultivated human cells. Effects of donor's age, tissue and genotype. *Laboratory Investigation*, **23**, 86–92.

MATERN, H., BETZ, H. & HOLZER, H. (1974). Compartmentation of inhibitors of proteinases A and B and carboxypeptidase Y in yeast. *Biochemical and Biophysical Research Communications*, **60**, 1051–1057.

MATILE, P. (1969). Plant Lysosomes. In *Lysosomes in Biology and Pathology*, ed. J. T. Dingle & H. B. Fell, pp. 3–42. Amsterdam and London: North-Holland Publishing Company.

MATILE, P., MOOR, H. & ROBINOW, C. F. (1969). Yeast cytology. In *The Yeasts*, **1**, ed. A. H. Rose & J. S. Harrison, pp. 219–302. New York and London: Academic Press.

MILLER, H. M., SULLIVAN, A. P. & SHEPHERD, M. G. (1974). Intracellular protein breakdown in thermophilic and mesophilic fungi. *Biochemical Journal*, **144**, 209–214.

MITCHISON, J. M. (1971). Synchronous cultures. In *The Biology of the Cell Cycle*, pp. 25–57. London: Cambridge University Press.

MIURA, T. & YANAGITA, T. (1972). Cellular senescence in yeast caused by carbon-source starvation. 1. Changes in activities of respiratory system and lipid peroxidation activity. *Journal of Biochemistry*, **72**, 141–148.

MORRIS, I. & SYRETT, P. J. (1965). The effect of N-starvation on the activity of nitrate reductase and other enzymes in *Chlorella*. *Journal of General Microbiology*, **38**, 21–28.

MORTON, A. G., DICKERSON, A. G. F. & ENGLAND, D. J. F. (1960). Changes in enzyme activity of fungi during nitrogen starvation. *Journal of Experimental Botany*, **11**, 116–128.

MOWAT, D., PEARLMAN, R. E. & ENGBERG, J. (1974). DNA synthesis following refeeding of starved *Tetrahymena pyriformis*. *Experimental Cell Research*, **84**, 282–286.

NG, A. M. L., SMITH, J. E. & MCINTOSH, A. F. (1973). Conidiation of *Aspergillus niger* in continuous culture. *Archiv für Mikrobiologie*, **88**, 119–126.

NILSSON, J. R. (1970). Cytolysosomes in *Tetrahymena pyriformis* GL. I. Synchronized cells dividing in inorganic salt medium. *Comptes Rendus Traveaux Laboratoire Carlsbergensis*, **38**, 87–106.

NOLAN, R. A. & BAL, A. K. (1974). Cellulase localization in hyphae of *Achlya ambisexualis*. *Journal of Bacteriology*, **117**, 840–843.

ORGEL, L. E. (1963). The maintenance of the accuracy of protein synthesis and its relevance to ageing. *Proceedings of the National Academy of Sciences USA*, **49**, 517–521.

PAGE, W. J. & STOCK, J. J. (1972). Isolation and characterization of *Microsporum gypseum* lysosomes: role of lysosomes in macroconidia germination. *Journal of Bacteriology*, **110**, 354–362.

PARK, D. & ROBINSON, P. M. (1966). Aspects of hyphal morphogenesis in fungi. In *Trends in Plant Morphogenesis*, ed. E. G. Cutler, pp. 27–44. London: Longmans, Green & Co. Ltd.

PARK, D. & ROBINSON, P. M. (1967). A fungal hormone controlling internal water distribution normally associated with cell ageing in fungi. *Symposium of the Society of Experimental Biology*, **21**, 323–336.

PINE, M. J. (1972). Turnover of intracellular proteins. *Annual Review of Microbiology*, **26**, 103–126.

PIRT, S. J. (1965). The maintenance energy of bacteria in growing cultures. *Proceedings of the Royal Society of London, Series B*, **163**, 224–231.

Pirt, S. J. (1966). A theory of the mode of growth of fungi in the form of pellets in submerged culture. *Proceedings of the Royal Society of London, Series B*, **166**, 369–373.

Pirt, S. J. (1975). *Principles of Microbe and Cell Cultivation*. Oxford: Blackwells.

Ricciardi, R. P., Holloman, D. W. & Gottlieb, D. (1974). Age dependent changes in fungi: ribosomes and protein synthesis in *Rhizoctonia solani* mycelium. *Archives of Microbiology*, **95**, 325–336.

Righelato, R. C., Trinci, A. P. J., Pirt, S. J. & Peat, A. (1968). The influence of maintenance energy and growth rate on the metabolic activity, morphology and conidiation of *Penicillium chrysogenum*. *Journal of General Microbiology*, **50**, 399–412.

Sando, N., Maeda, M., Endo, T., Oka, R. & Hayashibe, M. (1973). Induction of meiosis and sporulation in differentially aged cells of *Saccharomyces cerevisiae*. *Journal of General and Applied Microbiology*, **19**, 359–373.

Siekevitz, P. (1972). The turnover of proteins and the usage of information. *Journal of Theoretical Biology*, **37**, 321–334.

Smith-Sonneborn, J. & Klass, M. (1974). Changes in the DNA synthesis pattern of *Paramecium* with increased clonal age and interfission time. *Journal of Cell Biology*, **61**, 591–598.

Smith-Sonneborn, J., Klass, M. & Cotton, D. (1974). Parental age and life span versus progeny life span in *Paramecium*. *Journal of Cell Science*, **14**, 691–699.

Solomonson, L. P., Jetschmann, K. & Vennesland, B. (1973). Reversible inactivation of the nitrate reductase of *Chlorella vulgaris* Beijerinck. *Biochimica et Biophysica Acta*, **309**, 32–43.

Sols, A., Gancedo, C. & Delafuente, G. (1971). Energy yielding metabolism in yeasts. In *The Yeasts*, **2**, ed. A. H. Rose & J. S. Harrison, pp. 271–307. London and New York: Academic Press.

Subramanian, K. N. & Sorger, G. I. (1972). Regulation in nitrate reductase in *Neurospora crassa*: stability *in vivo*. *Journal of Bacteriology*, **110**, 538–546.

Suomalainen, H. & Oura, E. (1971). Yeast nutrition and solute uptake. In *The Yeasts*, **2**, ed. A. H. Rose & J. S. Harrison, pp. 3–74. London and New York: Academic Press.

Syrett, P. J. (1958). Respiration rate and internal ATP concentration in *Chlorella*. *Archives of Biochemistry and Biophysics*, **75**, 117–124.

Syrett, P. J. (1962). Nitrogen Assimilation. In *Physiology and Biochemistry of Algae*, ed. R. A. Lewin, pp. 171–188. London and New York: Academic Press.

Syrett, P. J. (1966). The kinetics of isocitrate lyase formation in *Chlorella*: evidence for the promotion of enzyme synthesis by photophosphorylation. *Journal of Experimental Botany*, **17**, 641–654.

Thacker, A. & Syrett, P. J. (1972). Disappearance of nitrate reductase activity from *Chlamydomonas reinhardi*. *New Phytologist*, **71**, 435–441.

Thornton, R. M. (1968). The fine structure of Phycomyces. I. Autophagic vesicles. *Journal of Ultrastructure Research*, **21**, 269–280.

Thurston, C. F. (1972). Disappearing enzymes. *Process Biochemistry*, **7**, 18–20.

Thurston, C. F., John, P. C. L. & Syrett, P. J. (1973). The effect of metabolic inhibitors on the loss of isocitrate lyase activity from *Chlorella*. *Archiv für Mikrobiologie*, **88**, 135–145.

Trinci, A. P. J. (1970). Kinetics of the growth of mycelial pellets of *Aspergillus nidulans*. *Archiv für Mikrobiologie*, **73**, 353–367.

Trinci, A. P. J. (1971). Influence of the width of the peripheral growth zone on the radial growth rate of fungal colonies on solid media. *Journal of General Microbiology*, **67**, 325–344.

TRINCI, A. P. J. & COLLINGE, A. J. (1973*a*). Structure and plugging of septa of wild type and spreading colonial mutants of *Neurospora crassa*. *Archiv für Mikrobiologie*, **91**, 355–364.

TRINCI, A. P. J. & COLLINGE, A. J. (1973*b*). Influence of L-sorbose on the growth and morphology of *Neurospora crassa*. *Journal of General Microbiology*, **78**, 179–192.

TRINCI, A. P. J. & COLLINGE, A. J. (1975). Hyphal wall growth in *Neurospora crassa* and *Geotrichum candidum*. *Journal of General Microbiology*, (in press).

TRINCI, A. P. J. & RIGHELATO, R. (1970). Changes in constituents and ultrastructure of hyphal compartments during autolysis of glucose-starved *Penicillium chrysogenum*. *Journal of General Microbiology*, **60**, 239–249.

VALADON, L. R. G., MYERS, A. & MANNERS, J. G. (1962). The behaviour of nucleic acids and other constituents in *Protomyces inundatus*. *Journal of Experimental Botany*, **13**, 378–389.

WATSON, T. G. (1970). Effect of sodium chloride on steady-state growth and metabolism of *Saccharomyces cerevisiae*. *Journal of General Microbiology*, **64**, 91–99.

WILLIAMS, F. M. (1971). Dynamics of microbial populations. In *Systems Analysis and Simulation in Ecology*, **1**, pp. 197–267. New York and London: Academic Press.

WITSCH, H. VON (1948). Physiologischer Zustand und Wachstumsintenität bei *Chlorella*. *Archiv für Mikrobiologie*, **14**, 128–141.

WRIGHT, B. E. & ANDERSON, M. L. (1960). Protein and amino acid turnover during differentiation in the slime mold. *Biochimica et Biophysica Acta*, **43**, 62–66.

YANAGITA, T. & KOGANÉ, F. (1962). Growth and cytochemical differentiation of mold colonies. *Journal of General and Applied Microbiology*, **8**, 201–213.

YANAGITA, T. & KOGANÉ, F. (1963). Cellular differentiation of growing mold colonies with special reference to phosphorus metabolism. *Journal of General and Applied Microbiology*, **9**, 313–330.

YANO, T., KODAMA, T. & YAMADA, K. (1961). Fundamental studies on the aerobic fermentation. VIII. Oxygen transfer within a mold pellet. *Agricultural and Biological Chemistry*, **25**, 580–584.

YENTSCH, C. S. (1962). Marine plankton. In *Physiology and Biochemistry of Algae*, ed. R. A. Lewin, pp. 771–798. New York and London: Academic Press.

YU, P. H., KULA, M. & TSAI, H. (1973). Studies on the apparent instability of *Neurospora* tryptophan synthase. Evidence for a protease. *European Journal of Biochemistry*, **32**, 129–135.

ZONNEVELD, B. J. M. (1972). Morphogenesis in *Aspergillus nidulans*. The significance of α-1,3-glucan of the cell wall and α-1,3-glucanase or cleistothecium development. *Biochimica et Biophysica Acta*, **273**, 174–187.

EXPLANATION OF PLATE

PLATE 1

Median transverse section through a pellet of *Basidiobolus ranarum* showing autolysis at the centre. Plate provided by Dr K. Gull.

PLATE I

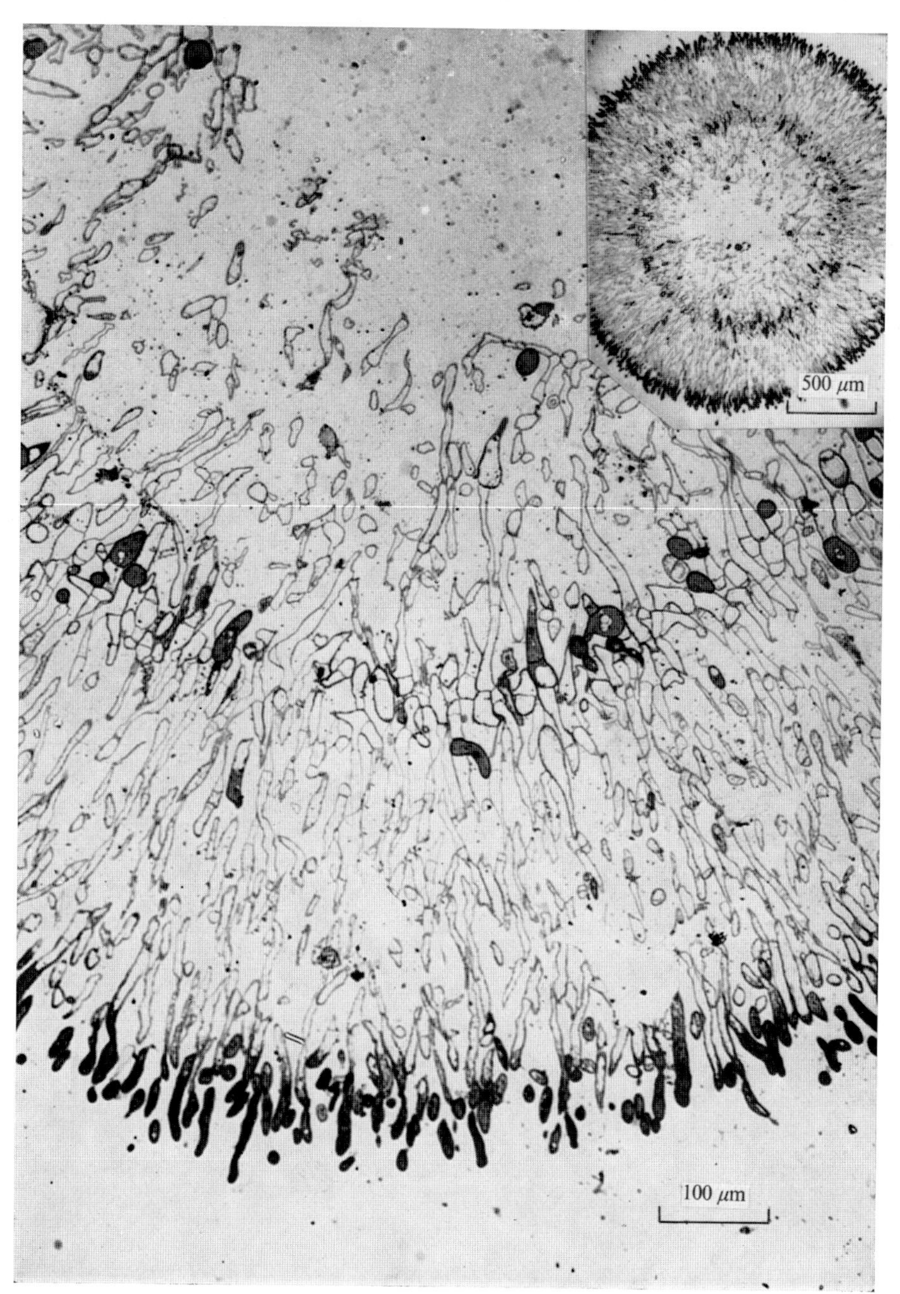

(*Facing page* 80)

COLD SHOCK AND FREEZING DAMAGE TO MICROBES

R. A. MACLEOD AND P. H. CALCOTT
Department of Microbiology, Macdonald Campus, McGill University, Montreal, Quebec

INTRODUCTION

It is now evident that a multiplicity of factors affect the response of microbial cells to chilling and freezing. Failure to appreciate this fact has often given rise to seemingly contradictory results. Out of the confusion, certain patterns of response are now becoming clear. When cells are frozen and thawed the rates of cooling and of warming both affect cell survival. The response to cooling and warming rates is similar for cells of widely differing types. The nature of the response was not readily predictable but once it was recognized, the effect of other variables on survival could be studied in a systematic way. The results of these studies reveal that different cooling and warming rates produce different kinds of damage to cells. The kind of damage produced can also vary depending on the chemical composition of the freezing menstruum. In this article, an attempt has been made to glean from the results of studies on cold shock, on freezing and thawing damage and on repair of freezing injury, information which will lead to a better understanding of the mechanisms leading to the injury and death of microbial cells on chilling and freezing.

COLD SHOCK

Cold shock is the injury or death to organisms caused by sudden chilling without freezing. First reported in *Escherichia coli* (Sherman & Albus, 1923) it is now known to occur in a variety of both Gram-positive (Ring, 1965; Smeaton & Elliott, 1967; Traci & Duncan, 1974) and Gram-negative bacteria (Gorrill & MacNeil, 1960; Strange & Dark, 1962). Cold shock has not been detected in yeast cells. The latter survive supercooling to −16 °C (Mazur, 1966).

Cold shock is accompanied by the release of a number of low molecular weight intracellular solutes including nucleotides, amino acids (Strange & Dark, 1962; Strange & Ness, 1963; Ring, 1965), K^+ (Haest *et al.*, 1972) and a low molecular weight protein (Smeaton &

Elliott, 1967). Loss of one or more of these solutes is often believed to be responsible for the death of the cells.

A number of factors can affect the sensitivity of cells to cold shock. The age of the culture is important. Cold shock usually occurs in cells harvested in the exponential but not the stationary phase of growth (Hegarty & Weeks, 1940; Meynell, 1958). An exception to this appears to be the release by cold shock of a low molecular weight protein, a ribonuclease inhibitor, from *Bacillus subtilis* grown to the stationary phase (Smeaton & Elliott, 1967). The composition of the growth medium can also have an effect. Cells grown on a simple, chemically defined medium were found to be more sensitive to cold shock than cells grown on a more complex medium (Farrell & Rose, 1968; Strange & Ness, 1963). The reverse, however, has also been observed. Meynell (1958) found that addition of peptone to a glucose salts medium increased the susceptibility of *E. coli* to sudden chilling.

The composition of the medium in which the cells are chilled has a profound effect on the extent of cold shock. Such divalent cations as Mg^{2+}, Ca^{2+} and Mn^{2+} substantially protect against the effect of chilling (Strange & Dark, 1962; Sato & Takahashi, 1968, 1969, 1970). Losses of intracellular solutes by cold shock are markedly sensitive to the relative osmolarities of the growth medium and the medium in which the cells are chilled. Leder (1972) showed that cold shock occurred if the two media were iso-osmolar but could be prevented by increasing the osmolarity of the medium used to chill the cells.

Loss of viability due to chilling is greater the smaller the cell population (Strange & Dark, 1962; Strange & Ness, 1963). Losses were insignificant in chilled suspensions of *Aerobacter aerogenes* containing more than 2.4×10^9 organisms ml^{-1}. The supernatant liquid separated from a chilled suspension containing a high density of exponential phase cells had a protective effect on bacteria subjected to chilling (Strange & Dark, 1962).

Both the rate of cooling and the temperature range over which cooling takes place affect the extent of cold shock (Sherman & Cameron, 1934; Smeaton & Elliott, 1967; Leder, 1972). Leder (1972) observed that *E. coli* cooled from 25 to 0 °C in 1 to 2 sec retained less than 5 per cent of permease-accumulated substrates but if cooled over the same range in 1 min, retained essentially 100 per cent of the substrates. Smeaton & Elliott (1967) determined that *B. subtilis* must be cooled through a critical temperature zone at 16 to 14 °C to obtain loss of solute from the cells. Sato & Takahashi (1968, 1969) observed two critical temperature zones through which *E. coli*, *B. subtilis* and

Pseudomonas fluorescens could be cooled to obtain cold shock. The position of the zones on the temperature scale was determined by the initial temperature of the cell suspension before the cold shock. They concluded that the magnitude of chilling rather than the rapid passage through a definite temperature zone is an essential requirement for cold shock.

Cold shock leads to a general increase in the permeability of bacterial cells. Strange & Postgate (1964) presented evidence that RNAase, anilino-naphthalene-8-sulphonate, hydrogen ions and hydroxyl ions more readily enter *Aerobacter aerogenes* subjected to cold shock or freezing and indicated that the permeability damage is sometimes reversible. The capacity of *Streptomyces hydrogenans* to take up thiourea was increased five-fold by sudden chilling from 30 to 0 °C (Ring, 1965). Cold shock increased the permeability of *B. subtilis* to a fluorescent dye (Smeaton & Elliott, 1967). Sudden chilling of a suspension of exponential phase cells of *Clostridium perfringens* resulted in a population unable to grow on a plating medium containing neomycin (Traci & Duncan, 1974). A study of conditions permitting recovery of the capacity of the cells to grow in the presence of neomycin threw light on the factors involved in the repair of injury caused by cold shock. Cells chilled suddenly from 37 to 10 °C recovered completely if incubated at 25 to 30 °C for 10 to 20 min in 0.01 per cent peptone. At 15 °C, repair was complete in 60 to 90 min. No recovery occurred if the cells were suspended in distilled water. Since neither nalidixic acid, chloramphenicol nor rifampin inhibited recovery, it was concluded that neither protein synthesis nor DNA-dependent RNA polymerase seemed to be involved in the repair of injury. When logarithmic-phase cells of *E. coli* or *P. fluorescens* were cold-shocked by suspending in cold Tris buffer, recovery from cold shock could be induced by adding Mg^{2+} to the suspension and incubating the cells at 30 °C (Sato & Takahashi, 1968, 1969). Ca^{2+} was almost as effective as Mg^{2+} in mediating recovery. For *B. subtilis* it was necessary to add an acid hydrolysate of casein along with Mg^{2+} to obtain recovery from cold shock. Studies with *E. coli* showed that the capacity of the cells to recover from cold shock was gradually lost the longer the cells were incubated in cold Tris buffer without added Mg^{2+}. The addition of 2,4-dinitrophenol (DNP), but not chloramphenicol, inhibited the Mg^{2+} mediated recovery of *E. coli*. The inhibition by DNP could be largely overcome by the further addition of NAD or ATP and nicotinamide to the suspension (Sato & Takahashi, 1969).

Sato & Takahashi (1970) found that the DNA of cold-shocked

E. coli cells contains more single-strand breaks (nicks) than unshocked cells or cells which had recovered, and proposed that cold shock leads to the loss of Mg^{2+} from the cells and hence to the loss of the DNA-ligase activity which would be required to cause the cells to recover from cold shock. The release of Mg^{2+}, however, occurred to almost the same extent in cells from both the exponential and the stationary phase of growth: cells in the stationary phase are not susceptible to cold shock.

The permeability changes resulting from cold shock arise as a result of a phase transition in the membrane lipids (Ring, 1965; Farrell & Rose, 1968; Haest *et al.*, 1972; Leder, 1972). Studies with *Mycoplasma laidlawii* have shown that when the membranes are cooled the lipids change from a liquid to a solid state (Steim *et al.*, 1969). The temperature of the phase transition varies with the degree of unsaturation and the chain length of the fatty acids in the membrane phospholipids (Steim *et al.*, 1969). It has been suggested that the rapid crystallization of membrane lipids caused by cold shock creates hydrophilic channels facilitating the escape of pool solutes. When the process of chilling is slow, the relative mobility of the lipid chains permits them to rearrange, thereby maintaining the integrity of the membrane barrier (Leder, 1972). Factors which would slow or prevent the phase transition would thus protect against cold shock.

Excellent support for the conclusion that cold shock is related to phase transition in the membrane lipids arises from the studies of Haest *et al.* (1972). Using an unsaturated fatty acid auxotroph of *E. coli* grown in the presence of different unsaturated fatty acids, these workers showed that the temperature at which the cold-shock-dependent release of K^+ from the cells starts depends on the fatty acid composition of the membrane. This suggests that the release of intracellular solutes starts at the temperature where phase transitions in the paraffin core of the membrane occur. If this is so, it is not immediately evident how various bacteria can have two critical temperature zones for cold shock (Sato & Takahashi, 1968, 1969).

DAMAGE BY FREEZING AND THAWING

It is now evident that many factors affect the response of cells to freezing. A list of some thirteen that have been recognized is given in Table 1.

Many theories to explain freezing and thawing damage have been proposed over the last 60 years but two have tended to predominate.

Table 1. *Factors known to affect survival of microbes from freezing and thawing*

Factor	Reference
Type and strain of organism	See Mazur, 1966
Population density	Major, McDougal & Harrison, 1955
Nutritional status	Calcott & MacLeod, 1974*b*
Growth phase and rate of growth	Davies, 1970; Calcott & MacLeod, 1974*b*
Composition of the freezing menstruum	Harrison, 1956; Postgate & Hunter, 1961
Cooling rate to the freezing point of the suspension	Sherman & Albus, 1923; Meynell, 1958
Cooling rate from the freezing point of the suspension	Mazur, 1966; Davies, 1970; Calcott & MacLeod, 1974*a*
Time held at low temperature	Haines, 1938; Harrison, 1956
Holding temperature	Haines, 1938; Straka & Stokes, 1959
Rate of warming to freezing point	Mazur, 1966; Calcott & MacLeod, 1974*a*
Diluting environment prior to viability determination	Bretz & Hartsell, 1959
Method of viability determination	Postgate, 1969
Medium used to assess viability	Straka & Stokes, 1959

The first one was presented by Keith (1913) who suggested that the formation of internal ice was responsible for cell death. The ice would cause the destruction of membrane components of the cells by mechanical means. The other popular idea was presented by Haines (1938). He envisaged that the high concentration of salts produced when ice was formed both intra- and extracellularly would lead to the denaturation of macromolecules such as proteins and/or to the breakdown of the barrier properties of membranes. At the time, it was considered that freeze-thaw death was the result of the action of a single factor. As information became available from mammalian cells and yeasts, it became clear that the death of cells was more likely the result of the interaction of at least two factors, one a direct function of cooling rate and the other an inverse function of cooling rate (Mazur, 1966; Mazur, Leibo & Chu, 1972). Recent studies with bacteria indicate that the situation may be even more complex.

When cells are cooled to below freezing temperature the rate of cooling and of warming can both affect cell survival. The pattern of response to cooling and warming rates is similar for cells of widely differing types. From a compilation of data by Mazur (1966) from various sources, a plot showing the influence of both cooling and warming rate on survival of *Saccharomyces cerevisiae* frozen in water suspension to −30 °C or below was produced (Fig. 1). As the cooling rate was increased from an infinitely slow rate the percentage survival of cells increased to a maximum and then decreased to a minimum before increasing again to a second maximum at ultra-rapid rates of

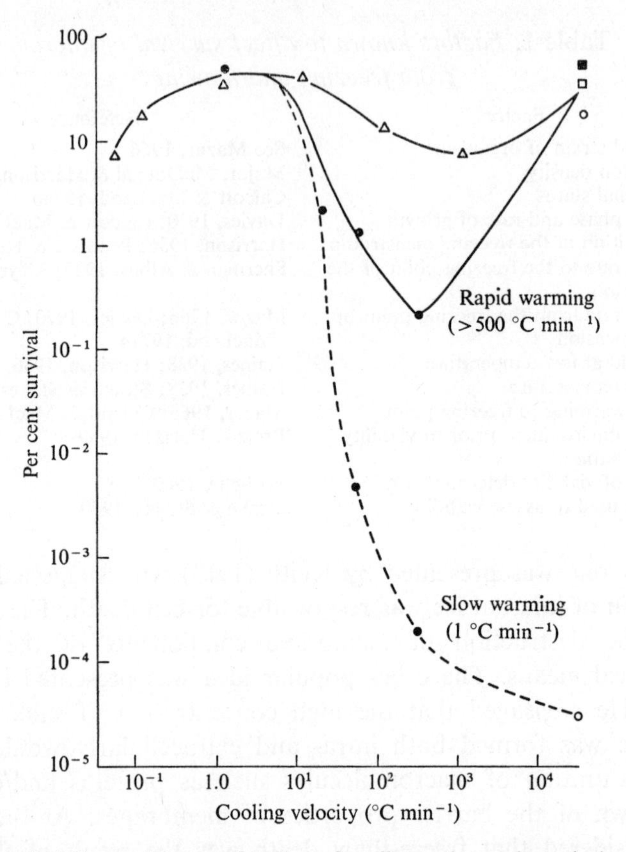

Fig. 1. Survival of *S. cerevisiae* as a function of cooling and warming velocity. Cells were suspended in water in most cases and cooled to −30 °C or below. Data compiled by Mazur (1966) from a number of sources.

Reproduced by kind permission of Academic Press, who hold the copyright for *Cryobiology*, ed. H. T. Meryman, 1966.

cooling. This response was obtained if the samples were thawed by rapid warming. For samples thawed by slow warming the response to increases in the cooling rate was the same until the maximum cooling rate for survival in the low cooling rate range was achieved. As the cooling rate was further increased, survival continued to be progressively reduced. Thus ultra-rapid warming and not rapid cooling is the major reason for high survival in the high cooling rate range.

The fact that there is an optimum cooling rate for survival in the low cooling rate range indicated that survival in this range was dependent on two categories of factors oppositely affected by cooling rate (Mazur, 1966; Mazur *et al.*, 1972). For reasons that will be considered later, the factor responsible for death at cooling rates

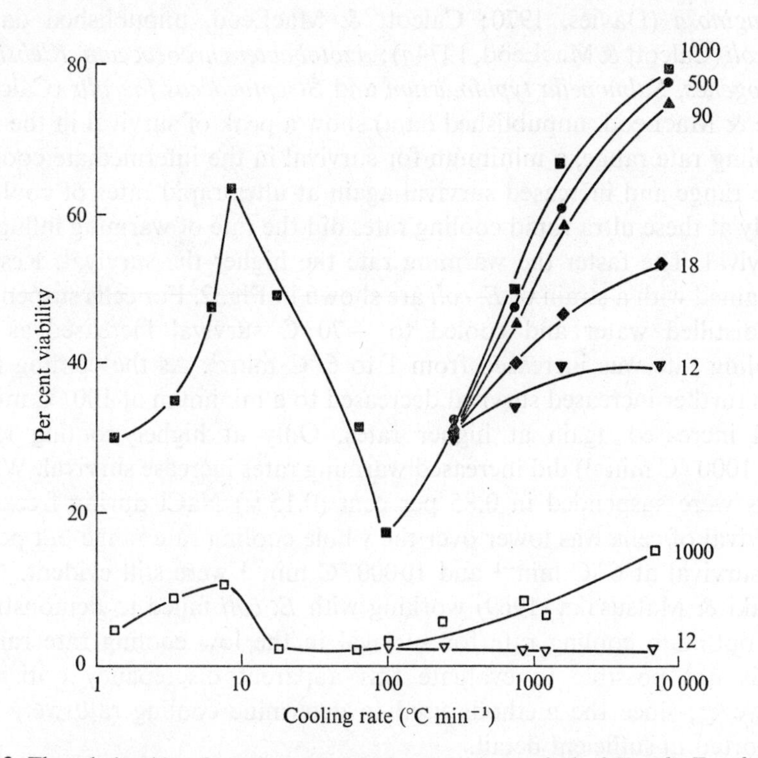

Fig. 2. The relationship of cooling and warming rates to survival of *E. coli*. *E. coli* was frozen by cooling to −70 °C at various cooling rates in either distilled water (closed symbols) or 0.85 per cent NaCl (open symbols) and thawed at a variety of rates (indicated at the right of the graph in °C min^{-1}). Viability was determined by slide culture.

Reproduced in part by permission of the National Research Council of Canada from the *Canadian Journal of Microbiology*, **20**, 671–681 (1974) (Calcott & MacLeod).

below the optimum for survival in the low cooling rate range was believed to be related to solution effects (e.g. high solute concentration) while the factor active at cooling rates above the optimum was associated with intracellular ice formation. Studies with cells as divergent as Chinese hamster tissue-culture cells, human red blood cells and mouse bone marrow stem cells have revealed that for each cell type, there is an optimum cooling rate for survival (Mazur *et al.*, 1970). This optimum cooling rate varies from 1.0 °C min^{-1} for stem cells to 3000 °C min^{-1} for red blood cells. Studies with Chinese hamster cells have shown that as in the case of yeast cells, survival depends on warming rate at cooling rates greater than the optimum (Mazur *et al.*, 1972).

Most bacteria examined respond to cooling and warming rates in a manner similar to other types of cells. Studies with *Pseudomonas aeruginosa* (Davies, 1970; Calcott & MacLeod, unpublished data); *E. coli* (Calcott & MacLeod, 1974*a*); *Azotobacter chroococcum*, *Klebsiella aerogenes*, *Salmonella typhimurium* and *Streptococcus faecalis* (Calcott, Lee & MacLeod, unpublished data) show a peak of survival in the low cooling rate range, a minimum for survival in the intermediate cooling rate range and increased survival again at ultra-rapid rates of cooling. Only at these ultra-rapid cooling rates did the rate of warming influence survival. The faster the warming rate the higher the survival. Results obtained with a strain of *E. coli* are shown in Fig. 2. For cells suspended in distilled water and cooled to -70 °C survival increased as the cooling rate was increased from 1 to 6 °C min^{-1}. As the cooling rate was further increased survival decreased to a minimum at 100 °C min^{-1} and increased again at higher rates. Only at higher cooling rates ($>$ 1000 °C min^{-1}) did increased warming rates increase survival. When cells were suspended in 0.85 per cent (0.15 M) NaCl during freezing, survival of cells was lower over the whole cooling rate range but peaks of survival at 6 °C min^{-1} and 10000 °C min^{-1} were still evident. Nei, Araki & Matsusaka (1969) working with *E. coli* failed to demonstrate an optimum cooling rate for survival in the low cooling rate range. It is not possible to evaluate this apparent discrepancy critically, however, since the methods used to determine cooling rate were not reported in sufficient detail.

Studies on the survival of a moderately halophilic bacterium on freezing and thawing (Deal, 1970) showed that survival was low at slow cooling rates, higher at intermediate rates and low again at faster rates. In contrast to other cells examined, however, the response to warming rate was the same over the whole cooling rate range. Significant survival was obtained only at fast warming rates.

Nature of freezing damage

Much evidence is available to show that both membrane and wall damage occur when cells are frozen and thawed. Conclusions regarding membrane damage arise from the finding that if microbes are frozen and thawed, leakage of materials occurs from the cells. In several cases a correlation has been reported between the extent of loss of internal solutes and percentage survival of the cells (cf. Lindeberg & Lode, 1963). Most of the lost materials are low molecular weight materials and include K^+, inorganic phosphate, phosphorylated sugars, sugars, fatty acids, esters and amino compounds (cf. Sakagami, 1959).

Other evidence of membrane damage arises from the increased penetrability of certain compounds into cells after freezing and thawing.

Strange & Postgate (1964) showed that a molecule as large as RNase can enter *A. aerogenes* after cold shock and also after freezing in the case of cells not sensitive to cold shock. *E. coli* became more sensitive to the toxic action of Cu^{2+} (MacLeod, Kuo & Gelinas, 1967). Postgate & Hunter (1963) observed that in a freeze-thawed population of cells showing only 5 per cent survival, 25 per cent of the cells were still osmotically active. This indicated that factors besides membrane damage must be responsible for loss of viability of the cells.

Evidence that freezing and thawing alter the cell wall as well as the membrane has been obtained. Kohn & Szybalski (1959) showed that if *E. coli* was frozen to −80 °C and thawed the cells could be converted to spheroplasts by treatment with lysozyme. This indicated that freezing and thawing increased the penetrability of the outer or cell wall membrane of this Gram-negative bacterium to lysozyme, permitting access of the enzyme to the underlying peptidoglycan layer. Ray, Jannsen & Busta (1972) confirmed these findings using *Salmonella anatum*. Morichi (1969) observed that freezing and thawing *E. coli* caused the release of the periplasmic enzyme alkaline phosphatase. Another periplasmic enzyme, cyclic phosphodiesterase, was released almost quantitatively from the cells of another strain of the same organism that had been frozen and thawed (Calcott & MacLeod, 1975). The release of periplasmic enzymes from the cells provides further evidence that freezing damage leads to an increase in the penetrability of the outer or cell wall membrane of this Gram-negative organism.

Relation of cell damage to cooling and warming rates

Most studies on cell damage caused by freezing and thawing have been conducted using cooling and warming rates which are ill-defined. Since the studies of Mazur and his co-workers have shown that death of cells by freezing and thawing can be caused by at least two categories of factors oppositely dependent on cooling rate, it is of interest to know how cell damage is related to cooling and warming rates.

A study has been made of the effect of cooling and warming rates during freezing and thawing on membrane damage in *E. coli* (Calcott & MacLeod, 1975). The effect was found to vary depending on whether the cells were frozen in water or 0.85 per cent NaCl. When cells were frozen in water there was no release of u.v.-absorbing material from the cells at cooling rates below the optimum for survival in the low cooling rate range (6 °C min^{-1}). Above the optimum cooling rate for survival the release of u.v.-absorbing material was inversely proportional

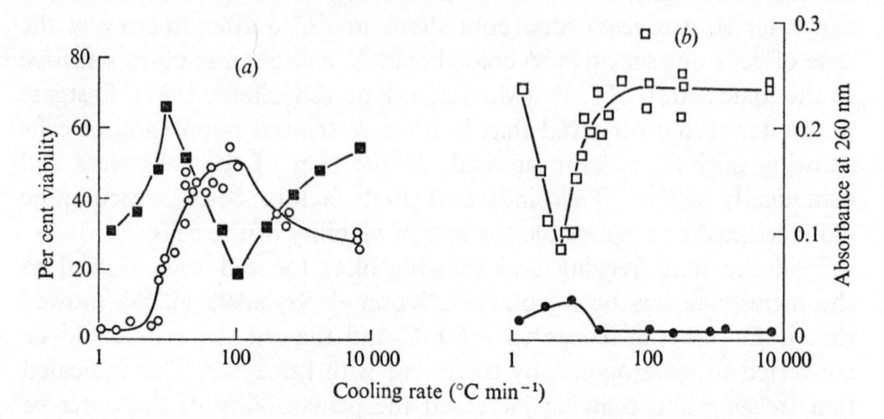

Fig. 3. Relation between survival of *Escherichia coli* and release of u.v.-absorbing material from the cells on freezing and thawing in water (*a*) and 0.85 per cent NaCl (*b*).

Cells were frozen by cooling to −70 °C at the various rates and thawed at 18 °C min^{-1}. Ultraviolet-absorbing material was assayed on the freezing menstruum after removal of cells (open symbols). Viability was determined by slide culture (closed symbols).

Reproduced by kind permission of the National Research Council of Canada from the *Canadian Journal of Microbiology*, **21** (1975) (Calcott & MacLeod).

to the survival of the population (Fig. 3*a*). A similar relationship was observed with K^+ release. Glucose-6-phosphate dehydrogenase (G6PDH), an intracellular enzyme, was not released by freezing and thawing but whole cell G6PDH activity increased, i.e. the substrates (glucose-6-phosphate and NADP) for this enzyme were able to enter the cell more easily. The shape of the curve relating enzyme accessibility to cooling and warming rates was similar to that relating release of u.v.-absorbing material and K^+ to cooling and warming rates. One may conclude then that, for *E. coli* suspended in water, death due to freezing at cooling rates less than the optimum in the low cooling rate range is not related to membrane damage, while at cooling rates above the optimum, death is associated with membrane damage.

For cells frozen in saline, death can be related to membrane damage even at cooling rates below the optimum for survival in the low cooling rate range. Fig. 3*b* shows the relation between the release of u.v.-absorbing material and cell survival for cells frozen in saline at different cooling rates. Release was high at the lowest cooling rate tested and decreased with increasing cooling rate to a minimum after which it increased to a maximum and levelled off. Similarly shaped curves were obtained for the release of K^+ and for the increase in accessibility of G6PDH to its substrates when the cells were frozen in saline. Thus for this organism, whether or not membrane damage is a factor influencing survival in the low cooling rate range during freezing depends on whether NaCl is present in the suspending solution.

We have conducted experiments (to be published) on the extent of damage to the wall and membrane of *E. coli* frozen in 0.15 M NaCl at a cooling rate (100 °C min^{-1}) which produces minimum cell survival. Thawing was by fast warming. Under these conditions, all K^+ and low molecular weight u.v.-absorbing material was released and sucrose, which normally penetrates the cell to the level of the cytoplasmic membrane, was able to permeate the cell fully. Two dextrans with molecular weights of 16000 and 60000 to 90000 respectively and inulin with a molecular weight of 5000, all of which were unable to penetrate the cells before freezing, were able to do so partially after freezing. The extent of penetration was roughly in proportion to the molecular weight of the penetrating compounds. This was interpreted to mean that damage to the cytoplasmic membrane was insufficient to permit penetration by any of the macromolecules but that some wall damage had occurred. Protectant studies with glycerol and Tween 80 indicated that while the former protected the cells against loss of viability it also prevented wall and membrane damage while the latter prevented membrane damage and loss of viability but was without effect on wall damage. This implied that death and membrane damage were related and that the wall damage which occurred on freezing and thawing was not detrimental to cell survival.

The effect of electrolytes on the survival of cells after freezing and thawing

Solutes of importance to cryosurvival can be divided into two groups, those which increase and those which decrease survival of cells which have been frozen and thawed.

In an effort to understand the action of solutes on cell survival it is necessary to consider some of the events which occur on freezing. When an aqueous solution is cooled below its freezing point, water freezes, the solute becomes concentrated and the freezing point of the remaining solution is lowered. Ultimately a temperature is reached, the eutectic temperature, below which a liquid state cannot exist and a eutectic mixture of ice and solute will form. Most electrolytes, e.g. NaCl, form crystalline eutectic mixtures. Many other compounds do not. Glycerol solutions, for instance, show progressive crystallization of ice with lowering temperature until a concentration of 60–70 per cent glycerol has been reached. The remaining solution then forms a glass (63.1 per cent glycerol). When glycerol and NaCl are both present, the mixture undergoes glass formation with no eutectic crystallization (Meryman, 1966).

One of the classic hypotheses of the mechanism of freezing damage is that the high concentration of electrolyte produced by freezing affects membrane lipids thus making the cells leaky. Lovelock (1953*a*, *b*) showed that red blood cells exposed to 0.8 M NaCl at room temperature, when returned to isotonic saline, were haemolysed to the same extent as cells frozen in isotonic saline and thawed. Harrison (1956) reported that the death rate of *E. coli* stored unfrozen at −22 °C in 4.6 M NaCl was similar to cells stored frozen in broth at the same temperature. He concluded that NaCl, not freezing *per se*, was the lethal agent. Lindeberg & Lode (1963) came to similar conclusions. Zimmerman (1964) obtained different results using *Serratia marcescens*. He found that the cells survived well in concentrated salt solutions at room temperature and concluded that the ice salt mixture had to pass through the eutectic point to be lethal to the cells.

Sato, Izaki & Takahashi (1972) report that the addition of 0.15 M NaCl to logarithmic phase *E. coli* at room temperature caused loss of cell viability. Viability of the cells could be restored by incubation with Mg^{2+}. In the study reported, the cell density used was very high (1×10^{10} cells ml^{-1}) and the cells were incubated under static conditions: the effect of NaCl increased progressively over a 30 min period of incubation. Studies in the authors' laboratory (Lee, Calcott & MacLeod, unpublished observations) have shown that when *E. coli* is suspended at room temperature at a cell density of 1×10^9 cells ml^{-1} under conditions of aeration, 0.15 M NaCl produced no significant loss of cell viability.

While many data are available to show that the inclusion of NaCl or other salts in the freezing menstruum decreases the survival of some micro-organisms after freezing and thawing, this is not universally true. The presence of 0.15 M NaCl in the suspending medium protects *Saccharomyces cerevisiae* from freezing damage to a small extent over the whole cooling rate range (Mazur, 1966). In a recent study in this laboratory (Lee, Calcott & MacLeod, unpublished observations) *E. coli*, *K. aerogenes*, *Salmonella typhimurium*, *Pseudomonas aeruginosa*, *A. chroococcum*, *B. subtilis* and *Staphylococcus aureus* were all sensitive to the presence of salt in the suspending medium during freezing. *Streptococcus faecalis* and *Lactobacillus casei* on the other hand, were not. One obvious way in which all the organisms in the former group differ from the two in the latter and the yeast *S. cerevisiae* is that the former all contain cytochrome pigments in their cytoplasmic membranes while the latter contain none. Although there may be other less obvious ways in which the membranes in the two groups of organisms may differ, the possibility exists that the presence of cytochrome pigments in the membrane renders the membranes sensitive to disruption by salt.

Deal (1974) has reported that in the case of a moderately halophilic bacterium, salts protect against freeze-thaw damage and for these cells the kinds and ratios of the ions present in the freezing menstruum can be critical for survival. For this organism, at all cooling rates tested, survival increased as the total molarity of mixtures of NaCl and KCl in the freezing menstruum were increased from 1.5 to 2.0 M. When cooling was slow, the ratio of K^+ to Na^+ was important for survival with high K^+ favouring survival. At intermediate or fast cooling rates Na^+ and K^+ appeared for the most part to be interchangeable. The addition of Mg^{2+} to the salt mixtures strongly protected the cells against freeze-thaw damage at all three cooling rates tested.

Cryoprotectants

A number of compounds serve as cryoprotectants for a wide range of cell types. Two groups can be distinguished, those which are penetrating, such as glycerol and dimethylsulphoxide and those which are not (Meryman, 1971). Among the latter are sucrose and polyvinylpyrrolidone. In the case of animal cells, the penetrating cryoprotectants such as glycerol appear to protect best in the low cooling rate range (Mazur, 1970). It has thus been proposed that they protect primarily against solution effects. They might do this by reducing the electrolyte concentration in the residual unfrozen solution in and around a cell at any given concentration. Non-penetrating cryoprotectants, on the other hand, have been shown to cause changes in the sodium and potassium content of erythrocytes frozen with these compounds. As a result, it has been suggested that these compounds confer protection by permitting a reversible influx and efflux of solute during freezing and thawing, thus enabling cells to avoid the effects of excessive osmotic gradients (Meryman, 1971).

Nash, Postgate & Hunter (1963) studied the effectiveness of a variety of compounds in protecting both *A. aerogenes* and red blood cells from freezing and thawing. Those compounds which protected one cell type also protected the other to a similar degree while those compounds which were ineffective with one type were also ineffective with the other. This was observed despite the fact that different freeze-

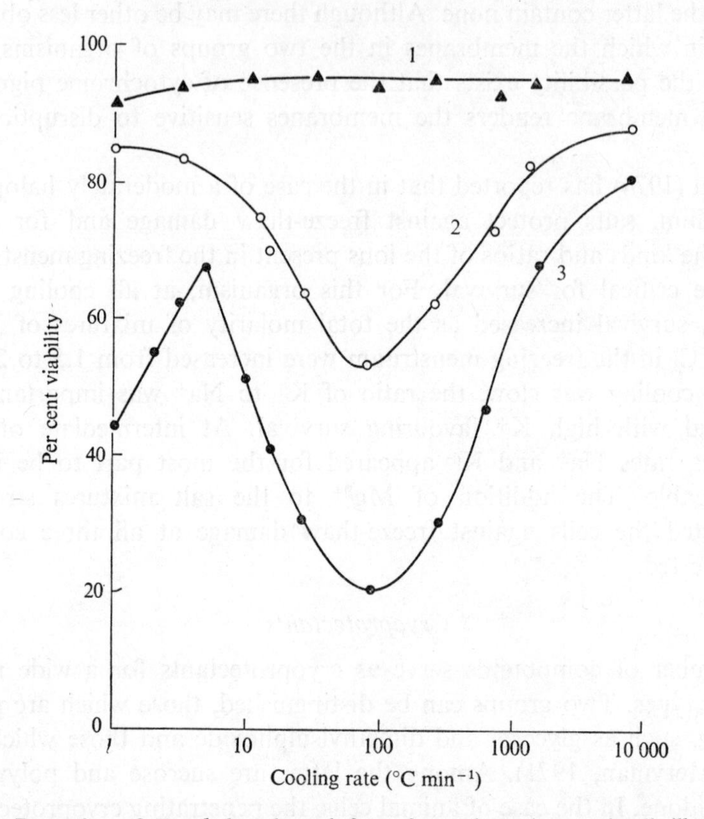

Fig. 4. Protection of *E. coli* by glycerol from freeze-thaw damage in distilled water Procedure as in Fig. 2 except organisms were frozen in distilled water + 3 per cent glycerol (curve 1), + 1 per cent glycerol (curve 2), or distilled water alone (curve 3) and thawed at 1000 °C min^{-1}.

Reproduced by kind permission of the National Research Council of Canada from the *Canadian Journal of Microbiology*, **20**, 671–681 (1974) (Calcott & MacLeod).

thaw techniques and methods of determining cell survival had to be used with the two types of cells.

A study has been made of the effect of cooling rate on the effect of representative penetrating and non-penetrating cryoprotectants on survival of *E. coli* from freezing and thawing (Calcott & MacLeod, 1974*a*). Glycerol, a penetrant, protected *E. coli* from death by freezing over the whole cooling rate range (Fig. 4). This occurred whether water or 0.15 M NaCl served as the freezing menstruum. Thus, for *E. coli*, glycerol does not only protect against 'solution effects'. Sucrose, a non-penetrant, also protected *E. coli* over the whole cooling rate range if the freezing menstruum was distilled water. If the freezing menstruum was 0.15 M NaCl, sucrose protected at cooling rates above

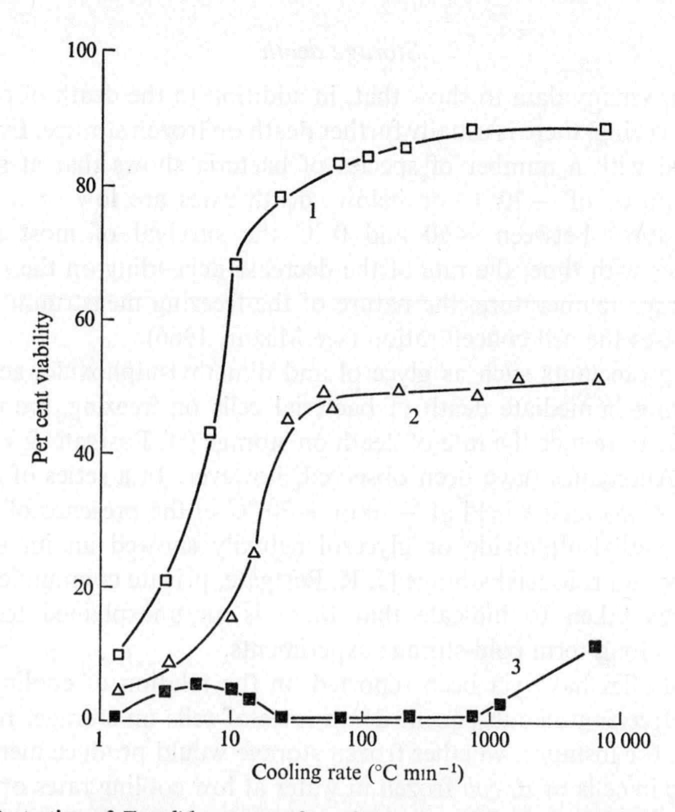

Fig. 5. Protection of *E. coli* by sucrose from freeze-thaw damage in NaCl. Procedure as in Fig. 2 except that organisms were frozen in saline + 5 per cent sucrose (curve 1), + 2.5 per cent sucrose (curve 2), or saline alone (curve 3).

Reproduced by kind permission of the National Research Council of Canada from the *Canadian Journal of Microbiology*, **20**, 671–681 (1974) (Calcott & MacLeod).

the optimum for survival but not below (Fig. 5). Tween 80, a non-penetrating compound, also protected the cells from death by freezing at cooling rates above the optimum for survival whether NaCl was present or not, but failed both in the presence and absence of NaCl to protect against freezing in the low cooling rate range. Both glycerol and Tween 80 prevent the release of K^+ from cells frozen at three different cooling rates, indicating that both compounds protect against the membrane damage produced by the conditions of freezing and thawing employed. Thus for *E. coli*, the non-penetrant Tween 80 does not confer protection by permitting a reversible influx and efflux of solute.

Storage death

There are many data to show that, in addition to the death of cells on initial freezing, there is usually further death on frozen storage. Evidence obtained with a number of species of bacteria shows that at storage temperatures of −70 °C or below, death rates are low or zero. At temperatures between −60 and 0 °C the survival of most species decreases with time, the rate of the decrease depending on the species, the storage temperature, the nature of the freezing menstruum and in some cases the cell concentration (see Mazur, 1966).

Cryoprotectants such as glycerol and dimethylsulphoxide, active in preventing immediate death of bacterial cells on freezing, are usually also able to reduce the rate of death on storage (cf. Postgate & Hunter, 1961). Anomalies have been observed, however. In a series of experiments, *A. aerogenes* held at −20 or −70 °C in the presence of 10 per cent dimethylsulphoxide or glycerol actually showed an increase in viability on prolonged storage (J. R. Postgate, private communication). This was taken to indicate that there is an unexplained technical pitfall in long-term cold-storage experiments.

No studies have yet been reported on the relation of cooling rates during freezing on subsequent death rates of cells on storage. It is not known, for instance, whether frozen storage would produce membrane damage in cells of *E. coli* frozen in water at low cooling rates or would lead to increased membrane damage in cells frozen in water at fast cooling rates. Until more information is available it is unproductive to speculate on the causes of cell death on storage.

Effect of nutritional status and growth rate on cryosensitivity

Earlier workers studying the influence of nutritional status on survival from freezing and thawing produced very conflicting reports. Certain groups showed that aerated cells were more resistant than non-aerated cells to freeze-thaw damage and that early log-phase cells of *E. coli* were more susceptible than stationary-phase cells to this stress. Others showed that phases of growth and degree of aerobiosis of *E. coli* were without effect on the cryosurvival of the cells (see Calcott & MacLeod, 1974*b*, for a review). More recent studies suggest that the apparent discrepancies in the results obtained may be explainable in terms of the differences in the composition of the growth media used and the different cooling and warming rates employed during freezing and thawing of the cells.

Nei *et al.* (1969) showed that aerated *E. coli* was more resistant to

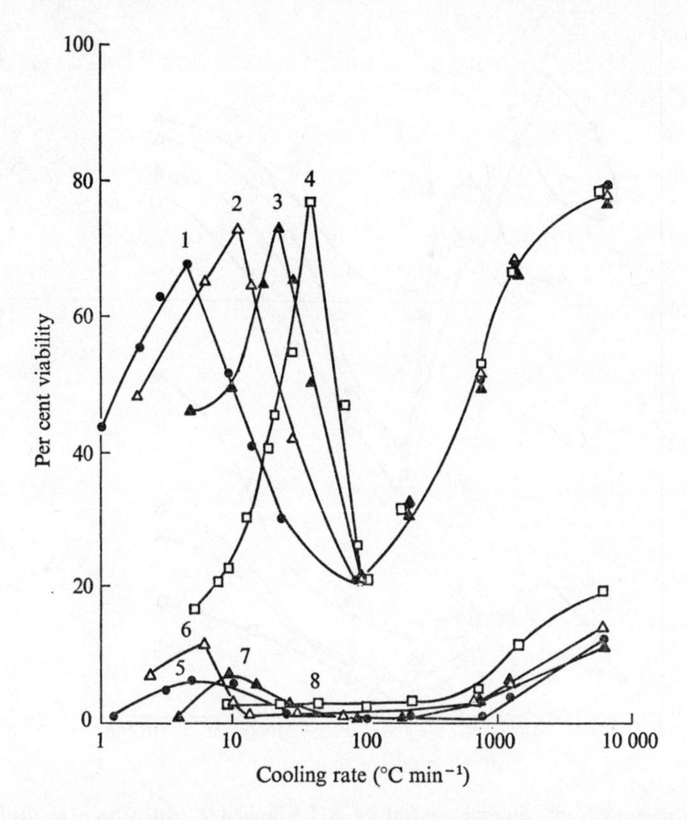

Fig. 6. Effect of growth rate on cryosurvival of glycerol-limited *E. coli* frozen in distilled water or NaCl. Twice-washed glycerol-limited *E. coli* cells grown at a variety of rates in a chemostat were cooled at varying rates and warmed rapidly (1000 °C min^{-1}). Growth rates: curves 1 and 5, $\mu = 0.10\ h^{-1}$; curves 2 and 6, $\mu = 0.20\ h^{-1}$; curves 3 and 7, $\mu = 0.39\ h^{-1}$; curves 4 and 8, $\mu = 0.60\ h^{-1}$. Freezing menstrua: curves 1 to 4, distilled water, curves 5 to 8, 0.85 per cent NaCl.

Reproduced by kind permission of the National Research Council of Canada from the *Canadian Journal of Microbiology*, **20**, 683–689 (1974) (Calcott & MacLeod).

freezing and thawing than non-aerated cells but the difference in response decreased as the cooling rate decreased. Davies (1970) presented evidence that not only did the phase of growth of a *Pseudomonas* species influence its cryosensitivity but so also did the medium used to grow the cells. Cells of the organism in the logarithmic phase of growth were more susceptible to freezing than cells in the stationary phase at some cooling rates while the reverse was true at other rates. Cells grown in Koser's citrate medium were markedly different in cryosensitivity than cells of the same organism grown in nutrient broth medium.

Calcott & MacLeod (1974*b*) showed that when *E. coli* was grown

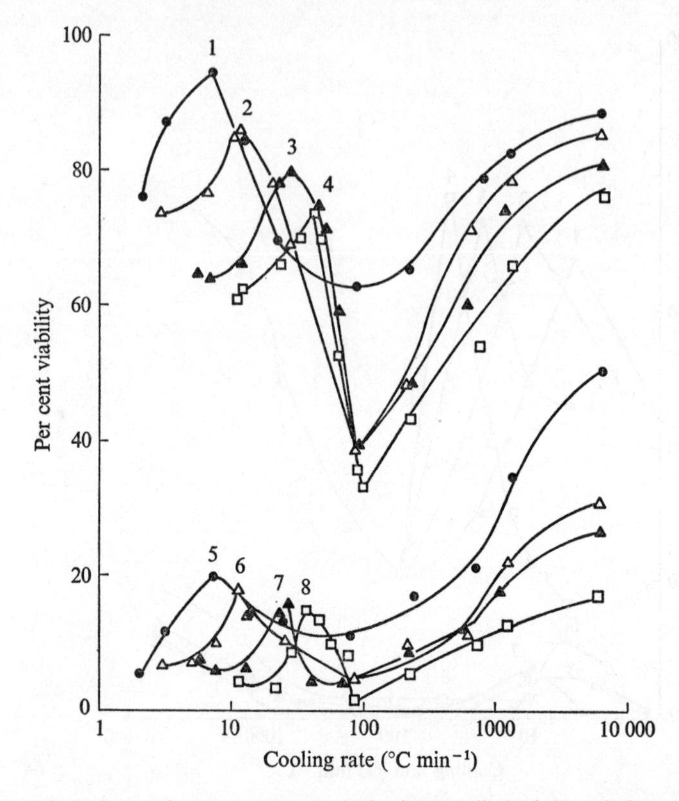

Fig. 7. Effect of growth rate on cryosurvival of NH_4^+-limited *E. coli* frozen in distilled water or NaCl. Procedures as for Fig. 6 except that NH_4^+-limited organisms were used. Growth rates: curves 1 and 5, $\mu = 0.11\ h^{-1}$; curves 2 and 6, $\mu = 0.218\ h^{-1}$; curves 3 and 7, $\mu = 0.40\ h^{-1}$; curves 4 and 8, $\mu = 0.54\ h^{-1}$. Freezing menstrua: curves 1 to 4, distilled water, curves 5 to 8, 0.85 per cent NaCl.

Reproduced by the kind permission of the National Research Council of Canada from the *Canadian Journal of Microbiology*, **20**, 683–689 (1974) (Calcott & MacLeod).

in a chemostat under carbon-limiting conditions, the peak of survival of cells frozen and thawed in the low cooling rate range shifted from 6 °C min^{-1} for slowly grown cells to 40 °C min^{-1} at higher growth rates (Fig. 6). At cooling rates above 100 °C min^{-1}, survival after freezing and thawing was independent of growth rate. In the case of NH_4^+-limited cells, the cooling rate for maximum survival of the cells also increased as the growth rate increased. For these cells, however, survival shifted downward with increasing growth rate of the cells over the whole cooling rate range (Fig. 7). In the case of the NH_4^+-limited cells, increased cryotolerance was associated with an increased carbohydrate content of the cells.

Gilliland & Speck (1974) observed a relation between the capacity of a variety of organisms grown in batch culture to survive freezing and thawing and the glucose content of the cells. In their studies, only one cooling rate was used to measure cryosensitivity.

Smittle *et al.* (1974) observed that *Lactobacillus bulgaricus* was more resistant to freezing in liquid nitrogen when the growth medium was supplemented with sodium oleate. All strains grown in broth containing sodium oleate contained larger amounts of octadecenoic and C_{19}-cyclopropane fatty acid than when grown without sodium oleate. Statistical analyses indicated that the stability of the lactobacilli in liquid nitrogen could be related to the per cent of C_{19}-cyclopropane fatty acid in the lipids extracted from the cells. Unfortunately, cryo-survival was measured at only one cooling rate.

REPAIR OF FREEZING DAMAGE

Straka & Stokes (1959) observed that when various species of bacteria were frozen and thawed the per cent survival varied depending upon the composition of the plating medium used to determine cell viability. Survival was always higher when a complex (Trypticase Soy) agar medium was employed than when a minimal (salts, glucose) agar medium was used. Cells unable to grow in either complex or minimal medium were considered to be killed, those able to grow on the complex but not the minimal medium were referred to as being metabolically injured while those able to grow on minimal medium only were believed to be unharmed. The percentages of injured, killed and unharmed cells in a population after freezing varied with the time and temperature of storage and with the nature and pH of the freezing menstruum. Metabolically injured cells were considered to represent that proportion of the population which had not been irreversibly damaged by freezing. It appeared that such cells had become nutritionally more exacting and provided the proper nutrients were supplied, could repair the damage caused by freezing. These observations were confirmed and extended by other workers using a variety of bacterial species (see Ray & Speck, 1973, for a review).

Since metabolically injured cells seem to be partially damaged it appeared possible that knowledge of the factors and conditions promoting recovery from the damage might provide clues to the mechanism of damage caused by freezing. Efforts have thus been made to identify the factors in complex medium responsible for promoting the growth of metabolically injured cells. Straka & Stokes (1959) observed that 1 to 2 per cent Trypticase added to minimal agar medium permitted as high a plate count as the complex Trypticase Soy agar medium. A supplement of acid-hydrolysed casein, a mixture of B-vitamins, purines and pyrimidines produced no repair. These workers proposed that the effectiveness of Trypticase was due to its content of specific peptides required for the resynthesis of essential enzymes or proteins destroyed by freezing. Moss & Speck (1966) isolated from Trypticase five closely related peptides which promoted the growth of *E. coli* metabolically injured by freezing. Using the same strain of *E. coli* as Moss & Speck and the same freezing conditions, Kuo & MacLeod (1969) observed that a mixture of crystalline amino acids, if added in a sufficiently low concentration, could replace Trypticase as a supplement to a minimal medium for obtaining repair of metabolically injured cells. Tests of the individual amino acids in the mixture showed that supplementing the minimal medium with serine reduced the plate count on a suspension of the cells after freezing but supplementation with glutamic acid, alanine or particularly aspartic acid increased the plate count. For this organism aspartic acid was as effective as Trypticase in promoting the growth of metabolically injured cells. For another organism, a strain of *A. aerogenes*, complex supplements such as Trypticase could be replaced by cysteine or the chelating agent ethylenediaminetetracetic acid for the growth of metabolically injured cells (MacLeod, Kuo & Gelinas, 1967). Evidence was presented indicating that these agents acted by detoxifying toxic traces of Cu^{2+} in the plating medium. It was concluded that freezing and storage damaged the cell membrane of the organism rendering it more penetratable by toxic trace elements.

When *S. anatum* or *E. coli* suspended in water were frozen 'rapidly', Ray *et al.* (1972; Ray & Speck, 1972) found over 90 per cent of the cells which survived had lost the capacity to form colonies on a plating medium containing Na-deoxycholate. These workers have determined the minimal conditions required to restore the capacity of the cells to form colonies on the deoxycholate agar. Incubating the cells in minimal broth medium restored their colony forming ability on the selective medium. Incubation of the cells in 0.25 per cent K_3PO_4 solution was almost as effective as incubation in the minimal medium while a combination of K_3PO_4, $MgSO_4$ and citrate was able to duplicate the action of the minimal medium. The development of resistance to deoxycholate by incubation in the phosphate solution was not prevented by actinomycin D, chloramphenicol or D-cycloserine but was markedly reduced by both cyanide and 2,4-dinitrophenol. It was thus concluded that RNA, protein and mucopeptide syntheses were not involved in the recovery process but that energy generation was required.

When *E. coli* was suspended in 0.15 M NaCl and frozen 'slowly', over 90 per cent of surviving cells failed to form colonies on a Trypticase Soy agar medium containing 0.55 M NaCl (Bunch & MacLeod, unpublished observations). Incubation of the cells after freezing in minimal medium containing aspartic acid restored their capacity to form colonies in the presence of NaCl. A solution containing a mixture of Na_3PO_4, K_3PO_4 and $MgSO_4$ could replace the minimal medium in restoring the capacity of the cells to grow in the presence of NaCl. The most important component of this recovery medium was Mg^{2+}. Failure to include Mg^{2+} in the recovery medium caused the cells to die rapidly as indicated by the loss of the ability of the cells to form colonies on the plating medium both in the presence and absence of added salt.

Recovery of the capacity of the freeze-damaged cells to grow in the presence of 0.55 M NaCl was not inhibited by the inclusion of cyanide, azide or chloramphenicol in the recovery medium. Thus neither protein nor energy synthesis appears to be involved in this recovery process.

The apparent discrepancy in the requirements for recovery from freezing damage reported by the two groups of workers probably can be attributed to two causes. One is the difference in freezing and thawing conditions used, the other is the difference in the agent employed to distinguish between injured and non-injured cells.

Neither group properly related the freezing and thawing conditions used to a curve relating cooling and warming rates to survival of the cells. The regimes employed might well have produced quite different kinds of damage to the cells.

There is evidence that the outer or cell wall membrane of Gram-negative bacteria serves as the barrier to the penetration of detergents (Unemoto & MacLeod, 1975) while the cytoplasmic membrane is the barrier to the penetration of NaCl. Thus conditions leading to the restoration of resistance to deoxycholate probably represent the requirements for the repair of the outer or cell wall membrane while those concerned with the development of resistance to NaCl, the repair of the cytoplasmic membrane. The conditions used by Bunch & MacLeod to restore the resistance of freeze-damaged *E. coli* to NaCl in the plating medium also restored resistance to sodium deoxycholate. The results obtained indicated, however, that the development of resistance to NaCl must precede the development of resistance to deoxycholate.

The studies with selective media indicate that most if not all of the cells that survived the conditions of freezing employed were in fact injured, even though many were still able to grow on minimal medium and all could grow and divide on a complex medium.

CONSIDERATION OF FACTORS CONTRIBUTING TO DEATH BY FREEZING

In a theoretical study, Mazur (1963) proposed a model to describe the events leading to extra- and intracellular ice formation in a suspension of cells during freezing. Mazur developed a differential equation based on this model which related cooling rate, surface volume ratio, membrane permeability to water and the temperature coefficient of the permeability factor. Solutions to this equation permitted him to predict the critical cooling rate above which intracellular ice would be expected to form in the cells. This critical cooling rate coincided well with the cooling rate for optimum survival of cells on freezing in the case of cell types having widely differing cooling rate optima. A number of electron microscope studies lend support to the conclusion that only at cooling rates greater than the optimum for survival does ice form intracellularly (Mazur, 1967; Bank & Mazur, 1973). Mazur thus concluded that death by freezing at cooling rates greater than the optimum was due to intracellular ice formation. Death by freezing at cooling rates below the optimum, since it occurred in cells which had become dehydrated before ice could form internally, was believed to be due to slowly concentrating solutes acting on the cells both internally and externally during the freezing process. Failure of cells to die when cooling and warming rates were ultra-rapid was attributed to the formation of small and hence non-lethal ice crystals. Decreased survival when cooling was ultra-rapid but warming slow, was ascribed to recrystallization and growth of large ice crystals from small ones (Mazur, 1966; Mazur *et al.*, 1972).

In the light of present knowledge let us consider whether the Mazur hypothesis requires amplification or modification in so far as it applies to the survival of a micro-organism such as *E. coli* after freezing.

When *E. coli* was suspended in water and cooled to very low temperatures at cooling rates less than the optimum for survival, the factor or factors causing death of the cells produced no membrane damage (Fig. 3*a*). It thus appears that the cell death that occurred in this cooling rate range was due to some factor operating internally such as high intracellular salt concentration or dehydration. It is difficult to explain, however, how a non-penetrating solute like sucrose could protect against the damage produced by a factor acting internally.

When *E. coli* was suspended in NaCl and cooled at rates less than the optimum, membrane damage occurred, survival was greatly reduced and the extent of survival was inversely proportional to the degree of membrane damage. Thus death of the cells could be related directly to membrane damage under these conditions. Since glycerol but not sucrose served as a cryoprotectant under these freezing conditions, the role of sucrose as a cryoprotectant has become even more obscure.

It would thus appear that when *E. coli* is frozen at cooling rates less than the optimum for survival two quite different factors can contribute to death of the cells. One is operative when NaCl is absent, both may be operative when NaCl is present. One causes membrane damage while the other does not. Both are similarly affected by cooling rate, their effect on cell survival decreasing as cooling rate increases.

When *E. coli* was frozen in water at cooling rates which increased from the rate for optimum to the rate for minimum survival, survival of cells decreased and membrane damage increased proportionally. Since sucrose could protect against this membrane damage the nature of the membrane damage must have been different from that produced when the cells were suspended in NaCl and cooled slowly. Thus the action of NaCl may be different in the two cooling rate ranges.

The membrane damage produced when cells were suspended in NaCl solution and cooled at rates above the optimum for survival was more extensive than for cells cooled at the same rate but suspended in water. At ultra-rapid rates of cooling, membrane damage occurred to cells suspended in NaCl but not in water. Thus the membrane damage produced at cooling rates above the optimum for survival in the presence of NaCl might well be different from that produced at these cooling rates in the absence of NaCl. Thus, at cooling rates above the optimum for survival, another two factors appear to act to decrease cell survival, one acting when NaCl is present in the suspending medium, the other acting in the presence or absence of NaCl. Both factors give rise to membrane damage. Over the range of cooling rates giving optimum to minimum survival, both factors are affected by cooling rate in a similar way, producing a decrease in cell survival as cooling rate increases.

To what extent does cold shock contribute to death by freezing? Since cold shock appears to act by making membranes leaky, cells of *E. coli* which have been suspended in water and frozen at cooling rates below the optimum for survival are not killed by cold shock since the barrier to the loss of u.v.-absorbing material from the cells remains intact under these conditions (Fig. 3*a*). When cells suspended in water are cooled at rates between those for optimum and minimum survival, the leaky membranes produced might well result from the manner in which phase transitions occur in the lipid components of the membranes. The presence of intracellular ice at these cooling rates could merely be fortuitous. The capacity of such cryoprotectants as sucrose and Tween 80 to protect against the membrane damage produced under these conditions could well be due to their capacity to modify the nature of the phase transitions. Sucrose is known to penetrate bacterial cells to the level of the cytoplasmic membrane (Postgate & Hunter, 1961; Buckmire & MacLeod, 1970) while Tween 80 appears to interact with the cell surface of *K. aerogenes* (Calcott & Postgate, 1971).

In the case of *E. coli*, a cell type markedly sensitive to NaCl in the external medium during freezing, the presence of NaCl leads to membrane damage over the whole cooling rate range. Membrane damage caused by cold shock (Sato & Takahashi, 1968, 1969) as well as membrane damage brought about by freezing cells in the presence of 0.15 M NaCl (Bunch & MacLeod, unpublished) can be repaired by incubating the cells with Mg^{2+}. Suspension of *E. coli* under certain conditions in the presence of NaCl even at room temperature leads to a loss of cell viability which could be restored by incubation with Mg^{2+} (Sato *et al.*, 1972). There is thus constantly recurring evidence of an antagonistic relationship between Na^+ and Mg^{2+} in the maintenance of membrane integrity in cells sensitive to NaCl during freezing. There is also much evidence that Na^+ can displace Mg^{2+} from the cell walls of bacteria (Strange & Shon, 1964; DeVoe & Oginsky, 1969; Rayman & MacLeod, 1975). It thus seems likely that when suspensions of cells in NaCl freeze, the increased concentration of Na^+ that results would lead to the displacement of Mg^{2+} from sites in the membrane where it would appear that Mg^{2+} is required to maintain membrane integrity.

Whenever membrane damage occurs, intracellular solutes, notably K^+ leak out. Deal (1974) has shown that in the case of a moderate halophile frozen at slow cooling rates, a high K^+ to Na^+ ratio in the freezing menstruum favoured survival. Since the ratio of K^+ to Na^+ inside cells is normally high this suggests that a drop in this ratio might well be one of the factors leading to death of cells by freezing particularly when slow cooling and slow warming rates are employed.

Freezing and thawing *E. coli* particularly in the presence of oxygen has been reported to produce single-strand breaks in the DNA (Swartz, 1971). Alur & Grecz (1975) showed a close correlation between the number of DNA single strand breaks and the extent of death of *E. coli* B/r due to freezing, cold storage and thawing. Double strand breaks have also been detected on freezing and thawing *E. coli* (Calcott & MacLeod, unpublished data). As mentioned earlier, the DNA of cold-shocked *E. coli* cells contains more single-strand breaks than unshocked cells or cells which had recovered (Sato & Takahashi, 1970). The latter authors have concluded that membrane damage leads to loss of Mg^{2+} from the cells which in turn is required for DNA-ligase activity and the repair of DNA. To determine whether DNA fragmentation could be an ultimate cause of cell death during freezing and thawing, the relation between the formation of DNA single- and double-strand breaks and cell survival should be studied over the whole cooling rate range.

CONCLUSION

It is now clear that when cells are frozen and thawed the pattern of response of cells to cooling and warming rates is similar for cells of widely differing types. Different cooling and warming rates have been shown to produce different kinds of damage to the cells. It also has been found that the kind of damage produced during freezing can vary depending on the chemical composition of the freezing menstruum.

To take full advantage of present knowledge and to make further meaningful contributions to our understanding of the mechanisms leading to cell death by freezing and thawing more precise control of freezing and thawing conditions will have to be exercised than has been customary in the past. It will be necessary not only to define the chemical composition of the freezing menstruum and provide the warming and cooling rates used, but also to show the position of these warming and cooling rates on a curve relating the effect of warming and cooling rates to cell survival for the particular cell system under investigation.

REFERENCES

ALUR, M. D. & GRECZ, N. (1975). Mechanism of injury of *Escherichia coli* by freezing and thawing. *Biochemical and Biophysical Research Communications*, **62**, 308–312.

BANK, H. & MAZUR, P. (1973). Visualization of freezing damage. *Journal of Cellular Biology*, **57**, 729–742.

BRETZ, H. W. & HARTSELL, S. E. (1959). Quantitative evaluation of defrosted *Escherichia coli*. *Food Research*, **24**, 369–381.

BUCKMIRE, F. L. A. & MACLEOD, R. A. (1970). Penetrability of a marine pseudomonad by inulin, sucrose and glycerol and its relation to the mechanism of lysis. *Canadian Journal of Microbiology*, **16**, 75–81.

CALCOTT, P. H. & MACLEOD, R. A. (1974*a*). Survival of *Escherichia coli* from freeze-thaw damage: a theoretical and practical study. *Canadian Journal of Microbiology*, **20**, 671–681.

CALCOTT, P. H. & MACLEOD, R. A. (1974*b*). Survival of *Escherichia coli* from freeze-thaw damage: influence of nutritional status and growth rate. *Canadian Journal of Microbiology*, **20**, 683–689.

CALCOTT, P. H. & MACLEOD, R. A. (1975). The survival of *Escherichia coli* from freeze-thaw damage: permeability barrier damage and viability. *Canadian Journal of Microbiology*, **21** (in press).

CALCOTT, P. H. & POSTGATE, J. R. (1971). Protection of *Aerobacter aerogenes* by nonionic detergents from freezing and thawing damage. *Cryobiology*, **7**, 238–242.

DAVIES, J. E. (1970). The role of peptides in preventing freeze-thawing injury. In *The Frozen Cell*, ed. G. E. W. Wolstenholme & M. O'Connor, Ciba Foundation Symposium, pp. 213–233. London: Churchill.

DEAL, P. H. (1970). Freeze-thaw behavior of a moderately halophilic bacterium as a function of salt concentration. *Cryobiology*, **7**, 107–112.

DEAL, P. H. (1974). Effect of freezing and thawing on a moderately halophilic bacterium as a function of Na^+, K^+ and Mg^{2+} concentration. *Cryobiology*, **11**, 13–22.

DEVOE, I. W. & OGINSKY, E. L. (1969). Antagonistic effect of monovalent cations in maintenance of cellular integrity of a marine bacterium. *Journal of Bacteriology*, **98**, 1355–1367.

FARRELL, J. & ROSE, A. H. (1968). Cold shock in a mesophilic and a psychrophilic pseudomonad. *Journal of General Microbiology*, **50**, 429–439.

GILLILAND, S. E. & SPECK, M. C. (1974). Relationship of cellular components to the stability of concentrated lactic streptococcus cultures at −17 °C. *Applied Microbiology*, **27**, 793–796.

GORRILL, R. H. & MCNEIL, E. M. (1960). The effect of cold diluent on the viable count of *Pseudomonas pyocyanea*. *Journal of General Microbiology*, **22**, 437–442.

HAEST, C. W. M., DEGIER, J., VANES, G. A., VERKLEIJ, A. J. & VANDEENAN, L. L. M. (1972). Fragility of the permeability barrier of *Escherichia coli*. *Biochimica et Biophysica Acta*, **288**, 43–53.

HAINES, R. B. (1938). The effect of freezing on bacteria. *Proceedings of the Royal Society of London, Series B*, **124**, 451–472.

HARRISON, A. P. JR (1956). Causes of death of bacteria in frozen suspension. *Antonie van Leeuwenhoek*, **22**, 407–421.

HEGARTY, C. P. & WEEKS, O. B. (1940). Sensitivity of *Escherichia coli* to cold shock during the logarithmic growth phase. *Journal of Bacteriology*, **39**, 475–480.

KEITH, S. JR (1913). Factors influencing the survival of bacteria at temperatures in the vicinity of the freezing point of water. *Science, New York*, **37**, 877–879.

KOHN, A. & SZYBALSKI, W. (1959). Lysozyme spheroplasts from thawed *Escherichia coli* cells. *Bacteriological Proceedings*, 126–127.

KUO, S. C. & MACLEOD, R. A. (1969). Capacity of aspartic acid to increase the bacterial count on suspensions of *Escherichia coli* after freezing. *Journal of Bacteriology*, **98**, 651–658.

LEDER, J. (1972). Interrelated effects of cold shock and osmotic pressure on the permeability of *Escherichia coli* membrane to permease accumulated solute. *Journal of Bacteriology*, **111**, 211–219.

LINDEBERG, G. & LODE, A. (1963). Release of ultra-violet absorbing material from *Escherichia coli* at sub-zero temperatures. *Canadian Journal of Microbiology*, **9**, 523–530.

LOVELOCK, J. E. (1953*a*). The haemolysis of human red blood cells by freezing and thawing. *Biochimica et Biophysica Acta*, **10**, 414–426.

LOVELOCK, J. E. (1953*b*). The mechanism of the protective action of glycerol against haemolysis by freezing and thawing. *Biochimica et Biophysica Acta*, **11**, 28–36.

MACLEOD, R. A., KUO, S. C. & GELINAS, R. (1967). Metabolic injury to bacteria. II. Metabolic injury induced by distilled water or Cu^{++} in the plating diluent. *Journal of Bacteriology*, **93**, 961–969.

MAJOR, C. P., MCDOUGAL, J. D. & HARRISON, A. P. JR (1955). The effect of initial cell concentration upon survival of bacteria at −22 °C. *Journal of Bacteriology*, **69**, 244–249.

MAZUR, P. (1963). Kinetics of water loss from cells at sub-zero temperatures and the likelihood of intracellular freezing. *Journal of General Physiology*, **47**, 347–369.

MAZUR, P. (1966). Physical and chemical basis of injury in single-celled microorganisms subjected to freezing and thawing. In *Cryobiology*, ed. H. T. Meryman, pp. 214–315. New York and London: Academic Press.

MAZUR, P. (1967). Physical-chemical basis of injury from intracellular freezing in yeast. In *Cellular Injury and Resistance in Freezing Organisms*, ed. E. Asahina. Proceedings of International Conference on Low Temperature Science, pp. 171–189.

MAZUR, P. (1970). Cryobiology: the freezing of biological systems. *Science, Washington*, **168**, 939–949.

MAZUR, P., LEIBO, S. P., FARRANT, J., CHU, E., HANNA, M. A. & SMITH, L. H. (1970). Interactions of cooling rate, warming rate and protective additive on the survival of frozen mammalian cells. In *The Frozen Cell*, ed. G. E. W. Wolstenholme & M. O'Connor, Ciba Foundation Symposium, pp. 69–88. London: Churchill.

MAZUR, P., LEIBO, S. P. & CHU, E. H. Y. (1972). A two-factor hypothesis of freezing injury. Evidence from Chinese hamster tissue-culture cells. *Experimental Cell Research*, **71**, 345–355.

MERYMAN, H. T. (1966). Review of biological freezing. In *Cryobiology*, ed. H. T. Meryman, pp. 1–114. New York and London: Academic Press.

MERYMAN, H. T. (1971). Cryoprotective agents. *Cryobiology*, **8**, 173–183.

MEYNELL, G. G. (1958). The effect of sudden chilling on *Escherichia coli*. *Journal of General Microbiology*, **19**, 380–389.

MORICHI, T. (1969). Metabolic injury in frozen *Escherichia coli*. In *Freezing and Drying of Microorganisms*, ed. T. Nei, p. 53–68. Tokyo: University of Tokyo Press.

MOSS, C. W. & SPECK, M. L. (1966). Identification of nutritional components in trypticase responsible for recovery of *Escherichia coli* injured by freezing. *Journal of Bacteriology*, **91**, 1098–1104.

NASH, T., POSTGATE, J. R. & HUNTER, J. R. (1963). Similar effects of various neutral solutes on survival of *Aerobacter aerogenes* and of red blood cells after freezing. *Nature, London*, **199**, 1113–1114.

NEI, T., ARAKI, T. & MATSUSAKA, T. (1969). Freezing injury to aerated and non-aerated cultures of *Escherichia coli*. In *Freezing and Drying of Microorganisms*, ed. T. Nei, p. 3. Tokyo: Tokyo University Press.

POSTGATE, J. R. (1969). Viable counts and viability. In *Methods in Microbiology*, **1**, ed. J. R. Norris & D. W. Ribbons, pp. 611–628. London: Academic Press.

POSTGATE, J. R. & HUNTER, J. R. (1961). On the survival of frozen bacteria. *Journal of General Microbiology*, **26**, 367–378.

POSTGATE, J. R. & HUNTER, J. R. (1963). Metabolic injury in frozen bacteria. *Journal of Applied Bacteriology*, **26**, 405–414.

RAY, B., JANSSEN, D. W. & BUSTA, F. F. (1972). Characterization of the repair of injury induced by freezing *Salmonella anatum*. *Applied Microbiology*, **23**, 803–809.

RAY, B. & SPECK, M. L. (1972). Metabolic process during the repair of freeze injury in *Escherichia coli*. *Applied Microbiology*, **24**, 585–590.

RAY, B. & SPECK, M. L. (1973). Freeze-injury in bacteria. *Chemical Rubber Company Critical Reviews, Clinical Laboratory Science*, **4**, 161–213.

RAYMAN, M. K. & MACLEOD, R. A. (1975). The interaction of Mg^{2+} with peptidoglycan and its relation to the prevention of lysis of a marine pseudomonad. *Journal of Bacteriology*, **122**, 650–659.

RING, K. (1965). The effect of low temperatures on permeability in *Streptomyces hydrogenans*. *Biochemical and Biophysical Research Communications*, **19**, 576–581.

SAKAGAMI, Y. (1959). Effects of freezing and thawing on growth and metabolism of yeast cells. *Low Temperature Science, Series B*, **17**, 105–124.

SATO, M. & TAKAHASHI, H. (1968). Cold shock of bacteria. I. General features of cold shock in *Escherichia coli*. *Journal of General Applied Microbiology*, **14**, 417–428.

SATO, M. & TAKAHASHI, H. (1969). Cold shock of bacteria. II. Magnesium-mediated recovery from cold shock and existence of two critical temperature zones in various bacteria. *Journal of General Applied Microbiology*, **15**, 217–229.

SATO, M. & TAKAHASHI, H. (1970). Cold shock of bacteria. IV. Involvement of DNA ligase reaction in recovery of *Escherichia coli* from cold shock. *Journal of General Applied Microbiology*, **16**, 279–290.

SATO, T., IZAKI, K. & TAKAHASHI, H. (1972). Recovery of cells of *Escherichia coli* from injury induced by sodium chloride. *Journal of General Applied Microbiology*, **18**, 307–317.

SHERMAN, J. M. & ALBUS, W. R. (1923). Physiological youth in bacteria. *Journal of Bacteriology*, **8**, 127–139.

SHERMAN, J. M. & CAMERON, G. M. (1934). Lethal environmental factors within the natural range of growth. *Journal of Bacteriology*, **27**, 341–348.

SMEATON, J. R. & ELLIOTT, W. H. (1967). Selective release of ribonuclease inhibitor from *Bacillus subtilis* cells by cold shock treatment. *Biochemical and Biophysical Research Communications*, **26**, 75–81.

SMITTLE, R. B., GILLILAND, S. E., SPECK, M. L. & WALTER, W. M. JR. (1974). Relationship of cellular fatty acid composition to survival of *Lactobacillus bulgaricus* in liquid nitrogen. *Applied Microbiology*, **27**, 738–743.

STEIM, J. M., TOURTELOTTE, M. E., REINERT, J. C., MCELHANEY, R. N. & RADER, R. L. (1969). Calorimetric evidence for the liquid crystalline state of lipids in a biomembrane. *Proceedings of the National Academy of Sciences, USA*, **63**, 104–109.

STRAKA, R. P. & STOKES, J. L. (1959). Metabolic injury to bacteria at low temperature. *Journal of Bacteriology*, **78**, 181–185.

STRANGE, R. E. & DARK, F. A. (1962). Effect of chilling on *Aerobacter aerogenes* in aqueous suspension. *Journal of General Microbiology*, **29**, 719–730.

STRANGE, R. E. & NESS, A. G. (1963). Effect of chilling on bacteria in aqueous suspension. *Nature, London*, **197**, 819.

STRANGE, R. E. & POSTGATE, J. R. (1964). Penetration of substances into cold-shocked bacteria. *Journal of General Microbiology*, **36**, 393–403.

STRANGE, R. E. & SHON, M. (1964). Effect of thermal stress on viability and RNA of *Aerobacter aerogenes* in aqueous suspension. *Journal of General Microbiology*, **34**, 99–114.

SWARTZ, H. M. (1971). Effect of oxygen on freezing damage: II. Physical chemical effects. *Cryobiology*, **8**, 255–264.

TRACI, P. A. & DUNCAN, C. L. (1974). Cold shock lethality and injury in *Clostridium perfringens*. *Applied Microbiology*, **28**, 815–821.

UNEMOTO, T. & MACLEOD, R. A. (1975). Capacity of the outer membrane of a Gram-negative marine bacterium in the presence of cations to prevent lysis by Triton X-100. *Journal of Bacteriology*, **121**, 800–806.

ZIMMERMAN, C. (1964). Freezing and freeze-drying of *Serratia marcescens* suspended in sodium chloride. *United States Army Biological Laboratory Technical Manuscript*, **187**, 1–11.

SURVIVAL OF DRIED AND AIRBORNE BACTERIA

R. E. STRANGE AND C. S. COX
Microbiological Research Establishment, Porton Down, Salisbury SP4 0JG, Wiltshire

INTRODUCTION

Aerosolization, freeze-drying and air-drying expose bacteria to composite stresses and some of the individual stresses involved are common to all three treatments. The major common stress is dehydration but to determine its effect on viability or virulence, the organisms are subjected to a further potential stress, rehydration.

Generation of aerosols by natural or artificial means usually exerts some degree of stress on microbes (Anderson & Cox, 1967; Dimmick, 1969). Injury is caused mainly by impaction and shear forces, and its extent varies with the method of generation, microbial species and physiological status of the organisms. Once airborne, the microbes are rapidly desiccated to an extent depending on the temperature and relative humidity of the atmosphere; under natural conditions at least they are exposed to various forms of radiation and the 'Open Air Factor' (Druett & May, 1968) which is toxic to micro-organisms at concentrations of a few parts per hundred million. Also, at certain relative humidities, oxygen is toxic to airborne microbes (Hess, 1965; Cox, Gagen & Baxter, 1974). Recovery of airborne microbes usually involves their collection into aqueous media when they may be subjected to impaction and shear forces, and osmotic shock. Rehydration conditions affect the viability of populations of *Escherichia coli* recovered from aerosols at certain critical relative humidities (Cox, 1966*a*, *b*).

During freeze-drying, the first potential stress imposed on microbes is rapid chilling ('cold shock') which may damage susceptible bacteria by interfering with permeability control mechanisms (Meynell, 1958; Strange & Dark, 1962). This is followed by freezing which causes various phase transitions of cellular water and a decrease in the water activity of the system (Gonda & Koga, 1973; Gonda, Ohtomo & Koga, 1973); these changes may result in physical, chemical and physiological injury to the organisms. Finally, dehydration occurs by sublimation of water molecules under low pressure and removal of 'bound water' which may be injurious (Webb, 1965). When dry, some microbes are killed by oxygen (Rogers, 1914; Lion, Kirby-Smith & Randolph, 1961; Benedict *et al.*, 1961; Cox & Heckly, 1973). As with airborne microbes, reconstitution of freeze-dried microbes exposes them to a degree of osmotic shock depending on the composition and method of addition of the suspending fluid.

The effects of dehydration and rehydration in the absence of other stresses imposed during aerosolization and freeze-drying have been studied by desiccating microbes on various supporting media including cellulose fibres (Annear, 1957, 1962; McDade & Hall, 1963*a*; Rountree, 1963), films (Maltman, 1965; Maltman, Orr & Hinton, 1960; Webb, 1960; McDade & Hall, 1964) and membrane filters (Webb, 1960).

An important advance in microbiology is recognition of the fact that microbes injured as a result of various stresses may recover fully if given a suitable environment. The phenomenon of 'reversible injury' is widespread and has been studied by many investigators (Straka & Stokes, 1959; Arpai, 1962; Moss & Speck, 1963, 1966*a*, *b*; Postgate & Hunter, 1963; Morichi, 1969; Webb, 1969*a*; Speck & Cowman, 1969; Speck & Ray, 1973; Cox, Bondurant & Hatch, 1971*a*, *b*). The practical implications of the phenomenon are obviously important in medical and veterinary science, and in food and pharmaceutical microbiology. Microbes may be reversibly injured by aerosolization and freeze-drying.

On the other hand, microbes may be damaged to a considerable extent by a given stress and yet remain fully viable. An example of this is provided by the investigations of Leon A. Heppel and his colleagues (see Heppel, 1967, 1971) on the effects of cold osmotic shock; a variety of microbial constituents (enzymes, antigens, transport factors, nucleotide components, lipopolysaccharide) may be released without significant loss of microbial viability. These findings should be of interest to those investigators of microbial survival who attempt to explain the lethal effect of stress on the basis of a single compositional, biochemical or physiological change they have detected in the stressed microbes.

In this review, the effects of aerosolization and freeze-drying on microbial viability and possible causes of injury and death, are discussed.

AEROSOLIZATION

Airborne microbes are partly responsible for the transmission of disease in man, animals and plants, and may infect water supplies, food, medicinal preparations and pharmaceutical products. The major objectives of studies in aerobiology are to define factors that affect the survival and infectivity of microbes in the airborne state and identify mechanisms responsible for microbial injury and death. To achieve these objectives, a large variety of techniques and apparatus have been developed for the generation, storage and collection of aerosols in the laboratory (Anderson & Cox, 1967; Dimmick, 1969). The methodology of aerosolization will not be comprehensively reviewed here and the emphasis will be on more recent developments.

Laboratory apparatus and techniques

Organism. Under strictly controlled conditions in the laboratory, the responses of different species of bacteria to aerosolization vary although different strains of a given species may respond similarly. For example, four strains of *E. coli* were all less stable at high than at low relative humidities when disseminated from aqueous suspension into nitrogen atmospheres (Cox, 1966*a*, 1968*b*); in contrast, dissemination of *E. coli* B and *Francisella tularensis* from both the wet and dried states gave survival patterns that varied with the method of dissemination and also with the organism (Cox, 1970, 1971; Cox & Goldberg, 1972). A summary of literature on microbial survival in aerosol was given by Anderson & Cox (1967).

Growth of micro-organisms. Growth conditions including composition of the growth medium (Brown, 1953; Strasters & Winkler, 1966; Dark & Callow, 1973), aeration (Brown, 1953) and growth phase (Brown, 1953; Goodlow & Leonard, 1961; Cox, 1966*a*; Cox *et al.*, 1971*a*; Dark & Callow, 1973), affect the subsequent survival properties of bacteria. The death rate of *E. coli* is highest with exponential-phase bacteria and resting-phase bacteria usually survive best. The survival in aerosols of continuously growing *E. coli* depends less on the growth-limiting substrate in a chemostat than on the growth rate; slow-growing bacteria survive much better than fast-growing bacteria (Dark & Callow, 1973).

Variation in microbial response to aerosolization, apparently under the same conditions, is a constant irritant to investigators and with long-term studies it is essential to have available a supply of organisms with reproducible properties. Since the inocula used may influence the properties of cultures, some workers store a quantity of organisms in the frozen state, sufficient for the entire programme (Cox, 1968*c*). For medium-term studies, continuously grown bacteria may be satisfactory.

Generation of aerosols. Most laboratory studies have been concerned with aerosols generated from aqueous suspensions (wet dissemination), but recently microbial aerosols produced from dry powders (dry dissemination) have also been investigated. Wet dissemination is achieved by atomization in a jet of air, with the spinning top, vibrating reed or ultrasound (Rosebury, 1947; Green & Lane, 1964; Anderson & Cox, 1967; Dimmick, 1969).

The method of aerosol generation from aqueous suspensions affects the range and distribution of particle sizes and since particle size influences survival in the airborne state, the aim is to produce homogeneous clouds containing one microbe per particle. Air jet atomizers provided with baffles (e.g. Collison, 1935) produce particles from bacterial suspensions in distilled water with a mass median diameter of 2 to 3 μm decreasing to 0.6 to 1 μm on evaporation but the output is often electrostatically charged (Chow & Mercer, 1971). The physical characteristics of clouds produced with other types of generators were mentioned by Anderson & Cox (1967) and recently an apparatus was described for producing airborne particles in the sub-micron range (Waite & Ramsden, 1971). Young, Larson & Dominik (1974) reported a modification to the May (1949) spinning top generator which they claim improves performance.

Various ways of generating aerosols from dry microbial powders have been described including explosions (Beebe, 1959) and air jets (Dimmick, 1959, 1960; Cox *et al.*, 1970), and recently a device for continuous dry generation was reported (Crider, Barkley & Strong, 1968). Powders are often electrostatically charged and ionizing radiation may have to be used to decrease physical losses due to electrostatic interactions (Cox *et al.*, 1970). De-agglomeration and fracture of particulate solids may present problems (Derr, 1965) and the residual water content of the powder should be controlled since it may affect the subsequent survival of the microbes in the airborne state (Nei, Araki & Souzu, 1965; Nei, Souzu & Araki, 1965; Greiff, 1970).

Microbes in aerosols dispersed from solids usually have survival properties different from those in aerosols disseminated from aqueous suspension (Beebe, 1959; Dimmick, 1960; Goodlow & Leonard, 1961; Ehrlich & Miller, 1971; Cox & Goldberg, 1972). For example, the relative humidity at which minimum survival of *F. tularensis* occurs is approximately 50 per cent for wet dissemination and 80 per cent for dry dissemination (Cox & Goldberg, 1972). This difference in behaviour probably depends on whether the direction of water flow is into (dry dissemination) or out of (wet dissemination) the aerosol particle (Cox & Goldberg, 1972).

The size of the particle which contains the micro-organism(s) can influence survival. Dunklin & Puck (1948) showed that the death rate of pneumococci in 1.6 μm particles was lower than that in 3.2 μm particles, whereas under conditions when the Open Air Factor (see later) is present, micro-organisms survive better in large rather than in small particles (Druett, 1970; Benbough & Hood, 1971).

Storage. Aerosol containers include rooms, the Henderson (1952) apparatus, the DATA system of Hatch & Dimmick (1965, 1966), wind tunnels (Druett & May, 1952) and the rotating drum (Goldberg *et al.*, 1958; Dimmick & Wang, 1969; Goldberg, 1970, 1971). Recently Hood (1971) described a system in which large, medium and small particles remain airborne for long periods of time. Basically, the apparatus consists of a 7 m diameter mild steel sphere and therefore is unlikely to become standard equipment for aerosol experiments. Its great advantage is that it allows studies of survival of aerosolized micro-organisms to be made under conditions equivalent to those in outside air.

Another recent technique is the use of microthreads consisting of ultra-fine spider threads wound onto stainless steel frames (May & Druett, 1968). The frames are loaded into a 'sow' through which the aerosol is passed and particles become attached to the spider thread producing a 'captive aerosol'. Gravitational fall-out of aerosol particles does not occur as with aerosols in storage vessels and particles of 1 to 20 μm size range can be studied. A further advantage of the 'captive aerosol' is that it can easily be exposed to different environments (e.g. rooms, open air etc.) for extended periods of time without the problem of physical losses of aerosol particles. Difficulties do occur with this technique, however, because retention of aerosol particles by spider thread varies from frame to frame and also some aerosol particles can stick to the stainless steel frame causing errors in viability estimations. In addition, problems arise with regard to control of relative humidity. Hood (1971) reported that the survival of *E. coli* was lower when held as an aerosol in a closed sphere than on microthreads in a sealed sow. On the other hand, Semliki Forest virus decays more rapidly on microthreads than in the airborne state (Benbough & Hood, 1971). Even so, the microthread technique provides an extremely useful tool for aerobiologists and its development was largely responsible for the discovery of the Open Air Factor (see later).

Survival of bacteria in large drops (about 100 μm diameter) can be conveniently studied using a technique which involves the support of drops on fine glass fibres (Silver, 1965). The method includes stress-free droplet formation, storage and collection and has been useful in determining causes of death (Cox, 1965; Silver, 1965). Other techniques for studying effects of dehydration are discussed later in the section on freeze-drying.

Collection of aerosols. To obtain information regarding the physical, chemical and biological state of microbes in an aerosol, it is necessary to collect the particles into liquid or onto a solid surface. The information required may include particle size distribution, the total concentration of organisms and the viable fraction, the proportion of organisms that are reversibly injured and the extent to which infectivity is affected. Many different types of collecting devices have been designed because no single sampling method will provide all the information required. If the aim is to determine microbial response to aerosolization, conditions of collection should be as bland as possible, like, for example, the animal respiratory system.

The principles of methods for sampling microbial aerosols include impingement into liquids, impaction onto solid surfaces, filtration, sedimentation, centrifugation and electrostatic and thermal precipitation (Green & Lane, 1964; Wolf *et al.*, 1959; Bachelor, 1960; Anderson & Cox, 1967; Noble, 1967; May, 1967, 1972; Akers & Won, 1969; Davies, 1971). Two recent improvements to collection techniques have been reported. The first is a method for slowing the rate of evaporation of water to prevent the drying out of the nutrient agar media used in slit-type samplers (May, 1969; Thomas, 1970*a*, *b*, *c*). The second is a method for rehydrating aerosol particles in a damp atmosphere before collection into an impinger. Aerosols are passed through a small chamber containing muslin wetted with water or saturated salt solutions and the survival of microbial populations subsequently collected into impinger fluid was often much higher than those of similar populations collected directly from the aerosol (Cox, 1966*b*, 1967, 1968*b*; Maltman & Webb, 1971). A similar technique was used by Hatch & Warren (1969) and Goldberg & Ford (1973).

Assessment of microbial populations recovered from aerosols. Experimental aerobiology depends on accurate reliable methods for determining the viability of microbial populations recovered from aerosols. Survival estimates based solely on colony or plaque counts are inaccurate because physical losses due to deposition on surfaces etc. decrease the total count to an unpredictable extent. Also, if successive samples are removed from an enclosed aerosol the effect of dilution must be allowed for.

Assessment methods were reviewed by Anderson & Cox (1967) and discussed briefly by Strange *et al.* (1972). Methods commonly used for determining total microbial numbers in samples recovered from aerosols, are based on tracer techniques. These involve adding material that undergoes only physical loss to the bacterial suspension used for generating the aerosol. Determination of tracer and viable microbes in spray fluids and collecting fluids enables the total and viable numbers of recovered micro-organisms and their viability to be estimated.

Fluorescent compounds, dyes, bacterial spores, radioisotopes, radioactively labelled microbes and enzymes have been used as tracers in aerosols. Recently, a method for determining the total number of bacteria (or bacteriophages) in aqueous suspension with radioactively labelled homologous antibodies was successfully applied to samples collected from microbial aerosols (Strange *et al.*, 1972). The slide culture technique of Postgate, Crumpton & Hunter (1961) can be used to determine the percentage viability of populations of certain bacteria recovered from aerosols (Cox & Baldwin, 1967; Strange *et al.*, 1972).

Factors affecting survival in aerosols

Relative humidity. Humidity influences the survival of microbes in aerosols, but determination of its intrinsic effect is complicated because many other factors may also affect survival.

Many workers, especially before the last decade, investigated survival at only three relative humidity values, high, medium and low. Their results indicated that survival is a 'smooth function' of relative humidity. More recently survival has been examined over small increments of relative humidity and the results indicate that for a variety of bacteria narrow zones of marked instability occur (Anderson, 1966; Cox, 1966*a*, *b*, 1967, 1968*a*, *b*, 1969, 1970, 1971; Cox & Goldberg, 1972; Benbough, 1967). An example of this phenomenon with aerosolized *E. coli* is shown in Fig. 1. Similar zones of instability have also been found for freeze-dried bacteria (Monk *et al.*, 1956; Monk & McCaffrey, 1957; Davis & Bateman, 1960; Bateman, 1968; Dewald *et al.*, 1967; Bateman *et al.*, 1961) and for viruses (Akers, Bond & Goldberg, 1966; Akers & Hatch, 1968; Benbough, 1971; Trouwborst, 1971; Trouwborst & de Jong, 1973; Trouwborst & Kuyper, 1974; Trouwborst, de Jong & Winkler, 1972).

Because a large number of other variables modify the apparent effect of humidity, the only general statement that can be made is that survival tends to be better at high and low than at mid-range humidities. An example of how another factor may affect the situation is provided by *F. tularensis*: survival (at a given relative humidity) was markedly

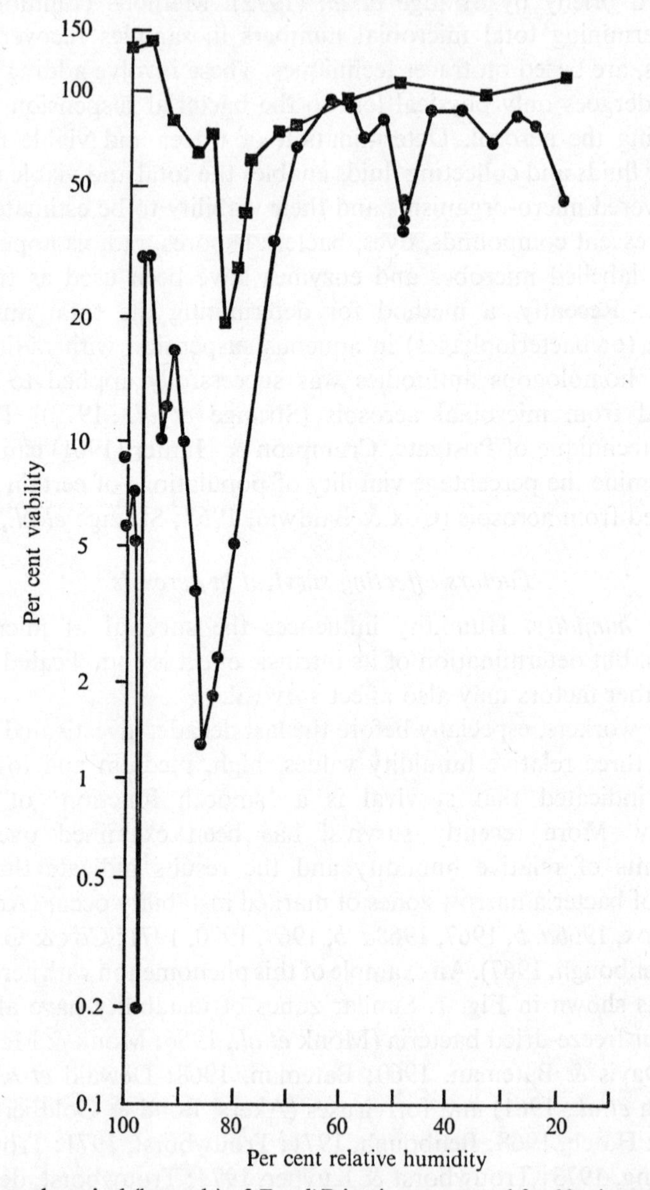

Fig. 1. Aerosol survival (log scale) of *E. coli* B in nitrogen, stored for 30 min and collected by impinger into phosphate buffer. Sprayed from distilled water (●); sprayed from 0.1 M raffinose solution (■). Reproduced by permission of the Society for General Microbiology, *Journal of General Microbiology*, **43**, 383–399 (1966) (Cox).

different with wet- and dry-disseminated aerosols (Cox & Goldberg, 1972).

Recently, C. S. Cox (unpublished) proposed a mathematical model based on first order denaturation kinetics that accounts for the influence of relative humidity and time upon survival of a variety of micro-organisms. This model is compatible with the findings of Trouwborst and colleagues who showed experimentally that micro-organisms can be killed by surface denaturation (Trouwborst, de Jong & Winkler, 1972, 1973; Trouwborst, Kuyper, de Jong & Plantigua, 1974). A model involving first order kinetics has also been proposed by Bateman (1973) but only the effect of time upon survival is considered.

Composition of the atmosphere. Air used in laboratory aerosol experiments is generally presumed to be free from toxic components and often is not submitted to purification procedures other than, perhaps, filtration. Rogers (1914) found that the presence of oxygen could cause loss of viability of freeze-dried bacteria. This result has since been confirmed by several investigators (Benedict *et al.*, 1961; Lion, 1963; Lion & Bergmann, 1961*a*, *b*; Dewald, 1966*b*; Cox & Heckly, 1973). One of the first reports that oxygen is toxic for bacteria in aerosols was by Ferry, Brown & Damon (1958) and later Levin & Cabelli (1963) reported that germination of *Bacillus subtilis* spores collected from aerosols depended upon the oxygen content of the atmosphere. More recently, a toxic action of oxygen apparent only at low relative humidity has been reported (Hess, 1965; Cox, 1966*a*, 1968*b*, 1970, 1971; Cox & Baldwin, 1966, 1967; Benbough, 1967, 1969; Webb, 1967*a*, 1969*a*; Cox *et al.*, 1971*a*, *b*; Goldberg & Ford, 1973). On the other hand Rosebury (1947) found that an inert atmosphere did not prolong survival of *F. tularensis* at low humidity. However, Cox (1971) showed that the composition of the spray fluid was important in this respect and that if the suspending fluid was cysteine broth (or distilled water), oxygen was toxic to this organism at low relative humidity values. Using other techniques for removing liquid water, such as freezing, freeze-drying or dehydration on membrane filters a toxic action of oxygen has also been detected (Schwartz, 1970, 1971; Cox & Heckly, 1973; Bateman, 1968).

Mathematical models have been developed that account for the toxic action of oxygen and the influence of relative humidity and time on this toxicity (Cox, 1973*a*; Cox, Baxter & Maidment, 1973; Cox, Gagen & Baxter, 1974). The models involve a carrier X which combines reversibly with oxygen to give XO_2 and this in turn reacts with an acceptor A to give AO_2; A but not AO_2 is the biologically active form

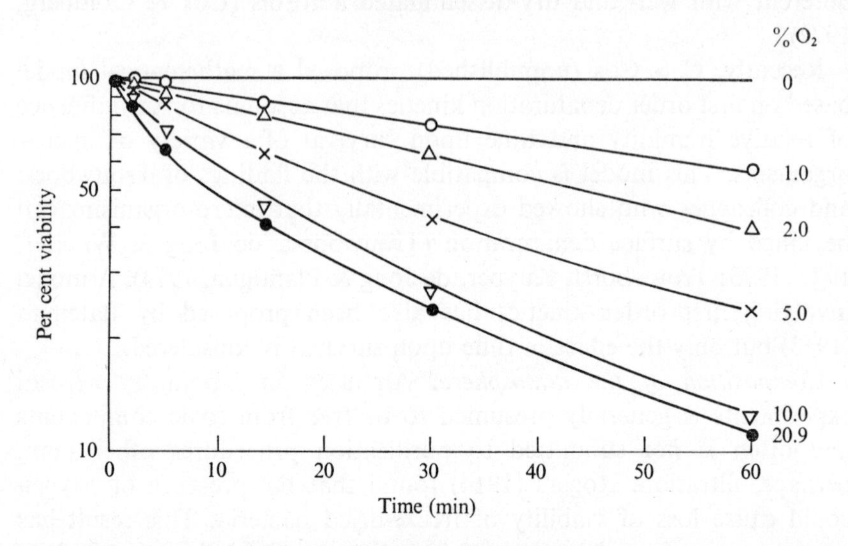

Fig. 2. Aerosol survival (log scale) of *Serratia marcescens* 8UK sprayed from suspension in distilled water into different nitrogen+oxygen mixtures at 30 per cent relative humidity and 25.0±0.1 °C. Points are experimental, lines are calculated. Reproduced by permission of the National Research Council of Canada from the *Canadian Journal of Microbiology* **20**, 1529–1534, (1974) (Cox, Gagen & Baxter).

of the acceptor. Calculated and experimental values for the survival of *Serratia marcescens* in different concentrations of oxygen are shown in Fig. 2.

A recent apparently contradictory finding with *E. coli* B (Cox, 1970) and *F. tularensis* (Cox, 1971) was that oxygen enhanced survival under certain conditions. With *F. tularensis*, a component of the fluid used for suspending the bacteria prior to aerosolization exerted a toxic action in the aerosol in the absence of oxygen, whereas this component lost its toxicity in the presence of oxygen (Cox, 1971). However, it is not understood why the presence of oxygen at high relative humidities enhance the survival of *E. coli* B when dry-disseminated (Cox, 1970).

Cox (1968*a*) found that the aerosol survival of wet-disseminated *E. coli* B was similar in atmospheres of nitrogen, argon and helium. In contrast, Greiff (1970) and Greiff & Rightsel (1969), using freeze-dried influenza virus, showed that the nature of the 'inert' atmosphere under which the virus was maintained, affected the survival of the virus. Helium was found to be best for maintaining survival. Such differences may have been due to the residual water content of the organisms and/or their different natures.

Comparisons showed that the survival of micro-organisms in outside air was much lower than that in enclosed air (Hood, 1971, 1973; Benbough & Hood, 1971; Southey & Harper, 1971) due to the toxic action of the Open Air Factor (Druett & May, 1969; May, Druett & Packman, 1969; Druett, 1970, 1973*a*; May, 1972; Harper, 1973; Hood, 1973; de Mik & de Groot, 1973; Nash, 1973).

Oxides of nitrogen and sulphur dioxide cause loss of microbial viability (Won & Ross, 1969; Lightheart, Hiatt & Rossano, 1971; Chatigny *et al.*, 1973) but it is thought that these do not constitute the Open Air Factor. More likely the factor is a product resulting from the reaction of olefins with ozone (Druett, 1970, 1973*a*; Dark & Nash, 1970; Nash, 1973). Olefins having a double bond in a ring structure when mixed with ozone seem to be the most bactericidal at very low concentrations (Dark & Nash, 1970; Nash, 1973). Recently the molecular weight of the Open Air Factor was estimated to be 50 to 150 (Hood, 1974).

A mechanism was proposed for the action of the Open Air Factor and a mathematical expression derived that apparently accounts for the effects of time and of different concentrations of the factor upon survival (Cox, Hood & Baxter, 1973). Using the expression it was estimated that the usual day-to-day variations in the concentration of the factor was within a seven- to eight-fold range.

Pressure changes. Druett (1973*b*) reported that rapid expansion and recompression of air under conditions which cause the condensation of water on, and re-evaporation from, aerosol particles rapidly killed *E. coli* contained within them. Electron microscope observations and the fact that the recovered bacteria were sensitive to lysozyme indicate that bacterial wall damage results from this stress.

Temperature. In general, survival of airborne microbes improves as the temperature of the atmosphere is lowered.

Radiation. The effects of sunlight, ultraviolet (u.v.) light and X-rays on the survival of airborne bacteria were discussed by Anderson & Cox (1967) and since then this aspect of aerobiology does not appear to have greatly concerned investigators. The survival of vegetative microbes following exposure to u.v. and ionizing radiations is discussed in this volume by B. A. Bridges.

Cloud age. The survival of vegetative bacteria in aerosols decreases with time, the rate of loss of viability depending on the other factors discussed in this chapter. In contained aerosols, there is usually an initial rapid death rate followed by a slower rate, although in the presence of the Open Air Factor and at high humidity there is often a lag period before any loss of viability occurs. Monk & Mattuck

(1956) and Bateman (1973) derived mathematical expressions to describe the survival characteristics of airborne bacteria but these only take account of time. Mathematical expressions that take account not only of time but also of oxygen concentration, humidity and concentration of the Open Air Factor were developed by Cox and coworkers (Cox, 1973*a*; Cox, Baxter & Maidment, 1973; Cox, Hood & Baxter, 1973; Cox, Gagen & Baxter, 1974). Given the conditions under which airborne bacteria are exposed, these expressions allow theoretical survival curves to be constructed and these curves agree well with those obtained from experimental data.

Protective additives. For most purposes, the reponse of bacteria sprayed from suspensions in water is the basis on which the effects of additives to spray fluids are assessed. The effects of additives to spray fluids have been studied both in attempts to improve survival and to identify lethal mechanisms. Much of the literature relating to this aspect was summarized by Anderson & Cox (1967, their table 2), most of which was concerned with aerosols stored in the dark and in the absence of the Open Air Factor.

Potential protecting additives include amino acids, antibiotics, aromatic compounds, dyes, metal-chelating agents, polyhydric alcohols, salts, spent growth media and sugars. Among these, inositol and sugars (particularly di- and tri-saccharides) appear to offer best protection over the widest range of conditions (Cox, 1965, 1966*a*; Webb, 1965; Goldberg & Ford, 1973).

Effects of collection. These were reviewed previously (Anderson & Cox, 1967) and since that time the main emphasis has been on rehumidification (or rehydration) before sampling. Cox (1966*b*, 1967, 1968*b*) showed that if an aerosol is passed into a rehumidification chamber (about 100 per cent relative humidity) before collection into an impinger, a marked increase in survival occurs with some strains of *E. coli* stored at certain relative humidities. On the other hand the process can decrease the survival of other strains (Cox, 1968*b*). Maltman & Webb (1971) and Goldberg & Ford (1973) have also demonstrated similar beneficial effects of rehumidification on *Klebsiella pneumoniae*. Recently it was shown that rehumidification enhances the survival of freeze-dried or aerosolized phage (Cox, Harris & Lee, 1974). These results and those brought about by shifts in relative humidity (Hatch & Dimmick, 1966; Hatch, Wright & Bailey, 1970) indicate that loss of viability often occurs during both dehydration and rehydration processes.

Factors affecting infectivity in aerosols

The importance of the spread of disease by the aerosol route was emphasized at a 'Conference on Airborne Infection' (1961). This view has not always been taken; for example, Chapin (1912) stated 'Most diseases are not likely to be dustborne and they are sprayborne only for two or three feet, a phenomenon which after all resembles contact infection more than it does aerial infection as ordinarily understood'. Chope & Smillie (1936) shared this view. Such beliefs were caused in part by the absence of suitable techniques for infectivity studies, but even so Lurie (1930) had reported the aerogenic transmission of tuberculosis among guinea pigs. Also, Wells (1933, 1934) advanced the theory of droplet nuclei for transmitting infectious microbes and developed techniques for quantitatively sampling microbial aerosols. The importance of the airborne route is now recognized (Langmuir, 1961) and recently was the topic of the fourth International Symposium on Aerobiology (1973).

Since survival of micro-organisms is a prerequisite to their infectivity, all the factors known to influence their survival must also effect their infectivity. In addition there are factors which modify infectivity *per se*, but these have received less attention than those influencing survival.

The part of the respiratory tract to which an aerosol particle can penetrate depends on its size. From 0.5 to 2 μm diameter the fraction of particles deposited in the lungs increases, while below 0.5 μm diameter the fraction decreases. From 2 to 15 μm diameter, the fraction increases and then falls off. Since some aerosol particles are hygroscopic, these rehydrate in the respiratory system resulting in an increased particle size. Particle size is important because this determines the landing site of the particle (which in turn can affect the form of the disease) and also because the larger the particle the greater is the number of micro-organisms required to produce infection of a host. Infection via the aerosol route can be produced by one organism (Furcolow, 1961; Hood, 1961; McCrumb, 1961; Tigertt, Benenson & Gochenour, 1961). The infective dose depends on aerosol age since infectivity may decline faster than does viability (Schlamm, 1960; Hood, 1961; Sawyer *et al.*, 1966). Also, the nature of the spray fluid can modify the decline of infectivity *per se* in the aerosol, since spray fluids may contain infectivity suppressors, such as the chloride ion (Hood, 1961).

Death mechanisms in aerosolized bacteria

Survival of bacteria in aerosol experiments is usually estimated on the basis of the number of survivors capable of forming colonies on the chosen growth medium. However, stresses, including aerosolization, may increase the mutation rate in a microbial population and such mutants may require extra nutrients for growth. In addition, reversible damage may be inflicted which requires complex molecules for repair, so that the bacteria appear to be non-viable on chemically defined media but viable when grown on complex media.

Causes of death during aerosol generation. Potential death mechanisms were discussed previously (Anderson & Cox, 1967) and recently it was shown that shear forces exerted during refluxing in a Collison spray (Collison, 1935) may cause detachment of phage tails resulting in loss of phage viability (Cox, Harris & Lee, 1974). This suggests that the Collison spray may physically damage other micro-organisms including bacteria.

Death during storage caused by dehydration and rehydration. Microbial aerosols stored in the dark in containers may be damaged by oxygen-dependent and/or oxygen-independent lethal mechanisms. Both processes may occur simultaneously unless suitable experimental conditions are chosen to prevent it. Ideally, to distinguish the separate effects of dehydration and rehydration, a completely inert atmosphere should be used.

The conditions of rehydration markedly influence survival. For some strains of *E. coli* and *K. pneumoniae* rehumidification before sampling decreases the loss of viability, indicating that rapid rehydration kills these bacteria or prevents reversal of dehydration and storage damage, whereas slow rehydration does not. Rehydration conditions also affect the viability of some aerosolized or freeze-dried phages.

Certain protective additives, e.g. raffinose, which do not penetrate bacterial walls, are able to protect bacteria (Cox, 1965, 1966*a*). The modes of action of such compounds may be either to slow down the rate of rehydration or to stabilize the wall against damage. That the bacterial wall can be damaged by aerosolization and collection was demonstrated by Hambleton (1970, 1971), Maltman & Webb (1971), Druett (1973*b*), Benbough & Hambleton (1973) and Hambleton & Benbough (1973). When the sensitivity of control and aerosolized bacteria to hydrolytic enzymes and antibiotics was determined, unstressed bacteria were insensitive while recovered aerosolized bacteria were sensitive, indicating damage to their walls. Furthermore, protecting additives diminished sensitivity to these hydrolytic enzymes (Hambleton,

1970), while with rehumidification before collection of the aerosol, the bacteria were completely insensitive to them (Maltman & Webb, 1971). In addition, the work of Benbough & Hambleton (1973) indicates that sub-lethal damage can occur to transport systems. In contrast, Maltman & Webb (1971) concluded that there was no positive correlation between viability loss and susceptibility to hydrolytic enzymes, and that such susceptibility was induced or enhanced by protecting additives. However, Maltman & Webb (1971) performed their experiments under conditions where both oxygen-dependent and oxygen-independent death mechanisms probably occurred simultaneously, unlike the work of Benbough, Cox and Hambleton (*loc. cit.*), where only oxygen-independent decay was studied.

Damage to microbial membranes is not necessarily lethal because mechanisms for membrane repair can operate (Hambleton, 1971; Hambleton & Benbough, 1973) and collecting fluids containing high concentrations of phosphate prevent such damage (Maltman & Webb, 1971).

Particles disseminated from the wet state are dehydrated in the atmosphere and rehydrated during collection but those disseminated from the dry state are exposed only to rehydration. Hence it should be possible to distinguish between the effects of rehydration and dehydration. Beebe (1959), Dimmick (1960), Goodlow & Leonard (1961), Cox (1970), Ehrlich & Miller (1971) and Cox & Goldberg (1972) showed that a variety of dry-disseminated organisms apparently die in aerosols, but the observed loss of viability may have occurred during collection and rehydration. Cox (1971) and Cox & Goldberg (1972) compared the survival of *F. tularensis* disseminated from the wet and dried states and found that both dehydration and rehydration affected viability. In contrast, the survival properties of wet- and dry-disseminated *E. coli* B in nitrogen atmospheres differed only slightly, suggesting that rehydration rather than dehydration was the major stress (Cox, 1966*a*, 1968*a*, 1970). This different behaviour of *E. coli* B and *F. tularensis* may be due to differences in the sensitivity to aerosol stress of certain components and/or structures in the two organisms, that is, different sites are involved in the loss of viability of the respective organisms (Cox, unpublished).

Results with *F. tularensis* (Cox & Goldberg, 1972) showed that the relative humidity for minimum survival was different for wet and dry dissemination. The difference was not caused by the freeze-drying process used to prepare the powder for dry dissemination. These results suggest that relative humidity alone does not control survival, because if it did the relative humidity for minimum survival should have been the same for wet- and dry-disseminated bacterial aerosols. A possible explanation (Cox, unpublished) is that key molecules in bacteria may exist in different hydration states (as does copper sulphate) that undergo a first-order inactivation process at a rate depending on the particular hydrate present. Because of hysteresis in water sorption isotherms for bacteria (Bateman *et al.*, 1962), the hydrate existing at a given relative humidity would be different for wet and dry dissemination. Similarly, a given hydrate would occur at different relative humidities when bacteria are wet- and dry-disseminated. Hence, the relative humidity for minimum survival would be different for wet- and dry-disseminated bacteria. Effects of protecting additives and collection conditions could be explained in terms of slowing down or reversing the above first-order inactivation process.

The ability of *E. coli* to synthesize β-galactosidase in the presence of a specific inducer is considerably impaired in the period immediately following collection from the aerosol; loss of ability to synthesize this protein appeared to precede death (Anderson, 1966). Similar results have been obtained by Webb, Dumasia & Singh Bhorjee (1965) and by Webb & Walker (1968). However, more recent studies (Benbough *et al.*, 1972) indicate that the active transport mechanism for uptake of inducer is impaired by aerosolization. Hence, it may be that the ability to synthesize β-galactosidase is not in fact impaired, but the means of transporting inducer is.

^{43}K-labelled populations of *E. coli* lose practically all of this isotope within a very short time after recovery from the aerosol, except at very high humidities and after short storage times. Loss of potassium was a consequence of aerosolization, but is not in itself lethal (Anderson & Dark, 1967). Such losses imply loss of control of other ions and substrates and also disruption of structures and functions dependent upon this cation. Cox (1969) found that RNA breakdown occurs in aerosolized *E. coli* K12HfrC, to an extent depending upon relative humidity; lowest survival was associated with greatest breakdown. This could have been due to loss of solutes necessary to maintain the structure of RNA. Also, the ability to synthesize RNA and protein was impaired in parallel with the extent of the losses of viability occurring at different relative humidity values. These experiments were performed in the absence of oxygen, as were those of Cox *et al.* (1971*a*, *b*) which indicated that, in similarly stressed organisms, DNA synthesis was also slightly impaired. The ability of the recovered organisms to metabolize oxygen was also decreased (Cox, 1969; Cox *et al.*, 1971*a*, *b*).

Hence, the oxygen-independent loss of viability induced by aerosolization, storage and collection, appears to be related to wall and membrane damage, metabolic injury, reduced DNA, RNA and protein synthesis, impaired active transport and a lack of energy-reserve compounds for biochemical synthesis.

Webb (1969*a*) suggested that loss of viability of aerosolized *E. coli* is due to DNA damage and the inability of the bacteria to repair this damage. Cox *et al.* (1971*a*, *b*) and Hatch, Bondurant & Ehresmann (1973) studying a wider range of bacteria found no evidence for DNA repair mechanisms in bacteria collected from aerosols. But their results are consistent with aerosol stability being genetically determined and inherited by progeny cells.

Death during storage caused by oxygen and by the Open Air Factor. The better survival of bacteria in dark, inert atmospheres rather than in air, indicates that the latter contains toxic components. In enclosed air the effect is attributed to oxygen, while in open air conditions the effect is attributed to oxygen and to the Open Air Factor.

Cox & Baldwin (1964, 1966) demonstrated that *E. coli* B killed by oxygen could still support replication of phage T7 i.e. oxygen must therefore operate on processes not involved in T7 replication in *E. coli* B. The most obvious of these are DNA synthesis directed by *E. coli* B DNA, cell wall synthesis and cell division. Cox *et al.* (1971*a*, *b*) and Benbough (1967) showed that DNA synthesis directed by *E. coli* DNA was not inhibited in these bacteria when they were killed by oxygen. Also, Cox *et al.* (1971*a*, *b*) showed that oxygen-killed *E. coli* B/r and *E. coli* B_{s-1} metabolized oxygen at an undiminished rate compared to controls (i.e. bacteria aerosolized into nitrogen). Likewise, Anderson & Dark (1967) found that the initial loss of ^{43}K by populations of *E. coli* recovered from aerosols was not altered by the presence of oxygen. Also, Cox & Baldwin (1966) found that *E. coli* B, collected after exposure to oxygen in an aerosol, very often grew on recovery medium into extremely long filaments. Hence, oxygen inhibited cell division processes in this organism but it is not known whether this is a general phenomenon. In this context, it may be significant that oxygen is not toxic to dried bacteriophages and viruses.

The mathematical model for oxygen toxicity proposed by Cox and co-workers (see above) cannot predict, and takes no account of, the chemical nature of moieties concerned with oxygen toxicity, but it does indicate a possible mechanism. The model also applies to freeze-dried bacteria killed by oxygen (Bateman *et al.*, 1961; Benedict *et al.*, 1961; Lion & Bergmann, 1961*a*, *b*; Lion, 1963; Heckly, Dimmick &

Windle, 1963; Heckly & Dimmick, 1968; Cox & Heckly, 1973). The possible involvement of free radicals in such oxygen toxicity is discussed below in the section on freeze-drying.

The cause of loss of viability induced by the Open Air Factor and other pollutants (e.g. oxides of nitrogen) is not understood.

Death caused by irradiation during storage. As we mentioned earlier, Anderson & Cox (1967) reviewed irradiation effects and, as far as the present authors know, no work specifically concerned with radiation as a potential cause of microbial death in aerosols has been reported since that review.

Death during collection. Loss of viability of bacteria during their recovery from aerosols is discussed above. Recently, Rechsteiner (1970) showed a significant loss of viability of virus caused by shear forces which operate in samplers of the impinger type. Cox, Harris & Lee (1974) demonstrated a higher viable recovery of phage in an impactor sampler compared to an impinger sampler and showed that the difference was due to the greater shear forces operating in the latter sampler. Evidently, shear and impaction forces imposed during collection may seriously damage micro-organisms.

FREEZE-DRYING

Freeze-drying is a technique used routinely not only for preserving microbes in an 'anhydrobiotic' or 'cryptobiotic' (Keilin, 1959) state but also to prepare them for subsequent qualitative and/or quantitative analysis. Here we shall be concerned with the effects of the process on the integrity and survival of microbes – not with the practical and theoretical aspects of the process itself (see Meryman, 1966*a*; Rey, 1964, 1966; Suzuki, Sawada & Obayashi, 1969; Lapage *et al.*, 1969).

With regard to the preservation of microbes, a major object of research is to establish conditions that allow the recovery of microbes with the minimum loss of their physical and biological properties including viability, biochemical properties, virulence and immunogenicity. Injury may be inflicted during one or more of the four stages of freeze-drying (freezing, dehydration, storage of dried organisms and rehydration), and the total damage caused depends on many factors. These factors are discussed by Fry (1966) and include bacterial species; growth conditions and growth phase; population density; suspending medium composition and added protective agents; rate of freezing; temperature of drying; atmosphere, radiation, temperature and residual moisture during storage; and rehydration conditions. To some extent, the severity of the different stresses imposed during the four stages of freeze-drying can be estimated by investigating each of the stresses in isolation. Thus, the extensive studies on the effects of freezing and thawing on microbes are relevant but obviously the treatment is not equivalent to freezing and drying followed by rehydration. Similarly, dehydration *in vacuo* of frozen microbes is not equivalent to desiccation of unfrozen microbes by air-drying. However, the effects of rehydration are possibly similar whether microbes are dried by freeze-drying, air-drying or aerosolization.

The literature concerning the influence of the various factors mentioned above on the survival of bacteria subjected to freeze-drying will not be comprehensively reviewed here. Briefly summarizing what appear to be generally accepted opinions:

(*a*) Bacterial spores are very resistant, Gram-positive bacteria are more resistant than Gram-negative bacteria and leptospira are among the least resistant organisms.

(*b*) Early stationary-phase bacteria are more resistant than bacteria harvested during other growth phases in batch cultures and early exponential-phase bacteria are the least resistant.

(*c*) The denser the population, the higher the survival rate, at least in the absence of added protective agents (Record & Taylor, 1953).

(*d*) The addition of certain sugars, amino acids, peptides, proteins or other substances, and various mixtures of them (e.g. *Mist. desiccans*, Fry & Greaves, 1951) to suspending fluids may afford protection to bacteria during freeze-drying.

(*e*) A controlled and relatively slow freezing rate improves survival. The possibility of intracellular ice formation resulting in bacterial death depends on the relationship between the rate of removal of liquid water due to freezing and the permeability of the bacterial membrane to water (Nei, Araki & Matsusaka, 1969).

(*f*) A controlled and relatively low drying temperature improves survival. However, the evidence is contradictory (Fry, 1966); when certain bacteria were dried from the frozen and unfrozen states, recoveries of viable bacteria were similar.

(*g*) Over-drying is deleterious; a residual moisture content of 0.5 to 1.5 per cent appears to be optimal for survival.

(*h*) During storage of freeze-dried bacteria, oxygen is toxic. The higher the storage temperature, the greater the rate of loss of viability.

(*i*) During rehydration, conditions that minimize osmotic shock and shear stress improve survival.

In terms of the cultural, biochemical, serological and pathogenic properties of survivors of freeze-drying, there is little evidence that significant changes occur either during the process itself or subsequent storage sometimes for several years (Fry, 1966). However, morphological and physiological changes do occur (Nei, 1960, 1962, 1973).

Many investigators consider that most of the loss of viability in populations of susceptible bacteria occurs during the drying stage. In this context, the high-vacuum drying technique of Dewald (1966*a*) and the 'convective procedure' of Wagman & Weneck (1963) which involves sublimation of ice at atmospheric pressure, are of interest since it is claimed that they decrease losses of viability compared with those that occur with conventional freeze-drying methods.

Injury caused during the freezing stage

During the first stages of freeze-drying, organisms are subjected to cold shock and freezing, stresses discussed in this volume by MacLeod & Calcott. The effect of freezing is similar to that of drying in that fluid water is removed from the system (Meryman, 1966*b*). With slow rates of freezing, intracellular water may not freeze but will be withdrawn from the organisms at a rate depending on the extent of freezing and tonicity of the extracellular environment. With cooling at moderate rates down to −20 °C or below, about 90 per cent of intracellular water is frozen leaving about 10 per cent which is usually referred to as 'bound'. There is controversy as to whether this residual water is bound or supercooled but Mazur (1966) considers it is definitely bound to cellular solids by forces of varying strength.

Certain effects of subjecting bacteria to sub-zero temperatures have been investigated directly on frozen material; for example, changes in the physical state of water with differential scanning calorimetry (Gonda *et al.*, 1973) and morphology with electron microscopy (Nei *et al.*, 1969; Nei, 1973). But assessment of the effects of freezing on viability and physiological activity requires that the frozen organisms are first subjected to the further potential stress of thawing. Here, only such damage as occurs during freezing and storage in the frozen state for a limited period is relevant.

In his excellent review of the basis of freezing injury in microorganisms, Mazur (1966) discusses the state of water in frozen microbes and describes several potentially damaging events that may occur. These events are extracellular ice formation, increase in the concentration of extracellular solutes, intracellular ice formation and increase in the concentration of intracellular solutes. Concentration of intracellular solutes may cause injury by speeding up reaction rates to an extent that more than compensates for the decreased rates due to lowered temperature, and/or by causing large changes in pH as certain components of buffer systems are preferentially precipitated. The effect of increased concentrations of certain reactants or large changes in pH could be denaturation of cellular macromolecules such as protein, DNA and RNA. With high cooling velocities, intracellular ice formation may occur at temperatures below −20 °C and cause mechanical injury. Nei *et al.* (1969) reported that fast freezing of *E. coli* to −150 °C produced small intracellular cavities in the bacteria that could be loci previously occupied by ice crystals and they suggested that bacterial death during fast freezing may be due primarily to intracellular ice formation.

During freezing, most bacteria usually remain physically intact but shrinking or swelling may occur (Nei, 1960, 1962, 1973; Mazur, 1966; Nei *et al.*, 1969).

Clearly, for maintenance of bacterial integrity and viability during this stage of freeze-drying, ideal conditions are those that cause freezing of the whole mass of material without the growth of large ice crystals that disrupt the organisms or their membranes and change ionic concentrations.

Injury caused during the drying stage

During the drying stage, ice sublimes at the surface of the ice crystal, water vapour is transferred to the specimen surface and from here it is removed to the condenser or desiccant. For maintenance of viability, the extent to which 'bound water' is removed from microbes appears to be critical.

Webb (1965) emphasized the importance of bound water in microbes and components isolated from them and complained that workers use the terms 'dry' and 'freeze-dried' without knowing how dry their materials are. According to Webb, approximately 80 per cent of the physiology of organisms depends on the movement, not of free water, but of bound water. The basis for this statement is obscure to us but, nevertheless, it appears to be established that over-drying organisms to below a certain moisture content, possibly lying between 0.5 and 1.5 per cent, is injurious (Fry, 1966). Suzuki *et al.* (1969) investigated the decrease in the survival rate of BCG in the course of drying relative to its dehydration curve. With products processed in 5 or 10 per cent sodium glutamate, a critical decrease in the survival rate occurred at a residual water content of approximately 0.1 per cent but in 10 per cent sodium glutamate + 5 per cent soluble starch, the survival rate gradually decreased in parallel with the residual water. Takano & Terui (1964, 1969) working with yeasts obtained data consistent with the assumption that liberation of water during secondary drying *in vacuo* until a vacuum of 10^{-3} to 10^{-5} Torr was reached, results from a chemical reaction involving a lethal change in some critical molecule. Nitrogen-starved yeasts with a depleted amino acid pool were more susceptible than yeasts with a normal pool but the addition of amino acids prior to freeze-drying protected them. It was suggested that the lethal change may involve amino-carbonyl reactions occurring at considerably lowered levels of water activity. Greiff (1969) found that the survival of freeze-dried influenza virus also depended on the residual level but Suzuki (1969) failed to confirm this effect with vaccinia virus freeze-dried with 5 per cent sodium glutamate; reduction of virus titre did not correlate with residual moisture contents between 1.42 and 3.71 per cent.

Desiccation is reported to have mutagenic effects on vegetative bacteria (Webb, 1965, 1967*b*) and bacterial spores (Zamenof, Eichhorn & Rosenbaum-Oliver, 1968).

It is generally accepted that drying from the frozen state is less injurious to microbes than drying from the liquid state but the evidence for this is not straightforward (Fry, 1966). The effect of the temperature of drying depends on the nature of the organism, the composition of the suspending medium and added protective agents. Some organisms tolerate drying temperatures above −20 °C whereas others lose viability. Various methods of drying bacteria without initially freezing them have been tested and the products compared with those obtained after drying from the frozen state. Annear (1957, 1958, 1962) dropped single drops of suspension onto cellulose or alginate tufts or onto the bottoms of the drying tubes, and evacuated them. Immediate survivals of *Staphylococcus aureus* and *Vibrio cholera* were very high; however, during storage, survival on cellulose tufts was good, but poor on bare glass.

The effect of desiccation on microbes air-dried on surfaces is important in the context of the transmission of infectious disease. McDade & Hall (1963*a*) investigated the survival of *S. aureus* on braided silk, cotton and dacron suture materials. Lengths were impregnated with bacterial suspension, partially dried and held in atmospheres at controlled temperatures and relative humidities. The materials were then incubated in nutrient broth with shaking. Results showed that survival was best at relative humidities of 95 to 98 per cent and worst at 53 to 59 per cent. Bacteria on treated strips were also used to infect mice which showed that a small number of bacteria remained viable and infective after exposure to the atmosphere for one week. The survival of this organism on surfaces of glass, ceramics, rubber, asphalt and polished steel was best at 25 °C and relative humidities of 11 to 33 per cent (McDade & Hall, 1963*b*). Rountree (1963) found that strains of *S. aureus* could be divided into two categories according to their ability to survive on pieces of textile. Organisms in the first category started dying immediately whereas those in the second survived for long periods which apparently gave these latter strains a selective advantage. Gram-negative bacteria, air-dried on various surfaces, survived best at 25 °C and a relative humidity of 11 per cent (McDade & Hall, 1964). Webb air-dried bacteria in thin films on Millipore filters in controlled atmospheres and showed that loss of viability in relatively sparse populations was much less in the presence of inositol. An interesting finding was that whereas desiccation decreased the ability of *E. coli* to oxidize amino acids, such activity was decreased to a much greater extent in inositol-protected bacteria. This phenomenon is discussed at length by Webb (1965).

Injury during storage

The use of freeze-drying for the preservation of vaccines and collections of type cultures depends on viability and DNA integrity being maintained not only during the process itself but also during long periods of storage. Factors that affect survival during storage include the atmosphere, temperature, residual moisture and possibly light (Fry, 1966).

Effect of the atmosphere. Obayashi, Ota & Arai (1961) found that freeze-dried *Lactobacillus bifida* survived better in air than *in vacuo* but, apart from this isolated finding, it seems to be established that oxygen is injurious to freeze-dried bacteria. The effect of oxygen on freeze-dried *Serratia marcescens* is shown in Fig. 3. The death rate of dried bacteria has been correlated with the production of free radicals since both events are more extensive when oxygen is present (Dimmick, Heckley & Hollis, 1961). Heckly *et al.* (1963) reported that addition of water vapour to the atmosphere of freeze-dried bacteria decreased free radical production. Heckly & Dimmick (1968) confirmed this and also showed that, in fact, raising the humidity increased the rate of formation of free radicals but decreased their half-lives; these authors suggested that a low molecular weight DNA and a substance similar to adrenaline or propyl gallate were both required to form the toxic free radical precursor. However, the kinetic studies of Cox & Heckly

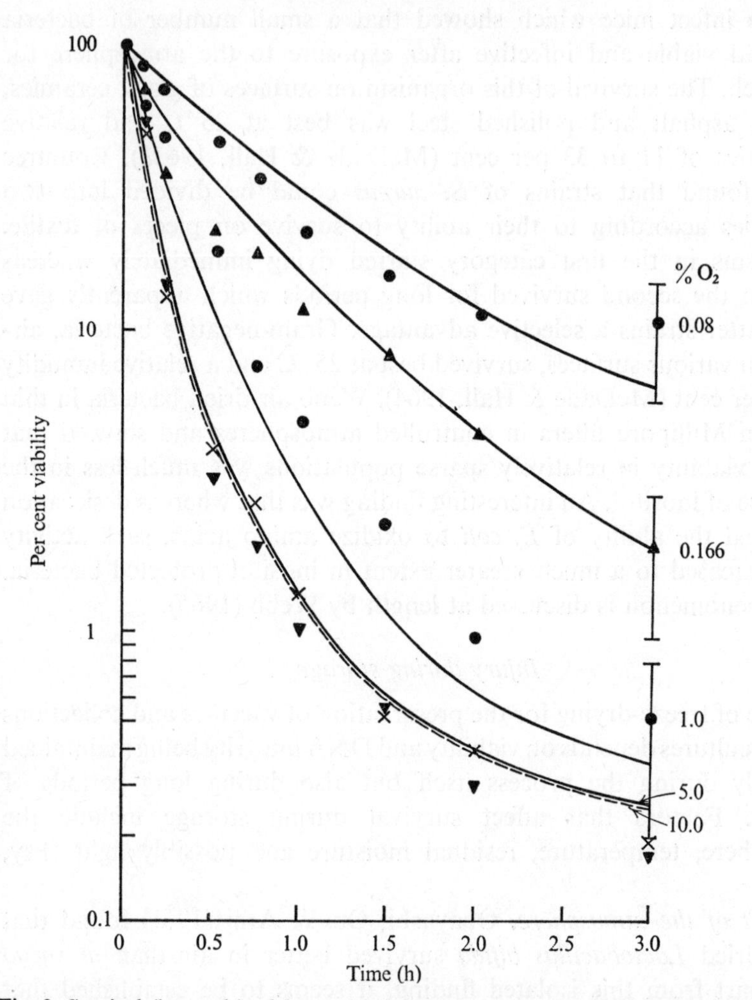

Fig. 3. Survival (log scale) of freeze-dried *Serratia marcescens* 8UK exposed to different oxygen concentrations at zero per cent relative humidity. Points are experimental data, lines are calculated. Reproduced by permission of the Society for General Microbiology, *Journal of General Microbiology* **75**, 179–185 (1973) (Cox, Baxter & Maidment).

(1973) led these authors to believe that the production of free radicals, detected by electron paramagnetic resonance, is not involved in the lethal effect of oxygen on dried bacteria. In support of this proposition, Cox, Gagen & Baxter (1974) provided a mathematical expression for oxygen-induced death in dehydrated bacteria that takes account of oxygen concentration, residual water and exposure time to oxygen without invoking production of free radicals. Cox (1973*b*) offered an explanation for the discrepancy between his and other workers' findings

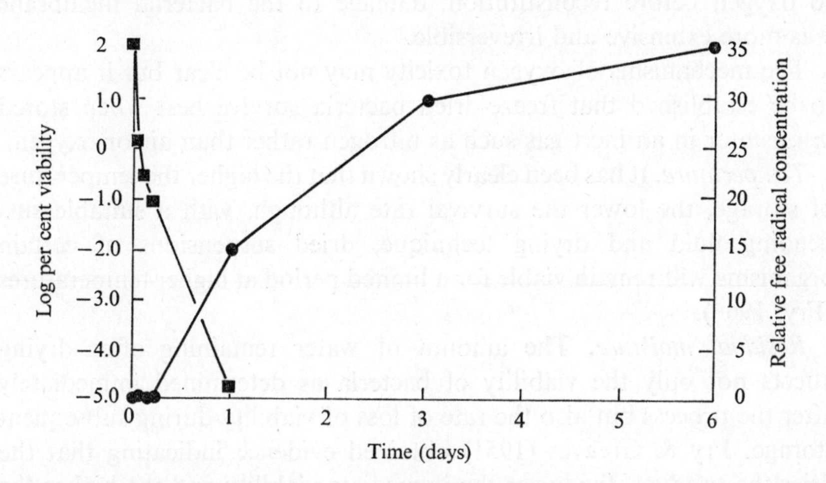

Fig. 4. Free radical formation (●) and loss of viability (■) as a function of time in freeze-dried *Serratia marcescens* 8UK/2 under 10 per cent oxygen at zero per cent relative humidity.

Reproduced by permission of the National Research Council of Canada from the *Canadian Journal of Microbiology*, **19**, 189–194 (1973) (Cox & Heckley).

based on the differences in exposure times of dried bacteria to oxygen. Cox's measurements were made immediately following a short exposure to oxygen whereas other workers' results were obtained after exposure of bacteria for much longer periods. Since, in Cox's experiments, the rate of loss of viability of *S. marcescens* in the presence of oxygen was much greater than the rate of production of free radicals, and most of the bacterial population was dead before free radicals were detectable, apparently the latter could not be responsible for bacterial death (Fig. 4). This conclusion must be weighed against evidence that certain additives (e.g. thiourea, sodium iodide) in drying fluids protect bacteria against the lethal effect of oxygen and decrease free radical formation (Lion *et al.*, 1961; Lion & Bergman, 1961*b*; Lion, 1963). However, these additives may be bimodal in action, behaving as free radical scavengers in one mode and protecting organisms against oxygen-induced loss of viability by another independent mode.

Israeli, Giberman & Kohn (1974) reported that the lethal effect of oxygen on freeze-dried *E. coli* is due to irreversible damage inflicted on the bacterial cytoplasmic membrane. Freeze-drying itself injured the transport system for *o*-nitrophenylgalactoside and potassium so as to make the membrane leaky to these substances. The damage was partially repaired on incubation of reconstituted bacteria in nutrient medium; however, if the bacteria were not held *in vacuo* but exposed to oxygen before reconstitution, damage to the bacterial membrane was more extensive and irreversible.

The mechanisms of oxygen toxicity may not be clear but it appears to be established that freeze-dried bacteria survive best when stored *in vacuo* or in an inert gas such as nitrogen rather than air or oxygen.

Temperature. It has been clearly shown that the higher the temperature of storage, the lower the survival rate although, with a suitable suspending fluid and drying technique, dried suspensions of certain organisms will remain viable for a limited period at higher temperatures (Fry, 1966).

Residual moisture. The amount of water remaining after drying affects not only the viability of bacteria as determined immediately after the process but also the rate of loss of viability during subsequent storage. Fry & Greaves (1951) obtained evidence indicating that the drier the product, the lower the immediate viability and the higher the rate of loss of viability during subsequent storage. Many investigators believe that substances such as sugars and amino acids which protect organisms during drying and storage, act by retaining the moisture necessary for microbial survival (Obayashi & Cho, 1957). The optimum residual water content varies with the composition of the fluid in which organisms are dried, the storage atmosphere (Scott, 1958) and probably with the species and physiological state of the organism (Fry, 1966).

The mutagenic effect of desiccation on *B. subtilis* spores was shown to continue during storage in the dried state (Zamenof *et al.*, 1968); after drying in high vacuum over P_2O_5 and storage for seven months, mutant frequencies increased 19-fold and 250-fold, respectively, over normal mutant frequencies.

Light. There is some controversy concerning the effect of light on dried bacteria but since there is rarely any reason why the dried product should not be stored in the dark, the problem is of limited interest. Fry (1966) compared the survival of paracolon bacilli, freeze-dried in *Mist. desiccans* and stored in both darkness and discontinuous sunlight. An increased death rate was evident after 13 days in sunlight; viabilities after eight years in the dark and sunlight were 21 and 0.03 per cent, respectively. Webb (1963) found that artificial sunlight decreased the survival of air-dried *Serratia marcescens* but the extent of the lethal effect depended on the relative humidity.

Protective additives in the drying fluid

A large number of simple and complex substances, alone or in mixtures, have been found to decrease the loss of viability of bacteria during freeze-drying and subsequent storage of the product (Fry, 1966). Complex protective additives include serum, serum proteins, gelatine, milk, broth, dextran, starch, polyethylene glycol, polyvinylpyrrolidone and peptone. Simple substances that may protect include glucose, sucrose, lactose, galactose, sodium glutamate and sodium aspartate. Favourite mixtures include serum (equine or human) + broth + glucose (*Mist. desiccans*, Fry & Greaves, 1951) and dextrin + ammonium chloride + thiourea + ascorbic acid (Naylor & Smith, 1946). Some of these additives have also been found to protect aerosolized bacteria (see above).

Sodium glutamate is used as a protective agent in freeze-dried vaccines (Obayashi *et al.*, 1961; Suzuki *et al.*, 1969) and its stabilizing effect is enhanced by colloids such as polyvinylpyrrolidone. Redway & Lapage (1974) obtained results which suggest that *meso*-inositol, non-reducing disaccharides and certain polyalcohols are the most suitable of the compounds they tested for incorporating into suspending media for use in the freeze-drying of bacteria. Sinha, Dudani & Ranganathan (1974) reported that addition of reconstituted skim milk with added ascorbic acid and thiourea gave the maximum survival of *Streptococcus lactis* populations during freeze-drying and storage.

Various mechanisms have been suggested to explain the protective activity of these different classes of compounds. Some of them (e.g. sugars and amino acids) are supposed to act by regulating the final moisture content of the product to near the optimum level for survival. However, Scott (1958) claimed that microbial death during drying is caused by reactions between carbonyl compounds and amino groups on microbial proteins that results in the blocking of amino acid side chains. Lea, Hannan & Greaves (1950) showed that when human plasma containing added glucose is freeze-dried, the amino-nitrogen content falls and up to 5 per cent by weight of glucose is attached to the proteins; insolubility and a brown colour eventually develop. If this reaction is a major cause of bacterial death during drying and storage, the protective activity of amino acids, peptides, broth and proteins may be due to the fact that these substances compete for the available carbonyl groups thus sparing microbial protein. In addition to exacerbating this potentially damaging reaction, added reducing sugars may cause structural damage to bacteria during rehydration.

Record, Taylor & Miller (1962) freeze-dried exponential-phase *E. coli* in 10 per cent glucose solution and found that rapid rehydration of the product led to the formation of spheroplasts. The authors' interpretation of this phenomenon was that glucose penetrated the bacterial wall and accumulated between it and the inner membrane; after drying and rapid rehydration, osmotic diffusion forces developed which caused the walls to rupture. Record *et al.* (1962) also offered an explanation of the protective effect of added colloids in the presence of glucose; high molecular weight colloids are unable to penetrate the walls but the osmotic pressure they exert causes the walls to collapse onto the inner membranes thus decreasing the space in which the sugar can accumulate.

More recently, Orndorff & MacKenzie (1973) suggested that additives such as glucose, sorbitol, dextran and Ficoll protect bacteria during freeze-drying by supersaturating and by providing inert amorphous matrices capable of retaining substances released from the bacteria.

Rehydration

During reconstitution of freeze-dried bacteria with water, intracellular and extracellular concentrations of solutes differ until equilibrium is reached and the resulting diffusion pressures may cause physical damage to the organisms. Leach & Scott (1959) found that rehydration with successive small volumes of water instead of adding the same total volume of water at once, or with relatively concentrated solutions of sugars or sugar alcohols, enhanced survival. Record *et al.* (1962) found that reconstitution of dried *E. coli* with 50 per cent glucose solution followed by dialysis of the suspension against the same solution that was stirred and slowly diluted with buffer for 6 h, markedly improved survival compared with rapid reconstitution with buffer.

'Osmotic shock' is often implicated as the cause of injury or death in bacteria and its lethal effect was demonstrated by Sherman & Cameron (1934) and Record *et al.* (1962). More recent systematic investigations of the phenomenon have been made under conditions that impose other overt stress. For example, the studies of Heppel and his colleagues (Neu & Heppel, 1965; Nossal & Heppel, 1966; Heppel, 1967, 1971) involved subjecting bacteria to a combination of stresses including exposure to ethylenediaminetetracetic acid, osmotic shock and cold shock. In the present context, these studies are not strictly relevant but it is of interest that Gram-negative coliform organisms survive these insults despite the fact that a variety of microbial enzymes, transport factors, nucleotides and lipopolysaccharide are released into the environment during the treatment. Morichi (1969) 'osmotically shocked' *E. coli* with the same procedure and found that some of the population were 'metabolically injured'.

Reversible injury in freeze-dried bacteria

It is well established that, when bacterial populations are subjected to stresses including aerosolization, cold shock, freezing and thawing, osmotic shock, radiation and mild heat stress, some bacteria may die, others are injured and the remainder are apparently unaffected. The injured bacteria are of particular interest since they may fail to grow on medium adequate for growth of unstressed bacteria but, given a suitable environment, are capable of full recovery. Reversible injury inflicted on bacteria by freezing and thawing (MacLeod & Calcott, this volume) has been investigated by many workers but the occurrence of the phenomenon in freeze-dried bacteria has received less attention. With a given organism, it cannot be assumed that the extent of reversible injury caused by freezing, drying and reconstitution is equivalent to that caused by freezing and thawing.

It is common knowledge that freeze-dried bacteria often need pampering with supplemented medium before they revert to their normal growth characteristics. Whole blood is a suitable supplement to recovery medium for freeze-dried *Pasteurella* species, BCG, *Chromobacterium lividum* and *Vibrio cholera* (see Fry, 1966).

Morichi, Okamoto & Irie (1973) and Morichi & Irie (1973) investigated the effect of peptone added to recovery medium on the apparent viability of freeze-dried *V. metschnikovii* IAM 1039, *E. coli* K-12, *E. coli* B and strains of *Streptococcus cremoris*. Washed bacteria were suspended in 0.33 M phosphate buffer (pH 7.0) and freeze-dried in buffer alone or containing sodium glutamate, sucrose, meat extract or skim milk as protective agents. After freeze-drying bacteria were rehydrated in 0.9 per cent NaCl solution and appropriate dilutions of the samples were plated on 18 to 19 different recovery media. Each medium consisted of the appropriate basal medium for the particular species or strain, with and without one of 18 different 'peptones' (included Casamino acids and lactalbumin hydrolysate). The extent of recovery depended on the organism, the presence of additive in the suspending medium and peptone in the recovery medium. Freeze-drying strains of *E. coli* resulted in the appearance of amino acid-requiring and peptide-requiring bacteria as found in frozen and thawed populations of *E. coli* by Morichi (1969). An injury requiring peptides for repair was also detected in freeze-dried populations of *S. cremoris* but only peptones made from casein (e.g. Trypticase and Tryptone) were effective. Addition of certain of the peptones to recovery media resulted in a marked decrease in the viable bacteria recovered from freeze-dried preparations of *V. metschnikovii* and *E. coli.*

Mechanisms of injury and death in freeze-dried microbes

During freeze-drying, storage in the dried state and reconstitution, microbes are subjected to several stresses, the injurious or lethal effect of each of which may be due to one or more of several different events. For example, freezing may damage bacteria structurally, physiologically (e.g. 'metabolic injury') and genetically. The difficulties associated with identifying major causes of injury and death are not relieved by the fact that a number of unrelated compounds may each afford protection to bacteria but the mechanisms of action of these substances are not fully understood. Possible causes of injury and death in freeze-dried bacterial populations are summarized below.

Cold shock and freezing injury

(*a*) Alteration of the bacterial membrane and interference with permeability control mechanisms during rapid chilling (Meynell, 1958; Strange & Dark, 1962) and freezing (Strange & Postgate, 1964; Moss & Speck, 1966*b*).

(*b*) Concentration of intracellular solutes resulting in pH changes and/or increased rates of reaction between cellular constituents even at very low temperatures (Mazur, 1966).

(*c*) Formation of intracellular ice crystals causing physical damage (Nei *et al.*, 1969).

(*d*) 'Metabolic injury' that may be repaired in the presence of certain peptides or amino acids in the recovery medium (Speck & Cowman, 1969; Morichi, 1969).

(*e*) Denaturation or aggregation of macromolecules with loss of their biological activity due to removal of freezable water from the system (Meryman, 1966*b*; Hanafusa, 1966, 1969, 1972, 1973; Souza, 1973*a*, *b*).

(*f*) Detachment of 'latent RNAase' from ribosomes may subsequently lead to degradation of RNA (Smeaton & Elliott, 1967).

(*g*) Damage to DNA that may be reparable (Schwartz, 1970; Takano, Sinskey & Baraldi, 1973).

Desiccation damage

(*a*) Changes in the configuration of macromolecules due to removal of water; this is likely to occur to a greater extent during drying than during freezing (Webb, 1965). Damage inflicted on certain proteins results in loss of enzymic activity; the activity *in vivo* of DNA may be affected and damage may be more severe in some regions of the chromosome than others (Webb, 1968, 1969*b*).

(*b*) Amino-carbonyl reactions may occur at very low levels of water activity resulting in loss of the biological activities of proteins (Takano & Terui, 1969).

(*c*) Injury to cytoplasmic membranes (Wagman, 1960).

Storage damage

(*a*) Oxygen is toxic to freeze-dried bacteria to an extent depending on the residual water content; the effect may be due to free radical formation and/or irreversible damage to bacterial membranes (Israeli *et al.*, 1974). Another explanation involving unidentified oxygen carrier and receptor components in bacteria but not free radicals has been proposed (Cox, 1973*b*).

(*b*) Decrease in the residual water content of bacteria to below a certain level increases the rate of loss of viability during storage. This may be due to the progressive effects of phenomena occurring during the drying stage causing increased damage to macromolecules.

Rehydration damage

Injury to microbes may be caused by osmotic shock and shear forces, the degree of which depends on the diffusion pressures that develop before equilibrium is reached. In severe cases, walls may be stripped off bacteria (Record *et al.*, 1962).

CONCLUSIONS

Elucidation of the mechanisms of injury and death in microbial populations is satisfying to the academic scientist and of importance to those concerned with practical applications of microbiology. However, the hope of many of the earlier investigators that the lethal effect of a given stress on microbes could be neatly explained on the basis of a universal primary event has, in general, not been realized. Rather, in most cases, injury and death appear to be due to one or more of several different events depending on the particular organism and the stress environment. With a composite stress such as aerosolization or freeze-drying, it is difficult to isolate and assess the effects of the individual stresses involved.

A considerable number of mechanisms have been proposed to explain reversible and lethal injury in airborne and freeze-dried bacteria. These have involved most of the known structural components, macromolecules, small molecular weight solutes including ions and the physiological activities of organisms. If practically all bacterial components are affected to a lesser or greater degree by aerosolization and desiccation, the chance of identifying the initial traumatic event is small. Nevertheless, recognition of potential causes of injury allows ameliorative action to be taken that may, for example, greatly improve microbial survival during freeze-drying.

With certain investigations in aerobiology, recovery of a sufficient number of organisms for particular assays is difficult and some workers have attempted to overcome the problem by air-drying relatively large microbial populations under controlled conditions that simulate those in aerosols. It appears that, whether bacteria are freeze-dried or air-dried in aerosols, films or on membrane filters, in general they respond similarly to temperature, humidity, oxygen and rehydration. Recent findings on the toxic effect of oxygen on dried bacteria conflict with the view that free radical formation is responsible for this lethal effect, although the latter view is supported by the fact that certain additives which depress free radical production also decrease loss of viability. Cox's mathematical expression (Cox, Gagen & Baxter, 1974) contains two components, the chemical identities of which are not known. Substantiation of his hypothesis requires that these components be identified and shown to be the only ones involved in the lethal effect.

In this review, philosophical attitudes about the phenomenon of anhydrobiosis – latent life induced by removal of water from the living system (Crowe, 1971) – have been ignored. It was suggested by Hinton (1968) that the phenomenon is characteristic of primitive protoplasm and according to this criterion, many different types of bacteria are primitive for they are capable of tolerating desiccation for long periods. Physiological and biochemical adaptations among organisms capable of survival in a dry state are poorly understood but it may be that replacement of 'bound water' in macromolecules by polar compounds is involved. For example, Sussman & Halvorson (1966) suggested that dipicolinic acid replaces 'bound water' in bacterial spores and perhaps the protective agent inositol, which Webb (1965) claims replaces 'bound water' in desiccated vegetative bacteria, mimics natural adaptation processes that equip organisms for survival in the dried state (Crowe, 1971).

REFERENCES

AKERS, A. B. & WON, W. D. (1969). Assay of living, airborne microorganisms. In *An Introduction to Experimental Aerobiology*, ed. R. L. Dimmick & A. B. Akers, pp. 59–99. New York, London, Sydney and Toronto: Wiley Interscience.

AKERS, T. G., BOND, S. & GOLDBERG, L. J. (1966). Effect of temperature and relative humidity on the survival of airborne Columbia SK group viruses. *Applied Microbiology*, **14**, 361–364.

AKERS, T. G. & HATCH, M. T. (1968). Survival of a picornovirus and its infectious RNA after aerosolization. *Applied Microbiology*, **16**, 1811–1813.

ANDERSON, J. D. (1966). Biochemical studies of lethal processes in aerosols of *Escherichia coli. Journal of General Microbiology* **45**, 303–313.

ANDERSON, J. D. & COX, C. S. (1967). Microbial Survival. *Symposia of the Society for General Microbiology*, **17**, 203–226.

ANDERSON, J. D. & DARK, F. A. (1967). Studies on the effects of aerosolization on the rates of efflux of ions from populations of *Escherichia coli. Journal of General Microbiology*, **46**, 95–105.

ANNEAR, D. I. (1957). The preservation of bacteria by drying on cellulose and alginate films. *Journal of Applied Bacteriology*, **20**, 17–20.

ANNEAR, D. I. (1958). Observations on drying bacteria from the frozen and from the liquid state. *Australian Journal of Experimental Biology and Medical Science*, **36**, 211–221.

ANNEAR, D. I. (1962). Recoveries of bacteria after drying on cellulose fibres. A method for the routine preservation of bacteria. *Australian Journal of Experimental Biology and Medical Science*, **40**, 1–8.

ARPAI, J. (1962). Non-lethal freezing injury to metabolism and motility of *Pseudomonas fluorescens* and *Escherichia coli. Applied Microbiology*, **10**, 297–301.

BACHELOR, H. W. (1960). Aerosol samplers. *Advances in Applied Microbiology*, **2**, 31–64.

BATEMAN, J. B. (1968). Long term bactericidal effects of reduced ambient water activity; use of membrane filter support for test organisms. *American Journal of Epidemiology*, **87**, 349–366.

BATEMAN, J. B. (1973). Calculation of inactivation curves for dehydrated organisms. In *Airborne Transmission and Airborne Infection*, ed. J. F. Ph. Hers & K. C. Winkler, pp. 117–124. Utrecht, The Netherlands: Oosthoek Publishing Company.

BATEMAN, J. B., MCCAFFREY, P. A., O'CONNOR, R. J. & MONK, G. W. (1961). Relative humidity and the killing of bacteria. The survival of damp *Serratia marcescens* in air. *Applied Microbiology*, **9**, 576–571.

BATEMAN, J. B., STEVENS, C. L., MERCER, W. B. & CARSTENSEN, E. L. (1962). Relative humidity and the killing of bacteria: the variation of cellular water content with external relative humidity or osmolality. *Journal of General Microbiology*, **29**, 207–219.

BEEBE, J. M. (1959). Stability of disseminated aerosols of *Pasteurella tularensis* subjected to solar radiation at various humidities. *Journal of Bacteriology*, **78**, 18–24.

BENBOUGH, J. E. (1967). Death mechanisms in airborne *Escherichia coli. Journal of General Microbiology*, **47**, 325–333.

BENBOUGH, J. E. (1969). Factors affecting the toxicity of oxygen towards airborne coliform bacteria. *Journal of General Microbiology*, **56**, 241–250.

BENBOUGH, J. E. (1971). Some factors affecting the survival of airborne viruses. *Journal of General Virology*, **10**, 209–220.

Benbough, J. E. & Hambleton, P. (1973). Structural, organisational and functional changes associated with envelopes of bacteria sampled from aerosols. In *Airborne Transmission and Airborne Infection*, ed. J. F. Ph. Hers & K. C. Winkler, pp. 135–137. Utrecht, The Netherlands: Oosthoek Publishing Company.

Benbough, J. E., Hambleton, P., Martin, M. L. & Strange, R. E. (1972). Effect of aerosolization on the transport of α-methyl glucoside and galactosides into *Escherichia coli. Journal of General Microbiology*, **72**, 511–520.

Benbough, J. E. & Hood, A. M. (1971). Viricidal activity of open air. *Journal of Hygiene, Cambridge*, **69**, 619–626.

Benedict, R. G., Sharpe, E. S., Corman, J., Meyers, G. B., Baer, E. F., Hall, H. H. & Jackson, R. W. (1961). Preservation of micro-organisms by freeze-drying. II. The destructive action of oxygen. Additional stabilizer for *Serratia marcescens*. Experiments with other organisms. *Applied Microbiology*, **9**, 256–262.

Brown, A. D. (1953). The survival of airborne microorganisms. II. Experiments with *Escherichia coli* near 0 °C. *Australian Journal of Biological Science*, **6**, 470–485.

Chapin, C. V. (1912). *The Sources and Modes of Infection*, 2nd edition. New York: John Wiley and Sons.

Chatigny, M.A., Wolochow, H., Lief, W. R. & Herbert, J. (1973). The toxicity of nitrogen oxides for airborne microbes: effects of relative humidity, test procedures and containment and composition of spray suspension. In *Airborne Transmission and Airborne Infection*, ed. J. F. Ph. Hers & K. C. Winkler, pp. 94–97. Utrecht, The Netherlands: Oosthoek Publishing Company.

Chope, H. D. & Smillie, W. G. (1936). Airborne infection. *Journal of Industrial Hygiene and Toxicology*, **18**, 780–792.

Chow, H. V. & Mercer, T. T. (1971). Charges on droplets produced by atomization of solutions. *American Industrial Hygiene Association Journal*, **32**, 247–255.

Collison, W. E. (1935). *Inhalation therapy technique*. London: William Heinemann (Medical Books).

Conference on Airborne Infection (1961). *Bacteriological Reviews*, **25**, 173–377.

Cox, C. S. (1965). Protecting agents and their mode of action, *First International Symposium on Aerobiology*, pp. 345–368. Oakland, California: Naval Biological Laboratory, Naval Supply Center.

Cox, C. S. (1966*a*). The survival of *Escherichia coli* atomized into air and into nitrogen from distilled water and from solutions of protecting agents as a function of relative humidity. *Journal of General Microbiology*, **43**, 383–399.

Cox, C. S. (1966*b*). The survival of *Escherichia coli* in nitrogen under changing conditions of relative humidity. *Journal of General Microbiology*, **45**, 283–288.

Cox, C. S. (1967). The aerosol survival of *Escherichia coli* Jepp sprayed from protecting agents into nitrogen atmospheres under shifting conditions of relative humidity *Journal of General Microbiology*, **49**, 109–114.

Cox, C. S. (1968*a*). The aerosol survival of *Escherichia coli* B in nitrogen, argon and helium atmospheres and the influence of relative humidity. *Journal of General Microbiology*, **50**, 139–147.

Cox, C. S. (1968*b*). The aerosol survival and cause of death of *Escherichia coli* K12. *Journal of General Microbiology*, **54**, 169–175.

Cox, C. S. (1968*c*). Method for the routine preservation of microorganisms. *Nature, London*, **220**, 1139.

Cox, C. S. (1969). The cause of loss of viability of airborne *Escherichia coli* K12. *Journal of General Microbiology*, **57**, 77–80.

Cox, C. S. (1970). Aerosol survival of *Escherichia coli* B disseminated from the dry state. *Applied Microbiology*, **19**, 604–607.

Cox, C. S. (1971). Aerosol survival of *Pasteurella tularensis* disseminated from the wet and dry states. *Applied Microbiology*, **21**, 482–486.

Cox, C. S. (1973*a*). A kinetic model for oxygen-induced death in dehydrated bacteria. In *Airborne Transmission and Airborne Infection*, ed. J. F. Ph. Hers & K. C. Winkler, pp. 108–110. Utrecht, The Netherlands: Oosthoek Publishing Company.

Cox, C. S. (1973*b*). Oxygen induced free radicals and viable decay in freeze-dried bacteria. In *Freeze-drying of Biological Materials*, pp. 55–59. Paris: Institut International du Froid.

Cox, C. S. & Baldwin, F. (1964). A method for investigating the cause of death of airborne bacteria. *Nature, London*, **202**, 1135.

Cox, C. S. & Baldwin, F. (1966). The use of phage to study causes of loss of viability of *Escherichia coli* in aerosols. *Journal of General Microbiology*, **44**, 15–22.

Cox, C. S. & Baldwin, F. (1967). The toxic effect of oxygen upon aerosol survival of *Escherichia coli* B. *Journal of General Microbiology*, **49**, 115–117.

Cox, C. S., Baxter, J. & Maidment, B. J. (1973). A mathematical expression for oxygen-induced death in dehydrated bacteria. *Journal of General Microbiology*, **75**, 179–185.

Cox, C. S., Bondurant, M. C. & Hatch, M. T. (1971*a*). Effects of oxygen on the aerosol survival of radiation sensitive and resistant strains of *Escherichia coli* B. *Journal of Hygiene, Cambridge*, **69**, 661–672.

Cox, C. S., Bondurant, M. C. & Hatch, M. T. (1971*b*). Absence of dark repair in selected mutants of airborne *Escherichia coli* in air or nitrogen atmospheres. *Bacteriological Proceedings*, A134.

Cox, C. S., Derr, J. S., Fleurie, E. G. & Roderick, R. C. (1970). An experimental technique for the study of aerosols of lyophilized bacteria. *Applied Microbiology*, **20**, 927–934.

Cox, C. S., Gagen, S. J. & Baxter, J. (1974). Aerosol survival of *Serratia marcescens* as a function of oxygen concentration, relative humidity and time. *Canadian Journal of Microbiology*, **20**, 1529–1534.

Cox, C. S. & Goldberg, L. J. (1972). Aerosol survival of *Pasteurella tularensis* and the influence of relative humidity. *Applied Microbiology*, **23**, 1–3.

Cox, C. S., Harris, W. J. & Lee, J. (1974). Viability and electron microscope studies of phages T3 and T7 subjected to freeze-drying, freeze-thawing and aerosolization. *Journal of General Microbiology*, **81**, 207–215.

Cox, C. S. & Heckly, R. J. (1973). Effects of oxygen upon freeze-dried and freeze-thawed bacteria: viability and free radical studies. *Canadian Journal of Microbiology*, **19**, 189–194.

Cox, C. S., Hood, A. M. & Baxter, J. (1973). Method for comparing concentrations of the Open Air Factor. *Applied Microbiology*, **26**, 640–642.

Crider, W. L., Barkley, N. P. & Strong, A. A. (1968). Dry powder aerosol dispersing device with long-time output stability. *Reviews of Scientific Instruments*, **39**, 152–155.

Crowe, J. H. (1971). Anhydrobiosis: an unsolved problem. *American Naturalist*, **105**, 563–574.

Dark, F. A. & Callow, D. S. (1973). The effect of growth conditions on the survival of airborne *Escherichia coli*. In *Airborne Transmission and Airborne Infection*, ed. J. F. Ph. Hers & K. C. Winkler, pp. 97–99. Utrecht, The Netherlands: Oosthoek Publishing Company.

Dark, F. A. & Nash, T. (1970). Comparative toxicity of various ozonized olefins to bacteria suspended in air. *Journal of Hygiene, Cambridge*, **68**, 245–252.

DAVIES, R. R. (1971). Air sampling for fungi, pollens and bacteria. In *Methods in Microbiology*, **4**, ed. J. R. Norris & D. W. Ribbons, pp. 367–404. London and New York: Academic Press.

DAVIS, M. S. & BATEMAN, J. B. (1960). Relative humidity and the killing of bacteria. II. Selective changes in oxidative activity associated with death. *Journal of Bacteriology*, **80**, 580–584.

DERR, J. S. (1965). Model for de-agglomeration and fracture of particulate solids. In *First International Symposium on Aerobiology*, pp. 227–261. Oakland, California: Naval Biological Laboratory, Naval Supply Center.

DEWALD, R. R. (1966*a*). Preservation of *Serratia marcescens* by high vacuum lyophilization. *Applied Microbiology*, **14**, 561–567.

DEWALD, R. R. (1966*b*). Kinetic studies on the destructive action of oxygen on lyophilized *Serratia marcescens*. *Applied Microbiology*, **14**, 568–572.

DEWALD, R. R., BROWALL, K. W., SCHAEFER, L. D. & MESSER, A. (1967). Effect of water vapour on lyophilized *Serratia marcescens* and *Escherichia coli*. *Applied Microbiology*, **15**, 1299–1302.

DIMMICK, R. L. (1959). Jet disperser for compacted powders in the one to ten micron range. *Archives of Industrial Health*, **20**, 8–14.

DIMMICK, R. L. (1960). Characteristics of dried *Serratia marcescens* in the airborne state. *Journal of Bacteriology*, **80**, 289–296.

DIMMICK, R. L. (1969). Production of biological aerosols. In *An Introduction to Experimental Aerobiology*, ed. R. L. Dimmick & A. B. Akers, pp. 22–45. New York and London: Wiley Interscience.

DIMMICK, R. L., HECKLY, R. J. & HOLLIS, D. P. (1961). Free radical formation during storage of freeze-dried *Serratia marcescens*. *Nature, London*, **192**, 776–777.

DIMMICK, R. L. & WANG, L. (1969). Rotating drum. In *An Introduction to Experimental Aerobiology*, ed. R. L. Dimmick & A. B. Akers, pp. 164–176. New York and London: Wiley Interscience.

DRUETT, H. A. (1970). The open air factor. In *Third International Symposium on Aerobiology*, ed. I. H. Silver, p. 212. London and New York: Academic Press.

DRUETT, H. A. (1973*a*). The open air factor. In *Airborne Transmission and Airborne Infection*, ed. J. F. Ph. Hers & K. C. Winkler, pp. 141–149. Utrecht, The Netherlands: Oosthoek Publishing Company.

DRUETT, H. A. (1973*b*). Effect on the viability of microorganisms in aerosols of the rapid rarefaction of the surrounding air. In *Airborne Transmission and Airborne Infection*, ed. J. F. Ph. Hers & K. C. Winkler, pp. 90–94. Utrecht, The Netherlands: Oosthoek Publishing Company.

DRUETT, H. A. & MAY, K. R. (1952). A wind tunnel for the study of airborne infections. *Journal of Hygiene, Cambridge*, **50**, 69–81.

DRUETT, H. A. & MAY, K. R. (1968). Unstable germicidal pollutant in rural air. *Nature, London*, **220**, 395–396.

DRUETT, H. A. & MAY, K. R. (1969). The Open Air Factor. *New Scientist*, **41**, 579–581.

DUNKLIN, E. W. & PUCK, T. T. (1948). The lethal effect of relative humidity on airborne bacteria. *Journal of Experimental Medicine*, **87**, 87–101.

EHRLICH, R. & MILLER, S. (1971). Effect of relative humidity and temperature on airborne Venezuelan equine encephalitis virus. *Applied Microbiology*, **22**, 194–199.

FERRY, R. M., BROWN, W. F. & DAMON, E. B. (1958). Loss of viability of bacterial aerosols. III. Factors affecting death rates of certain non-pathogens. *Journal of Hygiene, Cambridge*, **56**, 389–403.

FRY, R. M. (1966). Freezing and drying of bacteria. In *Cryobiology*, ed. H. T. Meryman, pp. 665–696. London and New York: Academic Press.

FRY, R. M. & GREAVES, R. I. N. (1951). The survival of bacteria during and after drying. *Journal of Hygiene, Cambridge*, **49**, 220–246.

FURCOLOW, M. L. (1961). Airborne histoplasmosis. *Bacteriological Reviews*, **25**, 301–309.

GOLDBERG, L. J. (1970). The Naval Biological Laboratory programmed environmental facility; theory and operation. In *Third International Symposium on Aerobiology*, ed. I. H. Silver, p. 268. London and New York: Academic Press.

GOLDBERG, L. J. (1971). The Naval Biomedical Research Laboratory programmed environmental aerosol facility. *Applied Microbiology*, **21**, 244–252.

GOLDBERG, L. J. & FORD, I. (1973). The function of chemical additives in enhancing microbial survival in aerosols. In *Airborne Transmission and Airborne Infection*, ed. J. F. Ph. Hers & K. C. Winkler, pp. 86–89. Utrecht, The Netherlands: Oosthoek Publishing Company.

GOLDBERG, L. J., WATKINS, H. M. S., BOERKE, E. E. & CHATIGNY, M. A. (1958). Use of a rotating drum for the study of aerosols over extended time intervals. *American Journal of Hygiene*, **68**, 85–93.

GONDA, K. & KOGA, S. (1973). Low temperature thermograms of *Saccharomyces cerevisiae*. *Journal of General and Applied Microbiology*, **19**, 393–395.

GONDA, K., OHTOMO, T. & KOGA, S. (1973). Differential scanning calorimetry of frozen microbial cells. In *Freeze-drying of Biological Materials*, pp. 19–28. Paris: Institut International du Froid.

GOODLOW, R. G. & LEONARD, F. A. (1961). Viability and infectivity of microorganisms in experimental airborne infection. *Bacteriological Reviews*, **25**, 182–187.

GREEN, H. L. & LANE, W. R. (1964). *Particulate Clouds: Dusts, Smokes and Mists*. London: E. and F. N. Spon Ltd.

GREIFF, D. (1969). The effects of residual moisture on the predicted stabilities of suspensions of viruses dried by sublimation of ice *in vacuo*. In *Freezing and Drying of Microorganisms*, ed. T. Nei, pp. 93–109. Tokyo: University of Tokyo Press.

GREIFF, D. (1970). Stabilities of suspensions of influenza virus dried by sublimation of ice *in vacuo* to different contents of residual moisture and sealed under different gases. *Applied Microbiology*, **20**, 935–938.

GREIFF, D. & RIGHTSEL, W. A. (1969). Stabilities of dried suspensions of influenza virus sealed in a vacuum or under different gases. *Applied Microbiology*, **17**, 830–835.

HAMBLETON, P. (1970). The sensitivity of Gram-negative bacteria recovered from aerosols to lysozyme and other hydrolytic enzymes. *Journal of General Microbiology*, **61**, 197–204.

HAMBLETON, P. (1971). Repair of wall damage in *Escherichia coli* recovered from an aerosol. *Journal of General Microbiology*, **69**, 81–88.

HAMBLETON, P. & BENBOUGH, J. E. (1973). Damage to envelopes of Gram-negative bacteria recovered from aerosols. In *Airborne Transmission and Airborne Infection*, ed. J. F. Ph. Hers & K. C. Winkler, pp. 131–134. Utrecht, The Netherlands: Oosthoek Publishing Company.

HANAFUSA, N. (1966). Denaturation of catalase by freeze-thawing and freeze-drying. *Contribution from Institute of Low Temperature Science*, Hokkaido University, Japan, B**24**, 57–66. (In Japanese with English summary.)

HANAFUSA, N. (1969). Denaturation of enzyme protein by freeze-thawing and freeze-drying. In *Freezing and Drying of Microorganisms*, ed. T. Nei, pp. 117–129. Tokyo: University of Tokyo Press.

HANAFUSA, N. (1972). Denaturation of enzyme protein by freeze-thawing and freeze-drying. *Contribution from Institute of Low Temperature Science*, Hokkaido University, Japan, **B17**, 1–38.

HANAFUSA, N. (1973). Freezing and drying of enzyme protein. In *Freeze-drying of Biological Materials*, pp. 9–18. Paris: Institut International du Froid.

HARPER, G. J. (1973). The influence of urban and rural air on the survival of microorganisms exposed on microthreads. In *Airborne Transmission and Airborne Infection*, ed. J. F. Ph. Hers & K. C. Winkler, pp. 151–154. Utrecht, The Netherlands: Oosthoek Publishing Company.

HATCH, M. T., BONDURANT, M. C. & EHRESMANN, D. W. (1973). Analysis of repair systems and the genetic basis of aerosol resistance in derivatives of *Escherichia coli* B. In *Airborne Transmission and Airborne Infection*, ed. J. F. Ph. Hers & K. C. Winkler, pp. 103–105. Utrecht, The Netherlands: Oosthoek Publishing Company.

HATCH, M. T. & DIMMICK, R. L. (1965). A study of dynamic aerosols of bacteria subjected to rapid changes in relative humidity. In *First International Symposium on Aerobiology*, pp. 265–281. Oakland, California: Naval Biological Laboratory, Naval Supply Center.

HATCH, M. T. & DIMMICK, R. L. (1966). Physiological responses of airborne bacteria to shifts in relative humidity. *Bacteriological Reviews*, **30**, 597–603.

HATCH, M. T. & WARREN, J. C. (1969). Enhanced recovery of airborne T3 coliphage and *Pasteurella pestis* bacteriophage by means of a pre-sampling humidification technique. *Applied Microbiology*, **17**, 685–689.

HATCH, M. T., WRIGHT, D. M. & BAILEY, G. D. (1970). Response of airborne *Mycoplasma pneumoniae* to abrupt changes in relative humidity. *Applied Microbiology*, **19**, 232–238.

HECKLY, R. J. & DIMMICK, R. L. (1968). Correlations between free radical production and viability of lyophilized bacteria. *Applied Microbiology*, **16**, 1081–1085.

HECKLY, R. J., DIMMICK, R. L. & WINDLE, J. J. (1963). Free radical formation and survival of lyophilized microorganisms. *Journal of Bacteriology*, **85**, 961–966.

HENDERSON, D. W. (1952). An apparatus for the study of airborne infection. *Journal of Hygiene, Cambridge*, **50**, 53–68.

HEPPEL, L. A. (1967). Selective release of enzymes from bacteria. *Science, Washington*, **156**, 1451–1455.

HEPPEL, L. A. (1971). The concept of periplasmic enzymes. In *Structure and Function of Biological Membranes*, ed. L. I. Rothfield, pp. 224–247. London and New York: Academic Press.

HESS, G. E. (1965). Effects of oxygen on aerosolized *Serratia marcescens*. *Applied Microbiology*, **13**, 781–787.

HINTON, H. E. (1968). Reversible suspension of metabolism and the origin of life. *Proceedings of the Royal Society of London, Series B*, **171**, 43–57.

HOOD, A. M. (1961). Infectivity of *Pasteurella tularensis* clouds. *Journal of Hygiene, Cambridge*, **59**, 497–504.

HOOD, A. M. (1971). An indoor system for the study of biological aerosols in open air conditions. *Journal of Hygiene, Cambridge*, **69**, 607–617.

HOOD, A. M. (1973). Open Air Factors in enclosed systems. In *Airborne Transmission and Airborne Infection*, ed. J. F. Ph. Hers & K. C. Winkler, pp. 149–151. Utrecht, The Netherlands: Oosthoek Publishing Company.

HOOD, A. M. (1974). Open Air Factors in enclosed systems. *Journal of Hygiene, Cambridge*, **72**, 53–60.

INTERNATIONAL SYMPOSIUM ON AEROBIOLOGY (FOURTH) (1973). *Airborne transmission and airborne infection*, ed. J. F. Ph. Hers & K. C. Winkler. Utrecht, The Netherlands: Oosthoek Publishing Company.

ISRAELI, E., GIBERMAN, E. & KOHN, A. (1974). Membrane malfunctions in freeze-dried *Escherichia coli*. *Cryobiology*, **11**, 473–477.

KEILIN, D. (1959). The problem of anabiosis or latent life: history and current concept. *Proceedings of the Royal Society of London, Series B*, **150**, 149–191.

LANGMUIR, A. D. (1961). Epidemiology of airborne infection. *Bacteriological Reviews*, **25**, 173–181.

LAPAGE, S. P., SHELTON, J. E., MITCHELL, T. G. & MACKENZIE, A. R. (1969). Culture Collections and the Preservation of Bacteria. In *Methods in Microbiology*, ed. J. R. Norris & D. W. Ribbons, **3A**, pp. 136–228. London and New York: Academic Press.

LEA, C. H., HANNAN, R. S. & GREAVES, R. I. N. (1950). The reaction between proteins and reducing surgars in the dry state. *Biochemical Journal*, **47**, 626–629.

LEACH, R. H. & SCOTT, W. J. (1959). The influence of rehydration on the viability of dried microorganisms. *Journal of General Microbiology*, **21**, 295–307.

LEVIN, M. A. & CABELLI, V. J. (1963). Germination of spores as consequence of aerosolization and collection. *Bacteriological Proceedings*, 26.

LIGHTHEART, B., HIATT, V. E. & ROSSANO, A. T. (1971). The survival of airborne *Serratia marcescens* in urban concentrations of sulphur dioxide. *Journal of Air Pollution Control Association*, **21**, 639–642.

LION, M. B. (1963). Quantitative aspects of the protection of freeze-dried *Escherichia coli* against the toxic effect of oxygen. *Journal of General Microbiology*, **32**, 321–329.

LION, M. B. & BERGMANN, E. D. (1961*a*). The effect of oxygen on freeze-dried *Escherichia coli*. *Journal of General Microbiology*, **24**, 191–199.

LION, M. B. & BERGMANN, E. D. (1961*b*). Substances that protect lyophilized *Escherichia coli* against the lethal effect of oxygen. *Journal of General Microbiology*, **25**, 291–296.

LION, M. B., KIRBY-SMITH, J. S. & RANDOLPH, M. C. (1961). Electron spin resonance signals from lyophilized bacterial cells exposed to oxygen. *Nature, London*, **192**, 34–36.

LURIE, N. B. (1930). Airborne contagion of tuberculosis in an animal room. *Journal of Experimental Medicine*, **51**, 743–751.

MALTMAN, J. R. (1965). Bacterial responses to desiccation and rehydration. In *First International Symposium on Aerobiology*, pp. 291–303. Oakland, California: Naval Biological Laboratory, Naval Supply Center.

MALTMAN, J. R., ORR, J. H. & HINTON, N. A. (1960). The effect of desiccation on *Staphylococcus pyogenes* with special reference to implications concerning virulence. *American Journal of Hygiene*, **72**, 335–350.

MALTMAN, J. R. & WEBB, S. J. (1971). The action of hydrolytic enzymes and vapour rehydration on semi-dried cells of *Klebsiella pneumoniae*. *Canadian Journal of Microbiology*, **17**, 1443–1450.

MAY, K. R. (1949). An improved spinning top homogeneous spray apparatus. *Journal of Applied Physics*, **20**, 932–938.

MAY, K. R. (1967). Physical aspects of sampling airborne microbes. *Symposia of the Society for General Microbiology*, **17**, 60–80.

MAY, K. R. (1969). Prolongation of microbiological air sampling by a monolayer agar gel. *Applied Microbiology*, **18**, 513–514.

MAY, K. R. (1972). Assessment of viable airborne particles. In *Assessment of Airborne Particles. Fundamentals, Applications and Implications to Inhalation*

Therapy, ed. T. T. Mercer, P. E. Morrow & W. Stöber, pp. 480–494. Springfield, Illinois: Charles C. Thomas, Publisher.

MAY, K. R. & DRUETT, H. A. (1968). A microthread technique for studying the viability of microbes in a simulated airborne state. *Journal of General Microbiology*, **51**, 353–366.

MAY, K. R., DRUETT, H. A. & PACKMAN, L. P. (1969). Toxicity of open air to a variety of microorganisms. *Nature, London*, **221**, 1146–1147.

MAZUR, P. (1966). Physical and chemical basis of injury in single-celled microorganisms subjected to freezing and thawing. In *Cryobiology*, ed. H. T. Meryman, pp. 213–315. London: Academic Press.

MCCRUMB, F. R. (1961). Aerosol infection with *Pasteurella tularensis*. *Bacteriological Reviews*, **25**, 262–267.

MCDADE, J. J. & HALL, L. B. (1963*a*). An experimental method to measure the influence of environmental factors on the viability and pathogenicity of *Staphylococcus aureus*. *American Journal of Hygiene*, **77**, 98–108.

MCDADE, J. J. & HALL, L. B. (1963*b*). Survival of *Staphylococcus aureus* in the environment. I. Exposure on surfaces. *American Journal of Hygiene*, **78**, 330–337.

MCDADE, J. J. & HALL, L. B. (1964). Survival of Gram-negative bacteria in the environment. I. Effect of relative humidity on surface exposed organisms. *American Journal of Hygiene*, **80**, 192–204.

MERYMAN, H. T. (1966*a*). Freeze-drying. In *Cryobiology*, ed. H. T. Meryman, pp. 609–663. London: Academic Press.

MERYMAN, H. T. (1966*b*). Review of biological freezing. In *Cryobiology*, ed. H. T. Meryman, pp. 1–114. London: Academic Press.

MEYNELL, G. G. (1958). The effect of sudden chilling on *Escherichia coli*. *Journal of General Microbiology*, **19**, 380–389.

MIK, G. DE & GROOT, I. DE (1973). The survival of *Escherichia coli* in the open air in different parts of the Netherlands. In *Airborne Transmission and Airborne Infection*, ed. J. F. Ph. Hers & K. C. Winkler, pp. 155–158. Utrecht, The Netherlands: Oosthoek Publishing Company.

MONK, G. W., ELBERT, M. L., STEVENS, C. L. & MCCAFFREY, P. A. (1956). The effect of water on the death rate of *Serratia marcescens*. *Journal of Bacteriology*, **72**, 368–372.

MONK, G. W. & MATTUCK, R. D. (1956). Biological cloud dynamics. *Bulletin of Mathematics and Biophysics*, **18**, 57–64.

MONK, G. W. & MCCAFFREY, P. A. (1957). Effect of sorbed water on the death rate of washed *Serratia marcescens*. *Journal of Bacteriology*, **73**, 85–88.

MORICHI, T. (1969). Metabolic injury in frozen *Escherichia coli*. In *Freezing and Drying of Microorganisms*, ed. T. Nei, pp. 53–68. Tokyo: University of Tokyo Press.

MORICHI, T. & IRIE, R. (1973). Factors affecting repair of sublethal injury in frozen or freeze-dried bacteria. *Cryobiology*, **10**, 393–399.

MORICHI, T., OKAMOTO, T. & IRIE, R. (1973). Effect of peptone added to the recovery medium on the viability of freeze-dried bacteria. In *Freeze-Drying of Biological Materials*, pp. 47–59. Paris: Institut International du Froid.

MOSS, C. W. & SPECK, M. L. (1963). Injury and death of *Streptococcus lactis* due to freezing and frozen storage. *Applied Microbiology*, **11**, 326–329.

MOSS, C. W. & SPECK, M. L. (1966*a*). Identification of nutritional components in Trypticase responsible for recovery of *Escherichia coli* injured by freezing. *Journal of Bacteriology*, **91**, 1098–1104.

MOSS, C. W. & SPECK, M. L. (1966*b*). Release of biologically active peptides from *Escherichia coli* at subzero temperatures. *Journal of Bacteriology*, **91**, 1105–1111.

NASH, T. (1973). Detection of presumed OAF using a nuclear counter. In *Airborne Transmission and Airborne Infection*, ed. J. F. Ph. Hers & K. C. Winkler, pp. 158–162. Utrecht, The Netherlands: Oosthoek Publishing Company.

NAYLOR, H. B. & SMITH, P. A. (1946). Factors affecting the viability of *Serratia marcescens* during dehydration and storage. *Journal of Bacteriology*, **52**, 565–573.

NEI, T. (1960). Effects of freezing and freeze-drying on microorganisms. In *Recent Research in Freezing and Drying*, ed. A. S. Parkes & A. U. Smith, pp. 221–228. Oxford: Blackwell.

NEI, T. (1962). Electron microscopic study of microorganisms subjected to freezing and drying: cinematographic observations of yeast and coli cells. *Experimental Cell Research*, **28**, 560–575.

NEI, T. (1973). Some aspects of freezing and drying of microorganisms on the basis of cellular water. *Cryobiology*, **10**, 403–408.

NEI, T., ARAKI, T. & MATSUSAKA, T. (1969). Freezing injury to aerated and non-aerated cultures. In *Freezing and Drying of Microorganisms*, ed. T. Nei, pp. 3–15. Tokyo: University of Tokyo Press.

NEI, T., ARAKI, T. & SOUZU, H. (1965). Studies of the effect of drying conditions on residual moisture content and cell viability in the freeze-drying of microorganisms. *Cryobiology*, **2**, 68–73.

NEI, T., SOUZU, H. & ARAKI, T. (1965). Effect of residual moisture content on the survival of freeze-dried bacteria during storage under various conditions. *Cryobiology*, **2**, 276–279.

NEU, H. C. & HEPPEL, L. A. (1965). The release of enzymes from *Escherichia coli* by osmotic shock and during the formation of spheroplasts. *Journal of Biological Chemistry*, **240**, 3685–3692.

NOBLE, W. C. (1967). Sampling airborne microbes; handling the catch. *Symposia of the Society for General Microbiology*, **17**, 81–108.

NOSSAL, N. G. & HEPPEL, L. A. (1966). The release of enzymes by osmotic shock from *Escherichia coli* in exponential phase. *Journal of Biological Chemistry*, **241**, 3055–3062.

OBAYASHI, Y. & CHO, C. (1957). Further studies on the adjuvant for dried BCG vaccine. *Bulletin World Health Organization*, **17**, 255–274.

OBAYASHI, Y., OTA, S. & ARAI, S. (1961). Some factors affecting the preservability of freeze-dried bacteria. *Journal of Hygiene, Cambridge*, **59**, 77–91.

ORNDORFF, G. R. & MACKENZIE, A. P. (1973). Function of the suspending medium during the freeze-drying preservation of *Escherichia coli*. *Cryobiology*, **10**, 475–487.

POSTGATE, J. R., CRUMPTON, J. E. & HUNTER, J. R. (1961). The measurement of bacterial viability by slide culture. *Journal of General Microbiology*, **24**, 15–24.

POSTGATE, J. R. & HUNTER, J. R. (1963). Metabolic injury in frozen bacteria. *Journal of Applied Bacteriology*, **26**, 405–414.

RECHSTEINER, J. (1970). The recovery of airborne respiratory syncytial virus. In *Third International Symposium on Aerobiology*, ed. I. H. Silver, p. 269. London and New York: Academic Press.

RECORD, B. R. & TAYLOR, R. (1953). Some factors affecting the survival of *Bacterium coli* on freeze-drying. *Journal of General Microbiology*, **9**, 475–484.

RECORD, B. R., TAYLOR, R. & MILLER, D. S. (1962). The survival of *Escherichia coli* on drying and rehydration. *Journal of General Microbiology*, **28**, 585–598.

REDWAY, K. F. & LAPAGE, S. P. (1974). Effect of carbohydrates and related compounds on the long term preservation of freeze-dried bacteria. *Cryobiology*, **11**, 73–79.

REY, L. (ed.) (1964). *Aspects Theoretiques et Industriels de la Lyophilisation*. Paris: Hermann.

REY, L. (ed.) (1966). *Lyophilisation Reserches et Applications Nouvelle*. Paris: Hermann.

ROGERS, L. A. (1914). The preparation of dried cultures. *Journal of Infectious Diseases*, **14**, 100–123.

ROSEBURY, T. (1947). *Experimental Airborne Infection*. Baltimore: The Williams and Wilkins Company.

ROUNTREE, P. M. (1963). The effect of desiccation on the viability of *Staphylococcus aureus*. *Journal of Hygiene, Cambridge*, **61**, 265–272.

SAWYER, W. D., JEMSKI, J. V., HOGGE, A. L., EIGELSBACH, H. T., ELWOOD, K., WOLFE, E. K., DANGERFIELD, H. G., GOCHENOUR, W. S. & CROZIER, D. (1966). Effect of aerosol age on the infectivity of airborne *Pasteurella tularensis* for *Macaca mulatta* and man. *Journal of Bacteriology*, **91**, 2180–2184.

SCHLAMM, N. A. (1960). Detection of viability in aged or injured *Pasteurella tularensis*. *Journal of Bacteriology*, **80**, 818–822.

SCHWARTZ, H. M. (1970). Effect of oxygen on freezing damage. I. Effect on survival of *Escherichia coli* B/r and *Escherichia coli* B_{S-1}. *Cryobiology* **6**, 546–551.

SCHWARTZ, H. M. (1971). Effect of oxygen on freezing damage. II. Physical-chemical effects. *Cryobiology*, **8**, 255–264.

SCOTT, W. J. (1958). The effect of residual water on the survival of dried bacteria during storage. *Journal of General Microbiology*, **19**, 624–633.

SHERMAN, J. M. & CAMERON, G. M. (1934). Lethal environmental factors within the natural range of growth. *Journal of Bacteriology*, **27**, 341–348.

SILVER, I. H. (1965). Viability of microbes using a suspended droplet technique. In *First International Symposium on Aerobiology*, pp. 319–333. Oakland, California: Naval Biological Laboratory, Naval Supply Center.

SINHA, R. N., DUDANI, A. T. & RANGANATHAN, B. (1974). Effect of individual ingredients of fortified skim milk as suspending media on survival of freeze-dried cells of *Streptococcus lactis*. *Cryobiology*, **11**, 368–370.

SMEATON, J. R. & ELLIOT, W. H. (1967). Isolation and properties of a special bacterial ribonuclease inhibitor. *Biochimica Biophysica Acta*, **145**, 547–560.

SOUTHEY, R. F. W. & HARPER, G. J. (1971). The survival of *Erwinia amylovora* in airborne particles: tests in the laboratory and in the open air. *Journal of Applied Bacteriology*, **34**, 547–556.

SOUZA, H. (1973*a*). Studies on the character of phospholipase C and its effect on survival of the cell after freeze-thawing or freeze-drying of yeast. In *Freeze-Drying of Biological Materials*, pp. 29–36. Paris: Institut International du Froid.

SOUZA, H. (1973*b*). The phospholipid degradation and cellular death caused by freeze-thawing or freeze-drying of yeast. *Cryobiology*, **10**, 427–431.

SPECK, M. L. & COWMAN, R. A. (1969). Metabolic injury to bacteria resulting from freezing. In *Freezing and Drying of Microorganisms*, ed. T. Nei, pp. 39–51. Tokyo: University of Tokyo Press.

SPECK, M. L. & RAY, B. (1973). Recovery of *Escherichia coli* after injury from freezing. In *Freeze-drying of Biological Materials*, pp. 37–46. Paris: Institut International du Froid.

STRAKA, R. P. & STOKES, J. L. (1959). Metabolic injury to bacteria at low temperatures. *Journal of Bacteriology*, **78**, 181–185.

STRANGE, R. E., BENBOUGH, J. E., HAMBLETON, P. & MARTIN, K. L. (1972). Methods for the assessment of microbial populations recovered from enclosed aerosols. *Journal of General Microbiology*, **72**, 117–125.

STRANGE, R. E. & DARK, F. A. (1962). Effect of chilling on *Aerobacter aerogenes* in aqueous suspension. *Journal of General Microbiology*, **29**, 719–730.

STRANGE, R. E. & POSTGATE, J. R. (1964). Penetration of substances into cold-shocked bacteria. *Journal of General Microbiology*, **36**, 393–403.

STRASTERS, K. C. & WINKLER, K. C. (1966). Viability of hospital staphylococci in air. *Bacteriological Reviews*, **30**, 674–677.

SUSSMAN, A. S. & HALVORSON, H. O. (1966). *Spores: Their Dormancy and Germination*, p. 74. New York and London: Harper & Row.

SUZUKI, M. (1969). Relation between reduction of vaccinia virus titre and residual moisture content. In *Freezing and Drying of Microorganisms*, ed. T. Nei, pp. 111–115. Tokyo: University of Tokyo Press.

SUZUKI, M., SAWADA, T. & OBAYASHI, Y. (1969). Reduction of the survival rate of BCG in the course of drying relative to its dehydration curve. In *Freezing and Drying of Microorganisms*, ed. T. Nei, pp. 81–92. Tokyo: University of Tokyo Press.

TAKANO, M., SINSKEY, A. J. & BARALDI, D. (1973). DNA damage in *Salmonella typhimurium* LT-2 by a combination of freezing and nutrition. In *Freeze-Drying of Biological Materials*, pp. 61–70. Paris: Institut International du Froid.

TAKANO, M. & TERUI, G. (1964). Some problems of secondary drying of living cells in the course of freeze-drying. *Journal of Fermentation Technology*, **42**, 173–180. (In Japanese with English summary.)

TAKANO, M. & TERUI, G. (1969). Correlation of dehydration with death of microbial cells in the secondary stage of freeze-drying. In *Freezing and Drying of Microorganisms*, ed. T. Nei, pp. 131–142. Tokyo: University of Tokyo Press.

THOMAS, G. (1970*a*). Sampling of airborne viruses. In *Third International Symposium on Aerobiology*, ed. I. H. Silver, p. 266. London and New York: Academic Press.

THOMAS, G. (1970*b*). An adhesive surface sampling technique for airborne viruses. *Journal of Hygiene, Cambridge*, **68**, 273–282.

THOMAS, G. (1970*c*). Sampling rabbit pox aerosols of natural origin. *Journal of Hygiene, Cambridge*, **68**, 511–517.

TIGERTT, W. D., BENENSON, A. S. & GOCHENOUR, W. S. (1961). Airborne Q fever. *Bacteriological Reviews*, **25**, 285–293.

TROUWBORST, T. (1971). Inactivating in aerosolen van microörganismen en macromoleculen. Thesis, University of Utrecht.

TROUWBORST, T. & DE JONG, J. C. (1973). Interaction of some factors in the mechanism of inactivation of bacteriophage MS2 in aerosols. *Applied Microbiology*, **26**, 252–257.

TROUWBORST, T., DE JONG, J. C. & WINKLER, K. C. (1972). Mechanism of inactivation in aerosols of bacteriophage T1. *Journal of General Virology*, **15**, 235–242.

TROUWBORST, T., DE JONG, J. C. & WINKLER, K. C. (1973). Inactivation of the enzyme trypsin in aerosols. *Journal of Colloid and Interface Science*, **45**, 198–208.

TROUWBORST, T. & KUYPER, S. (1974). Inactivation of bacteriophage T3 in aerosols: effect of prehumidification on survival after spraying from solutions of salt, peptone and saliva. *Applied Microbiology*, **27**, 834–837.

TROUWBORST, T., KUYPER, S., DE JONG, J. C. & PLANTIGUA, A. D. (1974). Inactivation of some bacterial and animal viruses by exposure to liquid air interfaces. *Journal of General Virology*, **24**, 155–165.

WAGMAN, J. (1960). Evidence of cytoplasmic membrane injury in the drying of bacteria. *Journal of Bacteriology*, **80**, 558–564.

WAGMAN, J. & WENECK, E. J. (1963). Preservation of bacteria by circulating-gas freeze-drying. *Applied Microbiology*, **11**, 244–248.

WAITE, D. A. & RAMSDEN, D. (1971). The production of experimentally labelled aerosols in the sub-micron range. *Journal of Aerosol Science*, **2**, 425–436.

WEBB, S. J. (1960). Factors affecting the viability of airborne bacteria. III. The role of bonded water and protein structure in the death of airborne cells. *Canadian Journal of Microbiology*, **6**, 89–105.

WEBB, S. J. (1963). The effect of relative humidity and light on air-dried organisms. *Journal of Applied Bacteriology*, **26**, 307–313.

WEBB, S. J. (1965). *The Role of Bound Water in the Maintenance of the Integrity of a Cell or Virus*. Springfield, Illinois: Charles C. Thomas, Publisher.

WEBB, S. J. (1967*a*). The influence of oxygen and inositol on the survival of semi-dried microorganisms. *Canadian Journal of Microbiology*, **13**, 733–742.

WEBB, S. J. (1967*b*). Mutation of bacterial cells by controlled desiccation. *Nature, London*, **213**, 1137–1139.

WEBB, S. J. (1968). Effect of dehydration on bacterial recombination. *Nature, London*, **217**, 1231–1234.

WEBB, S. J. (1969*a*). The effects of oxygen on the possible repair of dehydration damage by *Escherichia coli*. *Journal of General Microbiology*, **58**, 317–326.

WEBB, S. J. (1969*b*). Some effects of dehydration on the genetics of microorganisms. In *Freezing and Drying of Microorganisms*, ed. T. Nei, pp. 153–167. Tokyo: University of Tokyo Press.

WEBB, S. J., DUMASIA, M. D. & SINGH BHORJEE, J. (1965). Bound water, inositol and the biosynthesis of temperate and virulent bacteriophages by air dried *Escherichia coli*. *Canadian Journal of Microbiology*, **11**, 141–149.

WEBB, S. J. & WALKER, J. L. (1968). The influence of cell water content on the inactivation of RNA by partial desiccation and ultra-violet light. *Canadian Journal of Microbiology*, **14**, 565–572.

WELLS, W. F. (1933). Apparatus for the study of the bacterial behaviour of air. *American Journal of Public Health*, **23**, 58–59.

WELLS, W. F. (1934). Droplets and droplet nuclei. *American Journal of Hygiene*, **20**, 611–618.

WELLS, W. F. (1955). *Airborne Contagion and Air Hygiene*. Cambridge, Massachusetts: Harvard University Press.

WOLF, H. W., SKALIY, P., HALL, L. B., HARRIS, M. M., DECKER, H. M., BUCHANAN, L. M. & DAHLGREN, C. M. (1959). *Sampling Microbiological Aerosols. Public Health Service Publications, Washington*, no. 60.

WON, W. P. & ROSS, H. (1969). Reaction of airborne *Rhizobium meliloti* to some environmental factors. *Applied Microbiology*, **18**, 555–557.

YOUNG, H. W., LARSON, E. W. & DOMINIK, J. W. (1974). Modified spinning top homogeneous spray apparatus for use in experimental respiratory disease studies. *Applied Microbiology*, **28**, 929–934.

ZAMENOF, S., EICHHORN, H. H. & ROSENBAUM-OLIVER, D. (1968). Mutability of stored spores of *Bacillus subtilis*. *Nature, London*, **220**, 818–819.

OSMOTIC STRESS AND MICROBIAL SURVIVAL

A. H. ROSE

Zymology Laboratory, School of Biological Sciences, University of Bath, Claverton Down, Bath BA2 7AY, Avon

INTRODUCTION

The metabolic reactions which a micro-organism must carry out in order to maintain viability and the capacity to grow and divide require, for optimum activity, high concentrations of low molecular weight compounds and ions. Since these reactions take place at a variety of loci inside the micro-organisms, it follows that there exists a high concentration of low molecular weight compounds and ions inside the microbial plasma membrane. However, the majority of micro-organisms remain viable and grow in environments in which the concentrations of solutes, including nutrients, are much lower than the concentrations of solutes present inside the micro-organism. An appreciable osmotic pressure difference exists therefore across the surface layers (plasma membrane, cell wall and, where applicable, capsule and slime layer) of a micro-organism. It is the purpose of this review to evaluate the importance of this osmotic pressure difference in relation to the capacity of micro-organisms to grow and to survive, and to discuss the molecular basis of the responses of microbes to osmotic stress. Compared with other stresses to which microbial populations are subjected, and which form the subjects of other reviews in this volume, osmotic stress has a much less severe or adverse effect on members of the population. In general, microbes are endowed with a marked capacity to resist osmotic stress, and to recover from stresses that temporarily incapacitate organisms and prevent their multiplication.

OSMOSIS AND MICRO-ORGANISMS

Osmosis has been defined as the process whereby solvent spontaneously moves from one region in a solution where its activity is high to another region where its activity is lower (Morris, 1974). The process only takes place in practice when a selectively permeable membrane separates a solution either from pure solvent or from a solution with the same solvent but containing a different concentration of solutes.

The microbial plasma membrane is selectively permeable to solutes and, because of the higher concentration of solutes inside most micro-organisms compared with that in the environment, there is a spontaneous flow of water from the environment into the micro-organism. The major role of the strong non-deformable wall which encases the majority of micro-organisms is to limit expansion of the protoplast which results from the entry of water into the organism.

There are clearly two main factors with regard to osmosis across the surface layers of a micro-organism, namely the osmotic pressure of the contents of the organism, and that of the solution surrounding the microbe. The second section of this review describes the ways in which micro-organisms react to changes in the osmotic pressure of the environment, a subject on which there is no shortage of data.

However, information on the osmotic pressure of the contents of micro-organisms is by comparison very meagre. Two types of method have been used to measure the osmotic pressure of microbial contents. In the first method, micro-organisms are converted into protoplasts or sphaeroplasts, and the stability of these structures is assessed in solutions of different molarity. The osmotic pressure of the contents of the protoplast or sphaeroplast – and by inference that of the contents of the intact micro-organism – is taken to be equal to the osmotic pressure of the solution which just protects the structures from lysis. Using this method, Gebicki & James (1960) obtained values of $5–6 \cdot 10^5$ Pascals for the contents of *Aerobacter aerogenes*, while Hugo & Russell (1960) found values of $4–8 \cdot 10^5$ Pascals for the contents of *Escherichia coli*. Earlier, Mitchell & Moyle (1956) employed a similar technique which involved equilibrating populations of *Staphylococcus aureus* in solutions of different molarity, and assuming the osmotic pressure of the bacterial contents to be equal to that of the solution which just prevented plasmolysis of the bacteria (see p. 159). Applying this technique, they obtained a value of $20–25 \cdot 10^5$ Pascals for the contents of *S. aureus*.

The second method measures the osmotic pressure of microbial contents by determining the freezing point of the extruded contents. Two groups of workers have applied this cryoscopic technique to *S. aureus*, and have reported values of $20–30 \cdot 10^5$ Pascals (Mitchell & Moyle, 1956) and approximately $25 \cdot 10^5$ Pascals (Scott, 1953), which agree remarkably well. Conway & Armstrong (1961) extended the technique to *Saccharomyces cerevisiae*, and reported the contents of resting organisms to have a molarity of 0.59.

The meagre nature of data on the osmotic pressure of microbial contents reflects largely the conviction of many microbial physiologists that the value, although useful for assessing the overall response of a microbe to changes in the osmotic pressure of the surrounding solution, has very little physiological significance, since there must exist in a micro-organism numerous micro-loci each with a different concentration of dissolved solutes. The methods used to measure the value are also open to severe criticism. As indicated in the last section of this review, the capacity of the microbial plasma membrane to stretch is determined to some extent by its lipid composition, a finding which casts doubt on the validity of measuring the osmotic pressure of microbial contents by observing the behaviour of the plasma membrane.

MICROBIAL RESPONSES TO OSMOTIC STRESS

What constitutes an osmotic stress?

Before proceeding to discuss the manner in which micro-organisms respond to osmotic stress, it is pertinent to ask what constitutes an osmotic stress for a micro-organism. Many possible definitions can be arrived at, but probably the broadest and most applicable to the present review is that a micro-organism is osmotically stressed when the magnitude of the osmotic pressure difference across the surface layers of the organism is sufficiently low or high to retard the rate of growth of the organism and to affect its ability to reproduce.

The magnitude of an osmotic stress must be related to the osmotic pressure of the solution surrounding an organism, so that a stress can be quantified in this way. However, microbiologists have been interested mostly in the response of microbes to large osmotic stresses, mainly because of the applications which these have in preserving perishable materials, particularly foods. As a result, it has been found more convenient to express the magnitude of an osmotic stress imposed on a population of micro-organisms not by quoting the osmotic pressure of the solution in which the organisms are suspended but by referring to the *water activity* of that solution.

Water activity (a_w) is a measure of the amount of free water in a system. When a solute is dissolved in water, some of the water becomes bound to the solute, so that the amount of free water available is decreased. In thermodynamic terms, there is a decrease in entropy when a solute is dissolved in water. There is also a decrease in the vapour pressure, which can be described according to Raoult's Law:

$$P/P_0 = N_2/N_1+N_2$$

where P is the vapour pressure of the solution, P_0 the vapour pressure

Table 1. *Minimum water activity values for growth of micro-organisms*

Species	Minimum a_W value	Reference
Bacteria		
Aerobacter aerogenes	0.94	Christian & Scott (1953)
Bacillus cereus var. *mycoides*	0.99	Burcik (1950)
B. subtilis	0.90	Marshall, Ohye & Christian (1971)
Clostridium botulinum Type A	0.93	Marshall *et al.* (1971)
Lactobacillus viridescens	0.94	Wodzinski & Frazier (1961)
Microbacterium sp.	0.94	Brownlie (1966)
Micrococcus sp.	0.83	Marshall *et al.* (1971)
Pseudomonas fluorescens	0.97	Limsong & Frazier (1966)
Salmonella typhimurium	0.92	Clayson & Blood (1957)
Staphylococcus aureus	0.86	Clayson & Blood (1957)
S. albus	0.88	Marshall *et al.* (1971)
Vibrio costicolus	0.86	Kushner (1968)
Yeasts		
Candida pseudotropicalis	0.93	Battley & Bartlett (1966)
Endomyces vernalis	0.88	Burcik (1950)
Hansenula suaveolens	0.97	Battley & Bartlett (1966)
Saccharomyces cerevisiae	0.92	Onishi (1957)
S. rouxii	0.85	Onishi (1957)
Torula utilis	0.94	Burcik (1950)
Moulds		
Alternaria citri	0.84	Tomkins (1929)
Aspergillus glaucus	0.71	Scott (1957)
A. niger	0.84	Pitt & Christian (1968)
A. ruber	0.70	Snow (1949)
Botrytis cinerea	0.93	Snow (1949)
Mucor plumbeus	0.93	Snow (1949)
Penicillium martensii	0.79	Ayerst (1969)
Rhizopus nigricans	0.93	Snow (1949)
Xeromyces bisporus	0.61	Pitt & Christian (1968)

of water, N_1 the number of moles of solute, and N_2 the number of moles of solvent. The value P/P_0 is numerically equal to the value for the water activity of the solution, the water activity of pure water being 1.000. This is not the occasion on which to deal in depth with the concept of water activity except in so far as it provides a measure of the degree to which a micro-organism is stressed osmotically. The review by Scott (1957), who did much to develop the concept in relation to microbial activity, remains a rich source of information as does the more recent article by Corry (1973). The latter article, as well as the review by Smith (1971), describes ways in which water activity can be measured and controlled.

Microbial physiologists have been concerned mainly with the optimum and minimum a_w values for growth of micro-organisms, and sufficient data are now available to make certain generalizations concerning the responses of various groups of microbes to osmotic stress. Of main interest as far as survival of micro-organisms is concerned are the values for the minimum or limiting water activities, which are the lowest a_w values at which an organism is just capable of growth. A representative range of minimum a_w values for microbes is given in Table 1. In general, bacteria are less tolerant of low water activities than either yeasts or moulds, although exceptions are found among the Micrococcaceae, particularly the staphylococci. Micro-organisms which are capable of withstanding considerable osmotic stress have attracted the attention of physiologists for some years, none more so than the so-called osmophilic yeast, *Saccharomyces rouxii*, and more recently the even more resistant mould *Xeromyces bisporus*. These micro-organisms have acquired the ability to avoid being stressed osmotically compared with the majority of other micro-organisms; their physiological peculiarities are discussed later in this review.

Responses to hypertonic solutions

The responses of bacteria to osmotic stress were first observed some three quarters of a century ago by Fischer (1903). Pioneer workers on this subject borrowed the nomenclature used to describe the effects of solute concentration on the behaviour of plant cells, a phenomenon which had been studied fairly extensively in the latter part of the nineteenth century. Microscopic observations indicated that, when microbes are suspended in solutions that create an osmotic pressure difference across the surface layers of the microbe such that water tends to leave the organism, the plasma membrane in the microbe can, provided suitable microscopic techniques are used, be seen to separate from the inside of the cell wall. Unfortunately, this phenomenon, which following the practice of plant physiologists was termed *plasmolysis*, is somewhat poorly described in the microbiological literature. Moreover, it seems to have become a rather unpopular term. While one can regularly find reference to plasmolysis in many of the older basic microbiological texts, even when these have gone into several editions (Salle, 1961; Lamanna & Mallette, 1965), more recent texts refrain from using the term (Brock, 1970; Stanier, Doudoroff & Adelberg, 1970; Levy, Campbell & Blackburn, 1973).

Nevertheless, there have been spasmodic reports of plasmolysis in various micro-organisms, and of the effects of different conditions on the phenomenon, and these reports have led to certain generalizations concerning the process of plasmolysis. Not all microbes are affected to the same extent by a particular hypertonic solution, which is hardly

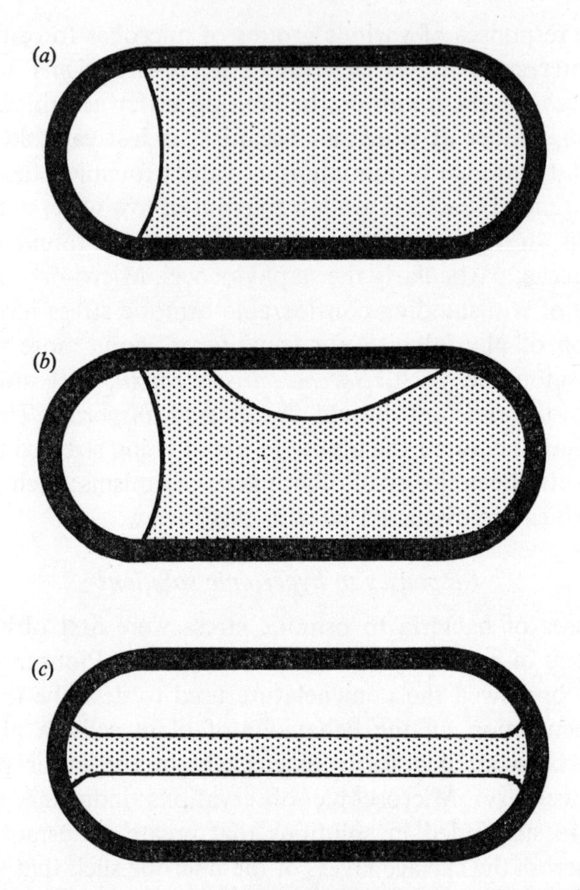

Fig. 1. Schematic representation of three different degrees of plasmolysis in a rod-shaped bacterium: (*a*) slight; (*b*) extensive; (*c*) severe. (After Scheie, 1969.)

surprising since the osmotic pressure difference created across the surface layers of a microbe will depend on the concentration of solutes inside the organism. Those organisms with a comparatively low concentration of intracellular solutes (Gram-negative bacteria in particular) are more susceptible to plasmolysis than organisms (Gram-positive bacteria, yeasts) containing a higher concentration of solutes (Hale, 1957). Moreover, although the plasma membrane becomes removed from the inside of the wall during plasmolysis, it nevertheless remains attached to the inside of the wall at certain points. On this aspect of plasmolysis, firm data are not plentiful, but it would seem that, after a particular hypertonic stress, contact of the plasma membrane with the inside of the wall is more extensive in Gram-positive bacteria and yeasts than in Gram-negative bacteria. Plasmolysis in *E. coli* has been studied by examining thin sections of plasmolysed organisms in the electron microscope (Cota-Robles, 1963; Scheie, 1969). Separation of the membrane from the inside of the wall generally starts at one pole of the bacterium and, as the magnitude of the stress is increased, takes place also in the median region of the cell and then extends to the other pole (Fig. 1). Unfortunately, electron microscope studies on microbial plasmolysis are few in number, and more studies in this area would considerably illuminate our understanding of the cytology of the process. It is worth noting, too, that students of the bacterial cell wall have largely ignored the possibility of parts of the plasma membrane being covalently linked to the inside of the wall. Conceivably, the points of attachment could, at least in Gram-positive bacteria, be where lipoteichoic acids are located in the wall, for there is evidence that these large amphipathic polymers may extend from the plasma membrane to the surface layers in bacteria (Wicken, Gibbens & Knox, 1973).

After plasmolysis has occurred, a gradual return of the plasma membrane to the inside of the wall – a process of *deplasmolysis* – may occur. The most plausible explanation of deplasmolysis is that, following the removal of water from the organism, there is a gradual increase in the concentration of solutes inside the organism as low molecular weight compounds are transported into the organism from the environment. Unlike water, these low molecular weight compounds pass across the plasma membrane by processes involving transport proteins, and the relative speed of these transport processes could explain the differences observed in the rates of deplasmolysis (Henneman & Umbreit, 1964; Record, Taylor & Miller, 1962; Scheie, 1969; Bayer, 1968). Interestingly, plasmolysis does not occur when microbes are in hypertonic solutions of glycerol (Robinow, 1960), a finding usually taken to indicate that glycerol passes across the microbial plasma membrane not by a transport protein-mediated process but by diffusion.

An important finding was that plasmolysis of *E. coli* leads to a loss of motility (Okrend & Doetsch, 1969). Bacterial flagella are attached to the plasma membrane by a basal structure (Smith & Koffler, 1971), which again focuses attention on the microbial plasma membrane as the organelle which is mainly responsible for the response of micro-organisms to osmotic stress.

There is evidence that many micro-organisms can adapt to an environment which nominally imposes an osmotic stress on the organism. Almost 70 years ago, Drabble, Drabble & Scott (1907) reported that the volumes of both bacteria and yeasts decrease when they are grown in media containing a hypertonic concentration of NaCl. More recent data from Lillehoj & Ottolenghi (1965) confirmed the effect for several species of *Saccharomyces*. These workers attributed the effect to an increase in the osmotic pressure of the cellular contents, but suggested that this arose not from uptake of solute from the medium but as a result of synthesis of low molecular weight compounds which are retained intracellularly.

Responses to hypotonic solutions

Although determinations have been made of the optimal a_w values for growth of micro-organisms (Scott, 1957; Corry, 1973), it is difficult to distinguish the stress that a micro-organism experiences following an increase in the water activity of the suspending solution above the optimal value from the stress that comes from being starved of essential nutrients. Micro-organisms differ tremendously in their capacity to remain viable when suspended in hypotonic non-nutrient solutions, which are often dilute buffers or just water. Survival depends on the environmental conditions employed (e.g. temperature, dissolved oxygen tension) and on the extent to which the organisms can call on endogenous reserve polymers as a source of energy. Organisms which contain these reserves are able to survive for much longer periods of time compared with those that do not or are not so generously endowed (Dawes & Senior, 1973; Dawes, this volume).

The importance of metabolic energy in maintaining the viability of micro-organisms placed under conditions of nutrient deficiency in hypotonic solutions has been placed on a quantitative basis by Atkinson and his colleagues. Adenylate charge, which is the value measured by these workers, is a linear measure of the amount of metabolically available energy stored in the adenylate pool in a micro-organism (Atkinson & Walton, 1967), and is numerically equal to the mole fraction of ATP plus half the mole fraction of ADP. Growing micro-organisms have an adenylate charge of 0.7–0.9 but, when placed in a hypotonic solution lacking a source of energy, the value for the adenylate charge begins to decline after a period of time which varies with different micro-organisms. In *E. coli* viability is seriously impaired when the value declines below about 0.5 (Chapman, Fall & Atkinson, 1971), although aerobically grown *S. cerevisiae* organisms remain viable even after the adenylate charge has declined to 0.15 or below (Ball & Atkinson, 1975). These findings represent a major contribution to our understanding of microbial survival in hypotonic solutions, since they allow the experimentor to predict quantitatively the likely capacity of starved organisms to multiply. At the same time they show

how very different are the responses of prokaryotic as compared with eukaryotic micro-organisms to the imposition of osmotic and energy stresses.

Cold osmotic shock

A number of different shock treatments are known to cause release from micro-organisms of intracellular constituents and to affect growth and viability of organisms. One of the first of these treatments to be examined in detail was referred to as cold shock, and involves the sudden chilling of a dilute suspension of exponential-phase bacteria in water or very dilute buffer, a process which leads to death of the majority of organisms in the population (Sherman & Albus, 1923; Hegarty & Weeks, 1940; Meynell, 1958; Gorrill & McNeil, 1960; Strange & Ness, 1963; Farrell & Rose, 1968). Cold shock has also been demonstrated in *Neurospora crassa* (Wiley, 1970) and *S. cerevisiae* (Patching & Rose, 1971*a*). The treatment involves imposition of an osmotic as well as a low-temperature stress on the organisms, although it is believed that the sudden chilling is primarily responsible for the decline in viability (see also MacLeod and Calcott, this volume).

A modification to the cold-shock procedure has been devised which embodies a much greater osmotic stress in addition to a low-temperature stress, and it has come to be known as *cold osmotic shock*. The technique was developed largely by Heppel, and involves the following series of operations. Vegetative micro-organisms are suspended in a hypertonic solution of a metabolically inert solute (such as 0.8 M sucrose with bacteria) containing 10 mM ethylenediaminetetracetic acid (EDTA). The organisms are immediately removed from the suspension and re-suspended in an ice-cold solution of $MgCl_2$ (10 mM). As a result of being subjected to this procedure, susceptible organisms release about 5 per cent of their total cell protein, which is made up of a mixture of enzymes and of proteins that are involved in transport processes. The effect on survival of the organisms is minimal in that organisms that have been subjected to cold osmotic shock, when incubated in complete medium, experience only a longer lag phase of growth compared with organisms which have not been so treated.

The enzymes which are released from micro-organisms after cold osmotic shock are believed to be located in a region between the outside of the plasma membrane and the inside of the cell wall, a region usually referred to as the *periplasm*. Among the first enzymes which were shown to be released from bacteria by cold osmotic shock was alkaline phosphomonoesterase (Brockman & Heppel, 1968), and others

Table 2. *Examples of periplasmic enzymes and solute-binding proteins released from bacteria by cold osmotic shock*

Periplasmic enzymes	Reference
Acid hexose phosphatase	Neu & Heppel (1964)
Alkaline phosphomonoesterase	Brockman & Heppel (1968)
Asparaginase II	Cedar & Schwartz (1967)
Deoxyriboaldolase	Munch-Petersen (1968)
Deoxyribonuclease I	Cordonnier & Bernardi (1965)
Cyclic phosphodiesterase	Neu & Heppel (1965)
5′-Nucleotidase	Neu & Heppel (1965)
Penicillinase	Datta & Richmond (1966)
Ribonuclease I	Neu & Heppel (1964)
Thymidine phosphorylase	Munch-Petersen (1967)
Solute-binding proteins	
L-arabinose-binding	Hogg & Englesberg (1969)
L-arginine-binding	Wilson & Holden (1969)
D-galactose-binding	Anraku (1968)
L-histidine-binding	Rosen & Vasington (1970)
Inorganic phosphate-binding	Medveczky & Rosenberg (1966)
Inorganic sulphate-binding	Pardee (1968)
L-leucine, L-valine and L-isoleucine-binding	Piperno & Oxender (1968)
L-phenylalanine-binding	Klein, Dahms & Boyer (1970)

are listed in Table 2. Further information on the release of periplasmic enzymes from bacteria by cold osmotic shock is given by Heppel (1967, 1969, 1971) and Simoni (1972). The technique is now widely exploited for isolating these enzymes, the physiological role of which would appear to be metabolism of solutes that cannot be transported directly into the bacteria.

Rather more attention has been given to the release by cold osmotic shock of proteins that have the capacity to bind with solutes. Kundig *et al.* (1966) were the first to show that submitting *E. coli* to the shock procedure caused the organisms to lose the ability to transport certain sugars and that proteins with affinities for the corresponding sugars were released into the dilute $MgCl_2$ solution. Since then numerous other reports have appeared showing that cold osmotic shock mainly of Gram-negative bacteria leads to release of solute-binding proteins, and examples of these are listed in Table 2. That the proteins which are released are involved in transport of solutes into the organisms was deduced from three lines of evidence. Firstly, submitting bacteria to cold osmotic shock leads to a sharp decrease in the ability of the organisms to transport solutes which bind to the released proteins with affinities (as judged by apparent K_m values) closely similar to the affinities which the solutes have for intact organisms. Secondly, transport-negative mutants of bacteria have been isolated which are unable to release the corresponding binding proteins. Finally, bacteria in which synthesis of proteins associated with active transport of certain sugars has been repressed are unable to release binding proteins when submitted to cold osmotic shock.

It is arguable whether or not the solute-binding proteins released by cold osmotic shock are bound to the plasma membrane and directly mediate transport of solutes across the membrane. Some physiologists argue that they are, and as such constitute the permeases of the classical solute-transport theory (Rickenberg *et al.*, 1956). They point out that proteins which mediate passage of solutes across the microbial plasma membrane would be expected to be loosely bound in the membrane. Others opine that the proteins released by cold osmotic shock are located on the outside of the plasma membrane, and that their physiological role is to trap solute molecules before these combine with the true transport molecules which lie tightly bound in the membrane.

The mechanism whereby solute-binding proteins are released from the bacterial plasma membrane after the organisms have been plasmolysed and then submitted to a rapid deplasmolysis in the cold remains a mystery. Conceivably, the proteins are extruded from the membrane when, following deplasmolysis at near-zero temperatures, the fatty-acyl chains of phospholipid molecules in the plasma membrane solidify and so lose their ability to retain the proteins. To test this hypothesis, Patching & Rose (1971*b*) grew *E. coli* at 15 °C instead of 37 °C in order to increase the degree of unsaturation in the fatty-acyl chains in membrane phospholipids (and hence lower their melting points), and examined the ability of bacteria grown at each temperature to release proteins that bind α-methylglucoside. No difference was detected, which suggests that freezing of the fatty-acyl chains in membrane phospholipids may be unimportant in cold osmotic shock. Bacteria grown at each of the two temperatures did, however, respond differently, when subjected to cold osmotic shock, to the presence of EDTA and to the osmotic stress, findings which are not easily explained in molecular terms. Further research on the mechanism of cold osmotic shock in bacteria would be welcome, not least because the procedure could be exploited more extensively in order to release plasma membrane-bound proteins from micro-organisms with a minimum of perturbation to the organism, and thereby a diminished possibility of denaturing the proteins.

AVOIDANCE OF OSMOTIC STRESSES

Some micro-organisms have acquired the ability to survive and grow in environments which have a very high osmotic pressure, or to reproduce in commonly encountered environments without, at the same time, being able to synthesize a cell wall to protect the plasma membrane from unrestricted expansion. These micro-organisms, because of their ability to grow in or under unusual conditions, have attracted a considerable amount of attention from microbial physiologists, arguably more than their ecological or economic importance justifies. Significantly, two of the three classes of such organisms have acquired the ability to grow under seemingly very stressful conditions, not as a result of alterations to the composition of the plasma membrane which might permit the stresses to be resisted, but as a result of other alterations to their physiological processes which cause the magnitude of the osmotic pressure difference across the plasma membrane to be minimized. In other words, these organisms have devised means of avoiding large osmotic stresses. The remainder of this section of the review deals with the three classes of organism, namely osmophiles, halophiles and wall-less organisms.

Osmophiles

The existence of micro-organisms which are able to thrive and multiply in environments with a high osmotic pressure or low water activity has been recognized for well over half a century (von Richter, 1912) and, as already noted, the property is confined to a small number of yeasts and moulds. The two groups of organisms have been referred to, respectively, as osmophilic yeasts and xerophilic moulds although neither epithet, in strictly physiological terms, is appropriate. Information on the behaviour of these organisms has come mainly as a result of their commercial importance, and is very largely confined to osmophilic yeasts and in particular strains of *Saccharomyces rouxii*. Yeasts capable of growing in concentrated sugar solutions have long been recognized as important contaminant organisms in industries which use sugar syrups or other types of concentrated sugar solutions (Scarr, 1951). The rather less extensively studied xerophilic moulds, usually *Xeromyces bisporus*, are found in extremely dry, sugary environments, especially on the surfaces of mummified fruits such as prunes (Pitt & Christian, 1968). The reviews by Scott (1957), Ingram (1957), Onishi (1963*a*) and Brown (1964) have described the earlier data reported on these organisms. The present account will concentrate on more recent data, and especially those which have illuminated the molecular basis of the osmophilic character in yeasts.

The fundamental problem when studying the physiology of osmophilic microbes is why they, unlike other micro-organisms, are not only able to survive but also to multiply rapidly in media which may have a water activity value as low as 0.65. An important breakthrough in research on osmophilic yeasts has recently come from Brown's laboratory in Australia. Following an initial survey of the growth patterns of a range of osmophilic yeasts, which included strains of *S. rouxii*, *S. mellis*, *S. rosei*, *S. acidifaciens*, *Zygosaccharomyces rugosus*, *Z. nectarophilus*, *Z. nussbaumeri* and *Torulopsis halonitratophila* (Anand & Brown, 1968), this group went on to show that several osmophilic strains of *S. rouxii* synthesize, and accumulate intracellularly, appreciable concentrations of arabitol, glycerol and several other unidentified polyols (Brown & Simpson, 1972; Brown, 1974). Non-osmophilic yeasts do not accumulate polyols. The concentration of arabitol in one strain of *S. rouxii* is approximately equivalent to 18.5 per cent of the dry weight of the yeast (about 1217 μmol g^{-1} dry weight). This discovery of polyol accumulation by osmophilic yeasts is of more than passing interest since these yeasts have for some time been known to be capable of secreting polyols, a phenomenon which had been exploited, on a pilot-plant scale, as a commercial source of these compounds (Spencer, 1968).

The conclusion which might be drawn from the discovery of intracellular polyol accumulation by osmophilic yeasts is that the accumulation of these hydrophilic compounds increases the concentration of bound water inside the yeast, and so decreases the osmotic pressure difference across the plasma membrane. This is a valid conclusion, and obviously vitiates von Richter's (1912) use of the term 'osmophile'. However, Brown and his colleagues have gone further and have stressed that the chemical nature of the accumulated solute is important. They developed the concept of a *compatible solute*, which can be defined as a solute which, in high concentration, allows an enzyme to function effectively (Brown & Simpson, 1972; Brown, 1974). The chemical nature of the compatible solute was shown to be important for the functioning of enzymes, including the NADP-specific isocitrate dehydrogenase from a strain of *S. rouxii*. These reports from Brown's laboratory have furnished the first insight into the physiological basis of osmophily in yeasts, and they set the scene for a wide-ranging study of the biochemical role of the compatible solutes synthesized by osmophilic yeasts.

Naturally occurring strains of osmophilic yeasts are facultative rather than obligate osmophiles under the majority of environmental conditions encountered. Onishi (1963*a*) did, however, report an obligate requirement for high concentrations of sugar by some yeasts at elevated temperatures, while temperature-conditional obligately osmophilic varieties of other strains of *S. rouxii* have been described by Lochhead & Heron (1929) and Spencer & Sallans (1956), and of *Torulopsis halonitratophila* by Onishi (1963*b*). With the aim of studying the physiological basis of the obligately psychrophilic habit, Koh (1975*a*) isolated an obligately osmophilic mutant of *S. rouxii* which is capable of growing only in media that contain at least 20 per cent (w/v) sucrose. In media containing 30 per cent (w/v) sucrose, the yeast grows in a filamentous form, and only when the sucrose concentration is raised to 60 per cent (w/v) does the yeast or single-celled form appear. Since the morphological form of a micro-organism is determined by the chemical composition of the wall, it must be deduced that the osmotic pressure of the medium in which this mutant is grown affects the activities of the plethora of cell wall-synthesizing enzymes located in the plasma membrane. A role for the composition of the plasma membrane in regulation of the morphological form of this mutant, when grown in media of different water activities, comes from the discovery (Koh, 1975*b*) that lipids from the obligately osmophilic mutant contain more unsaturated fatty-acyl residues than lipids from the facultatively osmophilic parent strain of *S. rouxii*.

Halophiles

Another group of micro-organisms which are able to grow in media with a low water activity are the extremely halophilic bacteria and algae. With their capacity to grow in the presence of almost saturated solutions of NaCl these organisms constitute a thoroughly intriguing group. With uncanny insight, Baas-Becking (1928) remarked that they seem to represent the 'borderland of physiological possibilities'. Since then, these organisms, and in particular the extremely halophilic bacteria, have attracted the attention of several groups of microbiologists, whose research has brought to light some fascinating physiological properties of these organisms. The microbiology of halophilic micro-organisms has been exhaustively reviewed by Larsen (1967, 1973) and Dundas (1976).

There are two quite separate groups of extremely halophilic bacteria. One is the halobacterium type, the organisms in this group all being assigned to the genus *Halobacterium*. The other is the halococcus group, members of which are grouped in the genus *Halococcus*. Extremely halophilic bacteria have the typical characteristics of pseudomonads, with the difference that they have a minimum requirement for growth of 2.5–3.0 M NaCl, and an optimum requirement of 4–5 M concentrations of this solute. These bacteria also require extremely high concentrations of Mg^{2+}, concentrations of the order of 0.1–0.5 M being necessary for optimal growth (Brown & Gibbons, 1955). Christian & Waltho (1962) were the first to show that extremely halophilic bacteria contain high concentrations of NaCl, a discovery which helped explain how these bacteria acquired the ability to grow in media of low water activity. It is clear too, following the concept developed by A. D. Brown and his colleagues (Brown & Simpson, 1972; Brown, 1974), that NaCl is acting as a compatible solute, since enzymes synthesized by these bacteria are able to catalyse reactions under conditions of salt concentration that would lead to denaturation of enzymes synthesized by non-halophilic bacteria. The compatible role for NaCl in extremely halophilic bacteria has now been demonstrated, e.g. by Holmes & Halvorson (1965), for well over 30 enzymes. It has, moreover, been shown that ribosomes in extremely halophilic bacteria are uniquely adapted to function in the presence of up to 4 M NaCl.

For the purposes of this review, the manner in which the envelope layers of extremely halophilic bacteria have adapted to deal with the high concentration of NaCl in which they are bathed is of more importance. In this respect, the two groups of extremely halophilic bacteria differ. The halococci have a relatively thick cell wall composed of a polysaccharide which contains residues of the relatively uncommon sugar gulosaminuronic acid (Reistad, 1975). This wall allows the bacteria to resist the large osmotic stress which is created across the envelope layers when the bacteria are placed in an environment with a relatively high water activity. Interestingly, it is not clear which wall polymers are responsible for the mechanical strength. Peptidoglycans of the structure found in the majority of bacteria would appear to be absent, since residues of amino acids and hexosamines account for only 7–15 per cent of the dry weight of the halococcal wall. The situation in the halobacteria is quite different, for they are completely devoid of a thick wall. Instead, the outside of the plasma membrane is covered with an acidic protein. Halobacteria are unable to withstand appreciable osmotic pressure differences across their envelope layers. Indeed, the relatively slight mechanical stress imposed by picking up a colony of these bacteria with a dry inoculum loop causes massive disruption of the bacteria.

Disappointingly, little has been reported on the composition of the plasma membrane in extremely halophilic bacteria. This dearth of information is particularly unfortunate, since it is known that extremely halophilic bacteria synthesize some very unusual lipids, and it would be very interesting to know how and to what extent these lipids are involved in the structure of the plasma membrane. Extreme halophiles have very little if any fatty-acyl residues in their lipids. Instead, the cell lipids contain predominantly di-*O*-alkyl analogues of phosphatidylglycerophosphate (Kates *et al.*, 1966). The major alcohol is dihydrophytyl alcohol, a compound which is unique to extremely halophilic bacteria.

The halophilic algae which have been studied – species of *Dunaliella* – avoid creating a large osmotic pressure difference across their envelope layers in a quite different manner from the extremely halophilic bacteria. There was for some years a difference of opinion among workers in this area as to whether *Dunaliella* species do or do not tolerate a high concentration of NaCl inside the plasma membrane. Trezzi, Galli & Bellini (1965) and Ginzburg (1969) were among workers who believed that, in this respect, the halophilic algae behaved similarly to the extremely halophilic bacteria. Others demurred (Johnson *et al.*, 1968), a view which is now accepted. The important finding which explained the ability of *Dunaliella* species to tolerate high concentrations of NaCl came first from Ben-Amotz & Avron (1973) and was later confirmed by Borowitzka & Brown (1974). Both of these groups of workers agree that Na^+ is effectively excluded from the contents of these algae. However, the organisms are able to synthesize large quantities of glycerol – a fact which has been known for some time (Craigie & McLachlan, 1964) – and the presence of high concentrations of this polyol inside the algae prevents formation of a large osmotic pressure difference across the envelope layers. Borowitzka & Brown (1974) worked with two strains of *Dunaliella* – *D. tertiolecta*, a marine isolate which is only moderately halophilic, and the more halophilic *D. viridis* – and they found that the amounts of glycerol retained intracellularly by these algae varied directly with the NaCl concentration in the environments. Thus, when *D. tertiolecta* was grown in a medium containing 1.36 M NaCl, it accumulated 1.4 mol kg^{-1} of glycerol; comparable data for *D. viridis* are an internal concentration of 4.4 mol kg^{-1} when grown in the presence of 4.25 M NaCl.

Brown & Simpson (1972) and Borowitzka & Brown (1974) have reported the compensatory nature of glycerol as an internal solute. The activities in crude preparations of two enzymes from *D. viridis* – glucose-6-phosphate dehydrogenase and the NADP-specific glycerol dehydrogenase – were found to be tolerant of concentrations of glycerol that are found in *D. viridis*. Comparable concentrations of NaCl and KCl, and of the non-electrolyte sucrose, caused appreciable inhibition of enzyme activity.

Wall-less micro-organisms

Another group of micro-organisms that avoid creating a large osmotic pressure difference across their surface layers are the wall-less mycoplasmas and L-forms. However, little has been reported on the content of dissolved solutes in these organisms, and the means whereby they avoid osmotic stress.

Recently, research in Guze's laboratory has thrown some light on the manner in which penicillin-induced L-forms of *Streptococcus faecalis*, which have been induced to grow in media lacking a stabilizing solute, avoid creating an osmotic pressure difference across their plasma membrane. In part, this is attributable to the accumulation of lower internal concentrations of Na^+ and K^+ (Montgomerie *et al.*, 1972), the mechanism for which has yet to be explained. The complexity of the adaptation is shown by the finding that L-forms of the bacterium grown in the absence of a stabilizing solute differ in fatty-acyl composition from those grown in media containing 0.5 M sucrose. Those grown in the absence of sucrose synthesize a higher proportion of $C_{18:1}$ acids and a lower proportion of C_{18}- and C_{19}-cyclopropane acids (Montgomerie, Kalmanson & Guze, 1973), which suggests that the composition of the plasma membrane is important in the process of osmoregulation in the L-forms.

RESPONSES OF THE MICROBIAL PLASMA MEMBRANE TO OSMOTIC STRESS

Fundamental to any understanding of the molecular basis of the response of micro-organisms to osmotic stress is the need to know how the nature of the lipids and proteins in the membrane influences the capacity of the membrane to contract (during plasmolysis) or to expand to a limit determined by the dimensions of the wall, while at the same time ensuring that the membrane continues to act as a continuous barrier round the protoplast. Few microbiologists can be unaware of the burgeoning literature on biomembranes that has trundled into libraries over the past decade, and they might be forgiven for assuming that much has been reported on the molecular basis of expansion and contraction in biological membranes. However, they would be wrong, for surprisingly few publications have appeared on this basic property of membranes.

Both the proteins and the lipids that occur in biological membranes are heterogeneous in composition. The proteins include transport proteins, proteins that catalyse synthesis of certain reactions that lead to cell-wall synthesis and, in prokaryotes, a variety of enzymes including those involved in genome replication and in electron transport. The most abundant class of lipids are the glycerophospholipids – usually referred to simply as phospholipids – and there are several different classes of these including phosphatidylcholine, phosphatidylethanolamine, phosphatidylserine, phosphatidylinositol, phosphatidylglycerol, and diphosphatidylglycerol. It has to be remembered, too, that each of these names is generic, and that many different phospholipids of each class can occur in a membrane, differing in the nature of the fatty-acyl groups in the molecule. In eukaryotic micro-organisms, the fatty-acyl chains are predominantly straight-chain saturated and mono-unsaturated residues, usually between 16 and 20 carbon atoms long (Erwin, 1973). Prokaryotic micro-organisms, however, have a greater variety of fatty-acyl residues in their membrane phospholipids; they include, in addition to those that occur in eukaryotic membranes, cyclopropane and branched-chain residues (Goldfine, 1972). Finally, in membranes of eukaryotic micro-organisms there are sterols, the nature of which varies to some extent in different organisms (Goodwin, 1974). It follows that the microbial plasma membrane contains a very wide range of different lipid molecules, and it is relevant to ask whether and to what extent the different lipids in membranes affect the ability of the membrane to contract and to stretch.

Since contraction and particularly stretching of a membrane will affect its stability, it is pertinent to turn to data which have been reported on the molecular basis of membrane stability. Many of these data have been obtained using lipid monolayers and artificial bi-layers including liposomes, and the generalization which has emerged from these studies is that stability is greatest in membranes in which the lipids interact most strongly with each other. Molecular interactions between lipids have been studied by measuring the mean molecular area occupied by a lipid molecule at an air–water interface as influenced by the presence of another type of lipid. The extent of interaction with phospholipids depends on many factors, including the length and degree of unsaturation in the fatty-acyl chains (Demel, Bruckdorfer & Van Deenen, 1972; Ghosh, Williams & Tinoco, 1973). Moreover, the stability of a phospholipid monolayer is considerably increased by the presence of sterol molecules, the magnitude of the effect depending on the structure of the sterol (Demel *et al.*, 1972; Ghosh & Tinoco, 1972).

The shortcomings in these studies are two-fold. Firstly, virtually no attention has been given to the phospholipids that occur in prokaryotic micro-organisms, so that the data are of little help in assessing the role of lipid composition in the stability of bacterial membranes. Secondly, because they have come from experiments using artificial lipid mixtures, the data do not permit an assessment to be made of the importance of proteins in stabilizing biomembranes. The only way round the second objection is to study plasma membranes from micro-organisms. The major problem here is that, with some notable exceptions, it is not possible to effect specific and, hopefully, stoicheiometric changes in the composition of the microbial plasma membrane, a facility which is absolutely essential if meaningful data are to be obtained on the role of lipid composition on membrane stability.

The most relevant data on the role of lipid composition on membrane stability in micro-organisms have come from experiments with organisms that are auxotrophic for one or more compounds that are incorporated into cell membranes. Among prokaryotes, there has been a very extensive exploitation of the fatty-acid and sterol requirements of mycoplasmas (Razin, 1973). A study by McElhaney, de Gier & Van Deenen (1970), using *Acholeplasma laidlawii* grown in the presence of different unsaturated fatty acids, compared the ability of organisms enriched in these different fatty-acyl residues to swell in solutions of glycerol at different temperatures. The rate of swelling was greater in mycoplasmas enriched with linoleic acid residues than with oleic acid residues, data which suggest that the introduction of additional unsaturation in the fatty-acyl chains causes the phospholipids to become less tightly packed, thereby decreasing the stability of the membrane.

Auxotrophy for membrane components in eukaryotic micro-organisms is not common, although mutants of *Saccharomyces cerevisiae* have been isolated that require either a saturated fatty acid (Schweizer & Bolling, 1970; Schweizer, Kühn & Castorph, 1971) or an unsaturated fatty acid (Resnick & Mortimer, 1966). Research in my laboratory has expoited the nutritional requirement in *S. cerevisiae* for a sterol and an unsaturated fatty acid induced by growing the organism under strictly anaerobic conditions (Andreasen & Stier, 1954). Both requirements are fairly non-specific (Light, Lennarz & Bloch, 1962; Proudlock *et al.*, 1968), which makes it possible to vary to an appreciable extent the sterol and fatty-acyl compositions of the lipids in the yeast plasma membrane.

The capacity of the yeast plasma membrane to stretch can be studied by examining the stability of sphaeroplasts in hypotonic solutions of a stabilizing solute such as mannitol or sorbitol. The importance of fatty-acyl unsaturation in conferring stability on the plasma membrane of *S. cerevisiae* NCYC 366 was studied by Alterthum & Rose (1973). This strain is one which is particularly susceptible to the action of a wall-dissolving β-glucanase enzyme derived from the basidiomycete QM 806. Sphaeroplasts were prepared from the yeast grown anaerobically in the presence of ergosterol, and oleic acid, linoleic acid or γ-linolenic acid. The degree of enrichment of the cell lipids with the fatty acid supplied in the medium ranged from 65 per cent of the total with yeast grown in the presence of oleic acid to 54 per cent in organisms grown in linolenic acid-containing medium. The stability of sphaeroplasts in hypertonic solutions of sorbitol, particularly in the concentration range 0.9–0.6 M, decreased as the proportion of unsaturation in the plasma membrane lipids was increased, suggesting that the increased motion of the unsaturated fatty acyl chains, a result of a much less close packing of the chains, restricts the capacity of the membrane to stretch.

While the findings reported by Alterthum & Rose (1973) were largely predictable from data reported on the role of lipid unsaturation on the area occupied by a phospholipid molecule at an air–water interface (Ghosh *et al.*, 1973), those obtained by Hossack & Rose (1976) were not. The latter workers grew *S. cerevisiae* NCYC 366 anaerobically in media containing oleic acid, and any one of several different sterols. These sterols were of two main types. The first type has a fully saturated side chain at C-17 in the sterol molecule and included cholesterol and sitosterol. In the second class of sterols, the side chain at C-17 contains a double bond at C-22, and these included ergosterol and stigmasterol. Organisms grown in the presence of any one of these sterols were enriched to the extent of 75–90 per cent with the exogenously supplied sterol. When sphaeroplasts were prepared from these organisms and subjected to osmotic lysis, greater stability, again over the concentration range 0.9–0.6 M sorbitol, was found with sphaeroplasts that had plasma membranes enriched in sterols that had unsaturation in the side chain (Fig. 2). The sterol side chain is thought to lie in the membrane juxtaposed to the terminal (C-10 onwards) part of the fatty-acyl chains in phospholipids, and it would seem that a decreased interaction of the unsaturated sterol side chain, as compared with a saturated chain, with the fatty-acyl chain enhances the capacity of the membrane to expand.

The osmotic stress imposed on sphaeroplasts by dilution into hypotonic solutions is drastic, and the ensuing damage to the plasma mem-

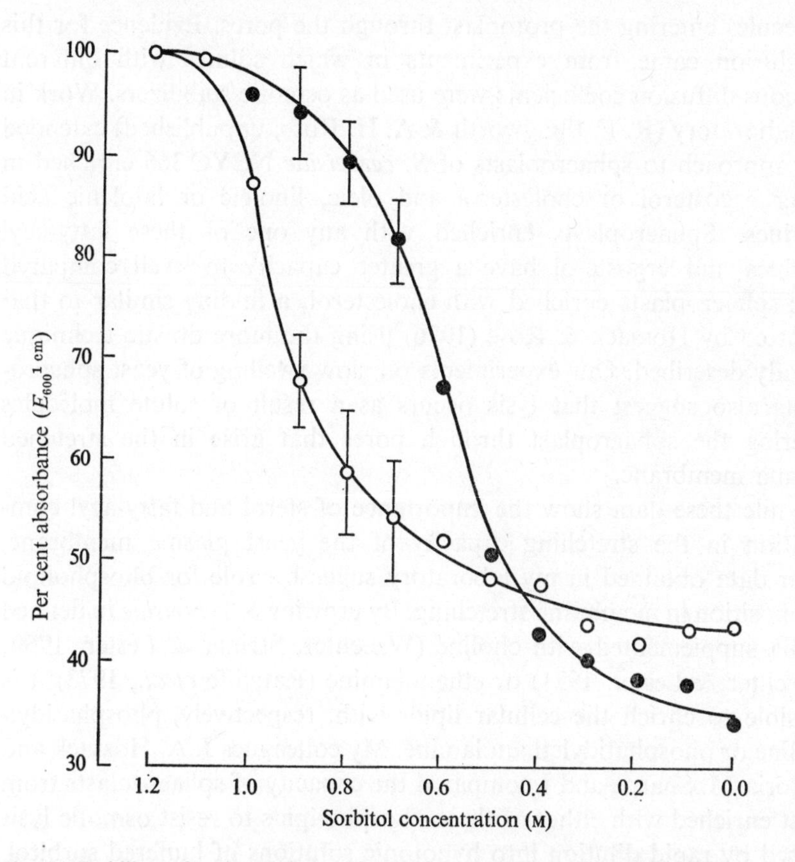

Fig. 2. Stability of sphaeroplasts of *Saccharomyces cerevisiae* NCYC 366 enriched in ergosterol (●) or cholesterol (○) when suspended in hypotonic solutions of buffered sorbitol. Portions (0.1 ml) of sphaeroplast suspension (containing 10 mg (dry wt equiv) ml^{-1} buffered 1.2 M sorbitol) were added to 2.9 ml portions of buffer containing concentrations of sorbitol ranging from 1.2 M to 0.0 M. The contents of each tube were then shaken gently, maintained at room temperature (18–22 °C) for 10 min, and the absorbance ($E_{600}{}^{1\,cm}$) of the suspension measured. Each value plotted is the average of at least three independent observations. The vertical bars indicate 95 per cent confidence limits on values. (From Hossack & Rose, 1976).

brane might be compared to brittle fracture of metals. A more gentle assessment of the capacity of protoplasts (and sphaeroplasts) to expand in volume was first used by Corner & Marquis (1969), and involves a slow addition of buffer to a suspension of protoplasts in a buffered solution of stabilizing solute, with periodic measurements of the average volume of the protoplasts. Using this technique with protoplasts of *Bacillus megaterium*, Corner & Marquis (1969) concluded that osmotic lysis of protoplasts occurs as a result of pores being formed in the bacterial plasma membrane as it stretches, and solute molecules entering the protoplast through the pores. Evidence for this conclusion came from experiments in which solutes with different aqueous diffusion coefficients were used as osmotic stabilizers. Work in my laboratory (R. F. Illingworth & A. H. Rose, unpublished) extended this approach to sphaeroplasts of *S. cerevisiae* NCYC 366 enriched in either ergosterol or cholesterol and oleic, linoleic or linolenic acid residues. Sphaeroplasts enriched with any one of these fatty-acyl residues and ergosterol have a greater capacity to swell compared with sphaeroplasts enriched with cholesterol, a finding similar to that reported by Hossack & Rose (1976) using the more drastic technique already described. Our experiments on slow swelling of yeast sphaeroplasts also suggest that lysis occurs as a result of solute molecules entering the sphaeroplast through pores that arise in the stretched plasma membrane.

While these data show the importance of sterol and fatty-acyl composition in the stretching capacity of the yeast plasma membrane, other data obtained in my laboratory suggest a role for phospholipid composition in membrane stretching. By growing *S. cerevisiae* in defined media supplemented with choline (Waechter, Steiner & Lester, 1969; Waechter & Lester, 1971) or ethanolamine (Ratcliffe *et al.*, 1973) it is possible to enrich the cellular lipids with, respectively, phosphatidylcholine or phosphatidylethanolamine. My colleagues J. A. Hossack and Victoria M. Sharpe and I compared the capacity of sphaeroplasts from yeast enriched with either of these phospholipids to resist osmotic lysis caused by rapid dilution into hypotonic solutions of buffered sorbitol. Those enriched in phosphatidylcholine were much more susceptible to osmotic lysis, again over the concentration range 0.9–0.6 M, compared with sphaeroplasts enriched with phosphatidylethanolamine. Phillips, Finer & Hauser (1972) showed that the polar headgroups in phosphatidylcholine and phosphatidylethanolamine are differently oriented in membranes, with the result that phosphatidylethanolamines, but not phosphatidylcholine, are difficult to hydrate. It is conceivable that the differences found in the stretching properties of yeast plasma membranes enriched in either of these phospholipids can be explained by these hydration effects.

I am deeply grateful to the many colleagues who have worked in my laboratory over the past seven years, and who have helped to formulate the views expressed in this review on the molecular basis of stretching of the microbial plasma membrane. Research carried out in my laboratory was financed by research grants B/SR/5724, B/RG/182 and B/SR/70610 from the Science Research Council.

REFERENCES

ALTERTHUM, F. & ROSE, A. H. (1973). Osmotic lysis of sphaeroplasts from *Saccharomyces cerevisiae* grown anaerobically in media containing different unsaturated fatty acids. *Journal of General Microbiology*, **77**, 371–382.

ANAND, J. C. & BROWN, A. D. (1968). Growth rate patterns of the so-called osmophilic and non-osmophilic yeasts in solutions of polyethylene glycol. *Journal of General Microbiology*, **52**, 205–212.

ANDREASEN, A. A. & STIER, T. J. B. (1954). Anaerobic nutrition of *Saccharomyces cerevisiae*. II. Unsaturated fatty acid requirement for growth in a defined medium. *Journal of Cellular and Comparative Physiology*, **43**, 27–281.

ANRAKU, Y. (1968). Transport of sugars and amino acids in bacteria. I. Purification and specificity of the galactoside- and leucine-binding proteins. *Journal of Biological Chemistry*, **243**, 3116–3122.

ATKINSON, D. E. & WALTON, G. M. (1967). Adenosine triphosphate conservation in metabolic regulation: rat liver citrate cleavage enzyme. *Journal of Biological Chemistry*, **242**, 3239–3241.

AYERST, G. (1969). The effects of moisture and temperature on growth and spore germination in some fungi. *Journal of Stored Products Research*, **5**, 127–141.

BAAS-BECKING, L. G. M. (1928). On organisms living in concentrated brine. *Tijdschrift der Nederlandsche dierkundige vereening*, Series III, **1**, 6–9.

BALL, W. J. & ATKINSON, D. E. (1975). Adenylate charge in *Saccharomyces cerevisiae* during starvation. *Journal of Bacteriology*, **121**, 975–982.

BATTLEY, E. H. & BARTLETT, E. J. (1966). A salt-concentration gradient method for the determination of the maximum salt concentration for microbial growth. *Antonie van Leeuwenhoek*, **32**, 256–260.

BAYER, M. E. (1968). Areas of adhesion between wall and membrane of *Escherichia coli*. *Journal of General Microbiology*, **53**, 395–404.

BEN-AMOTZ, A. & AVRON, M. (1973). The role of glycerol in the osmotic regulation of the halophilic alga *Dunaliella parva*. *Plant Physiology, Lancaster*, **51**, 875–878.

BOROWITZKA, L. J. & BROWN, A. D. (1974). The salt relations of marine and halophilic species of the unicellular alga, *Dunaliella*. The role of glycerol as a compatible solute. *Archives of Microbiology*, **96**, 37–52.

BROCK, T. D. (1970). *Biology of Micro-organisms*. Englewood Cliffs, N.J.: Prentice Hall Inc.

BROCKMAN, R. & HEPPEL, L. A. (1968). On the localization of alkaline phosphatase and cyclic phosphodiesterase in *Escherichia coli*. *Biochemistry, New York*, **7**, 2554–2562.

BROWN, A. D. (1964). Aspects of bacterial response to the ionic environment. *Bacteriological Reviews*, **28**, 296–329.

BROWN, A. D. (1974). Microbial water relations: features of the intracellular composition of sugar-tolerant yeasts. *Journal of Bacteriology*, **118**, 769–777.

BROWN, A. D. & SIMPSON, J. R. (1972). Water relations of sugar-tolerant yeasts: the role of intracellular polyols. *Journal of General Microbiology*, **72**, 589–591.

BROWN, H. J. & GIBBONS, N. E. (1955). The effect of magnesium, potassium and iron on the growth and morphology of red halophilic bacteria. *Canadian Journal of Microbiology*, **1**, 267–270.

BROWNLIE, L. E. (1966). Effect of some environmental factors on psychrophilic microbacteria. *Journal of Applied Bacteriology*, **29**, 447–454.

BURCIK, W. E. (1950). Über die Beziehungen zwischen Hydratur und Wachstum bei Bakterien und Hefen. *Archiv für Mikrobiologie*, **15**, 203–235.

CEDAR, H. & SCHWARTZ, J. H. (1967). Localization of two L-asparaginases in anaerobically grown *Escherichia coli. Journal of Biological Chemistry*, **242**, 3753–3755.

CHAPMAN, A. G., FALL, L. & ATKINSON, D. E. (1971). Adenylate energy charge in *Escherichia coli* during growth and starvation. *Journal of Bacteriology*, **108**, 1072–1086.

CHRISTIAN, J. H. B. & INGRAM, M. (1959). The freezing points of bacterial cells in relation to halophilism. *Journal of General Microbiology*, **20**, 27–31.

CHRISTIAN, J. H. B. & SCOTT, W. J. (1953). Water relations of salmonellae at 30 °C. *Australian Journal of Biological Sciences*, **6**, 565–573.

CHRISTIAN, J. H. B. & WALTHO, J. A. (1962). Solute concentrations within cells of halophilic and non-halophilic bacteria. *Biochimica et Biophysica Acta*, **65**, 506–508.

CLAYSON, D. H. F. & BLOOD, R. M. (1957). Food perishability: the determination of the vulnerability of food surfaces to bacterial infection. *Journal of the Science of Food and Agriculture*, **8**, 404–414.

CONWAY, E. J. & ARMSTRONG, W. McD. (1961). The total intracellular concentration of solutes in yeast and other plant cells and the distensibility of the plant-cell wall. *Biochemical Journal*, **81**, 631–639.

CORDONNIER, C. & BERNARDI, G. (1965). Localization of *E. coli* endonuclease I. *Biochemical and Biophysical Research Communications*, **20**, 555–559.

CORNER, T. R. & MARQUIS, R. E. (1969). Why do bacterial protoplasts burst in hypotonic solution? *Biochimica et Biophysica Acta*, **183**, 544–558.

CORRY, J. E. L. (1973). The water relations and heat resistance of micro-organisms. *Progress in Industrial Microbiology*, **12**, 73–108.

COTA-ROBLES, E. H. (1963). Electron microscopy of plasmolysis in *Escherichia coli. Journal of Bacteriology*, **85**, 499–503.

CRAIGIE, J. S. & McLACHLAN, J. (1964). Glycerol as a photosynthetic product in *Dunaliella tertiolecta. Canadian Journal of Botany*, **42**, 777–778.

DATTA, N. & RICHMOND, M. H. (1966). The purification and properties of a penicillinase whose synthesis is mediated by an R factor in *Escherichia coli. Biochemical Journal*, **98**, 204–209.

DAWES, E. A. & SENIOR, P. J. (1973). The role and regulation of energy reserve polymers in micro-organisms. *Advances in Microbial Physiology*, **10**, 135–266.

DEMEL, R. A., BRUCKDORFER, K. R. & VAN DEENEN, L. L. M. (1972). Structural requirements of sterols for the interaction with lecithin at the air–water interface. *Biochimica et Biophysica Acta*, **255**, 311–320.

DRABBLE, E., DRABBLE, H. & SCOTT, D. G. (1907). On the size of cells of *Pleurococcus* and *Saccharomyces* in solutions of a neutral salt. *Biochemical Journal*, **22**, 221–229.

DUNDAS, I. D. (1976). The physiology of extremely halophilic bacteria. *Advances in Microbial Physiology*, **14**, (in press).

ERWIN, J. (1973). Comparative biochemistry of fatty acids in eukaryotic microorganisms. In *Lipids and Membranes of Eukaryotic Micro-organisms*, ed. J. A. Erwin, pp. 41–143. New York: Academic Press.

FARRELL, J. & ROSE, A. H. (1968). Cold shock in mesophilic and psychrophilic pseudomonads. *Journal of General Microbiology*, **50**, 429–439.

FISCHER, A. (1903). *Vorlesungen Über Bakterien*, 2nd edition, p. 20. Jena: Fischer Verlag.

GEBICKI, J. M. & JAMES, A. M. (1960). The preparation and properties of spheroplasts of *Aerobacter aerogenes. Journal of General Microbiology*, **23**, 9–18.

GHOSH, D. & TINOCO, J. (1972). Monolayer interactions of individual lecithins with natural sterols. *Biochimica et Biophysica Acta*, **266**, 41–49.

GHOSH, D., WILLIAMS, M. A. & TINOCO, J. (1973). The influence of lecithin structure on thin monolayer behaviour and interactions with cholesterol. *Biochimica et Biophysica Acta*, **291**, 351–362.

GINZBURG, M. (1969). The unusual membrane permeability of two halophilic unicellular organisms. *Biochimica et Biophysica Acta*, **173**, 370–376.

GOLDFINE, H. (1972). Comparative aspects of bacterial lipids. *Advances in Microbial Physiology*, **8**, 1–58.

GOODWIN, T. W. (1974). Comparative biochemistry of sterols in eukaryotic micro-organisms. In *Lipids and Biomembranes of Eukaryotic Micro-organisms*, J. A. Erwin, pp. 1–40. New York: Academic Press.

GORRILL, R. H. & MCNEIL, E. M. (1960). The effect of cold diluent on the viable count of *Pseudomonas pyocyanea*. *Journal of General Microbiology*, **22**, 437–442.

HALE, C. M. F. (1957). A note on the relationship between the Gram reaction and plasmolytic effects in bacteria. *Experimental Cell Research*, **12**, 657–659.

HEGARTY, C. P. & WEEKS, O. B. (1940). Sensitivity of *Escherichia coli* to cold shock during the logarithmic growth phase. *Journal of Bacteriology*, **39**, 475–484.

HENNEMAN, D. H. & UMBREIT, W. W. (1964). Factors which modify the effect of sodium and potassium on bacterial cell membranes. *Journal of Bacteriology*, **87**, 1266–1273.

HEPPEL, L. A. (1967). Selective release of enzymes from bacteria. *Science, Washington*, **156**, 1451–1455.

HEPPEL, L. A. (1969). The effect of osmotic shock on release of bacterial proteins and on active transport. *Journal of General Physiology*, **54**, 95S–113S.

HEPPEL, L. A. (1971). The concept of periplasmic enzymes. In *Structure and Function of Biological Mebranes*, ed. L. I. Rothfield, pp. 223–247. New York: Academic Press.

HOGG, R. W. & ENGLESBERG, E. (1969). L-Arabinose binding protein from *Escherichia coli* B/r. *Journal of Bacteriology*, **100**, 423–432.

HOLMES, P. K. & HALVORSON, H. O. (1965). Purification of a salt-requiring enzyme from an obligately halophilic bacterium. *Journal of Bacteriology*, **90**, 312–315.

HOSSACK, J. A. & ROSE, A. H. (1976). Fragility of plasma membranes in *Saccharomyces cerevisiae* enriched with different sterols. *Biochimica et Biophysica Acta* (in press).

HUGO, W. B. & RUSSELL, A. D. (1960). Quantitative aspects of penicillin action on *Escherichia coli* in hypertonic medium. *Journal of Bacteriology*, **80**, 436–440.

INGRAM, M. (1957). Micro-organisms resisting high concentration of sugars or salts. *Symposia of the Society for General Microbiology*, **7**, 90–133.

JOHNSON, M. K., JOHNSON, E. J., MCELROY, R. D., SPEER, H. L. & BRAFF, B. S. (1968). Effects of salt on the halophilic alga *Dunaliella viridis*. *Journal of Bacteriology*, **95**, 1461–1468.

KATES, M., PALAMETA, B., JOO, C. N., KUSHNER, D. J. & GIBBONS, N. E. (1966). Aliphatic diether analogs of glyceride-derived lipids. IV. The occurrence of di-*O*-dihydrophytylglycerol ether containing lipids in extremely halophilic bacteria. *Biochemistry, New York*, **5**, 4092–4099.

KLEIN, W. L., DAHMS, A. S. & BOYER, P. D. (1970). The nature of the coupling of oxidative energy to amino acid transport. *Federation Proceedings. Federation of American Societies for Experimental Biology*, **29**, 341.

KOH, T. Y. (1975*a*). The isolation of obligate osmophilic mutants of the yeast *Saccharomyces rouxii*. *Journal of General Microbiology*, **88**, 184–188.

KOH, T. Y. (1975*b*). Studies on the 'osmophilic' yeast *Saccharomyces rouxii* and an obligate osmophilic mutant. *Journal of General Microbiology*, **88**, 101–114.

KUNDIG, W., KUNDIG, F. D., ANDERSON, B. & ROSEMAN, S. (1966). Restoration of active transport of glycosides in *Escherichia coli* by a component of a phosphotransferase system. *Journal of Biological Chemistry*, **241**, 3243–3246.

KUSHNER, D. J. (1968). Halophilic bacteria. *Advances in Applied Microbiology*, **10**, 73–99.

LAMANNA, C. & MALLETTE, M. F. (1965). *Basic Bacteriology: Its Biology and Chemical Background.* Baltimore: Williams & Wilkins Co.

LARSEN, H. (1967). Biochemical aspects of extreme halophilism. *Advances in Microbial Physiology*, **1**, 97–132.

LARSEN, H. (1973). The halobacteria's confusion to biology. *Antonie van Leeuwenhoek*, **39**, 383–396.

LEVY, J., CAMPBELL, J. J. R. & BLACKBURN, T. H. (1973). *Introductory Microbiology.* New York: John Wiley & Sons.

LIGHT, R. J., LENNARZ, W. J. & BLOCH, K. (1962). The metabolism of hydroxystearic acids in yeast. *Journal of Biological Chemistry*, **237**, 1793–1800.

LILLEHOJ, E. B. & OTTOLENGHI, P. (1965). Osmotic effects on yeast cells and protoplasts. *Symposium on Yeast Protoplasts*, pp. 145–512. Jena.

LIMSONG, S. & FRAZIER, W. C. (1966). Adaptation of *Pseudomonas fluorescens* to low levels of water activity produced by different solutes. *Applied Microbiology*, **14**, 899–901.

LOCHHEAD, A. G. & HERON, D. A. (1929). *Microbiological studies of honey. Bulletin of the Department of Agriculture, Canada*, **16** NS.

MARSHALL, B. J., OHYE, D. F. & CHRISTIAN, J. H. B. (1971). Tolerance of bacteria to high concentrations of NaCl and glycerol in the growth medium. *Applied Microbiology*, **21**, 363–364.

MCELHANEY, R. N., DE GIER, J. & VAN DEENEN, L. L. M. (1970). The effect of alterations in fatty acid composition and cholesterol content on the permeability of *Mycoplasma laidlawii* cells and derived liposomes. *Biochimica et Biophysica Acta*, **219**, 245–247.

MEDVECZKY, N. & ROSENBERG, H. (1970). The phosphate-binding protein of *Escherichia coli. Biochimica et Biophysica Acta*, **211**, 158–168.

MEYNELL, G. G. (1958). The sudden effect of chilling on *Escherichia coli. Journal of General Microbiology*, **19**, 380–389.

MITCHELL, P. D. & MOYLE, J. (1956). Osmotic structure and function in bacteria. *Symposium of the Society for General Microbiology*, **6**, 150–180.

MONTGOMERIE, J. Z., KALMANSON, G. M. & GUZE, L. B. (1973). Fatty acid composition of L-forms of *Streptococcus faecalis* cultured at different osmolalities. *Journal of Bacteriology*, **115**, 73–75.

MONTGOMERIE, J. Z., KALMANSON, G. M., HUBERT, E. H. & GUZE, L. B. (1972). Osmotic stability and sodium and potassium content of L-forms of *Streptococcus faecalis. Journal of Bacteriology*, **110**, 624–627.

MORRIS, J. G. (1974). *A Biologist's Physical Chemistry.* London: Edward Arnold.

MUNCH-PETERSEN, A. (1967). Thymidine breakdown and thymine uptake in different mutants of *Escherichia coli. Biochimica et Biophysica Acta*, **142**, 228–237.

MUNCH-PETERSEN, A. (1968). On the catabolism of deoxyribonucleosides in cells and cell extracts of *Escherichia coli. European Journal of Biochemistry*, **6**, 432–442.

NEU, H. C. & HEPPEL, L. A. (1964). The release of ribonuclease into the medium when *Escherichia coli* cells are converted to spheroplasts. *Journal of Biological Chemistry*, **239**, 3893–3900.

NEU, H. C. & HEPPEL, L. A. (1965). The release of enzymes from *Escherichia coli* by osmotic shock and during the formation of spheroplasts. *Journal of Biological Chemistry*, **240**, 3685–3692.

OKREND, A. G. & DOETSCH, R. N. (1969). Plasmolysis and bacterial motility; a method for the study of membrane function. *Archiv für Mikrobiologie*, **69**, 69–78.

ONISHI, H. (1957). Studies on osmophilic yeasts. I. Salt tolerance and sugar tolerance of osmophilic soy-yeasts. *Bulletin of the Agricultural Chemical Society of Japan*, **21**, 137–142.

ONISHI, H. (1963*a*). Osmophilic yeasts. *Advances in Food Research*, **12**, 53–94.

ONISHI, H. (1963*b*). Studies on osmophilic yeasts. XV. Effects of high concentrations of sodium chloride on polyalcohol production. *Agricultural Biological Chemistry*, **27**, 543–547.

PARDEE, A. B. (1968). Membrane transport proteins. *Science, Washington*, **162**, 632–637.

PATCHING, J. W. & ROSE, A. H. (1971*a*). Cold osmotic shock in *Saccharomyces cerevisiae*. *Journal of Bacteriology*, **108**, 451–458.

PATCHING, J. W. & ROSE, A. H. (1971*b*). Effect of growth temperature on cold osmotic shock in *Escherichia coli* ML 30. *Journal of General Microbiology*, **69**, 429–432.

PHILLIPS, M. C., FINER, E. G. & HAUSER, H. (1972). Differences between conformations of lecithin and phosphatidylethanolamine polar groups and their effects on interactions of phospholipid bilayer membranes. *Biochimica et Biophysica Acta*, **290**, 397–402.

PIPERNO, J. R. & OXENDER, D. L. (1968). Amino acid transport systems in *Escherichia coli* K12. H. *Journal of Biological Chemistry*, **243**, 5914–5920.

PITT, J. I. & CHRISTIAN, J. H. B. (1968). Water relations of xerophilic fungi isolated from prunes. *Applied Microbiology*, **16**, 1853–1858.

PROUDLOCK, J. W., WHEELDON, L. W., JOLLOW, D. J. & LINNANE, A. W. (1968). Role of sterols in *Saccharomyces cerevisiae*. *Biochimica et Biophysica Acta*, **152**, 434–437.

RATCLIFFE, S. J., HOSSACK, J. A., WHEELER, G. E. & ROSE, A. H. (1973). Modifications to the phospholipid composition of *Saccharomyces cerevisiae* induced by exogenous ethanolamine. *Journal of General Microbiology*, **76**, 445–449.

RAZIN, S. (1973). Physiology of mycoplasmas. *Advances in Microbial Physiology*, **10**, 1–80.

RECORD, B. R., TAYLOR, R. & MILLER, D. S. (1962). The survival of *Escherichia coli* on drying and rehydration. *Journal of General Microbiology*, **28**, 585–598.

RESNICK, M. A. & MORTIMER, R. K. (1966). Unsaturated fatty acid mutants of *Saccharomyces cerevisiae*. *Journal of Bacteriology*, **92**, 597–600.

REISTAD, R. (1975). Amino sugar and amino acid constituents of the cell walls of extremely halophilic cocci. *Archives of Microbiology*, **102**, 71–73.

RICKENBERG, H. V., COHEN, G. N., BUTTIN, G. & MONOD, J. (1956). La galactoside-permease d'*Escherichia coli*. *Annales de l'Institut Pasteur, Paris*, **91**, 829–857.

RICHTER, A. A. VON (1912). Über einen Osmophilien Organismus, den Hefepilz, *Zygosaccharomyces mellis acidi* sp. n. *Mykologia Zentralblatt*, **1**, 67–80.

ROBINOW, C. F. (1960). Outline of the visible organisation of bacteria. In *The Cell*, vol. VI, ed. J. Brachet & A. E. Mirsky, pp. 45–103. New York: Academic Press.

ROSEN, B. P. & VASINGTON, F. D. (1970). Relationship of the histidine binding protein and the histidine permease system in *S. typhimurium*. *Federation Proceedings, Federation of American Biological Societies*, **29**, 545.

SALLE, A. J. (1961). *Fundamental Principles of Bacteriology*. New York: McGraw-Hill Inc.

SCARR, M. P. (1951). Osmophilic yeasts in raw beet and cane sugars and intermediate sugar-refining products. *Journal of General Microbiology*, **5**, 704–713.

SCHEIE, P. O. (1969). Plasmolysis of *Escherichia coli* B/r with sucrose. *Journal of Bacteriology*, **98**, 335–340.

SCHWEIZER, E. & BOLLING, H. (1970). A *Saccharomyces cerevisiae* mutant defective in saturated fatty acid biosynthesis. *Proceedings of the National Academy of Sciences, USA*, **67**, 660–666.

SCHWEIZER, E., KÜHN, L. & CASTORPH, H. (1971). A new gene cluster in yeast: the fatty acid synthetase system. *Hoppe Seyler's Zeitschrift für Physiologische Chemie*, **352**, 377–384.

SCOTT, W. J. (1953). Water relations of *Staphylococcus aureus* at 30 °C. *Australian Journal of Biological Sciences*, **6**, 549–564.

SCOTT, W. J. (1957). Water relations of food spoilage micro-organisms. *Advances in Food Research*, **7**, 83–127.

SHERMAN, J. M. & ALBUS, W. R. (1923). Physiological youth in bacteria. *Journal of Bacteriology*, **8**, 127–139.

SIMONI, R. D. (1972). Macromolecular characterisation of bacterial transport systems. In *Membrane Molecular Biology*, ed. C. F. Fox & A. Keith, pp. 289–322. Stamford, Ct: Sinauer Associates Inc.

SMITH, P. R. (1971). The determination of equilibrium relative humidity on water activity in foods – a literature survey. *British Food Manufacturing Industries Research Association. Scientific and Technical Survey*, **70**.

SMITH, R. W. & KOFFLER, H. (1971). Bacterial flagella. *Advances in Microbial Physiology*, **6**, 219–339.

SNOW, D. (1949). The germination of mould spores at controlled humidities. *Annals of Applied Biology*, **36**, 1–13.

SPENCER, J. F. T. (1968). Production of polyhydric alcohols by yeasts. *Progress in Industrial Microbiology*, **7**, 1–42.

SPENCER, J. F. T. & SALLANS, H. R. (1956). Production of polyhydric alcohols by osmophilic yeasts. *Canadian Journal of Microbiology*, **2**, 72–79.

STANIER, R. Y., DOUDOROFF, M. & ADELBERG, E. A. (1970). *The Microbial World.* Englewood Cliffs, N.J.: Prentice-Hall Inc.

STRANGE, R. E. & NESS, A. G. (1963). Effect of chilling on bacteria in aqueous suspension. *Nature, London*, **197**, 819.

TOMKINS, R. G. (1929). Studies of the growth of moulds. I. *Proceedings of the Royal Society of London, Series B*, **105**, 375–402.

TREZZI, M., GALLI, M. & BELLINI, E. (1965). The resistance of *Dunaliella salina* to osmotic stresses. *Giornale Botanica*, **72**, 255–263.

WAECHTER, C. J. & LESTER, R. L. (1971). Regulation of phosphatidylcholine biosynthesis in *Saccharomyces cerevisiae*. *Journal of Bacteriology*, **105**, 837–843.

WAECHTER, C. J., STEINER, M. R. & LESTER, R. L. (1969). Regulation of phosphatidylcholine biosynthesis by the methylation pathway in *Saccharomyces cerevisiae*. *Journal of Biological Chemistry*, **244**, 3419–3422.

WICKEN, A. J., GIBBENS, J. W. & KNOX, K. W. (1973). Comparative studies on the isolation of membrane lipoteichoic acid from *Lactobacillus fermenti*. *Journal of Bacteriology*, **113**, 365–372.

WILEY, W. R. (1970). Tryptophan transport in *Neurospora crassa*: a tryptophan-binding protein released by cold osmotic shock. *Journal of Bacteriology*, **103**, 656–662.

WILSON, O. H. & HOLDEN, J. T. (1969). Stimulation of arginine transport in osmotically shocked *Escherichia coli* W. cells by purified arginine-binding protein fractions. *Journal of Biological Chemistry*, **244**, 2743–2749.

WODZINSKI, R. J. & FRAZIER, W. C. (1961). Moisture requirements of bacteria. III. Influence of temperature, pH, and malate and thiamine concentration on requirements of *Lactobacillus viridescens*. *Journal of Bacteriology*, **81**, 359–365.

SURVIVAL OF BACTERIA FOLLOWING EXPOSURE TO ULTRAVIOLET AND IONIZING RADIATIONS

B. A. BRIDGES

MRC Cell Mutation Unit, University of Sussex, Falmer, Brighton BN1 9QG, Sussex

INTRODUCTION

Irradiation has been a fact of biological life since the latter began. In particular, exposure to ultraviolet light (u.v.) from the sun was intense right up to the formation of the u.v.-absorbing ozone layer which was itself secondary to the creation of an oxygen-containing atmosphere by the photosynthetic metabolism of plants. The ability to cope with damage caused by u.v. must have been developed at very earliest times otherwise replication of nucleic acids with the conservation of enough genetic information to ensure biological continuity would have been impossible.

We now know that the defence mechanisms that have been developed by micro-organisms function not only against u.v. damage but also against that caused by ionizing radiation and many chemicals, both natural and man-made, and possibly even by heat. The results of experiments with radiations may thus have rather wider relevance than may appear at first sight. Even greater relevance to man is afforded by the realization that man himself is dependent upon similar defence mechanisms against the effects of radiations and radiomimetic chemicals.

Radiations

We are concerned here with two rather different types of radiation. Ionizing radiation causes the production of ionized molecules and free radicals when it passes through living matter. These are produced by the dissipation of packets of energy ('energy loss events') essentially randomly in all the constituents of a living cell and, being highly reactive, they rapidly initiate one or more chemical reactions. X- and γ-radiation and fast electrons are the types of ionizing radiation that will be considered. Other types are known (e.g. protons, neutrons, α-particles) and fuller details of the nature of all these radiations may be found in Dertinger & Jung (1970).

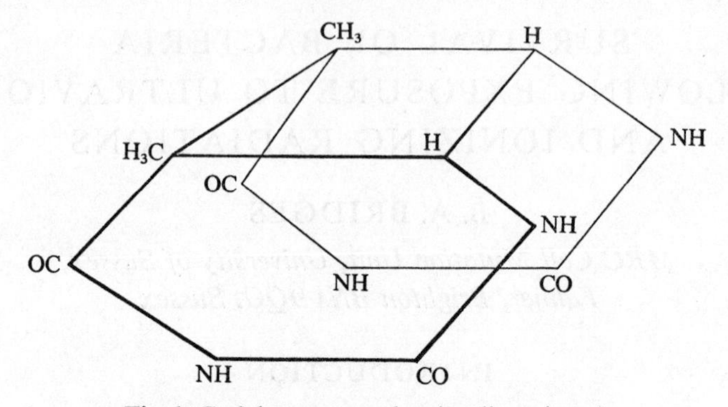

Fig. 1. Cyclobutane-type thymine dimer (*meso*).

Ultraviolet is electromagnetic radiation of wavelength between 200 and 300 nm and, in contrast to ionizing radiation, its energy is given up to an extent that depends almost entirely upon the chemical structure of the absorbing material. Although an absorbed photon gives up less energy than an 'energy loss event' of ionizing radiation, it is still capable of making or breaking several chemical bonds. A more detailed account of the properties of u.v. is given by Jagger (1967).

Although u.v. and ionizing radiations are completely different in their physical characteristics, modes of energy absorption, and types of chemical change that they initiate, they may be considered together because the most important site of damage for both is the genetic material (DNA) of the bacterium and because the bacterium reacts to these different types of damage in rather similar ways.

The reasons for the importance of DNA are different for the two types of radiation. Ultraviolet is absorbed most strongly by nucleic acids, particularly the wavelengths 250–260 nm which comprise the larger part of the output of most u.v. lamps. Ionizing radiation damage, on the other hand, is produced randomly throughout the organism so that the importance of DNA is solely due to the fact that each molecule is unique or nearly so within the bacterium and is irreplaceable (for review see Bridges & Munson, 1968).

The chief DNA photoproducts formed during u.v. irradiation are dimers between adjacent pyrimidines (e.g. Fig. 1). These have the merit for the experimentalist of being photoreversible (see below) which enables their involvement in many of the effects of u.v. to be ascertained. Another major product is cytosine hydrate although this is formed predominantly in single-stranded regions of DNA. It is unstable, reverting rapidly to cytosine, and its biological significance is unknown.

For further information see Setlow & Setlow (1972), Smith & Hanawalt (1969) and Kanazir (1969).

Damage to the bases in DNA is also found after exposure to ionizing radiation (Cerutti, 1974), although this does not include pyrimidine dimers. The most obvious lesion is the single-strand break, detectable with alkaline sucrose gradient sedimentation. Double-strand breaks, detectable with neutral sucrose gradients, are formed to a lesser extent and must be expected to be profoundly damaging to the integrity of the bacterial chromosome. It is generally accepted that *Escherichia coli*, the most studied bacterium in radiation biology, may not be able to survive most double-strand DNA breaks. Other micro-organisms, however, are considerably more resistant to ionizing radiation than *E. coli* and must be able to deal successfully with such breaks. For further information on the effects of ionizing radiation on biological macromolecules see Alexander & Lett (1967), Setlow & Setlow (1972), Blok & Loman (1973), and Bertinchamps *et al.* (1976).

The involvement of DNA damage in causing bacterial death following irradiation will be developed below. It should not be forgotten, however, that other components are also damaged and there will inevitably be situations where such damage may interact with and even mask the effects of that occurring in the DNA. For an account of the field up to 1968, including a discussion of high LET radiations (e.g. α particles) and of a modified target theory mathematical approach to cell killing (neither of which will be discussed further here) see Bridges & Munson (1968).

Death and survival

It is as well that the subject of this Symposium is survival rather than death, for our knowledge of the mechanisms for promoting bacterial survival far exceeds that of the ways in which bacteria die. Certainly death is not instantaneous and it is possible to alter the vitality of some bacteria by altering their environment many hours after irradiation (e.g. Alper & Gillies, 1958).

Following u.v. irradiation it is known that death is a result of damage to DNA because photoreactivating light, which splits pyrimidine dimers (see below), has a dramatic effect on survival. Exactly what causes the cell to die, however, is often unclear. There is currently an awareness that death may frequently be a consequence of inducible processes, the most obvious being the induction of integrated genomes such as prophages. Prophage induction does not even depend upon the bacterium's own DNA being damaged; even the introduction of a u.v.-irradiated episome may be sufficient (Borek & Ryan, 1958; Monk, 1967). In some bacteria death is characterized by the production of long filaments and although such bacteria may have a defect in the cell wall or membrane (James & Gillies, 1973) the filamentation process is characterized by many of the properties of prophage induction (cf. Witkin, 1967*a*).

Another apparently inducible response is acute respiratory deficiency and this may also be responsible for loss of vitality in certain instances (Swenson & Schenley, 1970, 1974). Inducible processes may also be involved in death and survival after exposure to ionizing radiation (cf. Marsden *et al.*, 1974).

In most cases, however, it must be admitted that the exact cause of death is not known and very often the most reasonable assumption is that death is a direct consequence of failure to replicate DNA so as to produce a viable and complete daughter chromosome. In the discussion that follows, attention will be concentrated upon those external factors which influence survival and those internal processes which have been developed by bacteria to promote their survival.

FACTORS MODIFYING THE SURVIVAL OF BACTERIA DURING IRRADIATION

The ability of bacteria to survive irradiation varies some 1000-fold from one species to another. Even within one species viability after a given exposure may vary considerably as a result of the influence of environmental factors. In this section we shall consider those factors which are effective immediately before or during irradiation. Such factors usually involve a modification of the initial (chemical) damage produced in the organism. Other factors affect the ability of the organism to respond to radiation-induced damage. Such factors usually involve changes in the physiological state of the cell and are usually operative after irradiation.

Chemical sensitizing agents

The most ubiquitous sensitizing agent for ionizing radiation is oxygen, which usually increases the sensitivity of bacteria between two- and four-fold (Fig. 2). The effect is general to most living things and was discovered by Holthusen in 1921, but it was many years before the phenomenon was rediscovered and studied quantitatively, notably by bacteriologists (e.g. Hollaender, Stapleton & Martin, 1951; Howard-Flanders & Alper, 1957). Typically the effect is ‘dose-modifying’, i.e.

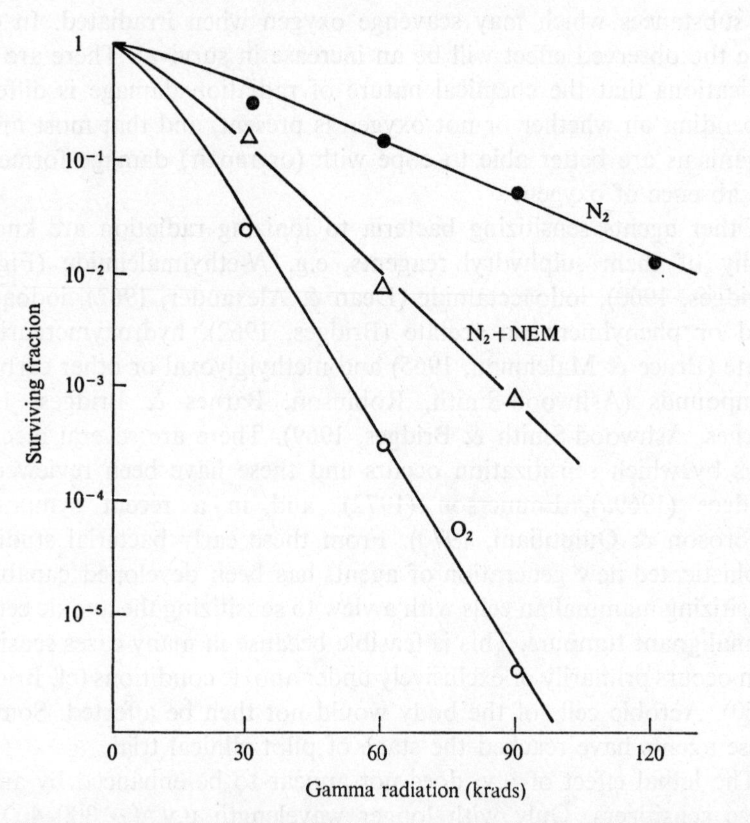

Fig. 2. Effect of oxygen and *N*-ethylmaleimide on the survival of *E. coli* B/r after exposure to gamma radiation. ●, Anxoic irradiation; △, anoxic irradiation in the presence of 10^{-3} M *N*-ethylmaleimide; ○, aerobic irradiation (from Bridges, 1960).

the effect may be described in terms of a factor by which the dose of radiation is effectively reduced. Thus the slope of the dose response curve is changed, but not its shape (see Fig. 2). The 'oxygen effect' is important in terms of human suffering since it is believed that the presence of anoxic necrotic centres to certain malignant tumours is in part responsible for their refractivity to radiotherapy. The breathing of hyperbaric oxygen during irradiation is one way in which it is hoped the degree of oxygenation of tumours may be increased.

So many factors can alter the intracellular oxygen concentration and it is necessary always to take this into consideration when studying the lethal effect of ionizing radiation. For example, dissolved oxygen may be depleted by cellular respiration (particularly when there are more than about 10^8 bacteria ml^{-1} or in the presence of nutrients), by the presence of autoxidizable chemicals (e.g. sulphydryl compounds), or by substances which may scavenge oxygen when irradiated. In each case the observed effect will be an increase in survival. There are also indications that the chemical nature of radiation damage is different depending on whether or not oxygen is present, and that most micro-organisms are better able to cope with (or repair) damage formed in the absence of oxygen.

Other agents sensitizing bacteria to ionizing radiation are known, many of them sulphydryl reagents, e.g. *N*-ethylmaleimide (Fig. 2) (Bridges, 1960), iodoacetamide (Dean & Alexander, 1962), iodoacetic acid or phenylmercuric acetate (Bridges, 1962), hydroxymercuribenzoate (Bruce & Malchman, 1965) and methylglyoxal or other carbonyl compounds (Ashwood-Smith, Robinson, Barnes & Bridges, 1967; Barnes, Ashwood-Smith & Bridges, 1969). There are several mechanisms by which sensitization occurs and these have been reviewed by Bridges (1969*a*), Emmerson (1972), and in a recent symposium (Moroson & Quintiliani, 1970). From these early bacterial studies a sophisticated new generation of agents has been developed capable of sensitizing mammalian cells with a view to sensitizing the anoxic centres of malignant tumours. This is feasible because in many cases sensitization occurs primarily or exclusively under anoxic conditions (cf. Bridges, 1960). Aerobic cells of the body would not then be affected. Some of these agents have reached the stage of pilot clinical trial.

The lethal effect of u.v. does not appear to be enhanced by any of these sensitizers. Only with longer wavelength u.v. (~ 300–400 nm) does one observe effects which are due to genuine photochemical sensitization. Coumarins, for example 8-methoxypsoralen, in the presence of 360 nm u.v., produce damage to bacterial DNA which, apart from being non-photoreversible, is similar to that produced by 254 nm u.v. alone in its effects (Igali *et al.*, 1970). It is known that both cross-links and adducts are formed in DNA by coumarin photosensitization and the treatment is highly mutagenic to bacteria.

Chemical protective agents

Early studies with ionizing radiation were bedevilled by the failure to control the dissolved oxygen tension and so oxygen depletion was often confused with genuine protection. Among the compounds capable of producing genuine protection even under anoxic conditions are glycerol, thiourea, dimethyl sulphoxide, and sulphydryl compounds such as cysteine. They all achieve a similar reduction (about two-fold) in the effectiveness of ionizing irradiation, both lethal and mutagenic (cf. Bridges, 1963). They are effective only when present during irradiation and act either by scavenging free radicals produced by radiolysis of intracellular water or by reacting with free radical centres on vital cell macromolecules and restoring them to their original state by proton donation (see Moroson & Quintiliani, 1970; Lion, 1976).

With 254 nm u.v., chemical protection is effected quite differently. Those chemicals which are effective (e.g. acridines, psoralens) appear to bind to the DNA double helix and act as energy sinks, diverting the energy of absorbed photons away from the normal sites of reaction and lowering the yield of photoproducts (Beukers, 1965; Setlow & Carrier, 1967; Bridges, 1971*b*). It is paradoxical that some of the chemicals act as photosensitizers with 360 nm u.v., greatly increasing the photoproduct yield (cf. Igali *et al.*, 1970).

Water content

Water is the largest single constituent of bacteria and it is not surprising that water content has a profound influence on radiation sensitivity. The classic work with ionizing radiation has been carried out with spores of *Bacillus megatherium* and has been comprehensively reviewed by Powers & Tallentire (1968). The effect is complex and involves a profound interaction with oxygen; the main findings may be summarized as follows. In the total absence of oxygen (including the stage of rehydration prior to viability determination), spores become slightly more resistant as the equilibrium water vapour pressure (EVP) is lowered from 8 to 5 mm mercury. Below this there is no further change down to 5×10^{-4} mm mercury. In the presence of oxygen, however, there is a substantial increase in sensitivity as the water vapour pressure is lowered from 10 mm to 5×10^{-4} mm mercury.

A rather similar picture has emerged with vegetative bacteria except that the damage in the absence of oxygen shows a much greater increase at high water contents. As such damage occurs even when oxygen-dependent damage is superimposed it results in a net increase in sensitivity even in the presence of oxygen. There is then a minimum sensitivity at a water content intermediate between 'wet' and 'dry'.

Removal of water also increases the u.v. sensitivity of *B. megatherium* spores (about three-fold) as the EVP is lowered from 20 to 10^{-5} mm mercury (see Powers & Tallentire, 1968). Vegetative bacteria appear to be affected to an even greater extent. For example, the sensitivity of *Serratia marcescens* to u.v. increased almost 20-fold when the EVP decreased from 23 to 8 mm mercury (Kaplan & Kaplan, 1956), a range which would still be regarded as 'wet' for spores. Enhanced u.v. sensitivity was also observed when vegetative bacteria were dried in aerosols

(Webb, 1965). The u.v. damage produced in 'dry' bacteria was not photoreversible and thus differed from that produced in wet bacteria (Kaplan & Kaplan, 1956; Webb & Tai, 1968).

Temperature

In general the sensitivity of bacteria to ionizing radiation is not greatly influenced by temperature. *Bacillus megatherium* spores become somewhat (< two-fold) more resistant as the temperature during irradiation is lowered from 22 to −148 °C; between −148 and −268 °C there is no further change in sensitivity (Webb, Ehret & Powers, 1958).

In contrast, most bacteria become considerably more sensitive to u.v. in the frozen state (Ashwood-Smith, Bridges & Munson, 1965; Ashwood-Smith & Bridges, 1967) due to the production of damage which is much less susceptible to the bacterial repair systems (Bridges, Ashwood-Smith & Munson, 1967). In the frozen state the yield of cyclobutane pyrimidine dimers is greatly decreased and another photoproduct (5-thyminyl-5,6-dihydrothymine) is produced (Rahn & Hosszu, 1968; Stafford & Donnellan, 1968).

It is interesting that the extremely radiation-resistant bacterium *Micrococcus radiodurans* can apparently deal with this photoproduct as well as the pyrimidine dimer since it is no more sensitive at −196 °C than at room temperature (Ashwood-Smith & Bridges, 1967).

One of the apparent results of the decreased repairability of the low temperature photoproduct is that the survivors of a dose of u.v. given at −79 °C contain a much higher proportion of mutants than the survivors of the same dose given at room temperature (Ashwood-Smith & Bridges, 1966). This is also true of the u.v. component of sunlight and it has been speculated that mutations resulting from low temperature solar irradiation may be concerned in speciation in bacteria (Ashwood-Smith, Copeland & Wilcockson, 1967). The effect of sunlight is even more dramatic on frozen bacteria than that of 254 nm u.v. since not only is the u.v. sensitivity of the bacteria greater but the chief photoproduct formation is not photoreversible by visible light, a process which normally partially mitigates the effect of the u.v. component of sunlight on bacteria.

RADIATION SENSITIVE AND RESISTANT MUTANTS

One of the more profitable approaches to the elucidation of any biochemical pathway is the isolation and study of mutants blocked at various stages. The isolation of the first radiation sensitive strain of

Table 1. *Some radiation-sensitivity genes in* Escherichia coli

Gene	Linkage	Approximate increase in sensitivity to u.v.	to aerobic ionizing radiation	Other phenotypic peculiarities	Biochemical deficiency (if known)
uvrA	*malB dnaB*	×15	×1	Unable to host-cell reactivate phages T1 and λ	Lacks u.v.-specific endonuclease
uvrB	*bio*	×15	×1	Unable to host-cell reactivate phages T1 and λ	Lacks u.v.-specific endonuclease
uvrC	*his*	×10	×1	Unable to host-cell reactivate phages T1 and λ	
uvrD	*ilv metE*	×10	×2	Unable to host-cell reactivate phages T1 and λ, dominant	
polA	*ilv metE*	×10	×3	Unable to host-cell reactivate some phages	Lacks DNA polymerase I activity
res	*ilv metE*	×10	×3	Tends to degrade its DNA	Lacks DNA polymerase I and associated exonuclease activity
uvr502	*ilv metE*	×10		Able to host-cell reactivate, mutator	
mutU	*ilv metE*	×2	×1.4	Able to host-cell reactivate, mutator	
ras	*lac purE*	×10	×1.5	High u.v. mutation induction	
recA	*pheA*	×20	×3	No genetic recombination, no radiation mutagenesis	
recB	*thyA*	×10	×3	Lowered genetic recombination	Lacks ATP-dependent exonuclease V
recC	*thyA*	×10	×3	Lowered genetic recombination	Lacks ATP-dependent exonuclease V
lex	*metA malB*	×15	×3	Probably same gene as *exrA*, no radiation mutagenesis	
exrA	*metA malB*	×15	×3	No radiation mutagenesis, dominant	
exrB	*metA malB*	×15	×5	No radiation mutagenesis, prone to filamentation	
phr	*gal*	×1	×1	Lacks ability to photoreactivate	Lacks photoreactivating enzyme
lon	*proC*	×10	×3	Medium dependent, filamentation death	

E. coli B (termed B_{s-1}) by Hill (1958) not only showed that the parent strain had means of coping successfully with radiation damage but also indicated the way in which this process might be studied. Since then a large number of mutants have been isolated showing various degrees of sensitivity, some of the more important are shown in Table 1, together with the corresponding biochemical defects where known. The vast majority have been found to be involved in a complex interacting set of pathways for the repair of damaged DNA. Almost all these genes are inessential for survival under 'normal' laboratory conditions but since the wild-type alleles are usually found in naturally occurring isolates one assumes that DNA damage, and the need for its repair, is one of the facts of life outside the laboratory. Even in the laboratory, some of the mutants are rather sick, particularly those which are also defective in genetic recombination.

One of the genes present in *E. coli* B itself confers radiation sensitivity in a manner which appears not to involve an interference with DNA repair. *E. coli* B is in fact more sensitive than most other strains of *E. coli* and the gene responsible, designated *lon*, appears to affect the link between DNA replication and cell division so that characteristic long filaments are formed which, unless other processes intervene, are unable to divide and form colonies. Both u.v., ionizing radiation, and many chemical mutagens cause a slowing down in the rate of DNA synthesis and so trigger filament formation in a manner which has many of the properties of an inducible process, cf. prophage induction (Witkin, 1967*a*). On the other hand, any treatment which retards RNA and protein synthesis while allowing DNA repair to occur both inhibits filament formation and eliminates the sensitivity to radiation. Because of this, *lon* strains are highly susceptible to modification of their radiation sensitivity by environmental influences (e.g. Alper & Gillies, 1958).

The first radiation-resistant mutant, *E. coli* B/r (Witkin, 1947), was subsequently found to have a suppressor for *lon*, designated *sul*, which maps near *azi* (Donch *et al.*, 1971). Filament-forming organisms with properties similar to the *lon* mutants of *E. coli* may not be uncommon in nature (cf. the work of Grula & Grula (1962) on *Erwinia carotovora*), although probably the majority of *E. coli* strains are phenotypically Lon$^+$.

'WILD-TYPE' SENSITIVITY

Wild-type *E. coli* are neither the most sensitive nor the most resistant of bacteria. Even within this species, independent isolates may vary enormously in their radiation sensitivity. Excluding *E. coli*, there has been little work on the frequency of natural occurrence of highly sensitive organisms. A *Pseudomonas* species originally isolated from a chicken carcass was almost as sensitive to gamma radiation as the most sensitive repair-deficient mutants of *E. coli* (Bridges, 1962) although it was not unduly sensitive to u.v. Many bacteria are more resistant than *E. coli*, often exhibiting survival curves with pronounced shoulders. Species of micrococci are pertinent examples.

The distinction of being the most radiation-resistant bacteria seems to belong to a group of organisms, usually red pigmented, of which the type species is *Micrococcus radiodurans*. This bacterium may survive doses in excess of 10^4 ergs mm^{-2} or 10^6 rads. Given that the comparatively sensitive organism *E. coli* possesses several well-defined pathways of DNA repair, it is apparent that *M. radiodurans* must possess repair ability to an exceedingly high degree. Certainly there is an extremely efficient system for excision repair (see below) of regions of DNA damaged by either ionizing radiation or u.v. (Boling & Setlow, 1966; Dean, Little & Serianni, 1970; Bonura & Bruce, 1974). As yet we do not know why the excision process in *M. radiodurans* should be so much more effective than that in most other micro-organisms, nor do we know what other repair pathways may operate.

As with *E. coli*, the isolation of repair-deficient radiation-sensitive mutants points the way to further advances. Moseley, Mattingly & Copland (1972) showed that two temperature-sensitive DNA synthesis mutants became sensitive to u.v. and ionizing radiation when incubated at restrictive temperatures before irradiation. There was a concomitant loss of their ability to effect recombination as measured by transformation which suggested that genetic recombination processes might be involved in reconstructing viable genomes. Other radiation-sensitive mutants showing some temperature sensitivity have been reported by Suhadi *et al.* (1972). In contrast to wild-type *M. radiodurans*, these mutants were almost totally unable to rejoin double-strand DNA scissions after gamma irradiation.

One of the earliest u.v.-sensitive mutants of *M. radiodurans* to be isolated was UV17 (Moseley, 1967). Moseley & Mattingly (1971) have claimed that this is equivalent to an *exr* or *lex* mutant of *E. coli*. (The *exr* mutant enhances u.v. and ionizing radiation sensitivity, chiefly through blocking an error-prone pathway active in post-replication repair – see below.) Bonura & Bruce (1974), however, have challenged this interpretation as they have shown UV17 to be defective in the repair of single-strand DNA breaks during incubation in buffer. In *E. coli* this type of repair depends on the control of nuclease, polymerase and ligase activity and is unaffected by the *exr* mutation. Whatever mechanisms finally turn out to be involved in the maintenance of genetic integrity in *M. radiodurans*, we can be sure that in order to survive such high doses other cell components than DNA must need to be repaired or replaced, for example the cell membrane, and that these pathways of cytoplasmic repair must also be under genetic control.

DNA REPAIR PROCESSES

This section can do no more than summarize the present knowledge in this field and must inevitably be selective; its bias doubtless reflects the author's own preoccupations and prejudices. The topic has been thoroughly covered in the proceedings of a recent symposium (Hanawalt & Setlow, 1975), in Town, Smith & Kaplan (1973) and in Lehmann (1976).

Photoreactivation

One of the most easily demonstrated chemical lesions induced in DNA by u.v. (but not ionizing radiation) is the cyclobutane pyrimidine dimer in which neighbouring cytosine or thymine moieties are joined together by a cyclobutane ring (see Fig. 1). Many bacteria possess a highly specific enzyme capable of splitting such dimers and restoring the original base sequence (Rupert, 1960, 1962*a*, *b*). The enzyme has an apparently strict requirement for activation by light. Organisms possessing this enzyme show a dramatic restoration of viability when exposed to visible or near-visible light after u.v. irradiation (Plate 1). The phenomenon is called photoreactivation and was discovered independently for *Streptomyces griseus* by Kelner (1949*a*, *b*) and for bacteriophage (where a host enzyme is used) by Dulbecco (1949).

Photoreactivation has been a most useful tool in studying the response of bacteria to u.v. for two reasons. Firstly, enzymic photoreversal of any biological effect shows the essential involvement of pyrimidine dimers. There are few other agents or treatments whose effective initial lesions are known with such precision. Secondly, photoreactivation may be used to remove pyrimidine dimers at any time during the period that the biological response of the organism is being developed. Because of this, pyrimidine dimers are unique in being the only lesion that can be inserted and removed at will from the DNA of a growing bacterium.

Photoreactivating enzyme is extremely widely distributed in nature, including most plants and most members of the animal kingdom. For further reading see Setlow (1967), Jagger (1967), and Sutherland (1975).

Excision-repair

The fact that DNA *in vivo* consists of two strands containing complementary base pairs (and thus complementary genetic information) means that damage to one strand need not in principle be lethal since

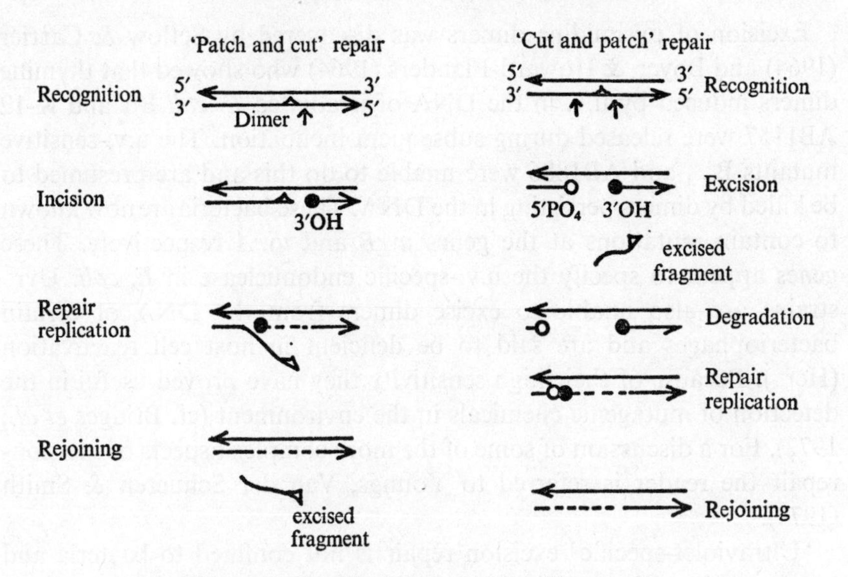

Fig. 3. Two hypothetical modes of excision repair.

the correct information is still present in the complementary strand. One of the ways in which bacteria take advantage of this is in the process called excision repair.

After u.v. irradiation many bacteria possess one or more types of 'u.v.-specific' endonuclease, an enzyme which can recognize a pyrimidine dimer and make a nick in the sugar-phosphate backbone near to it. A short region of the strand containing the dimer is then excised either by the exonucleolytic activity of DNA polymerase I ('Kornberg polymerase') or by a u.v.-specific exonuclease (Grossman *et al.*, 1968; Takagi *et al.*, 1968; Kaplan, Kushner & Grossman, 1969; Kelly *et al.*, 1969). The resulting single-strand gap is then filled in by a DNA polymerase (usually I) using the complementary strand as a template (Kanner & Hanawalt, 1970) and reconnection to the rest of the DNA strand is effected by a polynucleotide ligase (Pauling & Hamm, 1968; Gellert and Bullock, 1970). The mechanism thus described has been termed 'cut and patch' but it must be admitted that *in vivo* we are unable to distinguish it from a 'patch and cut' model utilizing the same enzymes (Fig. 3). Besides pyrimidine dimers, a number of DNA lesions produced by chemical mutagens are also recognized by 'u.v.-specific' endonucleases which may therefore recognize distortions in the double helix rather than specific chemical alterations (Hanawalt & Haynes, 1965). The designation 'u.v.-specific' is thus something of a misnomer.

Excision of pyrimidine dimers was discovered by Setlow & Carrier (1964) and Boyce & Howard-Flanders (1964) who showed that thymine dimers induced by u.v. in the DNA of irradiated *E. coli* B/r and K-12 AB1157 were released during subsequent incubation. The u.v.-sensitive mutants B_{s-1} and AB1886 were unable to do this and are presumed to be killed by dimers persisting in the DNA. These bacteria are now known to contain mutations at the genes *uvrB* and *uvrA* respectively. These genes appear to specify the u.v.-specific endonuclease in *E. coli.* Uvr^- strains are also unable to excise dimers from the DNA of certain bacteriophages and are said to be deficient in host cell reactivation (Hcr^-). Because of their high sensitivity they have proved useful in the detection of mutagenic chemicals in the environment (cf. Bridges *et al.*, 1972). For a discussion of some of the more complex aspects of excision-repair the reader is referred to Youngs, Van der Schueren & Smith (1974).

'Ultraviolet-specific' excision repair is not confined to bacteria and has been demonstrated in many other organisms including cultured human cells. The knowledge obtained with bacteria has helped towards an understanding of the human disease xeroderma pigmentosum in which the skin is abnormally sensitive to sunlight and prone to develop malignant tumors. Such patients are in fact Uvr^- and Hcr^- mutants (see Cleaver (1974) for review).

Ionizing radiation does not produce pyrimidine dimers and there is relatively little damage susceptible to the action of 'u.v.-specific' endonuclease (Howard-Flanders, Boyce & Theriot, 1966; Bridges & Munson, 1966). Excision repair nevertheless occurs, some of it initiated by other endonucleases (see Cerutti, 1974) but much of it initiated by the single-strand breaks produced by ionizing radiation. About 80 % of these breaks are subject to exonuclease action to convert them into small gaps and then very rapidly repaired by DNA polymerase I (Town, Smith & Kaplan, 1971) and (presumably) polynucleotide ligase.

RecA⁺-dependent repair

Clark & Margulies (1965) isolated mutants of *E. coli* K-12 that were unable to give rise to recombinants when used as recipients in genetic crosses and found that they were u.v. sensitive. Shortly afterwards Howard-Flanders & Theriot (1966) found that some X-ray sensitive mutants were recombination deficient. In most of these strains there was found to be a deficiency at a gene mapping between *pheA* and *cysC* and designated *recA*. Other Rec^- genes have since been discovered but *recA* is the most dramatic in its effects, which are several.

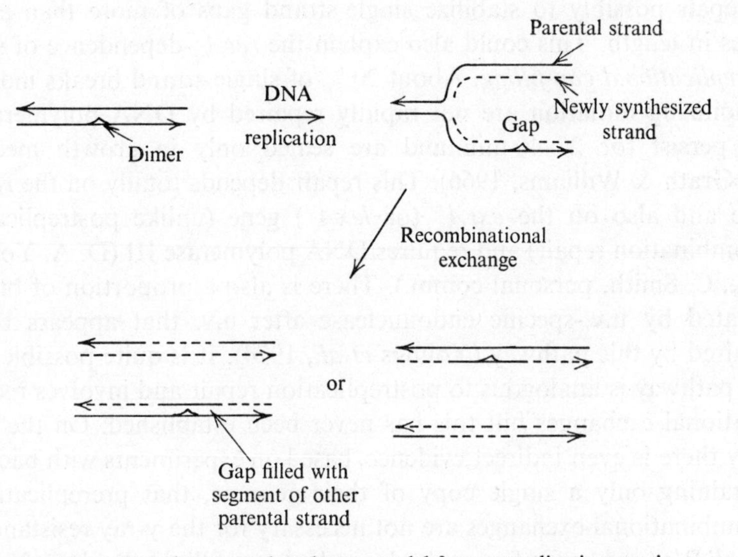

Fig. 4. Daughter-strand exchange model for postreplication repair.

RecA strains are deficient in at least three distinguishable pathways of repair. Probably the best characterized is *postreplicational recombination repair* of u.v. damage. This is most clearly seen in bacteria which cannot excise pyrimidine dimers. When such bacteria are given a low dose of u.v., sufficient to induce several dozen pyrimidine dimers in the chromosome, DNA synthesis continues but the daughter strands of DNA are of lower than normal molecular weight. In 10 to 20 min the molecular weight returns to normal (Rupp & Howard-Flanders, 1968). There is now convincing evidence that the daughter strands have a gap opposite each pyrimidine dimer in the parental strands and that this gap is filled by a recombinational exchange with the other daughter chromosome (Rupp *et al.*, 1971; Ganesan, 1974) (Fig. 4). Some synthesis *de novo* is also involved (Ley, 1973) and either DNA polymerase I or DNA polymerase III is required (Sedgwick & Bridges, 1974). In *recA* bacteria, daughter strand gaps are formed (Smith & Meun, 1970) but are never sealed, even if the opposing pyrimidine dimer is subsequently removed by photoreversal (Bridges & Sedgwick, 1974). Postreplication recombination repair has not been convincingly demonstrated after ionizing irradiation. After doses at which daughter DNA has low molecular weight, the template also is of low molecular weight due to single strand breakage and the population contains many cells with inviable chromosomes.

It seems most likely that the *recA*+ gene product is required very early in repair possibly to stabilize single-strand gaps of more than a few bases in length. This could also explain the *recA*+-dependence of some *prereplicational gap-filling*. About 20 % of single-strand breaks induced by ionizing radiation are not rapidly repaired by DNA polymerase I but persist for 20–40 min and are sealed only in growth medium (McGrath & Williams, 1966). This repair depends totally on the *recA*+ gene and also on the *exrA*+ (or *lexA*+) gene (unlike postreplication recombination repair) and requires DNA polymerase III (D. A. Youngs & K. C. Smith, personal comm.). There is also a proportion of breaks initiated by u.v.-specific endonuclease after u.v. that appears to be repaired by this pathway (Youngs *et al.*, 1974). It is quite possible that this pathway is analogous to postreplication repair and involves recombinational exchanges but this has never been established. On the contrary there is even indirect evidence, based on experiments with bacteria containing only a single copy of their genome, that prereplicational recombinational exchanges are not necessary for the γ-ray resistance of *E. coli* B/r (and hence for *recA*+-dependent gap-filling) (Bridges, 1971*a*).

A third, and perhaps most important *recA*+-dependent pathway, is *error-prone repair* (for reviews see Witkin, 1969; Bridges, 1969*b*; Doudney, 1975). One of the characteristic effects of ionizing radiation and u.v. on bacteria is that mutants are induced among the survivors. It is now clear that these mutants arise as errors during repair of DNA damage. Apart from the *recA*+ gene, the *exrA*+ (*lexA*+) gene is required for the operation of this pathway (Witkin, 1967*b*; Bridges, Law & Munson, 1968) and the current working hypothesis is that *exrA*+-dependent mutagenesis can occur during the sealing of single-strand gaps in DNA (Bridges, *et al.*, 1968; Witkin, 1969), and the more persistent the gap the more likely is repair to occur in an error-prone manner (Bridges, 1975).

It has not been possible to distinguish formally whether error-prone repair is an optional step in other *recA*+-dependent repair pathways or whether it is an independent pathway, although there is evidence to suggest the latter. Certainly it is possible to arrange situations either genetically or physiologically where *exrA* bacteria are able to carry out filling of single-strand breaks after γ-irradiation and filling of daughter-strand gaps after u.v. but where there is still no induced mutation (Bridges *et al.*, 1973). There are also conditions in which *exrA*+ bacteria fail to form mutations even when gap-filling (as determined biochemically) does occur (Sedgwick, 1975; Eyfjord, Green & Bridges, 1975).

Ultraviolet-induced mutagenesis in bacteriophage lambda–*E. coli* complexes is also dependent upon the *lexA*$^+$ and *recA*$^+$ genes and is associated with an inducible repair process (Weigle, 1953). It has been suggested (Defais *et al.*, 1971) that the same inducible error-prone process is reponsible for u.v. mutagenesis in bacteria themselves and is one of a cluster of inducible functions dependent upon the *recA*$^+$ and *exrA*$^+$ genes which also includes filament formation in *lon* strains of *E. coli* and prophage induction. There is probably more circumstantial evidence for this hypothesis (Witkin & George, 1973; Radman, 1975) than against (Bridges, Rothwell & Green, 1973).

The mechanism, both of induction and of error-prone repair itself, is still not clear but certain deductions may be made. There is no obligatory requirement for recombinational exchanges since (*a*) there is at least one situation where u.v. mutagenesis of bacteria appears to take place without the possibility of exchanges (Bridges & Mottershead, 1971); (*b*) *exrA*$^+$-dependent mutagenesis in lambda is distinct from *recA*$^+$-dependent recombination repair which does not require *exrA*$^+$ (Blanco & Devoret, 1973); and (*c*) inducible *recA*$^+$-dependent mutagenesis occurs in the single-stranded DNA phage ϕX174 after u.v. irradiation with low multiplicities of infection where phage recombination can be excluded.

There would appear to be at least three possible mechanisms for filling gaps in an error-prone way, considering particularly the insertion of bases in a daughter-strand gap opposite a pyrimidine dimer where normal polymerases appear unable to function. Firstly, bases may be inserted randomly by a terminal nucleotidyl transferase (possibly inducible) which does not require a template (Witkin, 1967*b*; Bridges, Dennis & Munson, 1967). Secondly, there may be an unknown (inducible) polymerase which is able to insert bases randomly opposite pyrimidine dimers (Radman, personal communication 1973; 1975). Thirdly, there may be a co-factor, possibly inducible, that is able to overide the specificity of one of the three known DNA polymerases and enable it to insert bases at random opposite pyrimidine dimers. On the two latter models the polymerase would be effectively functioning as a terminal nucleotidyl transferase enzyme where the template was inadequate.

Recent experiments in the author's laboratory have indicated that DNA polymerase III (but not polymerases I or II) is obligatory for mutation fixation so the possibility exists that there may be an inducible function which interacts with polymerase III to cause error-prone repair polymerization.

Correlation with heat sensitivity

We have already seen that among the genes necessary for repair of radiation-induced DNA damage are some necessary for genetic recombination and DNA polymerization, and some involved in the repair of DNA damaged by chemical mutagens. Both the advantages of genetic recombination and the need to repair damage caused by chemical mutagens may in many instances have been involved in the selective pressures that have led to the evolution or retention of mechanisms for radiation resistance in wild-type organisms, rather than the need to survive actual radiation exposure. Another factor may have been the need to deal with single-strand breaks arising as a result of thermal stress. A correlation between heat (52 °C) sensitivity and ionizing radiation sensitivity was observed by Bridges *et al.* (1969) and Matsumoto & Kagami-Ishi (1970) for a number of strains of *E. coli* carrying various radiation sensitivity alleles, and has also been found for yeast. The production of single-strand breaks was observed during mild heating (Bridges *et al.*, 1969) and their repair in radiation resistant but not radiation sensitive bacteria (Woodcock & Grigg, 1972). More recently Pauling & Beck (1975) have implicated polynucleotide ligase in the repair of DNA breaks induced by 52 °C treatment as well as by irradiation.

DNA strand breakage after mild heating is a strain-dependent phenomenon and greatly influenced by physiological factors (Sedgwick & Bridges, 1972 and unpublished observations) and it is unclear whether or not the correlation is a general phenomenon. The existence of some sort of cross-sensitivity, however, suggests that the ability of bacteria to repair or bypass radiation damage may, at least in some instances, have arisen or been retained during evolution as a response to thermal selective pressures.

Repair inhibitors

Several chemicals have been postulated to inhibit repair processes and so render bacteria more sensitive to radiation. Caffeine, for example, has been reported to inhibit excision of u.v. photoproducts (Sideropoulos & Shankel, 1968) and also the error-prone system (Witkin & Farquharson, 1969). A number of DNA-binding dyes and psoralens have also been assumed to be excision inhibitors on the basis of their effect on survival of *uvr*$^+$ but not *uvr*$^-$ strains. Biochemical data are, however, available only for acriflavine (Setlow, 1964) and not all workers are satisfied that the action of acriflavine is explicable solely in terms of decreased excision-repair (Alper & Hodgkins, 1969; Witkin, 1969, p. 544; Bridges, 1971*b*).

Sulphydryl binding agents have been discussed above as sensitizers of bacteria to ionizing radiation and in some instances it appears that inhibition of repair processes may play some part in this effect (see, e.g. Shenoy *et al.*, 1970).

The inhibition of 'recombination' repair has been recently reviewed by Smith (1971, 1975). Chromatographically pure hydroxyurea partially and reversibly inhibited the sealing of single-strand DNA breaks after X-irradiation of *E. coli* but had no effect on survival. Some commercial preparations of hydroxyurea, however, contained an impurity which irreversibly blocked single-strand break repair and markedly enhanced the killing of *rec*$^+$ but not *recA* bacteria by X-radiation. Another inhibitor of *recA*$^+$-dependent repair is quinacrine (Smith, 1971).

Modification of survival by postirradiation treatment

There are numerous reports of changes in growth conditions affecting irradiated bacteria (for reviews see Rupert & Harm, 1966; Stapleton, 1960; Smith, 1971, 1974). Perhaps the most well known of post-irradiation phenomena is 'liquid holding recovery' which occurs when *recA* or *lon* strains of *E. coli* are held in buffer after u.v. Both *lon* and *recA* strains die (for different reasons) as a consequence of the replication of their damaged DNA. Incubation in buffer both prevents this and permits slow excision of photoproducts thus reducing the risk of death when replication is resumed (Jagger, Wise & Stafford, 1964; Rupert & Harm, 1966; Ganesan & Smith, 1969). In the light of more recent knowledge of repair processes it is now quite feasible to interpret a number of other effects in terms of modification of DNA repair (cf. Friesen *et al.*, 1970; Youngs *et al.*, 1974). Actual evidence to support these interpretations is, however, often inferential and for this reason they will not be considered here. Suffice it to say that there is often a great variation between different organisms in their response to post-irradiation treatments and generalizations are unprofitable.

REFERENCES

Alexander, P. & Lett, J. T. (1967). Effects of ionizing radiations on biological macromolecules. In *Comprehensive Biochemistry*, **27**, ed. M. Florkin & E. H. Stolz, pp. 267–356. Amsterdam: Elsevier.

Alper, T. & Gillies, N. E. (1958). 'Restoration' of *Escherichia coli* strain B after irradiation: its dependence on suboptimal growth conditions. *Journal of General Microbiology*, **18**, 461–472.

ALPER, T. & HODGKINS, B. (1969). 'Excision repair' and dose-modification; questions raised by radiobiological experiments with acriflavine. *Mutation Research*, **8**, 15–23.

ASHWOOD-SMITH, M. J. & BRIDGES, B. A. (1966). Ultraviolet mutagenesis in *Escherichia coli* at low temperatures. *Mutation Research*, **3**, 135–144.

ASHWOOD-SMITH, M. J. & BRIDGES, B. A. (1967). On the sensitivity of frozen micro-organisms to ultraviolet radiation. *Proceedings of the Royal Society of London, Series B*, **168**, 194–202.

ASHWOOD-SMITH, M. J., BRIDGES, B. A. & MUNSON, R. J. (1965). Ultraviolet damage to bacteria and bacteriophage at low temperatures. *Science, Washington*, **149**, 1103–1105.

ASHWOOD-SMITH, M. J., COPELAND, J. & WILCOCKSON, J. (1967). Sunlight and frozen bacteria. *Nature, London*, **214**, 33–35.

ASHWOOD-SMITH, M. J., ROBINSON, D. M., BARNES, J. H. & BRIDGES, B. A. (1967). Radiosensitization of bacterial and mammalian cells by substituted glyoxals. *Nature, London*, **216**, 137–139.

BARNES, J., ASHWOOD-SMITH, M. J. & BRIDGES, B. A. (1969). Radiosensitization of bacterial cells by carbonyl compounds. *International Journal of Radiation Biology*, **15**, 285–288.

BERTINCHAMPS, A. J., HÜTTERMAN, J., TEOULE, R. & KÖHNLEIN, W. (ed.) (1976). *Effects of Ionizing Radiation on Nucleic Acids*. Berlin: Springer (in press).

BEUKERS, R. (1965). The effect of proflavine on UV-induced dimerization of thymine in DNA. *Photochemistry and Photobiology*, **4**, 935–937.

BLANCO, M. & DEVORET, R. (1973). Repair mechanisms involved in prophage reactivation and UV reactivation of UV-irradiated phage λ. *Mutation Research*, **17**, 293–305.

BLOK, J. & LOMAN, H. (1973). The effects of gamma radiation in DNA. *Current Topics in Radiation Research Quarterly*, **9**, 165–245.

BOLING, M. E. & SETLOW, J. K. (1966). The resistance of *Micrococcus radiodurans* to ultraviolet radiation. III. A repair mechanism. *Biochimica et Biophysica Acta*, **123**, 26–33.

BONURA, T. & BRUCE, A. K. (1974). The repair of single-strand breaks in a radio-sensitive mutant of *Micrococcus radiodurans. Radiation Research*, **57**, 260–275.

BOREK, E. & RYAN, A. (1958). The transfer of irradiation-elicited induction in a lysogenic organism. *Proceedings of the National Academy of Sciences, U.S.A.* **44**, 374–377.

BOYCE, R. P. & HOWARD-FLANDERS, P. (1964). Release of ultraviolet light induced thymine dimers from DNA in *E. coli* K-12. *Proceedings of the National Academy of Sciences, U.S.A.* **51**, 293–300.

BRIDGES, B. A. (1960). Sensitization of *Escherichia coli* to gamma radiation by *N*-ethylmaleimide. *Nature, London*, **188**, 415.

BRIDGES, B. A. (1962). The chemical sensitization of *Pseudomonas* species to ionizing radiation. *Radiation Research*, **16**, 232–242.

BRIDGES, B. A. (1963). Effect of chemical modifiers on inactivation and mutation induction by gamma radiation in *Escherichia coli. Journal of General Microbiology*, **31**, 405–412.

BRIDGES, B. A. (1969*a*). Sensitization of organisms to radiation by sulphydryl-binding agents. In *Advances in Radiation Biology*, **3**, ed. L. G. Augenstein, R. Mason & M. Zelle, p. 123. New York: Academic Press.

BRIDGES, B. A. (1969*b*). Mechanisms of radiation mutagenesis in cellular and sub-cellular systems. *Annual Review of Nuclear Science*, **19**, 139–178.

Bridges, B. A. (1971*a*). RecA⁺-dependent repair of gamma-ray damage to *Escherichia coli* does not require recombination between existing homologous chromosomes. *Journal of Bacteriology*, **108**, 944–945.

Bridges, B. A. (1971*b*). Genetic damage induced by 254 nm ultraviolet light in *Escherichia coli*: 8-methoxypsoralen as protective agent and repair inhibitor. *Photochemistry and Photobiology*, **14**, 659–662.

Bridges, B. A. (1975). Genetic effects of UV on *Escherichia coli*: a model for prokaryotes. *Proceedings of the Fifth International Congress of Radiation Research, Seattle* (in press).

Bridges, B. A., Ashwood-Smith, M. J. & Munson, R. J. (1967). On the nature of the lethal and mutagenic action of ultraviolet light on frozen bacteria. *Proceedings of the Royal Society of London, Series B*, **168**, 203–215.

Bridges, B. A., Ashwood-Smith, M. J. & Munson, R. J. (1969). Correlation of bacterial sensitivities to ionizing radiation and mild heating. *Journal of General Microbiology*, **58**, 115–124.

Bridges, B. A., Dennis, R. E. & Munson, R. J. (1967). Differential induction and repair of ultraviolet damage leading to true reversions and external suppressor mutations of an ochre codon in *Escherichia coli* B/r WP2. *Genetics*, **57**, 897–908.

Bridges, B. A., Gray, W. J. H., Green, M. H. L., Rothwell, M. A. & Sedgwick, S. G. (1973). Genetic and physiological separation of the repair and mutagenic functions of the *exrA* gene in *Escherichia coli. Genetics Supplement*, **73**, 123–129.

Bridges, B. A., Law, J. & Munson, R. J. (1968). Mutagenesis in *Escherichia coli.* II. Evidence for a common pathway for mutagenesis by ultraviolet light, ionizing radiation and thymine deprivation. *Molecular and General Genetics*, **103**, 266–273.

Bridges, B. A. & Mottershead, R. P. (1971). RecA⁺-dependent mutagenesis occurring before DNA replication in UV- and γ-irradiated *Escherichia coli. Mutation Research*, **13**, 1–8.

Bridges, B. A., Mottershead, R. P., Rothwell, M. A. & Green, M. H. L. (1972). Repair-deficient bacterial strains suitable for mutagenicity screening: tests with the fungicide captan. *Chemico-Biological Interactions*, **5**, 77–84.

Bridges, B. A. & Munson, R. J. (1966). Excision-repair of DNA damage in an auxotrophic strain of *Escherichia coli. Biochemical and Biophysical Research Communications*, **22**, 268–273.

Bridges, B. A. & Munson, R. J. (1968). Genetic radiation damage and its repair in *Escherichia coli.* In *Current Topics in Radiation Research*, **4**, ed. M. Ebert & A. Howard, pp. 95–188. Amsterdam: North Holland.

Bridges, B. A., Rothwell, M. A. & Green, M. H. L. (1973). Repair processes and dose response curves in UV mutagenesis of bacteria. *Anais da Academia Brasileira de Ciências*, **45** *Supplement*, 203–209.

Bridges, B. A. & Sedgwick, S. G. (1974). Effect of photoreactivation on the filling of gaps in deoxyribonucleic acid synthesized after exposure of *Escherichia coli* to ultraviolet light. *Journal of Bacteriology*, **117**, 1077–1081.

Bruce, A. K. & Malchman, W. H. (1965). Radiation sensitization of *Micrococcus radiodurans*, *Sarcina lutea*, and *Escherichia coli* by *p*-hydroxymercuribenzoate. *Radiation Research*, **24**, 473–481.

Cerutti, P. A. (1974). Excision repair of DNA base damage. *Life Sciences*, **15**, 1567–1575.

Clark, A. J. & Margulies, A. D. (1965). Isolation and characterization of recombination-deficient mutants of *Escherichia coli* K-12. *Proceedings of the National Academy of Sciences, U.S.A.* **53**, 451–458.

Cleaver, J. E. (1974). Repair processes for photochemical damage in mammalian cells. *Advances in Radiation Biology*, **4**, 1–75.

DEAN, C. J. & ALEXANDER, P. (1962). Sensitization of radioresistant bacteria to X-rays by iodoacetamide. *Nature, London*, **196**, 1324–1326.

DEAN, C. J., LITTLE, J. G. & SERIANNI, R. W. (1970). The control of post-irradiation DNA breakdown in *Micrococcus radiodurans*. *Biochemical and Biophysical Research Communications*, **39**, 126–134.

DEFAIS, M., FAUQUET, P., RADMAN, M. & ERRERA, M. (1971). Ultraviolet reactivation and ultraviolet mutagenesis of λ in different genetic systems. *Virology*, **43**, 495–503.

DERTINGER, H. & JUNG, H. (1970). *Molecular Radiation Biology*. Heidelberg: Springer. (English translation, London: Longmans, Green & Co.)

DONCH, J. J., CHUNG, Y. S., GREEN, M. H. L., GREENBERG, J. & WARREN, G. (1971). Genetic analysis of *sul* mutants of *Escherichia coli* B. *Genetical Research*, **17**, 185–193.

DOUDNEY, C. O. (1975). Mutation in ultraviolet light-damaged microorganisms. In *Photochemistry and Photobiology of Nucleic Acids*. Vol. II. *Photobiology*, ed. S. Y. Wang New York: Academic Press. (In Press.)

DULBECCO, R. (1949). Reactivation of ultra-violet-inactivated bacteriophage by visible light. *Nature, London*, **163**, 949–950.

EMMERSON, P. T. (1972). X-ray damage to DNA and loss of biological function: effect of sensitizing agents. In *Advances in Radiation Chemistry*, **3**, ed. M. Burton & J. L. Magee, pp. 209–270. New York: Wiley-Interscience.

EYFJORD, J. E., GREEN, M. H. L. & BRIDGES, B. A. (1975). Mutagenic DNA repair in *Escherichia coli*. Conditions for error-free filling of daughter strand gaps. *Journal of General Microbiology* (in press.)

FRIESEN, B. S., IYER, P. S., BAPTIST, J. G., MEYN, R. & RODGERS, J.-M. (1970). Glucose-induced resistance to gamma-rays in *Escherichia coli*. *International Journal of Radiation Biology*, **18**, 159–172.

GANESAN, A. K. (1974). Persistence of pyrimidine dimers during post-replication repair in ultraviolet light irradiated *Escherichia coli* K-12. *Journal of Molecular Biology*, **87**, 103–119.

GANESAN, A. K. & SMITH, K. C. (1969). Dark recovery processes in *Escherichia coli* irradiated with ultraviolet-light. II. Effect of *uvr* genes on liquid holding recovery. *Journal of Bacteriology*, **97**, 1129–1133.

GELLERT, M. & BULLOCK, M. C. (1970). DNA ligase mutants of *Escherichia coli*. *Proceedings of the National Academy of Sciences, U.S.A.* **67**, 1580–1587.

GROSSMAN, L., KAPLAN, J., KUSER, S. & MAHLER, I. (1968). Enzymes involved in the early stages of repair of ultraviolet-irradiated DNA. *Cold Spring Harbor Symposium on Quantitative Biology*, **33**, 229–234.

GRULA, E. A. & GRULA, M. M. (1962). Cell division in a species of *Erwinia*. III. Reversal of inhibition of cell division caused by D-amino acids, penicillin and ultraviolet light. *Journal of Bacteriology*, **83**, 981–988.

HANAWALT, P. C. & HAYNES, R. H. (1965). Repair replication of DNA in bacteria: irrelevance of chemical nature of base defect. *Biochemical and Biophysical Research Communications*, **19**, 462–467.

HANAWALT, P. C. & SETLOW, R. B. (1975). *Molecular Mechanisms for the Repair of DNA*. New York: Plenum Press.

HILL, R. F. (1958). A radiation-sensitive mutant of *Escherichia coli*. *Biochimica et Biophysica Acta*, **30**, 636–637.

HOLLAENDER, A., STAPLETON, G. E. & MARTIN, F. L. (1951). X-ray sensitivity of *E. coli* as modified by oxygen tension. *Nature, London*, **167**, 103–104.

HOWARD-FLANDERS, P. & ALPER, T. (1957). The sensitivity of micro-organisms to irradiation under controlled gas conditions. *Radiation Research*, **7**, 518–540.

HOWARD-FLANDERS, P., BOYCE, R. P. & THERIOT, L. (1966). Three loci in *Escherichia coli* K-12 that control the excision of pyrimidine dimers and certain other mutagen products from DNA. *Genetics*, **53**, 1119–1136.

HOWARD-FLANDERS, P. & THERIOT, L. (1966). Mutants of *Escherichia coli* K-12 defective in DNA repair and in genetic recombination. *Genetics*, **53**, 1137–1150.

IGALI, S., BRIDGES, B. A., ASHWOOD-SMITH M. J. & SCOTT, B. R. (1970). Mutagenesis in *Escherichia coli*. IV. Photosensitization to near ultraviolet light by 8-methoxypsoralen. *Mutation Research*, **9**, 21–30.

JAGGER, J. (1967). *Introduction to Research in Ultraviolet Photobiology*. Englewood Cliffs, N.J.: Prentice Hall. Inc.

JAGGER, J., WISE, W. C. & STAFFORD, R. S. (1964). Delay in growth and division induced by near ultraviolet-radiation in *Escherichia coli* B and its role in photoprotection and liquid holding recovery. *Photochemistry and Photobiology*, **3**, 11–24.

JAMES, R. & GILLIES, N. E. (1973). The sensitivity of suppressed and unsuppressed *lon* strains of *Escherichia coli* to chemical agents which induce filamentation. *Journal of General Microbiology*, **76**, 429–436.

KANAZIR, D. T. (1969). Radiation-induced alterations in the structure of deoxyribonucleic acid and their biological consequences. *Progress in Nucleic Acid Research and Molecular Biology*, **9**, 117–222.

KANNER, L. & HANAWALT, P. (1970). Repair deficiency in a bacterial mutant defective in DNA polymerase. *Biochemical and Biophysical Research Communications*, **39**, 149–155.

KAPLAN, J. C., KUSHNER, S. R. & GROSSMAN, L. (1969). Enzymatic repair of DNA. I. Purification of two enzymes involved in the excision of thymine dimers from ultraviolet-irradiated DNA. *Proceedings of the National Academy of Sciences, U.S.A.* **63**, 144–151.

KAPLAN, R. W. & KAPLAN, C. (1956). Influence of water content on UV-induced S-mutation and killing in *Serratia*. *Experimental Cell Research*, **11**, 378–392.

KELLY, R. B., ATKINSON, M. R., HUBERMAN, J. A. & KORNBERG, A. (1969). Excision of thymine dimers and other mismatched sequences by DNA polymerase of *Escherichia coli*. *Nature, London*, **224**, 495–501.

KELNER, A. (1949*a*), Effect of visible light on the recovery of *Streptomyces griseus* conidia from ultraviolet irradiation injury. *Proceedings of the National Academy of Sciences, U.S.A.* **35**, 73–79.

KELNER, A. (1949*b*). Photoreactivation of ultraviolet-irradiated *Escherichia coli* with special reference to the dose-reduction principle and to ultraviolet-induced mutation. *Journal of Bacteriology*, **58**, 511–532.

LEHMANN, A. R. (1976). Repair processes of radiation-induced DNA alterations. In *Effects of Ionizing Radiation on Nucleic Acids*, ed. A. J. Bertinchamps, J. Hütterman, R. Teoule & W. Kohnlein. Springer: Berlin. (In press.)

LEY, R. D. (1973). Post-replication repair in an excision-defective mutant of *Escherichia coli*: ultraviolet light-induced incorporation of bromodeoxyuridine into parental DNA. *Photochemistry and Photobiology*, **18**, 87–95.

LION, M. (1976). Modification of the radiation effects. In *Effects of Ionizing Radiation on Nucleic Acids*, ed. A. J. Bertinchamps, J. Hüttermann, R. Teoule & W. Kohnlein. Springer: Berlin. (In press.)

MARSDEN, H. S., POLLARD, E. C., GINOZA, W. & RANDALL, E. P. (1974). Involvement of *recA* and *exr* genes in the *in vivo* inhibition of the *recBC* nuclease. *Journal of Bacteriology*, **118**, 465–470.

MATSUMOTO, S. & KAGAMI-ISHI, Y. (1970). The temperature dependence of mortality rate of radiosensitive strains of *E. coli* and *S. cerevisiae*. *Japanese Journal of Genetics*, **45**, 153–160.

McGrath, R. A. & Williams, R. W. (1966). Reconstruction *in vivo* of irradiated *Escherichia coli* deoxyribonucleic acid; the rejoining of broken pieces. *Nature, London*, **212**, 534–535.

Monk, M. (1967). Observations on the mechanism of indirect induction by mating with ultraviolet-irradiated Col I donors. *Molecular and General Genetics*, **100**, 264–274.

Moroson, H. L. & Quintiliani, M. (ed.) (1970). *Radiation Protection and Sensitization*. London: Taylor and Francis.

Moseley, B. E. B. (1967). Repair of ultraviolet radiation damage in sensitive mutants of *Micrococcus radiodurans*. *Journal of Bacteriology*, **97**, 647–652.

Moseley, B. E. B. & Mattingly, A. (1971). Repair of irradiated transforming deoxyribonucleic acid in a wild type and a radiation-sensitive mutant of *Micrococcus radiodurans*. *Journal of Bacteriology* **105**, 976–983.

Moseley, B. E. B., Mattingly, A. & Copland, H. J. R. (1972). Sensitization to radiation by loss of recombination ability in a temperature-sensitive DNA mutant of *Micrococcus radiodurans* held at its restrictive temperature. *Journal of General Microbiology*, **72**, 329–338.

Pauling, C. & Beck, L. A. (1975). Role of DNA ligase in the repair of single-strand breaks induced in DNA by mild heating of *Escherichia coli*. *Journal of General Microbiology*, **87**, 181–184.

Pauling, C. & Hamm, L. (1968). Properties of a temperature-sensitive radiation-sensitive mutant of *Escherichia coli*. *Proceedings of the National Academy of Sciences, U.S.A.* **60**, 1495–1502.

Powers, E. L. & Talentire, A. (1968). The roles of water in the cellular effects of ionizing radiations. In *Actions Chimiques et Biologiques des Radiations*, **12**, ed. M. Haissinsky, pp. 3–67. Paris: Masson et Cie.

Radman, M. (1975). SOS repair: an inducible mutagenic DNA repair. In *Molecular Mechanisms for the Repair of DNA*. ed. P. C. Hanawalt & R. B. Setlow. New York: Plenum Press. (In press.)

Rahn, R. O. & Hosszu, J. L. (1968). Photoproduct formation in DNA at low temperatures. *Photochemistry and Photobiology*, **8**, 53–63.

Rupert, C. S. (1960). Photoreactivation of transforming DNA by an enzyme from bakers' yeast. *Journal of General Physiology*, **43**, 573–595.

Rupert, C. S. (1962*a*). Photoenzymatic repair of ultraviolet damage in DNA. I. Kinetics of the reaction. *Journal of General Physiology*, **45**, 703–724.

Rupert, C. S. (1962*b*). Photoenzymatic repair of ultraviolet damage in DNA. II. Formation of an enzyme-substrate complex. *Journal of General Physiology*, **45**, 725–741.

Rupert, C. S. & Harm, W. (1966). Reaction after photobiological damage. *Advances in Radiation Biology*, **2**, 1–8.

Rupp, W. D. & Howard-Flanders, P. (1968). Discontinuities in the DNA synthesized in an excision-defective strain of *Escherichia coli* following ultraviolet irradiation. *Journal of Molecular Biology*, **31**, 291–304.

Rupp, W. D., Wilde, C. E., Reno, D. L. & Howard-Flanders, P. (1971). Exchanges between DNA strands in ultraviolet-irradiated *Escherichia coli*. *Journal of Molecular Biology*, **61**, 25–44.

Sedgwick, S. G. (1975). Evidence for inducible error-prone repair in *Escherichia coli*. *Proceedings of the National Academy of Sciences, U.S.A.* (in press).

Sedgwick, S. G. & Bridges, B. A. (1972). Evidence for indirect production of DNA strand scissions during mild heating of *Escherichia coli*. *Journal of General Microbiology*, **71**, 191–193.

SEDGWICK, S. G. & BRIDGES, B. A. (1974). Requirement for either DNA polymerase I or DNA polymerase III in post-replication repair in excision-proficient *Escherichia coli. Nature, London*, **249**, 348–349.

SETLOW, J. K. (1967). The effects of ultraviolet radiation and photoreactivation. In *Comprehensive Biochemistry*, **27**, ed. M. Florkin & E. H. Stolz, pp. 157–209. Amsterdam: Elsevier.

SETLOW, R. B. (1964). Physical changes and mutagenesis. *Journal of Cellular and Comparative Physiology*, **64**, supplement 1, 51–68.

SETLOW, R. B. & CARRIER, W. L. (1964). The disappearance of thymine dimers from DNA: an error-correcting mechanism. *Proceedings of the National Academy of Sciences, U.S.A.* **51**, 226–231.

SETLOW, R. B. & CARRIER, W. L. (1967). Formation and destruction of pyrimidine dimers in polynucleotides by ultraviolet irradiation in the presence of proflavine. *Nature, London*, **213**, 906–907.

SETLOW, R. B. & SETLOW, J. K. (1972). Effect of radiation on polynucleotides. *Annual Reviews of Biophysics and Bioengineering*, **1**, 293–346.

SHENOY, M. A., JOSHI, D. S., SINGH, B. B. & GOPAL-AYENGAR, A. R. (1970). Role of bacterial membranes in radiosensitization. *Advances in Biological and Medical Physics*, **13**, 255–271.

SIDEROPOULOS, A. S. & SHANKEL, D. M. (1968). Mechanism of caffeine enhancement of mutations induced by sublethal ultraviolet dosages. *Journal of Bacteriology*, **96**, 198–204.

SMITH, K. C. (1971). The roles of genetic recombination and DNA polymerase in the repair of damaged DNA. In *Photophysiology*, **6**, ed. A. C. Giese, pp. 209–278. New York: Academic Press.

SMITH, K. C. (1975). The inhibition of DNA repair processes. In *Proceedings of the Eleventh International Cancer Congress*, Florence. (In press.)

SMITH, K. C. & HANAWALT, P. C. (1969). *Molecular Photobiology, Inactivation and Recovery*. New York: Academic Press.

SMITH, K. C. & MEUN, D. H. C. (1970). Repair of radiation-induced damage in *Escherichia coli*. I. Effect of *rec* mutations on post-replication repair of damage due to ultraviolet radiations. *Journal of Molecular Biology*, **51**, 459–472.

STAFFORD, R. S. & DONNELLAN, J. E. (1968). Photochemical evidence for conformation changes in DNA during germination of bacterial spores. *Proceedings of the National Academy of Sciences, U.S.A.* **59**, 822–828.

STAPLETON, G. E. (1960). Protection and recovery in bacteria and fungi. In *Radiation Protection and Recovery*, ed. A. Hollaender, pp. 87–116. Oxford: Pergamon Press.

SUHADI, F., KITAYAMA, S., OKAZAWA, Y. & MATSUYAMA, A. (1972). Isolation and some radiobiological properties of mutants of *Micrococcus radiodurans* sensitive to ionizing radiations. *Radiation Research*, **49**, 197–211.

SUTHERLAND, B. M. (1975). Photoreactivation in animal cells. *Life Sciences*, **16**, 1–6.

SWENSON, P. A. & SCHENLEY, R. L. (1970). Evidence for the control of respiration by DNA in ultraviolet-irradiated *Escherichia coli* B/r cells. *Mutation Research*, **9**, 443–453.

SWENSON, P. A. & SCHENLEY, R. L. (1974). Respiration, growth and viability of repair-deficient mutants of *Escherichia coli* after ultraviolet irradiation. *International Journal of Radiation Biology*, **25**, 51–60.

TAKAGI, Y., SEKIGUCHI, M., OKUBO, S., NAKAYAMA, H., SHUNADA, A. K., YASUDA, S., MISHIMOTO, T. & YOSHIHARA, H. (1968). Nucleases specific for ultraviolet light irradiated DNA and their possible role in dark repair. *Cold Spring Harbor Symposium on Quantitative Biology*, **33**, 219–227.

TOWN, C. D., SMITH, K. C. & KAPLAN, H. S. (1971). DNA polymerase required for rapid repair of X-ray-induced DNA strand breaks *in vivo*. *Science, Washington*, **172**, 851–854.

TOWN, C. D., SMITH, K. C. & KAPLAN, H. S. (1973). Repair of X-ray damage to bacterial DNA. *Current Topics in Radiation Research Quarterly*, **8**, 351–399.

WEBB, R. B., EHRET, C. F. & POWERS, E. L. (1958). A study of the temperature dependence of radiation sensitivity of dry spores of *Bacillus megaterium* between 5 °K and 309 °K. *Experientia*, **14**, 324–326.

WEBB, S. J. (1965). *Bound Water in Biological Integrity*. Springfield, Illinois: C. Thomas.

WEBB, S. J. & TAI, C. C. (1968). Lethal and mutagenic actions of 3200–4000 Å light. *Canadian Journal of Microbiology*, **14**, 727–735.

WEIGLE, J. J. (1953). Induction of mutation in a bacterial virus. *Proceedings of the National Academy of Sciences, U.S.A.* **39**, 628–636.

WITKIN, E. M. (1947). Genetics of resistance to radiation in *Escherichia coli*. *Genetics*, **32**, 221–248.

WITKIN, E. M. (1967*a*). The radiation sensitivity of *Escherichia coli* B: a hypothesis relating filament formation and prophage induction. *Proceedings of the National Academy of Sciences, U.S.A.* **57**, 1275–1279.

WITKIN, E. M. (1967*b*). Mutation-proof and mutation-prone modes of survival in derivatives of *Escherichia coli* B differing in sensitivity to ultraviolet light. *Brookhaven Symposia in Biology*, **20**, 17–53.

WITKIN, E. M. (1969). Ultraviolet-induced mutation and DNA repair. *Annual Review of Genetics*, **3**, 525–552.

WITKIN, E. M. & FARQUHARSON, E. L. (1969). Enhancement and diminution of ultraviolet-light-initiated mutagenesis by post-treatment with caffeine in *Escherichia coli*. In *Mutation as Cellular Process*, ed. G. E. W. Wolstenholme & M. O'Connor, pp. 36–49. London: Churchill.

WITKIN, E. M. & GEORGE, D. L. (1973). Ultraviolet mutagenesis in *polA* and *uvrA polA* derivatives of *Escherichia coli* B/r: Evidence for an inducible error-prone repair system. *Genetics*, Suppl. **73**, 91–108.

WOODCOCK, E. & GRIGG, G. W. (1972). Repair of thermally induced DNA breakage in *Escherichia coli*. *Nature New Biology*, **237**, 76–79.

YOUNGS, D. A., VAN DER SCHUEREN, E. & SMITH K. C. (1974). Separate branches of the *uvr*-gene dependent excision repair process in ultraviolet irradiated *Escherichia coli* K-12 cells: their dependence upon growth medium and the *polA*, *recA*, *recB* and *exrA* genes. *Journal of Bacteriology*, **117**, 717–725.

EXPLANATION OF PLATE

PLATE 1

Photoreversal of u.v. killing of *E. coli* WP2 *uvrA*. About 10^7 bacteria were spread over the surface of the agar and irradiated with a sufficient dose of u.v. to reduce survival to less than 1 in 10^7 (except at one edge where there was some shielding). The areas within the letters PRL were exposed to visible light for 20 min before incubation of the plates at 37 °C resulting in a dramatic restoration (photo: Colin Atherton).

PLATE I

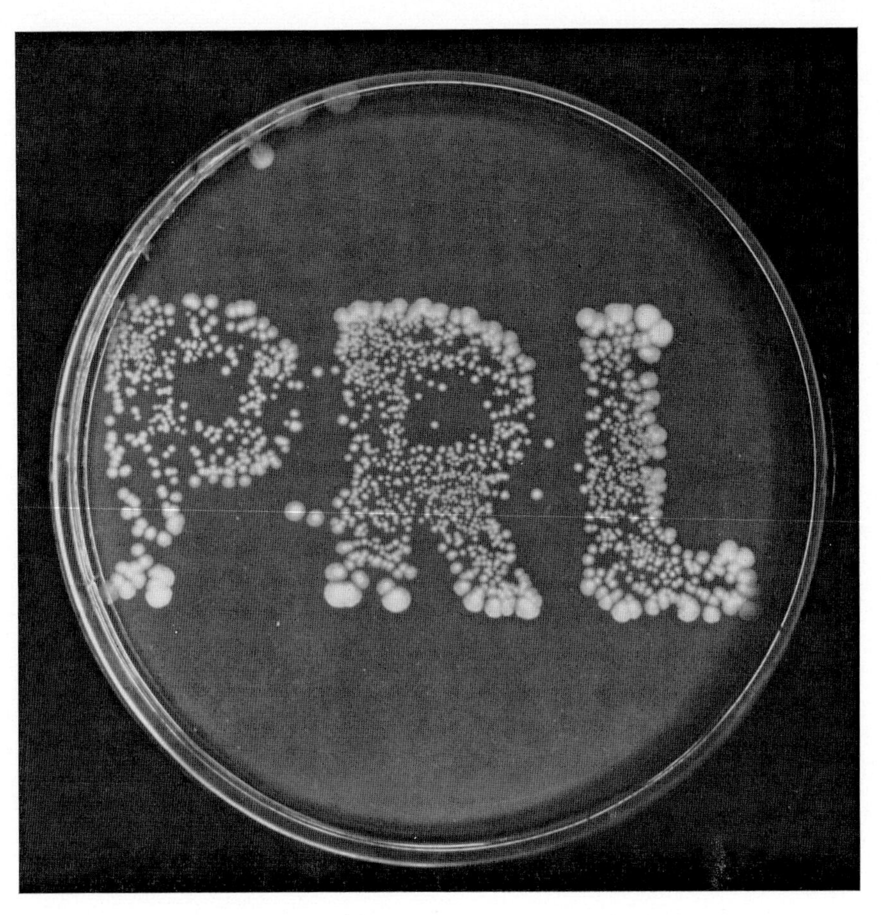

(*Facing page* 208)

CELLULAR DAMAGE INITIATED BY VISIBLE LIGHT

N. I. KRINSKY

Department of Biochemistry and Pharmacology,
Tufts University School of Medicine,
136 *Harrison Avenue, Boston, Massachusetts* 02111 *USA*

INTRODUCTION

In the seventh *Symposium of the Society for General Microbiology*, Stanier & Cohen-Bazire (1957) offered a most lucid discussion of the role of light in the microbial world. The authors focused primarily on positive aspects of light as seen from the world of the microbe. They dealt with photosynthesis, phototropism, and phototaxis. In that chapter, Stanier & Cohen-Bazire (1957) first presented the results of the very exciting work dealing with a new function for carotenoid pigments in photosynthetic bacteria and conducted at their laboratory. Griffiths *et al.* (1955) had reported that light had a lethal effect on a carotenoid-deficient mutant strain of the purple sulphur bacterium, *Rhodopseudomonas spheroides*. These observations were later extended (Sistrom, Griffiths, & Stanier, 1956) and the suggestion was made that the carotenoid pigments were protecting these mutant cells from the potentially harmful effects of light. The requirement for light, photosensitizer, and oxygen led Sistrom *et al.* (1956) to propose that the phenomenon they observed was a photodynamic effect.

The photodynamic effect, or the sensitization of living tissue by light, a suitable sensitizing pigment, and oxygen, had been known since the initial reports by Raab (1900). Raab had demonstrated that acridine could be used to kill paramecia in the presence of light. The history of the discovery of the photodynamic effect and a comprehensive bibliography up to 1940 can be found in the monograph of Blum (1941). Blum collected a vast amount of information on photosensitization phenomena in various organisms. For more recent surveys, the reader can refer to several extensive reviews (Harrison, 1967; Spikes & Straight, 1967; Krinsky, 1968; Spikes, 1968; Spikes & Livingston, 1969; Giese, 1971).

Although this chapter will also discuss photosensitization phenomena, it will be primarily concerned with the damage initiated by visible light and the mechanisms that cells have evolved to protect themselves against light. The protective effect of carotenoid pigments in nature has been extensively documented by this author earlier (Krinsky, 1968) but the molecular mechanisms have only been elucidated more recently. In order to understand fully the biological response called photosensitization or the photodynamic effect, it is first necessary to understand the basic photochemical reactions which initiate this phenomenon. Only through an understanding of the photochemical reactions will it be possible to learn how cells are damaged by light, and what mechanisms nature has evolved to offer protection. The ability of cells to protect themselves against light has served as a model for generating ideas on light protection that are applicable not only to unicellular organisms but also to humans.

PHOTOCHEMISTRY

At the time photobiologists were observing more and more examples of the photodynamic effect (Blum, 1941), a number of photochemists were investigating the chemical changes associated with photosensitized oxidations. These reactions are initiated by light (hν) exciting a suitable sensitizer (S) forming the first electronically excited species of the sensitizer, referred to as the singlet excited species (^{1}S). This electronically excited species has an extremely short lifetime ($< 10^{-11}$ sec) and can dissipate its energy through (1) interaction with the solvent, (2) emission of a photon in the form of fluorescence, or (3) with certain sensitizers may be converted through an intersystem crossing to a metastable excited species called the triplet sensitizer (^{3}S). This series of reactions, depicted in equation 1, produces an electronically excited species which has a sufficiently long lifetime to interact with other chemicals and initiate photochemical reactions.

$$\mathrm{S} \xrightarrow{h\nu} {}^1\mathrm{S} \rightsquigarrow {}^3\mathrm{S} \qquad (1)$$

The understanding of how ^{3}S can initiate photosensitized oxidations has gradually evolved as new techniques and new experimental evidence have been presented by a variety of photochemists working in this field. One of the earliest hypotheses was presented by Kautsky *et al.* (1931) who devised an experiment in which they physically separated the sensitizer from a suitable substrate molecule and obtained a photosensitized oxidation upon illumination. They postulated that the active intermediate was an excited species of oxygen formed by the reaction of ^{3}S and ground state oxygen. This reaction (equation 2) should proceed efficiently since oxygen is a paramagnetic molecule which exists as a triplet species in its ground state. Therefore, the interaction of ^3S and oxygen should result in spin-conserved reaction that would be formally allowed.

$$^3S + {}^3O_2 \longrightarrow S + {}^1O_2. \quad (2)$$

However, other workers have observed that different sensitizers could give rise to different products using the same substrates, which argued against a single, universal intermediate in photosensitized oxidations. In some cases, the sensitizer itself would be oxidized or reduced through hydrogen or electron transfer in the photosensitization process which could occur in the absence of oxygen. Such observations have argued strongly against the Kautsky hypothesis that the $^1O_2$ was produced in photosensitized oxidations.

This controversial area was greatly clarified by the reports of Foote & Wexler (1964*a*) and Corey & Taylor (1964) who reported that $^1O_2$, generated either chemically or by use of radiofrequency discharge apparatus, resulted in products which were identical to those formed in photosensitized oxidation using dye such as methylene blue as the photosensitizer. On the basis of these results, Foote & Wexler (1964*b*) proposed that the Kautsky hypothesis of $^1O_2$ production during photosensitized oxidations had been confirmed. These experiments have since been repeated and confirmed on numerous occasions, particularly from the laboratories of Wilson (1966), Foote (1968*a*, *b*), Kasha & Kahn (1970) and Kearns (1971) and they have all demonstrated $^1O_2$ as a major intermediate in photosensitized oxidations.

It would appear then that the ability of a photosensitizer to generate $^1O_2$ or to participate in a redox reaction involving hydrogen or electron abstraction depends on the chemical nature of the sensitizer, the environment, and the availability of oxygen. Gollnick & Schenck (1967) have classified these two different types of photosensitized reactions as: Type I, the redox reactions which do not involve an initial interaction of sensitizer with oxygen and Type II, those reactions in which the excited sensitizer interacts directly with molecular oxygen. The two types are described schematically in Fig. 1. In both types, ^3S is generated as described in equation 1 and can then react with a suitable substrate (AH, A) leading either to hydrogen abstraction or electron transfer. In both cases, the reactants can proceed in photoreactions either with oxygen or with other species to yield oxidized products. In Type II, ^3S reacts directly with $^3O_2$ yielding primarily $^1O_2$. A small percentage of the oxygen molecules reacting with ^3S can apparently undergo an electron transfer, yielding the superoxide anion radical, $O_2^{\overline{\cdot}}$. The extent

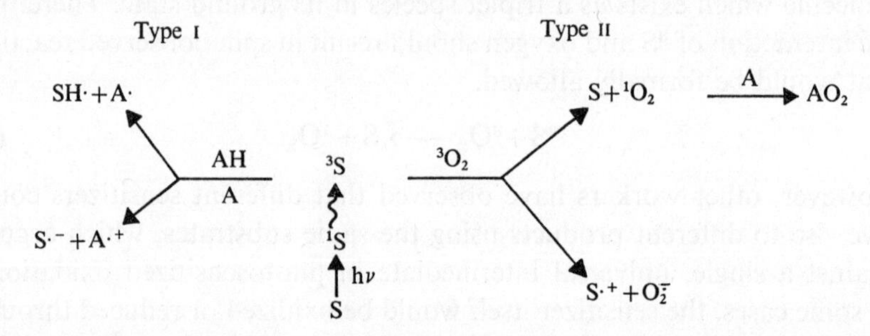

Fig. 1. Schematic representation of reactions involved in Type I and Type II photosensitized oxidations. The sensitizer (S), when excited to its triplet state (^3S) can react directly with substrates (AH, A) in Type I reactions, or can react directly with O_2 yielding primarily singlet oxygen ($^1O_2$) in a Type II reaction. This $^1O_2$ can react further with substrates (A) yielding oxidized products.

of this reaction has not yet been quantified. $^1O_2$ can then undergo various reactions in which the excited oxygen functions primarily as a dienophile (Gollnick & Schenck, 1967; Foote, 1975).

Almost every biological molecule that absorbs light in the region of the electromagnetic spectrum ranging from the near ultraviolet (320 nm) through the visible region to the near infrared (900 nm) has been proposed as a potential photosensitizing compound. In the biological systems to be discussed in the next section, it is frequently difficult to assign precisely the endogenous pigment responsible for a particular photobiological or photosensitized effect. For example, porphyrins, which are ubiquitously distributed in living tissue with the exception of some obligate anaerobes, are very effective photosensitizers (Spikes, 1975). In addition to endogenous sensitizers, many organisms can be sensitized to the effects of light following the application of exogenous sensitizers. These exogenous sensitizers can be administered experimentally while studying photodynamic effects, administered inadvertently while studying other effects of these compounds without realizing that they have the capacity to act as photosensitizers, or they can be presented to organisms through environmental changes. In many cases, once the sensitizer pigment is identified, it is possible to predict the type of chemical reaction that will take place. The prediction is based on whether the sensitizer, in the conditions obtainable within the organism, proceeds via a Type I reaction or Type II reaction. Even though the chemical consequences of these two types of reactions, which are either radical reactions or direct additions of oxygen, may be similar, it is frequently difficult to correlate the biological effects with the chemical reactions. Presumably this effect is due to the complexity of the biological response and the fact that it frequently represents a multitude of photosensitized reactions culminating in a given response of the organism.

BIOLOGICAL EFFECTS

Inasmuch as the counting of viable cells of a bacterial culture following a perturbation is a relatively simple laboratory operation, both the death of cells and failure to produce colonies have been used as the most frequent criteria for assessing the extent of photodynamic damage. Although such results are frequently clear cut, they represent the culmination of a series of complicated phenomena. It is impossible to tell the cause of death or the failure to multiply by counting the number of survivors. Any attempt to understand the molecular mechanisms that are involved in the lethal effects of photodynamic action necessitates looking at more refined alterations in the cell than the process of death. Toward this end, various investigators have looked at direct effects on nucleic acids following a photosensitized reaction. These nucleic acid effects include the development of mutants characterized by altered forms, altered biochemical properties, or increased ability to resist phage infection. All these effects have been used to characterize DNA involvement in the photodynamic effect. In addition, the effect of altered membrane function has also been used as a means of identifying the site and nature of the photosensitized effect. Therefore, studies have been carried out measuring changes in membrane-associated functions such as respiration, permeability, transport phenomena, and membrane-associated enzymes. However, even these effects which are not necessarily lethal can have complicated molecular mechanisms.

Failure to form colonies (death)

In addition to the monograph by Blum (1941), there are several recent reviews which deal with the lethal effects of photodynamic action, particularly with respect to micro-organisms (Harrison, 1967; Eisenstark, 1971; Krinsky, 1968, 1971).

Although the lethal effects of sunlight had been observed in the nineteenth century (Downes & Blunt, 1887), one of the earliest examples demonstrating that artificial light could also have bactericidal activity was that of Buchbinder, Solowey & Phelps (1941). These workers studied the survival rates of several strains of streptococci and observed that not only direct and indirect sunlight were effective in killing these micro-organisms, but that artificial illumination was also effective. They

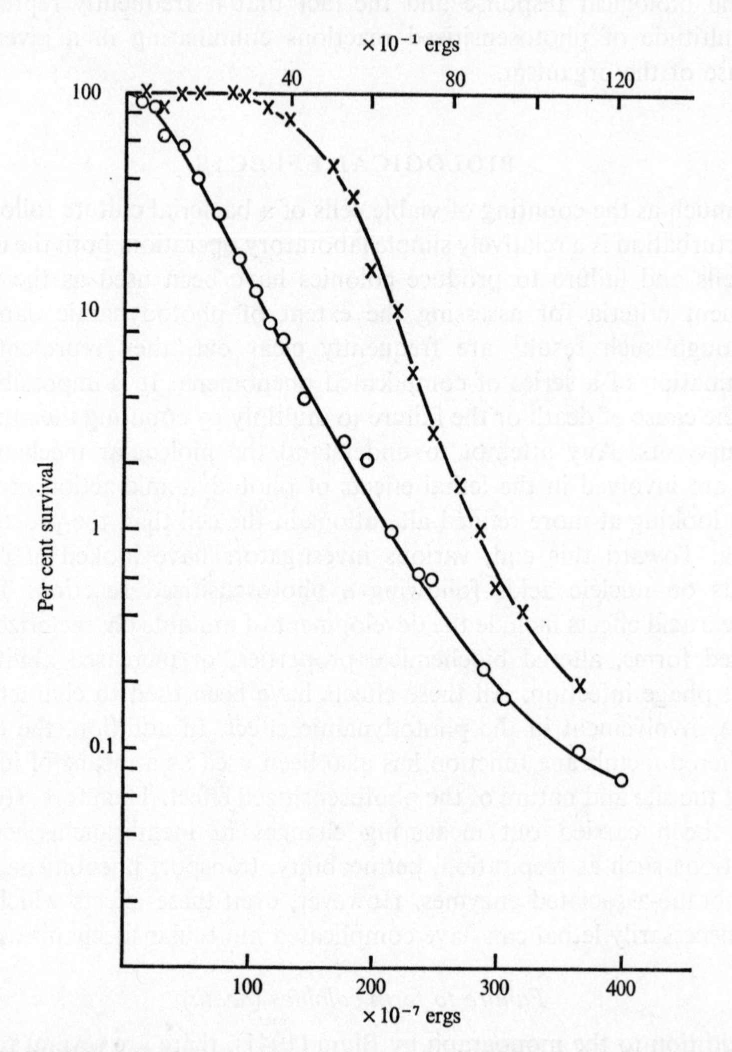

Fig. 2. Survival ratio plotted against energy per organism for *Escherichia coli* in liquid culture irradiated either with u.v. light at 265 nm (○) or visible light at 350–490 nm (×). (After Hollaender, 1943.)

suggested that daylight and direct sunlight might play an important role in suppressing various infections of the upper respiratory tract. The first quantitative experiment on the effect of visible light as opposed to u.v. light was carried out by Hollaender (1943) using *Escherichia coli* as the test organism and suitable light sources and filters to isolate different regions of the spectrum. As seen in Fig. 2, Hollaender was able to demonstrate that visible light (350–490 nm) was effective in bringing about the death of *Escherichia coli.* In his discussion, he pointed out differences between the lethal effect of visible radiation (350–490 nm) and that caused by u.v. irradiation (218–295 nm). He suggested that, since nucleic acids had no absorption in the visible region of the spectrum, compounds such as riboflavin might serve as the photosensitizing pigment.

Since the report of Hollaender (1943) differentiating the lethal effects of ultraviolet light from visible light, there have been many reports comparing the actions of these two portions of the electromagnetic spectrum. Much of this work was collected in an excellent review by Eisenstark (1971) who carefully tabulated many diverse effects of u.v. light (218–300 nm), near u.v.–visible light (300–490 nm) as well as the effects of photodynamic action on many different bacterial species. In his review, Eisenstark points out not only the difference between the effects of u.v. light and near u.v.–visible light, but also indicates the many similarities between the effects of photodynamic action and near u.v.–visible light. It is important to note that under both of these conditions it is necessary that oxygen is present in order to see a biological effect. Eisenstark makes a strong argument that the effects caused by near u.v.–visible light are due to the presence of natural photosensitizers which behave in a fashion analogous to that brought about by the addition of exogenous photosensitizers in a photodynamic action. Although all three types of light can attack and alter nucleic acids, only u.v. light does this by direct interaction whereas near u.v.–visible light and photodynamic action require the intervention of a natural or exogenous photosensitizing pigment.

A number of important articles have appeared recently dealing with this distinction between u.v. light and near u.v. and visible light effects. Webb & Lorenz (1970) studied lethality in a repairable strain and an excision repair deficient strain of *Escherichia coli* at wavelengths of 254, 313, 365 and 390–750 nm. They observed an oxygen enhancement of killing only at 365 nm and 390–750 nm. The oxygen-dependent damage induced by near ultraviolet light (365 nm) could be partially repaired by the excision-repair system, whereas there was very little repair in the cells treated with visible light (390–750 nm). Ferron, Eisenstark, & Mackay (1972) have studied the effects of visible and near ultraviolet light on recombinationless mutants of *Salmonella typhimurium.* They found that these two types of light gave very different effects which could be attributed to the direct action of u.v. light (< 295 nm) on DNA. However, near u.v. light (> 295 nm) makes use of a different sensitizer. In particular, these authors observed that DNA was rapidly degraded in these mutants following u.v. exposure and that pyrimidine dimers were also formed. Neither of these events took place following the near u.v. illumination. They also observed that phage were able to continue multiplying in the recombinationless mutants that were killed by u.v. light, whereas they could not multiply in cells exposed to near u.v. and visible light. They conclude that the death of the recombinationless mutants brought about by near u.v. and visible light is not associated with DNA degradation. The presence of a repair system following illumination with near u.v. light was also demonstrated by Cabrera-Juarez & Espinosa-Lara (1974) using *Hemophilus influenzae*. Their conclusion was based on the fact that the lethal effects of the irradiation with light from 325–400 nm showed a large shoulder indicative of the presence of a repair system.

Finally, the report of Kuhn & Starr (1970) merits attention. These workers studied various bacteria growing on nutrient agar in microscope growth chambers and found that irradiation in the visible range of the spectrum caused retardation of growth, morphological alterations and death. Part of this effect might be due to the increased temperature in the growth chambers during illumination. However, they found that limiting illumination to wavelengths above 420 nm did not result in significant increases in temperature and yet still produced killing in the chamber. Although they could find no correlation between pigmentation and sensitivity, they did conclude that dormant spores are more resistant to killing by visible light under these conditions than are vegetative cells.

Induction of mutations

Much of the work on mutagenesis initiated by visible light using either internal sensitizers or by means of a photodynamic effect has been reviewed earlier (Zelle & Hollaender, 1955; Zetterberg, 1964; Harrison, 1967; Spikes, 1968; Eisenstark, 1971). Some of the earliest observations include those made by Kaplan who observed that photodynamic effects could alter the colony form of *Serratia marcescens* (Kaplan, 1950*a*), develop both resistance to phage T7 (Kaplan, 1950*b*) and change the nutritional requirements in *Escherichia coli* (Kaplan, 1950*c*) and finally it could alter the morphology of colonies of *Penicillium notatum* (Kaplan, 1950*d*). Among the many recent examples of the induction of mutations by visible light are those of Mathews (1963) who observed that 8-methoxypsoralen photosensitization resulted in the development of many penicillin-resistant mutants in *Sarcina lutea*. Nutritional mutants in *E. coli* have been observed by Bellin, Lutwick &

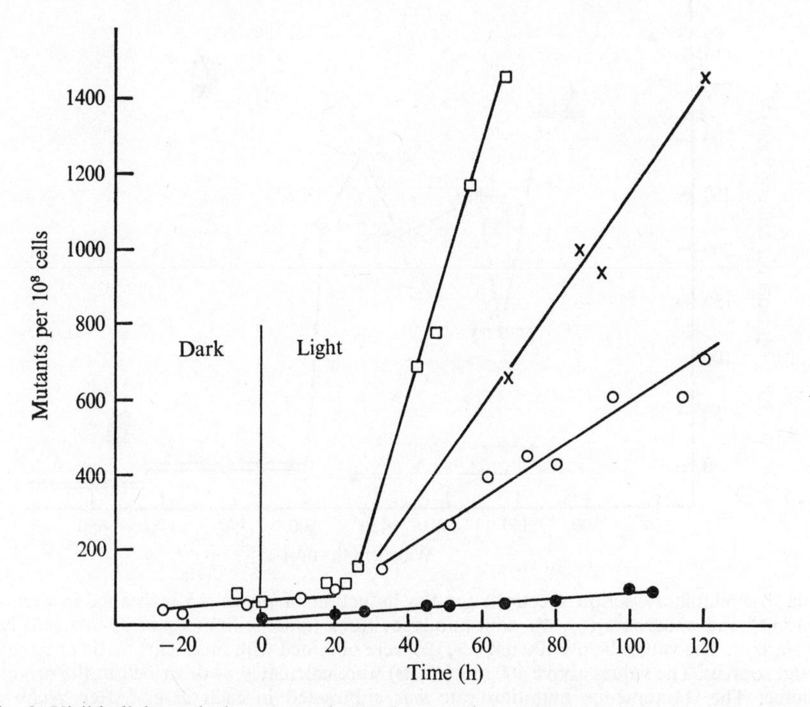

Fig. 3. Visible light and photodynamic (10^{-5} M acridine orange) mutation rates (resistance to phage T5) in glucose-limited chemostat cultures of *E. coli* B/r/1, *try*⁻ at several levels of irradiance by visible light. Growth rates were 0.18 to 0.22 division per hour except for the acridine orange chemostat, which had a rate of 0.10 division per hour. Included in this figure are dark controls (●), acridine orange-treated cells exposed to 5.3 μW mm^{-2} (□), and two visible light-exposed cultures, at either 96.5 μW mm^{-2} (×) or 47.0 μW mm^{-2} (○). (After Webb & Malina, 1967. Copyright © 1967 by the American Association for the Advancement of Science.)

Jonas (1969) using both neutral red and methylene blue as the photosensitizers. In contrast, rose bengal, which could be used to demonstrate lethal photodynamic effects did not elicit mutations in this organism. Webb & Tai (1969) illuminated various strains of *E. coli* with light from 320–400 nm without an external sensitizer and observed both the development of mutations to a variety of growth factors and the development of antibiotic resistance. Similar observations have been made by Cabrera-Juarez & Espinosa-Lara (1974) using similar wavelengths of light on *Haemophilus influenzae*. These workers were able to demonstrate that this irradiation could cause a marked increase in the development of mutants that were resistant to streptomycin.

Some of the most interesting results of the study of light-induced mutation comes from the laboratory of R. B. Webb. He has studied

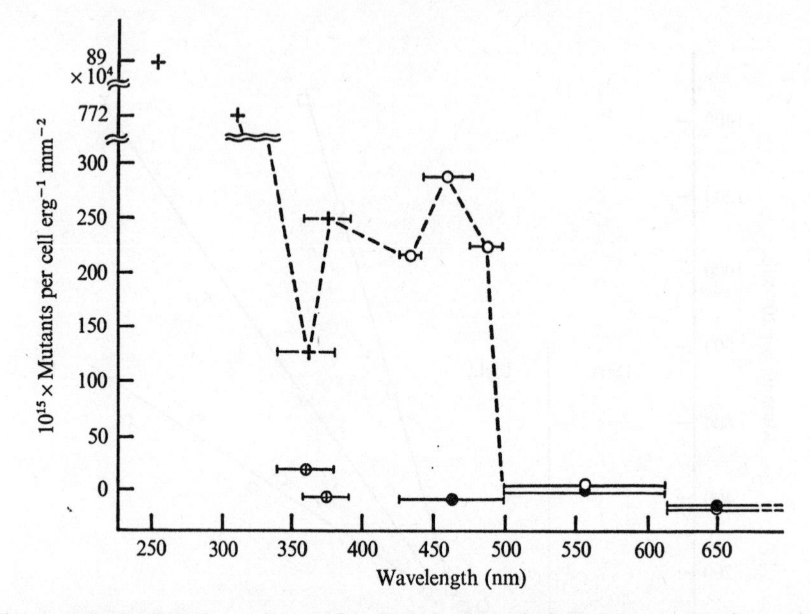

Fig. 4. Preliminary action spectrum for the induction of phage T5 resistance in aerobic (+, ○) and anaerobic (⊕, ●) continuous cultures (glucose-limited) of *Escherichia coli* B/r/l, *try*⁻. The values below 400 nm (+, ⊕) were obtained with monochromatic or narrow band sources. The values above 400 nm (○, ●) were calculated as described in the original source. The spontaneous mutation rate was subtracted in each case. (After Webb & Malina, 1970.)

the mutagenic effects of near u.v. and visible light either in the presence of sensitizers (Webb & Kubitschek, 1963) or using only endogenous photosensitizers (Webb & Malina, 1967; Webb & Malina, 1970; Webb, 1972). The mutation studied most frequently by the above authors is the development of resistance to bacteriophage T5 in chemostat cultures of *E. coli*. Very similar observations of this type of mutation were also reported by Kubitschek (1967). The development of mutants to bacteriophage T5 is a function of the light intensity as seen in Fig. 3. Under these experimental conditions, the authors were able to observe mutation rates more than 18 times greater than the spontaneous dark rate using intensities of visible light that were not lethal (Webb & Malina, 1967). The photodynamic effect on the development of this mutation was clearly demonstrated with the observation that oxygen was required to observe the development of phage resistance (Webb & Malina, 1970). An action spectrum for this phenomenon was determined by Webb & Malina (1970) and is seen in Fig. 4. The shape of this action spectrum is similar to that seen with many blue-light effects in biology and is suggestive of either a flavin or carotenoid involvement in the

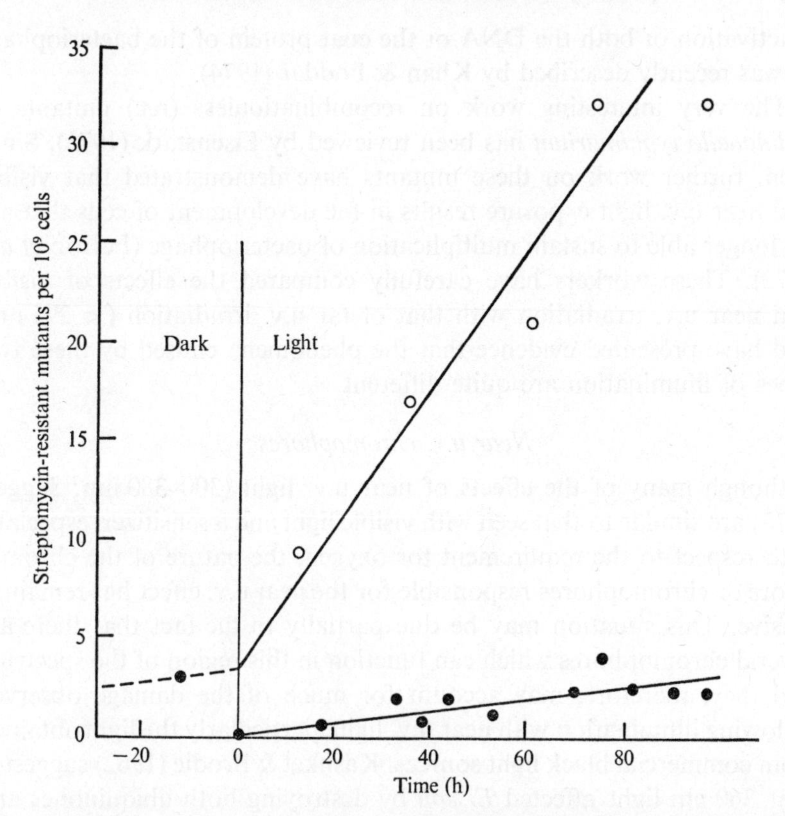

Fig. 5. Induction of streptomycin-resistant mutants in *E. coli* B/r/l, *try*$^-$ by cool-white fluorescent visible light (○) at 145 erg mm^{-2} sec^{-1} in glucose-limited chemostats compared to dark controls (●). Growth rate 0.25 generations per hour. (After Webb, 1972.)

process (Krinsky, 1971). A similar effect using relatively low light intensities from a cool-white fluorescent bulb has been reported by Webb (1972) in relation to the development of streptomycin resistance (Fig. 5).

At this time, it is still impossible to decide the nature of the chromophore that results in the induction of mutations following a photodynamic effect. There are a number of molecules which may serve as prime candidates although it is still not clear whether or not the effect is due directly to an interaction of a sensitizer associated with DNA or merely reflects the end result of a process initiated by a photosensitizer not associated with DNA. Numerous reports dealing with the inactivation or alteration of bacteriophage have also appeared (Welsh & Adams, 1954; Ritchie, 1964; Calberg-Bacq, Delmelle & Duchesne, 1968). Such bacteriophage mutation, however, can be a result of photodynamic inactivation of both the DNA or the coat protein of the bacteriophage as was recently described by Khan & Poddar (1974).

The very interesting work on recombinationless (*rec*) mutants of *Salmonella typhimurium* has been reviewed by Eisenstark (1971). Since then, further work on these mutants have demonstrated that visible and near u.v. light exposure results in the development of cells that are no longer able to sustain multiplication of bacteriophage (Ferron *et al.*, 1972). These workers have carefully compared the effects of visible and near u.v. irradiation with that of far u.v. irradiation (< 295 nm) and have presented evidence that the phenomena caused by these two types of illumination are quite different.

Near u.v. chromophores

Although many of the effects of near u.v. light (300–380 nm; Jagger, 1973) are similar to that seen with visible light and a sensitizer, especially with respect to the requirement for oxygen, the nature of the chromophore or chromophores responsible for the near u.v. effect has remained elusive. This situation may be due partially to the fact that there are several chromophores which can function in this region of the spectrum and they, therefore, may account for much of the damage observed following illumination with near u.v. light, particularly the light obtained from commercial black light sources. Kashket & Brodie (1962) suggested that 360 nm light affected *E. coli* by destroying both ubiquinones and naphthoquinones which are both essential to this organism's oxidative metabolism. This sensitivity of ubiquinone to near u.v. irradiation has been puzzling inasmuch as several workers observed inhibition at specific electron transport sites in organisms such as *E. coli* which had not been considered to require ubiquinone. Part of this phenomenon has been clarified by Bragg (1971) who has demonstrated that near u.v. light interferes with two segments of the respiratory chain of *E. coli*, but that the interruption at the second site is due to the destruction of cytochrome a_3 and not ubiquinone.

Swenson & Setlow (1970) have suggested that other chromophores may be involved in the expression of u.v. damage. These workers studied the induction of tryptophanase in *E. coli* and found that this process could be inhibited by near u.v. irradiation, again using a black light source with peak emission at 360 nm. They measured an action spectrum for the inhibition and found a peak around 334 nm which corresponded to the peak observed for growth delay. However, the tryptophanase inhibition is more sensitive to near u.v. irradiation than the other harmful effects that they could measure. They suggested that the chromophore for this process may be pyridoxal phosphate which both absorbs in the same region of the spectrum and also serves as a co-factor for tryptophanase. Ramabhadran, Fossum & Jagger (1975) have presented evidence that 4-thiouridine in transfer RNA serves as the chromophore for near u.v.-induced growth delay in *E. coli*.

Another chromophore has been suggested by Peters & Pauling (1974) based on the loss of the capacity of irradiated cells to replicate DNA. On the basis of their observations, they suggest that ribonucleotide diphosphate (RDP) reductase may be the sensitizer molecule. This enzyme has an absorption maximum at 360 nm. Several experiments indicate the deficiency of deoxynucleoside derivatives following 365 nm irradiation and this observation could be explained by a deficiency in RDP reductase.

It is becoming increasingly clear that in certain cases irradiation of the growth medium is sufficient to lead to inhibition of growth and division. This has been demonstrated clearly by Webb & Lorenz (1972) and Yoakum & Eisenstark (1972). The latter workers have also demonstrated that black light illumination of either the medium or L-tryptophan results in the production of a photoproduct which is toxic for recombinationless (*rec*) mutants of *Salmonella typhimurium*. The above phenomenon is a photodynamic effect, for the toxic product is only observed when L-tryptophan is irradiated in the presence of air. With regard to the actual mechanism of phototoxicity, Yoakum (1975) has demonstrated that the photoproducts of L-tryptophan have a capacity to sensitize bacterial deoxyribonucleic acid to black light illumination, particularly the light emitted at 365 nm. The effect is observed as a greater than 10-fold increase in the number of deoxyribonucleic acid strand breaks in the presence of L-tryptophan photoproducts and 365 nm illumination. Similar results have been observed in mammalian cells in culture by Wang, Stoien & Landa (1974) and Stoien & Wang (1974).

Membrane damage

Although the following effects may not be associated exclusively with light-induced alterations to membranes, most of them appear to be related phenomena that are believed to occur in or on membranes (Krinsky, 1974*b*). The ability of visible light to inhibit the respiration of various bacterial systems has been known for many years (Rubenstein, 1931) but the mechanisms have only recently begun to be understood. It is now apparent that some of the co-factors in enzymes involved in respiration are sensitive to both near u.v. and visible light and when these co-factors are damaged, an inhibition of respiration results. The effect

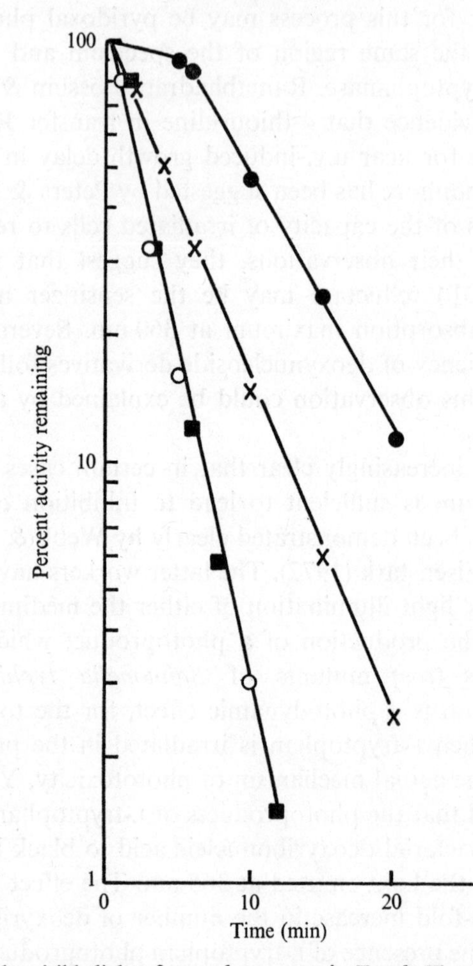

Fig. 6. Inhibition by visible light of several processes in *E. coli*. These processes include respiration (■) and uptake of either glycine (○), methyl-thio-β-D-galactoside (×), or phenylalanine (●). (After D'Aoust *et al.*, 1974.)

of near u.v. light has already been discussed. These observations, which have been reviewed recently (Epel, 1973), indicate that visible light, at intensities which do not kill cells, has the capacity to inhibit one component of cytochrome a_3. This phenomenon is not limited to bacterial systems but also occurs in yeast and in animal mitochondria (Ninnemann, Butler & Epel, 1970*a*, *b*). The ability of visible light to destroy haem proteins is not surprising inasmuch as Mitchell & Anderson (1965) demonstrated that visible light could inactivate catalase from either *Sarcina lutea*, corn seedlings or even the crystalline enzyme from beef liver. As will be discussed later, the effect of light can be attenuated by the presence of carotenoid pigments. D'Aoust *et al.* (1974), while studying the effects of visible light on active transport on *E. coli*, observed an inhibition of respiration which could be correlated with the inhibition of the active uptake of glycine. They had reported earlier (Barran *et al.*, 1974; Schneider *et al.*, 1974) that visible light inhibits various transport systems in *E. coli* but at different rates. The correlation between the inhibition of both respiration and glycine uptake is clearly demonstrated in Fig. 6. Other transport systems, however, such as those for phenylalanine and methyl-thio-β-D-galactoside are inactivated at very different rates. These authors suggest that the differential effects of visible light on active transport might serve as a means of dissecting out the actual molecular mechanisms that are involved in transport processes.

As mentioned above, visible light can inactivate enzymes both in solution and in living organisms, as exemplified by the inactivation of catalase (Mitchell & Anderson, 1965). In addition, Mathews & Sistrom (1960) have demonstrated that, in a carotenoidless strain of *Sarcina lutea*, visible light in the presence of a photosensitizer rapidly inactivates two membrane enzymes, succinic dehydrogenase and diphosphopyridine nucleotide oxidase. Under the same conditions, the permeability of the cells was altered as they observed an increase in 260 nm absorbing material released in the medium following illumination in air. Using isolated membranes of *Acholeplasma laidlawii*, Rottem, Gottfried & Razin (1968) demonstrated a destruction of the membrane enzyme adenosine triphosphatase in a dye-sensitized reaction. Under photodynamic conditions using an exogenous dye, Cooney & Krinsky (1972) demonstrated that light had a lethal effect upon this organism.

Photodynamic effects have also been observed in many non-bacterial systems. Bellin & Ronayne (1968) demonstrated that the cell membrane of *Euglena gracilis* became more permeable to dye ions following injury brought about by photodynamic action. Goldstein & Harber (1972) demonstrated the inactivation of the enzyme, acetylcholinesterase in red blood cells from erythropoietic protoporphyric patients. In this case the excess production of protoporphyrin results in the production of light-sensitive red blood cells which can be damaged following exposure to visible light. The ultimate effect is a photohaemolysis of the red blood cell. There is an extensive literature beyond the scope of this review, dealing with the harmful effects of visible light on the rod outer segments of the eye (Dowling & Sidman, 1962; Noell *et al.*, 1966). The possibility of using light deprivation as a therapeutic tool in certain degenerative diseases of the retina has been discussed by Berson (1973).

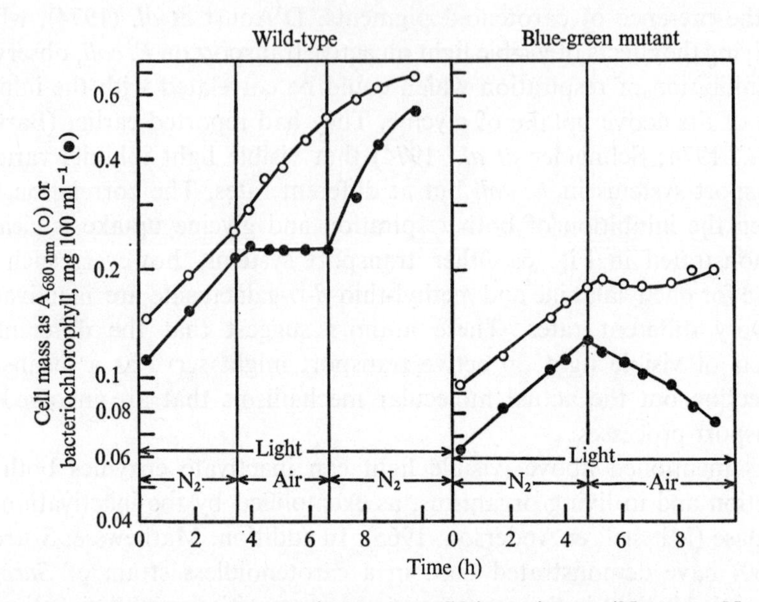

Fig. 7. The effect of air (95 per cent air–5 per cent CO_2), or nitrogen (95 per cent N_2–5 per cent CO_2) in the presence of light on growth (○) and pigment synthesis (●) by the wild-type strain and the blue-green mutant strain of *Rhodopseudomonas spheroides* at 30 °C. (After Sistrom *et al.*, 1956.)

PROTECTION AGAINST PHOTODYNAMIC EFFECTS

For some time now it has been appreciated that one of the major protective devices in bacterial systems for preventing harmful photodynamic effects consists of the presence of coloured carotenoid pigments (reviews by Krinsky, 1968, 1971, 1974*a*). The work on carotenoid-deficient mutants of the purple sulphur bacterium, *Rhodopseudomonas spheroides* (Griffiths *et al.*, 1955; Sistrom *et al.*, 1956) provided some of the earliest data on carotenoid protection. These workers studied both the wild-type strain and a blue-green mutant strain which lacked coloured carotenoid pigments. Under photosynthetic conditions, both strains displayed identical growth behaviour, but in the presence of light and air the growth of the mutant strain ceased, its bacteriochlorophyll was bleached and the organism was killed (Fig. 7). Sistrom *et al.* (1956) proposed that the coloured carotenoid pigments served as protective agents against a photodynamic effect initiated by the endogenous pigment bacteriochlorophyll. Shortly thereafter, Kunisawa & Stanier (1958), working with the chemoheterotrophic bacterium *Corynebacterium poinsettiae*, demonstrated the protective function of carotenoids in non-photosynthetic bacteria. In order to show this effect, these workers prepared a carotenoidless white mutant strain by exposing the wild-type strain to u.v. light. High intensity visible light had no effect on these two strains but a photodynamic effect could be initiated by adding an exogenous photosensitizer such as toluidine blue and exposing the organisms to visible light. Under these conditions the white mutant strain was readily killed without any apparent effect on the pigmented wild-type strain. This phenomenon was a true photodynamic effect because no killing could be observed when the two strains were illuminated under nitrogen in the presence of an exogenous photosensitizer.

Although the studies of Kunisawa & Stanier (1958) indicated that the presence of coloured carotenoid pigments protected bacterial cells from a photodynamic effect initiated by the addition of an exogenous sensitizer, they differed from the studies of Sistrom *et al.* (1956). In the latter case an endogenous pigment served as the sensitizer. It remained for the experiments of Mathews & Sistrom (1959) to demonstrate the ecological importance of coloured carotenoid pigments in bacteria. These workers used a yellow coccus, *Sarcina lutea*, and a white mutant strain as their test organism and exposed both strains to light under aerobic and anaerobic conditions. Their results, shown in Fig. 8, clearly indicate that a photodynamic effect occurs in the white mutant strain whereas the carotenoid-containing wild-type strain was not affected by this treatment. They concluded that organisms which normally exist in an aerobic environment with the potential for exposure to light would be protected by the presence of carotenoid pigments, whereas organisms lacking these pigments would undergo lethal effects. These studies in *Sarcina lutea* have been continued by Mathews and her collaborators (Mathews & Sistrom, 1960; Mathews, 1963, 1964; Mathews-Roth & Krinsky, 1970*a*, *b*).

Since then, there have been many photodynamic studies using carotenoid-containing bacteria and carotenoidless strains, either mutant strains or the wild-type strains rendered colourless by chemical inhibition of carotenoid biosynthesis. In almost all cases the presence of coloured carotenoid pigments has either delayed the onset of killing or given complete protection against the harmful effects of visible light. This phenomenon is not limited to bacteria but has been observed in corn seedlings (Koski & Smith, 1951; Anderson & Robertson, 1960), sun flowers (Wallace & Schwarting, 1954; Wallace & Habermann, 1959), algae (Claes, 1954; Sager & Zalokar, 1958) and plants that have been treated with herbicides which inhibit carotenoid biosynthesis (Burns, Carter & Buchanan, 1971). There have been several examples where

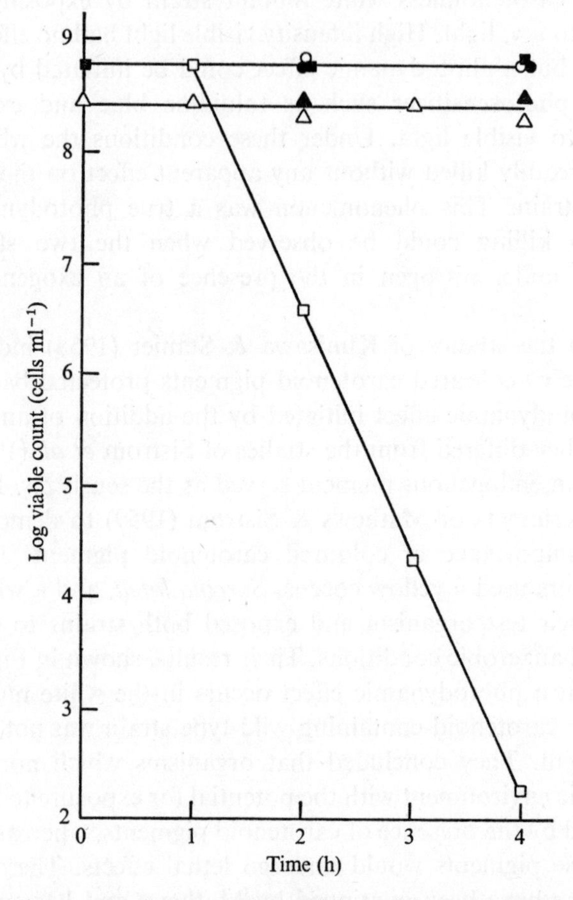

Fig. 8. Effect of exposure of carotenoid-containing wild-type *Sarcina lutea* (○) and a carotenoidless mutant strain (□, △) to direct sunlight in an atmosphere of air (○, □) or nitrogen (△). The dark controls are indicated by the corresponding filled symbols (●, ■, ▲). In all cases, the appropriate gases were bubbled through the cell suspension during the experiment. Light intensities measured with a Weston meter were 120000 lux at 0 h, 100000 lux at 2 h, and 50000 lux at 4 h. (After Mathews & Sistrom, 1959.)

the presence of coloured carotenoids is not associated with protection against the harmful effects of visible light. Maxwell, Macmillan & Chichester (1966) and Chichester & Maxwell (1969) presented evidence that carotenoid pigments do not protect the yeast, *Rhodotorula glutinis* against photosensitization initiated by visible light in the presence of an exogenous sensitizer. These workers studied both a pigmented and colourless strain of this yeast and were not able to observe protection in the carotenoid-containing wild-type strain in comparison to the colourless strain. One possible explanation for this observation is that the distribution of pigment in yeast is more complicated than that observed in prokaryotes and the carotenoids may not be associated with the site where the external photosensitizer initiates its lethal action. Cooney & Krinsky (1972) studied the photodynamic killing of *Acholeplasma laidlawii* and were able to demonstrate a clear photodynamic effect using either an exogenous sensitizer such as toluidine blue or an endogenous sensitizer under conditions of high light intensity. However, despite varying the carotenoid content per cell by factor of five, they were not able to demonstrate any difference in protection as the carotenoid concentration decreased. This may be due to the fact that even the low concentrations of carotenoids were above the threshold necessary to demonstrate protection. Huang & Haug (1974) have also studied the effect of varying the carotenoid content in *A. laidlawii*. They studied the effect of a 50-fold variation in carotenoid content as it affected membrane lipid fluidity. From their results they concluded that the carotenoid pigments serve to rigidify the plasma membrane of this organism. Schwartzel & Cooney (1974) studied the effects of light on *Micrococcus roseus* as a function of carotenoid content using both a wild-type strain, pigmented mutant strains, and a carotenoid-deficient strain induced by the addition of diphenylamine to the growth medium. They saw little difference in the photosensitivity of the wild-type and pigmented strains. In fact, the cells grown in the medium containing diphenylamine which did not contain significant quantities of coloured pigments were not as sensitive to photodynamic killing as was the wild-type strain. They concluded that carotenoids do not serve as protective agents against photodynamic effects in *Micrococcus roseus*.

The protective action of carotenoid pigments in bacteria, when present, can be observed not only in the intact organism but also in fractions obtained from whole cells. Rottem *et al.* (1968) demonstrated that the isolated membranes of *A. laidlawii* could be subjected to a photodynamic effect in the presence of toluidine blue. These workers observed that adenosine triphosphatase was destroyed in the membranes from carotenoid-deficient strains whereas the membranes from pigmented cells were protected. Similarly, Prebble & Huda (1973) studied the respiratory system of isolated cell membranes of *Sarcina lutea* from both wild-type and a carotenoid-deficient strain. They demonstrated that the carotenoid-containing strain was significantly less sensitive to a photodynamic effect than were the membranes from the carotenoid-deficient strain. They observed that several sites of the respiratory chain could be inactivated by visible light with an exogenous sensitizer and that brief periods of illumination, although resulting in damage to the respiratory system, could be reversed upon subsequent dark incubation.

Another factor which seems to determine the ability of carotenoid pigments to serve as protective agents is the number of conjugated double bonds or the chromophore length of the particular pigment. This phenomenon was first demonstrated by Claes and her collaborators (Claes & Nakayama, 1959; Claes, 1960, 1961). She had been studying the ability of carotenoid pigments to prevent the photobleaching of chlorophyll *a* and found that the carotenoid protection was a function of the number of conjugated double bonds in the polyene chain. It was Stanier (1959) who first pointed out the relationship between the biological effectiveness of carotenoid pigments in serving as protective agents and their chromophore length. Claes (1954) had reported earlier that a minimum of eleven conjugated double bonds was essential for biological protection whereas she was able to demonstrate (Claes & Nakayama, 1959; Claes, 1960, 1961) that nine conjugated double bonds would protect against chlorophyll photobleaching. Crounse, Feldman & Clayton (1963) observed that protection in mutant strains of *Rhodopseudomonas spheroides* was a relative phenomenon and they concluded that a minimum of nine conjugated double bonds was essential for protecting these mutant cells against photo-oxidations. Mathews-Roth & Krinsky (1970*a*) working with mutant strains of *Sarcina lutea*, also concluded that a minimum of nine conjugated bonds was essential for carotenoid pigments to exhibit protection. In a subsequent publication, Mathews-Roth & Krinsky (1970*b*) studied a mutant strain of *S. lutea* in which the major carotenoid pigment contained only eight conjugated double bonds. In this case, they found no protection against a photodynamic effect initiated by an exogenous photosensitizer.

The significance of the chromophore length appears to be due directly to the mechanism of action of carotenoid pigments in protecting cells against harmful photodynamic effects. In 1968, Foote & Denny presented the first evidence that carotenoid pigments could interact with $^1O_2$ generated in a Type II photochemical reaction and thereby interfere with the process of photosensitized oxidations. They observed that, in a system *in vitro*, low concentrations of β-carotene (10^{-4} M) inhibited the methylene blue-sensitized oxidation of 2-methyl-2-pentene by 95 per cent. They proposed that a direct transfer of energy occurred between $^1O_2$ and the carotenoid to yield the triplet carotenoid (^{3}Car) and ground state oxygen, also present as a triplet species. The reaction is described below:

$$^1O_2 + \text{Car} \longrightarrow {}^3\text{Car} + {}^3O_2. \qquad (3)$$

This reaction could only proceed effectively if the triplet energy level of β-carotene was near or below that of the $^1\Delta g$ state of $^1O_2$ which is 22.5 kcal mol^{-1}. The triplet energy of β-carotene was estimated to be 17.5 kcal mol^{-1} by Mathis & Kleo (1973). Based on measurements of quenching constants for the reaction of β-carotene and $^1O_2$, the triplet energy level of β-carotene was shown to be less than 23 kcal mol^{-1} by Farmilo & Wilkinson (1973). The experiments of these latter workers confirmed beyond any doubt the suggestion of Foote, Chang & Denny (1970*a*) that the quenching of $^1O_2$ by β-carotene was due to electronic energy transfer with the resulting production of the triplet state of β-carotene. Carotenoids can dissipate this energy directly to the solvent, and therefore can function in a cyclic fashion.

In order, therefore, for carotenoid pigments to quench $^1O_2$ effectively, the energy of the transition from the ground state to the first excited triplet state of the carotenoid must be equal to or less than that of the singlet–triplet transition of the excited oxygen. As mentioned above, this latter value is 22.5 kcal mol^{-1} and, according to the calculations of Mathis & Kleo (1973), polyenes with nine or more conjugated double bonds would be effective quenchers of $^1O_2$ whereas polyenes containing seven or less conjugated double bonds would not be effective. This reflects precisely the observations made in biological systems by Claes (Claes & Nakayama, 1959; Claes, 1960, 1961) and by Foote *et al.* (1970*a*) with respect to carotenoid quenching of $^1O_2$ in chemical systems. Foote *et al.* (1970*a*) have plotted the relationship they found between the quenching constants of polyenes of varying chain length and the observations reported earlier by Claes on protection against chlorophyll *a* oxidation. These data are presented in Fig. 9. An interesting observation related to this phenomenon was reported by Mathews-Roth & Krinsky (1970*a*) who found that a carotenoid pigment containing only eight conjugated double bonds from a mutant strain of *Sarcina lutea* did not protect the micro-organism from harmful oxidations initiated with an exogenous photosensitizer. This carotenoid pigment, as well as other naturally occurring carotenoids from both *S. lutea* and plant sources have now been tested by Mathews-Roth *et al.* (1974) for their ability to quench $^1O_2$. These authors found that the major carotenoid pigment of *S. lutea* which contains nine conjugated double bonds, is as effective a quencher of $^1O_2$ as is either β-carotene or two oxygen-containing carotenoids, isozexanthin and lutein containing 11 and 10 conjugated double bonds, respectively. On the other hand, the carotenoid pigment containing eight conjugated double bonds isolated from the mutant of *S. lutea* was significantly less effective in

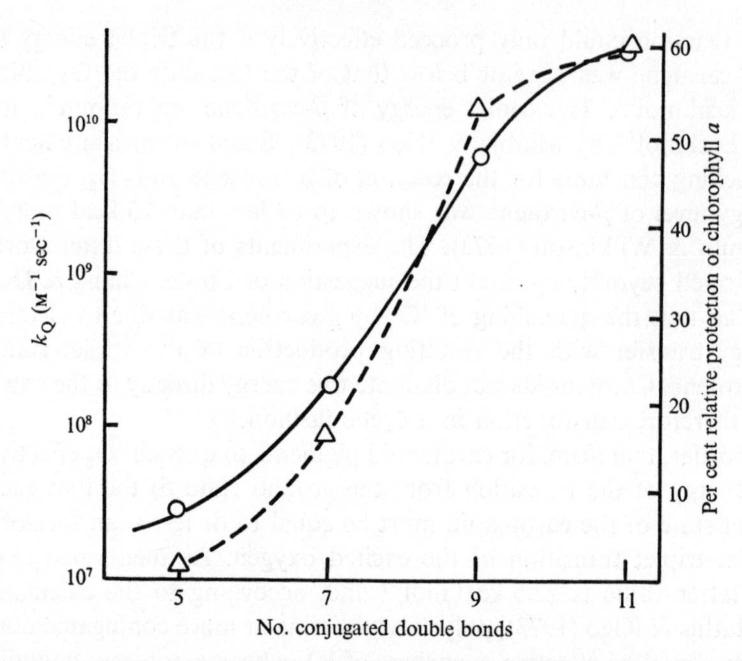

Fig. 9. The relationship between the length of the conjugated double bond system and either $^1O_2$ quenching rates k_Q (○), or protective action against photobleaching of chlorophyll *a* (△) (Claes, 1960; Claes & Nakayama, 1959). (After Foote *et al.*, 1970*a*. Copyright by the American Chemical Society.)

quenching $^1O_2$. Mathews-Roth *et al.* (1974) also suggest that other factors besides the number of conjugated double bonds may play a role in determining the ability of carotenoid pigments within cells to protect against the harmful results of a photodynamic effect. These other factors include the concentration of the pigment and the location of the pigment in relation to the location of the photosensitizer and the sensitive cellular sites for exhibiting photosensitized damage.

Another factor which seems to be characteristic of carotenoid quenching of $^1O_2$ is the unidirectional isomerization of *cis*-carotenoids to *trans*-carotenoids. This was first demonstrated in biological systems by Claes & Nakayama (1959) and Claes (1960) working with mutant strains of the algae *Chlorella pyrenoidosa*. Foote *et al.* (1970*b*) confirmed this observation and were able to demonstrate the fact that *cis* ⟶ *trans* isomerization was sensitized by $^1O_2$. In addition, Krinsky & Jong (1975) have been able to demonstrate that this isomerization also occurs in an artificial membrane system where carotenoids have been demonstrated to exert a protective effect against harmful photosensitized oxidations.

The use of artificial membrane systems (liposomes) as analogues for cells undergoing photosensitized oxidations was introduced by

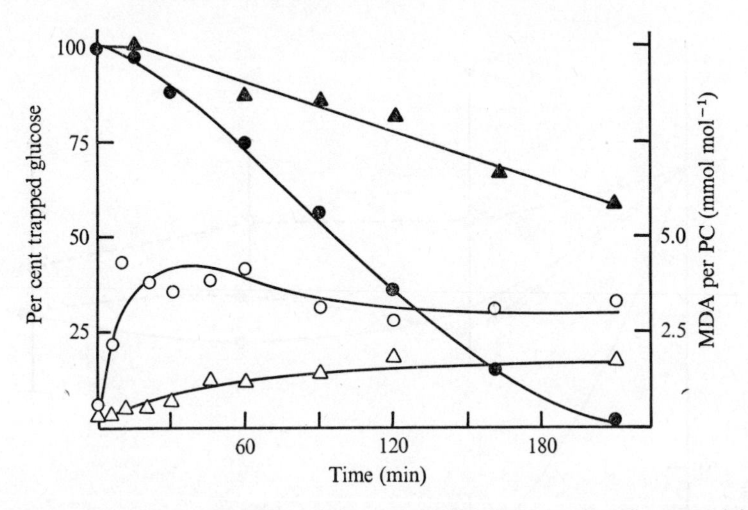

Fig. 10. Simultaneous measurement of glucose released from irradiated control (●) and canthaxanthin-containing liposomes (▲) and the malondialdehyde production (MDA) from control (○) and canthaxanthin-containing liposomes (△). The latter preparation contained 10 mmol canthaxanthin per mol phosphatidyl choline (PC). Irradiation was carried out at 10 °C, using white light > 520 nm at an irradiance of 5×10^3 W m^{-2}. Toluidine blue concentration was 6×10^{-5} M. (After Anderson & Krinsky, 1973.)

Anderson & Krinsky (1973) and Anderson *et al.* (1974). These workers studied the effects of illuminating artificial membrane systems exposed to photosensitizing dyes. They determined photodynamic damage either by the production of lipid peroxides or by the release of trapped solutes. As seen in Fig. 10, the incorporation of a diketo-carotenoid, canthaxanthin, into liposomes resulted in the formation of a liposome which was more resistant to the harmful effects of light in the presence of a photosensitizing dye. Lipid peroxides were not formed as rapidly as in the carotenoidless liposome and the release of the solute, glucose, was considerably slower in the carotenoid-containing liposomes. In an analogous fashion, Lamola, Yamane & Trozzolo (1973) demonstrated that the incorporation of β-carotene into red blood cells which had been treated with the photosensitizer, protoporphyrin, resulted in protection against a light-induced haemolysis normally exhibited by such sensitized red blood cells.

The ability of carotenoid pigments in bacteria to serve as protective agents against harmful photosensitized oxidations has been extended to a study of other types of oxidations that are harmful to bacteria. In particular, the polymorphonuclear leukocyte has a principal function in animals of ingesting and destroying bacteria using oxidative mechanisms (Klebanoff, 1975). Krinsky (1974*c*) has recently looked at the

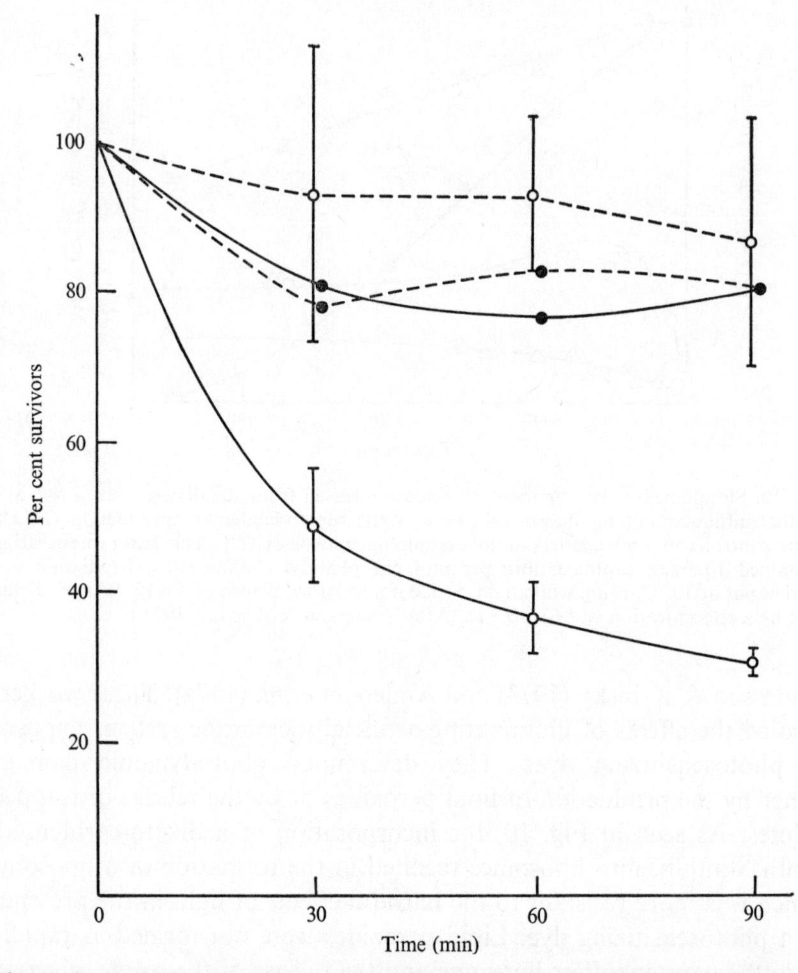

Fig. 11. Survival of carotenoid-containing wild-type *Sarcina lutea* (●) and a pigmentless mutant strain 93A (○) exposed to human polymorphonuclear leukocytes. Control samples (-----) contained no pooled human serum, thus preventing phagocytosis. Each data point represents three experiments; vertical lines denote one standard error and are plotted only for mutant strain 93A. No significant difference was found between the control and experimental (—) tubes for the wild-type pigmented strain. (After Krinsky, 1974*c*.)

ability of human polymorphonuclear leukocytes to kill wild-type and carotenoidless strains of *S. lutea* and has observed a difference in the killing rate between these two strains. As seen in Fig. 11, the carotenoid-less mutant strain is killed rapidly by human polymorphonuclear leukocytes, whereas the carotenoid containing wild-type strain is not.

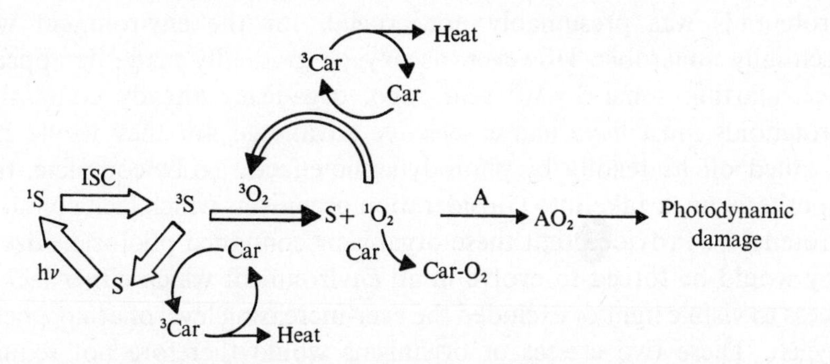

Fig. 12. Mechanisms of carotenoid protection. A sensitizer molecule (S) after excitation by light to its first excited state (^{1}S) undergoes intersystem crossing (ISC) to form the triplet species (^{3}S). This species can be quenched by carotenoids (Car) in an energy transfer reaction or can transfer its excitation energy to oxygen to yield singlet oxygen ($^1O_2$). This latter species is also effectively quenched by carotenoids reverting to ground state oxygen ($^3O_2$). In addition, carotenoids can react chemically with $^1O_2$ yielding oxidized species. The remaining $^1O_2$ can then react with substrates (A) and ultimately give rise to photodynamic effects. The width of the arrows indicates approximately the extent of these reactions.

CONCLUSION

It appears clear that one of the major functions of carotenoids in nature is to protect cells against potentially harmful or lethal photodynamic effects. What still requires clarification is the mechanism *in vivo*. In addition to quenching $^1O_2$, carotenoids can intercept the photosensitization reaction at an even earlier stage, by effectively quenching ^{3}S (Fujimori & Livingston, 1957) and thereby preventing the formation of $^1O_2$. In addition to quenching $^1O_2$ physically, carotenoids can also react chemically with $^1O_2$ forming carotenoid oxidation products which further decrease the $^1O_2$ available for initiating photodynamic damage. All of these protective mechanisms are depicted schematically in Fig. 12. According to the calculations of Foote (1975), carotenoids can quench 90 per cent of the ^{3}S in the case where chlorophyll is the sensitizing pigment. Of the remaining 10 per cent ^{3}S which can react with oxygen, carotenoids can quench 99.999 per cent of the $^1O_2$ formed physically. Some of the $^1O_2$ which survives these defences can then react chemically with carotenoids as a final protective stage prior to $^1O_2$ interaction with other cellular constituents.

It is difficult to conceive of other situations where naturally occurring compounds use three separate mechanisms to carry out a single function in the cell. This function is, of course, the preservation of life in an illuminated, aerobic environment. During the earliest period of evolution, before the appearance of the blue-green algae, the presence of carotenoids was presumably not crucial, for the environment was essentially anaerobic. However, as oxygen gradually made its appearance, starting some 3×10^9 years ago, organisms already containing carotenoids must have had a selective advantage, for they would not be killed off as readily by photodynamic effects. To be complete, this hypothesis must take into consideration organisms which lack coloured carotenoids. Provided that these organisms contained photosensitizers, they would be forced to evolve in an environment which either lacked access to visible light or excluded the ever-increasing level of atmospheric oxygen. These two classes of organisms would therefore not require the protective mechanisms exemplified by the formation of coloured carotenoid pigments.

REFERENCES

Anderson, I. C. & Robertson, D. S. (1960). Role of carotenoids in protecting chlorophyll from photodestruction. *Plant Physiology*, **35**, 531–534.

Anderson, S. M. & Krinsky, N. I. (1973). Protective action of carotenoid pigments against photodynamic damage to liposomes. *Photochemistry and Photobiology*, **18**, 403–408.

Anderson, S. M., Krinsky, N. I., Stone, M. S. & Clagett, D. C. (1974). Effect of singlet oxygen quenchers on oxidative damage to liposomes initiated by photosensitization or by radiofrequency discharge. *Photochemistry and Photobiology*, **20**, 65–69.

Barran, L. R., D'Aoust, J. Y., Labelle, I. L., Martin, W. G. & Schneider, H. (1974). Differential effects of visible light on active transport in *E. coli*. *Biochemical and Biophysical Research Communications*, **56**, 522–528.

Bellin, J. S., Lutwick, L. & Jonas, B. (1969). Effects of photodynamic action on *E. coli*. *Archives of Biochemistry and Biophysics*, **132**, 157–164.

Bellin, J. S. & Ronayne, M. E. (1968). Effects of photodynamic action on the cell membrane of *Euglena*. *Physiologia Plantarum*, **21**, 1060–1066.

Berson, E. L. (1973). Experimental and therapeutic aspects of photic damage to the retina. *Investigative Ophthalmology*, **12**, 35–44.

Blum, H. F. (1941). *Photodynamic Action and Diseases Caused by Light*. New York: Reinhold Publishing Co. (reprinted 1964, New York: Hafner Publishing Co.).

Bragg, P. D. (1971). Effect of near-ultraviolet light on the respiratory chain of *Escherichia coli*. *Canadian Journal of Biochemistry*, **49**, 492–495.

Buchbinder, L., Solowey, M. & Phelps, E. B. (1941). Studies on microorganisms in simulated room environments. III. The survival rates of streptococci in the presence of natural, daylight and sunlight, and artificial illumination. *Journal of Bacteriology*, **42**, 353–366.

Burns, E. R., Carter, M. C. & Buchanan, G. A. (1971). Inhibition of carotenoid synthesis as a mechanism of action of amitrole, dichlormate, and pyriclor. *Plant Physiology*, **47**, 144–148.

Cabrera-Juarez, E. & Espinosa-Lara, M. (1974). Lethal and mutagenic action of black light (325 to 400 nm) on *Haemophilus influenzae* in the presence of air. *Journal of Bacteriology*, **117**, 960–964.

Calberg-Bacq, C. M., Delmelle, M. & Duchesne, J. (1968). Inactivation and mutagenesis due to the photodynamic action of acridines and related dyes on extracellular bacteriophage T_4B. *Mutation Research*, **6**, 15–24.

CHICHESTER, C. O. & MAXWELL, W. A. (1969). The effects of high intensity visible and ultraviolet light on the death of microorganisms. *Life Science Space Research*, **7**, 11–18.

CLAES, H. (1954). Analyse der biochemischen Synthesekette für Carotenoide mit Hilfe von *Chlorella*-mutanten. *Zeitschrift für Naturforschung*, **9B**, 461–470.

CLAES, H. (1960). Interaction between chlorophyll and carotenes with different chromophoric groups. *Biochemical and Biophysical Research Communications*, **3**, 585–590.

CLAES, H. (1961). Energieübertrangung von angeregtem Chlorophyll auf C_{40}-Polyene mit verschiedenen chromophoren Gruppen. *Zeitschrift für Naturforschung*, **16B**, 445–454.

CLAES, H. & NAKAYAMA, T. O. M. (1959). Das photoxydative ausbleichen von Chlorophyll *in vitro* in gegenwart von Carotinen mit verschieden chromophoren Gruppen. *Zeitschrift für Naturforschung*, **14B**, 746–747.

COONEY, J. J. & KRINSKY, N. I. (1972). Photodynamic killing of *Acholeplasma laidlawii*. *Photochemistry and Photobiology*, **16**, 523–526.

COREY, E. J. & TAYLOR, W. C. (1964). A study of the peroxidation of organic compounds by externally generated singlet oxygen molecules. *Journal of the American Chemical Society*, **86**, 3881–3882.

CROUNSE, J. B., FELDMAN, R. P. & CLAYTON, R. K. (1963). Accumulation of polyene precursors of neurosporene in mutant strains of *Rhodopseudomonas spheroides*. *Nature, London*, **198**, 1227–1228.

D'AOUST, J. Y., GIROUX, J., BARRAN, L. R., SCHNEIDER, H. & MARTIN, W. G. (1974). Some effects of visible light on *Escherichia coli*. *Journal of Bacteriology*, **120**, 799–804.

DOWLING, J. E. & SIDMAN, R. L. (1962). Inherited retinal dystrophy in the rat. *Journal of Cell Biology*, **14**, 73–109.

DOWNES, A. & BLUNT, T. P. (1887). Researches on the effect of light upon bacteria and other organisms. *Proceedings of the Royal Society of London, Series B*, **26**, 488–500.

EISENSTARK, A. (1971). Mutagenic and lethal effects of visible and near-ultraviolet light on bacterial cells. *Advances in Genetics*, **16**, 167–198.

EPEL, B. L. (1973). Inhibition of growth and respiration by visible and near-visible light. In *Photophysiology*, **8**, ed. A. L. Giese, pp. 209–229. New York: Academic Press.

FARMILO, A. & WILKINSON, F. (1973). On the mechanism of quenching of singlet oxygen in solution. *Photochemistry and Photobiology*, **18**, 447–450.

FERRON, W. L., EISENSTARK, A. & MACKAY, D. (1972). Distinction between far- and near-ultraviolet light killing of recombinationless (*rec-A*) *Salmonella typhimurium*. *Biochimica et Biophysica Acta*, **277**, 651–658.

FOOTE, C. S. (1968*a*). Mechanisms of photosensitized oxidation. *Science, Washington*, **162**, 963–970.

FOOTE, C. S. (1968*b*). Photosensitized oxygenations and the role of singlet oxygen. *Accounts of Chemical Research*, **1**, 104–110.

FOOTE, C. S. (1975). Photosensitized oxidation and singlet oxygen: consequences in biological systems. In *Free Radicals and Biological Systems*, ed. W. Pryor. New York: Academic Press. (In press.)

FOOTE, C. S., CHANG, Y. C. & DENNY, R. W. (1970*a*). Chemistry of singlet oxygen. X. Carotenoid quenching parallels biological protection. *Journal of the American Chemical Society*, **92**, 5216–5218.

FOOTE, C. S., CHANG, Y. C. & DENNY, R. W. (1970*b*). Chemistry of singlet oxygen. XI. *Cis–trans* isomerization of carotenoids by singlet oxygen and a probable quenching mechanism. *Journal of the American Chemical Society*, **92**, 5218–5219.

FOOTE, C. S. & DENNY, R. W. (1968). Chemistry of singlet oxygen. VII. Quenching by β-carotene. *Journal of the American Chemical Society*, **90**, 6233–6235.

FOOTE, C. S. & WEXLER, S. (1964*a*). Olefin oxidations with excited singlet molecular oxygen. *Journal of the American Chemical Society*, **86**, 3879–3880.

FOOTE, C. S. & WEXLER, S. (1964*b*). Singlet oxygen. A probable intermediate in photosensitized autoxidations. *Journal of the American Chemical Society*, **86**, 3880–3881.

FUJIMORI, E. & LIVINGSTON, R. (1957). Interaction of chlorophyll in its triplet state with oxygen, carotene, etc. *Nature, London*, **180**, 1036–1038.

GIESE, A. C. (1971). Photosensitization by natural pigments. In *Photophysiology*, **6**, ed. A. C. Giese, pp. 77–129. New York: Academic Press.

GOLDSTEIN, B. D. & HARBER, L. C. (1972). Erythropoietic protoporphyria: lipid peroxidation and red cell membrane damage associated with photohemolysis. *Journal of Clinical Investigation*, **51**, 892–902.

GOLLNICK, K. & SCHENCK, G. O. (1967). Oxygen as a dienophile. In 1,4-*Cycloaddition Reactions*, ed. J. Hamer, pp. 255–344. New York: Academic Press.

GRIFFITHS, M., SISTROM, W. R., COHEN-BAZIRE, G. & STANIER, R. Y. (1955). Function of carotenoids in photosynthesis. *Nature, London*, **176**, 1211–1215.

HARRISON, A. P. (1967). Survival of bacteria. *Annual Review of Microbiology*, **21**, 143–156.

HOLLAENDER, A. (1943). Effect of long ultraviolet and short visible radiation (3500 to 4900 Å) on *Escherichia coli*. *Journal of Bacteriology*, **46**, 531–541.

HUANG, G. L. & HAUG, A. (1974). Regulation of membrane lipid fluidity in *Acholeplasma laidlawii*: effect of carotenoid pigment content. *Biochimica et Biophysica Acta*, **352**, 361–370.

JAGGER, J. (1973). The realm of the ultraviolet. *Photochemistry and Photobiology*, **18**, 353–354.

KAPLAN, R. W. (1950*a*). Mutationsauslösung bei *Bacterium prodigiosum* durch sichtbares Licht nach vital Färbung mit Erythrosin. *Archiv für Mikrobiologie*, **15**, 152–175.

KAPLAN, R. W. (1950*b*). Auslösung von Phagenresistenzmutationen bei *Bacterium coli* durch Erythrosin mit und ohne Belichtung. *Naturwissenschaften*, **37**, 308.

KAPLAN, R. W. (1950*c*). Mutation und Keimtötung bei *Bact. coli* Histidinless durch uv und Photodynamie. *Naturwissenschaften*, **37**, 547.

KAPLAN, R. W. (1950*d*). Photodynamische Auslösung von Mutationen in den Sporen von *Penicillium notatum*. *Planta*, **38**, 1–11.

KASHA, M. & KHAN, A. U. (1970). The physics, chemistry and biology of singlet molecular oxygen. *Annals of the New York Academy of Sciences*, **171**, 5–23.

KASHKET, E. R. & BRODIE, A. F. (1962). Effects of near-ultraviolet irradiation on growth and oxidative metabolism of bacteria. *Journal of Bacteriology*, **83**, 1093–1100.

KAUTSKY, H., DE BRUIJN, H., NEUWIRTH, R. & BAUMEISTER, W. (1933). Energieunwandlung an Grenzflachen. VII. Photosensibilisierte Oxydation als Wirkung eines Aktiven, metastabilen Zustandes des Sauerstoff-moleküls. *Chemische Berichte*, **66**, 1588–1600.

KEARNS, D. R. (1971). Physical and chemical properties of singlet molecular oxygen. *Chemical Reviews*, **71**, 395–427.

KHAN, N. C. & PODDAR, R. K. (1974). Photodynamic inactivation of antigenic determinants of single-stranded DNA bacteriophage ϕX174. *Journal of Virology* **13**, 997–1000.

KLEBANOFF, S. J. (1975). Antimicrobial mechanisms in neutrophilic polymorphonuclear leukocytes. *Seminars in Hematology*, **12**, 117–142.

KOSKI, V. M. & SMITH, J. H. C. (1951). Chlorophyll formation in a mutant, white seedling-3. *Archives of Biochemistry and Biophysics*, **34**, 189–195.

KRINSKY, N. I. (1968). The protective functions of carotenoid pigments. In *Photophysiology*, **3**, ed. A. C. Giese, pp. 123–195. New York: Academic Press.

KRINSKY, N. I. (1971). Function. In *Carotenoids*, ed. O. Isler, pp. 669–716. Basel: Birkhaüser Verlag.

KRINSKY, N. I. (1974*a*). The protective functions of carotenoid pigments against aerobic photosensitivity. In *Progress in Photobiology*, ed. G. O. Schenck, Proceedings of the VIth International Congress on Photobiology, Bochum, 21–25 Aug., 1972, no. 011. Frankfurt: Deut. Ges. für Lichtforschung.

KRINSKY, N. I. (1974*b*). Membrane photochemistry and photobiology. *Photochemistry and Photobiology*, **20**, 532–535.

KRINSKY, N. I. (1974*c*). Singlet excited oxygen as a mediator of the antibacterial action of leukocytes. *Science, Washington*, **184**, 563–564.

KRINSKY, N. I. & JONG, P. (1975). Unidirectional *cis* → *trans* isomerization of β-carotene in artificial membrane systems catalyzed by singlet oxygen ($^1O_2$). *Abstracts of the 4th International Symposium on Carotenoids*, Berne, 1975.

KUBITSCHEK, H. E. (1967). Mutagenesis by near-visible light. *Science, Washington*, **155**, 1545–1546.

KUHN, D. A. & STARR, M. P. (1970). Effects of microscope illumination on bacterial development. *Archiv für Mikrobiologie*, **74**, 292–300.

KUNISAWA, R. & STANIER, R. Y. (1958). Studies on the role of carotenoid pigments in a chemoheterotrophic bacterium, *Corynebacterium poinsettiae*. *Archiv für Mikrobiologie*, **31**, 146–156.

LAMOLA, A. A., YAMANE, T. & TROZZOLO, A. M. (1973). Cholesterol hydroperoxide formation in red cell membranes and photohemolysis in erythropoietic protoporphyria. *Science, Washington*, **179**, 1131–1133.

MATHEWS, M. M. (1963). Comparative study of lethal photosensitization of *Sarcina lutea* by 8-methoxypsoralen and by toluidine blue. *Journal of Bacteriology*, **85**, 322–328.

MATHEWS, M. M. (1964). The effect of low temperature on the protection by carotenoids against photosensitization in *Sarcina lutea*. *Photochemistry and Photobiology*, **3**, 75–77.

MATHEWS, M. M. & SISTROM, W. R. (1959). Function of carotenoid pigments in non-photosynthetic bacteria. *Nature, London*, **184**, 1892–1893.

MATHEWS, M. M. & SISTROM, W. R. (1960). The function of the carotenoid pigments of *Sarcina lutea*. *Archiv für Mikrobiologie*, **35**, 139–146.

MATHEWS-ROTH, M. M. & KRINSKY, N. I. (1970*a*). Studies on the protective function of the carotenoid pigments of *Sarcina lutea*. *Photochemistry and Photobiology*, **11**, 419–428.

MATHEWS-ROTH, M. M. & KRINSKY, N. I. (1970*b*). Failure of conjugated octaene carotenoids to protect a mutant of *Sarcina lutea*. *Photochemistry and Photobiology*, **11**, 555–557.

MATHEWS-ROTH, M. M., WILSON, T., FUJIMORI, E. & KRINSKY, N. I. (1974). Carotenoid chromophore length and protection against photosensitization. *Photochemistry and Photobiology*, **19**, 217–222.

MATHIS, P. & KLEO, J. (1973). The triplet state of β-carotene and of analog polyenes of different length. *Photochemistry and Photobiology*, **18**, 343–346.

MAXWELL, W. A., MACMILLAN, J. D. & CHICHESTER, C. O. (1966). Function of carotenoids in protection of *Rhodotorula glutinis* against irradiation from a gas laser. *Photochemistry and Photobiology*, **5**, 567–577.

MITCHELL, R. L. & ANDERSON, I. C. (1965). Photoinactivation of catalase in carotenoidless tissues. *Crop Science*, **5**, 588–591.

NINNEMANN, H., BUTLER, W. L. & EPEL, B. L. (1970*a*). Inhibition of respiration in yeast by light. *Biochimica et Biophysica Acta*, **205**, 499–506.

NINNEMANN, H., BUTLER, W. L. & EPEL, B. L. (1970*b*). Inhibition of respiration and destruction of cytochrome a_3 by light in mitochondria and cytochrome oxidase from beef heart. *Biochimica et biophysica Acta*, **205**, 507–512.

NOELL, W. K., WALKER, V. S., KANG, B. S. & BERMAN, S. (1966). Retinal damage by light in rats. *Investigative Ophthalmology*, **5**, 450–472.

PETERS, J. & PAULING, C. (1974). Mechanism of near ultraviolet light irradiation inactivation of *Escherichia coli. Radiation Research*, **59**, 235.

PREBBLE, J. & HUDA, A. S. (1973). Sensitivity of the electron transport chain of pigmented and non-pigmented *Sarcina* membranes to photodynamic action. *Photochemistry and Photobiology*, **17**, 255–264.

RAAB, O. (1900), Über die Wirkung fluoresierender Stoffe auf Infusorien. *Zeitschrift für Biologie*, **39**, 524–546.

RITCHIE, D. A. (1964). Mutagenesis with light and proflavine in phage T_4. *Genetical Research, Cambridge*, **5**, 168–169.

RAMABHADRAN, T. V., FOSSUM, T. & JAGGER, J. (1975). The chromophore and target for near-UV-induced growth delay in *E. coli. Abstracts, 3rd Annual Meeting, American Society for Photobiology*, p. 92, June 22–26, 1975, Louisville, Kentucky.

ROTTEM, S., GOTTFRIED, L. & RAZIN, S. (1968). Carotenoids as protectors against photodynamic inactivation of the adenosine triphosphatase of *Mycoplasma laidlawii* membranes. *Biochemical Journal*, **109**, 707–708.

RUBENSTEIN, B. B. (1931). Decrease in rate of oxygen consumption under the influence of visible light on *Sarcina lutea. Science, New York*, **74**, 419–420.

SAGER, R. & ZALOKAR, M. (1958). Pigments and photosynthesis in a carotenoid-deficient mutant of *Chlamydomonas. Nature, London*, **182**, 98–100.

SCHNEIDER, H., D'AOUST, J. Y., BARRAN, L. R. & MARTIN, W. G. (1974). Selective and differential photodynamic effects on *E. coli* membrane processes and their inhibition. *Radiation Research*, **59**, 22.

SCHWARTZEL, E. H. & COONEY, J. J. (1974). Action of light on *Micrococcus roseus. Canadian Journal of Microbiology*, **20**, 1015–1021.

SISTROM, W. R., GRIFFITHS, M. & STANIER, R. Y. (1956). The biology of a photosynthetic bacterium which lacks colored carotenoids. *Journal of Cellular and Comparative Physiology*, **48**, 473–515.

SPIKES, J. D. (1968). Photodynamic action. In *Photophysiology*, **3**, ed. A. C. Giese, pp. 33–64. New York: Academic Press.

SPIKES, J. D. (1975). Porphyrins and related compounds as photodynamic sensitizers. *Annals of the New York Academy of Sciences*, **244**, 496–508.

SPIKES, J. D. & LIVINGSTON, R. (1969). The molecular biology of photodynamic action sensitized oxidations in biological systems. *Advances in Radiation Biology*, **3**, 29–121.

SPIKES, J. D. & STRAIGHT, R. (1967). Sensitized photochemical processes in biological systems. *Annual Reviews of Physical Chemistry*, **18**, 409–436.

STANIER, R. Y. (1959). Formation and function of the photosynthetic pigment system in purple bacteria. *Brookhaven Symposia in Biology*, **11**, 43–53.,

STANIER, R. Y. & COHEN-BAZIRE, G. (1957). The role of light in the microbial world: some facts and speculation. *Symposia of the Society for General Microbiology*, **7**, 56–89.

STOIEN, J. D. & WANG, R. J. (1974). Effect of near-ultraviolet and visible light on mammalian cells in culture. II. Formation of toxic photo-products in tissue culture medium by blacklight. *Proceedings of the National Academy of Sciences, U.S.A.* **71**, 3971–3965.

SWENSON, P. A. & SETLOW, R. B. (1970). Inhibition of the induced formation of tryptophanase in *Escherichia coli* by near-ultraviolet radiation. *Journal of Bacteriology*, **102**, 815–819.

WALLACE, R. H. & HABERMANN, H. M. (1959). Genetic history and general comparisons of two albino mutations of *Helianthus annuus*. *American Journal of Botany*, **46**, 157–162.

WALLACE, R. H. & SCHWARTING, A. E. (1954). A study of chlorophyll in a white mutant strain of *Helianthus annuus*. *Plant Physiology, Lancaster*, **29**, 431–436.

WANG, R. J., STOIEN, J. D. & LANDA, F. (1974). Lethal effect of near-ultraviolet irradiation on mammalian cells in culture. *Nature, London*, **247**, 43–44.

WEBB, R. B. (1972). Photodynamic lethality and mutagenesis in the absence of added sensitizers. In *Research Progress in Organic, Biological and Medicinal Chemistry*, vol. III (part II), ed. U. Gallo & L. Santamaria, pp. 511–530. New York: American Elsevier.

WEBB, R. B. & KUBITSCHEK, H. E. (1963). Mutagenic and antimutagenic effects of acridine orange in *Escherichia coli*. *Biochemical and Biophysical Research Communications*, **13**, 90–94.

WEBB, R. B. & LORENZ, J. R. (1970). Oxygen dependence and repair of lethal effects of near ultraviolet and visible light. *Photochemistry and Photobiology*, **12**, 283–289.

WEBB, R. B. & LORENZ, J. R. (1972). Toxicity of irradiated medium for repair-deficient strains of *Escherichia coli*. *Journal of Bacteriology*, **112**, 649–652.

WEBB, R. B. & MALINA, M. M. (1967). Mutagenesis in *Escherichia coli* by visible light. *Science, Washington*, **156**, 1104–1105.

WEBB, R. B. & MALINA, M. M. (1970). Mutagenic effects of near ultraviolet and visible radiant energy on continuous cultures of *Escherichia coli*. *Photochemistry and Photobiology*, **12**, 457–468.

WEBB, S. J. & TAI, C. C. (1969). Physiological and genetic implications of selective mutation by light at 320–400 nm. *Nature, London*, **224**, 1123–1125.

WELSH, J. N. & ADAMS, M. H. (1954). Photodynamic inactivation of bacteriophage. *Journal of Bacteriology*, **68**, 122–127.

WILSON, T. (1966). Excited singlet molecular oxygen in photooxidation. *Journal of the American Chemical Society*, **88**, 2898–2902.

YOAKUM, G. H. (1975). Tryptophan photoproduct(s): sensitized induction of strand breaks (or alkali-labile bonds) in bacterial deoxyribonucleic acid during near-ultraviolet irradiation. *Journal of Bacteriology*, **122**, 199–205.

YOAKUM, G. & EISENSTARK, A. (1972). Toxicity of L-tryptophan photoproduct on recombinationless (*rec*) mutants of *Salmonella typhimurium*. *Journal of Bacteriology*, **112**, 653–655.

ZELLE, M. R. & HOLLAENDER, A. (1955). Effects of radiation on bacteria. In *Radiation Biology*, ed. A. Hollaender, pp. 365–430. New York: McGraw-Hill.

ZETTERBERG, G. (1964). Mutagenic effects of ultraviolet and visible light. In *Photophysiology*, **2**, ed. A. C. Giese, pp. 247–281. New York: Academic Press.

MICROBIAL ENDURANCE AND RESISTANCE TO HEAT STRESS

N. E. WELKER
Department of Biochemistry and Molecular Biology, Northwestern University, Evanston, Illinois 60201, *USA*

INTRODUCTION

It is surprising at a time when biochemists and molecular biologists are rapidly expanding our knowledge of the structure, function, and the biosynthetic processes of the bacterial cell, that the primary damage in a vegetative non-sporulating bacterial cell subjected to sub-lethal or mild (moist) heat is still unknown. The use of moist or dry heat as a method of sterilization or pasteurization has been used for a considerable period of time. The lack of data relating to the biochemical mechanisms of heat inactivation probably reflects the effectiveness of this method of sterilization, which is generally interpreted as being caused by the coagulation or denaturation of critical proteins in the cell. There is no doubt that protein denaturation does occur, but it is not unreasonable to propose that there are other more delicate and critical biochemical changes occurring in the cell before the denaturation of cell proteins becomes apparent.

Temperature is one of the most important environmental factors that affects the activities of a living organism. In the field of microbiology, micro-organisms are capable of growing over a relatively broad range of temperatures. Micro-organisms are usually placed into arbitrary groups on the basis of whether they grow at 'normal' temperatures or at either the low or high temperature extremes. Mesophiles grow at temperatures between 20 and 50 °C, with optimal growth between 35–42 °C; thermophiles show optimal growth at 60–75 °C, or higher in the case of the extreme thermophiles, and do not grow below 40 or 50 °C. Facultative thermophiles have a maximum temperature for growth between 50 and 65 °C, and can grow at mesophilic temperatures. Some facultative thermophiles will grow at temperatures as low as 30 or 35 °C. Micro-organisms which grow optimally at temperatures below 20 °C are designated psychrophiles and are discussed elsewhere in this volume (see Morita, this volume).

Studies on micro-organisms that grow at relatively high temperatures were initiated quite independently of those concerning the lethal action of heat on vegetative mesophiles. Most of the early studies were concerned with the survival of bacteria heated for a period of time (reviewed by Allwood & Russell, 1970). Generally, the death of a population of vegetative bacteria exposed to heat stress occurs exponentially, although deviations do occur. The biochemical basis for this phenomenon was difficult to explain in terms of single events occurring in the cell. The composition of the medium in which the cells were suspended during the heat treatment, culture age, as well as the composition of the growth medium and conditions used for the quantification of viable cells, all affect the survival under heat stress. Resistance to such stress was also found to vary considerably with different strains.

Studies on micro-organisms that grow at high temperatures were mainly concerned with the solution of the interesting biological problem of how these organisms manage to live at temperatures where most ordinary proteins and other cell constituents are inactivated or denatured. The initial discovery of a thermophilic bacterium is usually attributed to Miquel (1888). Since then thermophilic micro-organisms have been the centre of immense scientific and commercial interest. Over the years there have been numerous reports on the isolation and characterization of thermophilic bacteria (Gaughran, 1947; Gordon & Smith, 1949; Allen, 1953; Marsh & Larsen, 1953; Brock & Freeze, 1969), algae (Brock, 1967), actinomycetes (Cross, 1968), and fungi (Cooney & Emerson, 1964). Although most of the early reports were descriptive in nature, many studies were presented more completely so that theories of the origin and nature of these unique organisms became numerous.

There is considerable information regarding the effect of heat on mesophilic and thermophilic micro-organisms so that one can now gain some insight into the biochemical mechanisms underlying the thermostability of microbial constituents.

The principle aim of this article, therefore, is to review the most significant information from each of these areas in order to summarize our present knowledge concerning the biochemical mechanisms of endurance and resistance of vegetative non-sporulating bacterial cells to heat stress.

THERMOSENSITIVITY OF VARIOUS BACTERIAL CONSTITUENTS

Cell walls

The bacterial cell wall is responsible for maintaining the rigidity and shape of the cell and for protecting the cell form osmotic lysis. There is relatively little evidence that the cell wall plays a significant role in protecting the cell constituents from heat inactivation.

Lysozyme and trypsin have been used as enzymatic probes to test the effect of heat on cells of *Escherichia coli* (Allwood & Russell, 1970). The outer membrane of these cells was stripped off at temperatures between 75 and 100 °C. These cells were found to be sensitive to trypsin but were insensitive to lysozyme. There is evidence that boiled cells retain their cell wall antigenicity and the cell walls of Gram-negative bacteria were unaffected by high temperatures (Lennox, 1960). In contrast, the lipopolysaccharide component of the cell wall of *Proteus vulgaris* was irreversibly lost after cell walls were heated at either 50 or 100 °C. Cell walls of staphylococci are unaffected by heating at 100 °C. Similar results were obtained by Allwood & Russell (1969) who examined thin sections of whole cells of *Staphylococcus aureus* which had been heated.

The temperatures used in these investigations, however, are much higher than those normally used to examine the effect of temperature on cell injury.

Large surface blebs were observed in cells of *E. coli* heated at a temperature (55 °C) at which colony-forming ability was lost (Scheie & Ehrenspeck, 1973). They postulated that the initial event in thermal injury in *E. coli* is the denaturation of protein(s) in the outer membrane of the cell wall leading to a weakening of the rigid peptidoglycan layer.

Forrester & Wicken (1966) reported that the synthesis of cell wall peptidoglycan and teichoic acid in *Bacillus stearothermophilus* (thermophile) and *B. coagulans* (facultative thermophile) is influenced by the temperature at which the cultures of these organisms are grown. In both organisms, there was an increase in the peptidoglycan and a decrease in the teichoic acid content in the cell wall of cells grown at 55 °C as compared to the cell wall of cells grown at 37 °C. Novitsky *et al.* (1974) compared the amino acid and amino sugar composition of the cell wall peptidoglycan of two facultative thermophiles grown at 37 and 55 °C. With the exception of alanine, all the cell wall components were present in a higher proportion in cells grown at 55 °C. These results suggest that qualitative and quantitative changes in the various cell wall components may influence the susceptibility of an organism to heat. In effect the cell wall may act as an insulator to protect the underlying membrane from thermal injury.

The chemical composition of the cell wall peptidoglycan of various strains of *B. stearothermophilus* has been reported by Salton & Pavlik (1960), Forrester & Wicken (1966), and Sutow & Welker (1967). Although there were minor variations in the relative proportions of the cell wall amino acids and amino sugars, it was proposed that the cell walls of these thermophilic bacilli had an identical chemical composition (Sutow & Welker, 1967). Welker (1971) using a phage-induced lytic enzyme, showed that the cell wall peptidoglycan of *B. stearothermophilus* was similar in structure to that reported for the peptidoglycan of *E. coli*.

From the available data it appears unlikely that the cell wall peptidoglycan plays a significant role in protecting the cell from thermal injury. In the absence of an adequate assay system, it is impossible to determine whether the cell wall of thermophiles is more heat stable than the corresponding structure of mesophiles.

Deoxyribonucleic acid

The production of single-strand breaks in deoxyribonucleic acid (DNA) by heating may be the primary lethal event in the thermal inactivation of mesophilic bacteria (see also Bridges, this volume). Bridges, Ashwood-Smith & Munson (1969) demonstrated that, after incubation of *E. coli* for several minutes at 52 °C the DNA had single-strand breaks similar to those produced by ionizing radiations. The damage to DNA in each case parallels the observed loss of viability and the degree of sensitivity is correlated with the genetic makeup of the strain. Bacteria may use similar mechanisms to repair the damage to DNA caused by ionizing radiation and heat. Ionizing radiation is not usually found in natural environments at a high enough level to constitute a hazard to the bacterial cell and there is some speculation that the recovery or repair mechanisms might be involved in normal metabolism (e.g. genetic recombination). An alternative explanation is that these repair mechanisms are directed against other environmental hazards, all of which produce damage to the DNA similar to ionizing radiation.

Woodcock & Grigg (1972) estimated that DNA of *E. coli* B Thy$^-$ cells contain approximately 100 double-strand breaks after 15 min of incubation at 52 °C. When these same cells were heated at 52 °C for 15 min and subsequently incubated in phosphate buffer supplemented with thymidine, the molecular weight of the DNA increased reaching its initial value after 30 min. During this incubation period there was also a considerable increase in cell viability. They concluded that the repair of the damage caused by the heat treatment was taking place. The 'melting apart' (denaturation) of the two strands of DNA at elevated temperatures has been shown to be characteristic for each species of DNA and depends primarily on the salt concentration (Doty *et al.*, 1959) and secondarily on the guanine and cytosine (G+C) content of the DNA (Marmur & Doty, 1959). Since heating of DNA at 52 °C does not result in denaturation or strand breakage *in vitro*, and since the breakage depends on the cells being in the logarithmic phase of growth, they postulated that the DNA breakage was enzymatic. Evidence in support of this comes from studies involving mutants of *E. coli* deficient in repair enzymes.

Sedgwick & Bridges (1972) examined the introduction by heat of single-strand breaks in the DNA of *E. coli* carrying a mutation at *polA* which affects DNA polymerase I, an important component in the repair of DNA. The pol^+ and *polA* derivatives of *E. coli* K12 showed a similar pattern of single-strand breaks in DNA upon heating at 52 °C, whereas derivatives of *E. coli* B carrying res^+ or *resA* alleles (repair synthesis) were both insensitive to the introduction of strand breaks by this treatment. They concluded that the introduction of single-strand breaks in DNA following heat treatment is a consequence of strain-specific characteristics such as nucleases rather than a direct physical effect of heat on the DNA.

Pauling & Beck (1975) have examined the effect of heating on the repair of single-strand breaks in the DNA of a DNA ligase-deficient mutant (temperature sensitive), *E. coli* 15 T^-. The temperature sensitive (ts) mutant was much more sensitive to heating at 52 °C than the wild-type strain. After 20 min at 52 °C, 31 per cent of the wild-type bacteria remained viable, whereas only 0.1 per cent of the ts cells survived. Substantial single-strand breakage occurred in the DNA of the ts cells but not in the DNA of wild-type cells. They concluded that single-strand breakage occurs in both strains, probably upon exposure at all temperatures, but the damage to the DNA of the wild-type is repaired, whereas the ts mutant cannot function at the higher temperature and such breaks are now sensitive to further degradation by exonucleases.

The recovery of *E. coli* strains after exposure to mild heat treatment is greatly influenced by the culture conditions (Sedgwick & Bridges, 1972). *Salmonella typhimurium* LT-2 was shown to be less susceptible to recovery after heat treatment when suspended in complex medium (Gomez & Sinskey, 1973). This phenomenon, designated as minimal-medium recovery, was also observed in an excision-deficient mutant of *E. coli* K12 in which a higher survival after u.v. irradiation is observed if the mutant is plated on minimal medium instead of on a complex medium. Gomez & Sinskey (1973) showed that the loss of viability of *S. typhimurium* cells was correlated with the production of single-strand breaks in the DNA. Incubation of the heated cells in a minimal medium for 15 min at 37 °C did not change the viability or the extent of DNA breakage. In contrast, a marked increase in DNA single-strand breakage accompanied by a loss of viability was observed after a similar incubation in a nutritionally complex medium. If the heated bacteria were incubated in minimal medium before incubation in the complex medium, there was a decrease in the single-strand breaks in the DNA and an increase in viability. Gomez & Sinskey (1975) have shown that the repair of breaks in single-strand DNA in *S. typhimurium*, caused by sequential induction with heat and exposure to nutritionally complex media, requires air.

The loss of viability of cells exposed to mild heat can be correlated with the introduction of single-strand breaks in the cell DNA. It has been proposed that the damage to the DNA is enzymatic as a result of an activation or release of a nuclease. The ability of a cell to repair this damage, and to recover viability, depends upon the physiological and genetic state of the cell.

Most investigations with thermophiles have been concerned with the thermostability of their DNA by comparing the melting temperature (T_m) of DNA isolated from thermophilic strains of the genus *Bacillus* with that of DNA isolated from mesophilic strains (usually *E. coli*). The thermostability of DNA appears to bear no apparent relationship to the ability of organisms to grow at elevated temperatures. The base composition and T_m values of DNA preparations obtained from thermophiles are similar to those of DNA from mesophiles (Marmur, 1960; Welker & Campbell, 1965; Saunders & Campbell, 1966). Stenesh, Roe & Snyder (1968) pointed out that intergeneric differences cannot be ruled out in these studies and therefore made similar studies using mesophilic and thermophilic strains of the genus *Bacillus*. They compared the base composition, thermal melting profile, molecular weight, and viscosity of DNA preparations from mesophilic and thermophilic strains. The DNA of thermophiles showed a consistently higher G+C content (53 per cent G+C) than the DNA of mesophiles (45 per cent G+C). As expected, the thermal melting profiles showed that the thermophilic DNA was more heat stable than the DNA of mesophiles with T_m values of approximately 92 and 88 °C, respectively. This is not unusual in the sense that T_m values are related to the $G+C$ content of the DNA and vice versa.

From the available data, one must conclude that there is no apparent relationship between the thermostability of DNA and the ability of cells to grow at high temperatures.

Ribonucleic acid and ribosomes

Early studies were mostly concerned with the loss of 260 nm-absorbing material from heat-treated cells (see review by Allwood & Russell, 1970). A study on the effect of mild heat (47 °C) on ribonucleic acid (RNA) breakdown in *Aerobacter aerogenes* was made by Strange & Shon (1964). Since RNA degradation preceded the loss of viability, they concluded that damage to cell RNA was not the primary cause of cell death at this temperature. Although most of the early studies were concerned with the degradation of total cell RNA, recent investigations have reported the degradation of messenger RNA (mRNA) and ribosomal RNA (rRNA) as a result of heat shock.

Since rRNA comprises the bulk of the total cell RNA, most investigations have measured the damage to the ribosomes or rRNA as a result of heat treatment. The structure of the ribosome is complex. There are two ribosomal subunits, named according to their sedimentation coefficient. In prokaryotic cells, the smaller 30 S subunit consists of one 16 S RNA molecule, and about 20 protein molecules. The larger, 50 S subunit consists of one 23 S RNA molecule, one 5 S molecule, and about 20 to 35 protein molecules.

Rosenthal & Iandolo (1970) have shown that exposure of *S. aureus* cells to a temperature of 55 °C causes the degradation of ribosomes. Only the 30 S subunit of heated cells appears to be selectively damaged. Specifically, the 16 S RNA was destroyed and the secondary structure of the 23 S RNA was altered. They could find no evidence of an activation of a bound, inactive ribonuclease to account for their findings. No information was available as to what happens to the ribosomal proteins.

Evidence does exist that there may be a correlation between the degradation of cell RNA (most likely rRNA) and the loss of viability of cells exposed to high temperatures. Resynthesis of RNA must have to occur before damaged cells can recover. There have been too few studies of this nature to allow one to postulate that cell RNA is the primary target for thermal injury.

A number of investigations have been made concerning the thermostability of the various species of RNA in thermophiles.

Using a strain of *B. stearothermophilus*, Saunders & Campbell

(1966) showed that the base composition of mRNA was almost identical with the values obtained for DNA from this organism, and since one would also expect the base composition and T_m values of mesophile mRNA to be almost identical with those of its DNA, they concluded that the thermostability of mRNA probably does not play an important role in the cell's ability to grow at high temperatures. Indirectly, the thermostability of mRNA was studied by determining the endogenous amino acid incorporation in a cell-free system of *B. stearothermophilus* (Friedman & Weinstein, 1966). The incorporation at 65 °C was greater than at 37 °C and, with a similar system from *E. coli*, endogenous incorporation at 65 °C was less than 12 per cent of the value at 37 °C. For obvious technical reasons, no experiments were run using the amino acid incorporation system from the mesophile and mRNA from the thermophile (or vice versa). Bubela & Holdsworth (1966) demonstrated that the turnover rate for mRNA in *B. stearothermophilus* at 40 and 63 °C was approximately 1 min, compared with 5–6 min for *E. coli* at 40 °C. They suggested that the very rapid turnover of mRNA in the thermophile was a mechanism for providing a high rate of protein synthesis to facilitate replacement of cell constituents which are denatured at 63 °C. Friedman (1968), however, reported that the rate of protein turnover was 1–2 min at this temperature. He suggested that the anomalous results could be reconciled by postulating the presence of a proteolytic enzyme having a high heat of activation and a mRNA-degrading enzyme which does not have an elevated heat of activation.

The base composition of tRNA from several strains of *B. stearothermophilus* is similar to that reported for *E. coli* (Mangiantini *et al.*, 1965; Saunders & Campbell, 1966; Stenesh & Holazo, 1967). In addition, the thermal melting profiles of tRNA from *B. stearothermophilus* and *E. coli* are almost identical. Zeikus, Taylor & Brock (1970) demonstrated that tRNA from *Thermus aquaticus* has a T_m of 86° C where as *E. coli* tRNA has a T_m value of 80 °C. The guanine and cytosine content of the tRNA of *T. aquaticus* was higher (63.5 per cent G+C) than the corresponding component of either *E. coli* (59.5 per cent G+C) or *B. stearothermophilus* (58 per cent G+C). *Thermus aquaticus* is capable of growth at 79 °C which is 11 to 17 deg C higher than the maximum growth temperature reported for strains of *B. stearothermophilus* (Welker, unpublished data). They suggested that the enhanced thermostability of the *T. aquaticus* tRNA may be a reflection of the increased G+C content. It would be interesting to see how tRNA of extreme thermophiles which grow between 80 and 95 °C compare to *T. aquaticus* in terms of thermal melting profiles and base composition.

In this connection, Watanabe *et al.* (1974) reported finding 5-methyl-2-thiouridine (m^5s^2U) in the sequence G-m^5s^2U-ψ-C-G of tRNA of an extreme thermophile. Most of the thymidine normally present in the GTψC region (probable ribosomal surface-orientating site) was replaced by m^5s^2U. They proposed that the m^5s^2U is probably important for the capacity of the tRNA to function in protein biosynthesis at high temperatures.

Agris, Koh & Soll (1973) reported that the tRNA of *B. stearothermophilus* grown at 70 °C has 1.4 times as many methyl groups as the tRNA from cultures grown at 50 °C. This was due to a three-fold increase in the 2-*O*-methylribose moieties of the tRNA. The nature of the methyl nucleotides in tRNA was not affected by the growth temperature. The significance of the ribose methylation is not immediately obvious since the thermal melting profiles of tRNA from cells grown at 50 or 70 °C were similar indicating a similar degree of secondary and tertiary structure.

Studies with leucyl- and phenylalanyl-tRNA synthetase from *B. stearothermophilus* showed that when the thermophile enzymes were tested with thermophile tRNA, both synthetases had temperature optima between 55 and 75 °C (Friedman & Weinstein, 1966). The same results were obtained using the thermophile enzymes coupled with tRNA from *E. coli* and vice versa. The interchangeability of amino acyl-tRNA synthetases and tRNA from *B. stearothermophilus* and *E. coli* was also shown by Arca *et al.* (1964).

Comparative studies on the incorporation of lysine and phenylalanine with poly-U and poly-A as templates, respectively, with crude extracts of *E. coli* and *B. stearothermophilus* showed that the ribosomes of *E. coli* lose 90 per cent of their activity after 5 min at 65 °C, whereas the ribosomes of *B. stearothermophilus* showed a slight increase (10 per cent) in activity when subjected to similar treatment (Algranati & Lengyel, 1966). Studies by a number of investigators (reviewed by Farrell & Campbell, 1969) have demonstrated that the ribosomes isolated from *B. stearothermophilus* undergo thermal denaturation at much higher temperatures than ribosomes from *E. coli* and other mesophiles or psychrophiles. Stenesh & Holazo (1967) studied rRNAs from mesophilic and thermophilic strains of the genus *Bacillus*. Their data revealed an average mole percentage G+C value of 55.1 for the mesophiles and 59.8 for the thermophiles. Campbell & Pace (1968) pointed out, however, that the thermal melting profiles of thermophile and mesophile rRNA preparations are very similar, and that the slight difference observed in their profiles and in the base composition were not sufficiently great to account for thermostability of thermophile ribosomes. Pace & Campbell (1967) studied the thermostability of the ribosomes of 19 different micro-organisms (temperature growth optima ranging from 20° to 70 °C). They found a positive correlation between the thermostability of the ribosomes and the maximum growth temperature of the organism. Generally the G+C content tends to increase with increasing growth temperature of most of the organisms studied. Irwin, Akagi & Himes (1973) investigated the ribosomes and rRNA of psychrophilic, mesophilic, and thermophilic clostridia. The thermal melting temperature (T_m) of ribosomes of the psychrophile, mesophile, and thermophile was 64.4, 63.8 and 68.7–69.6 °C, respectively. The T_m values of the rRNAs were similar for all three groups (66.2–67.9 °C). Zeikus, Taylor & Brock (1970) have shown that the T_m values of ribosomes and rRNAs of *T. aquaticus* were higher than the T_m values of the corresponding components from *E. coli*.

Altenburg & Saunders (1971) prepared hybrid ribosomes formed from the subunits of mesophilic and thermophilic bacteria. The *B. stearothermophilus* 50 S subunit was more thermostable than the *E. coli* 50 S subunit. They proposed that, at elevated temperatures, the thermophile 50 S subunit protects the mesophile 30 S subunit from inactivation by helping to maintain its proper structural configuration. Zeikus *et al.* (1970) have demonstrated that total rRNA of *T. aquaticus* is more thermostable than *E. coli* rRNA; however this increased thermostability appears to reside only in the 23 S RNA and not in the 16 S RNA.

From the data presented, one is tempted to conclude that the nucleic acids of thermophiles do not appear to possess any unique structure or composition that would serve as an important factor in the thermostability of the various cellular functions and ultimately to the growth of bacteria at high temperatures. In most cases it has been established that the proteins (e.g. ribosomal proteins, amino acyl-tRNA synthetase, etc.) or even membrane (DNA attachment site) play a predominant role in the structural and functional integrity of nucleic acids at elevated temperatures.

Membrane

The bacterial cytoplasmic membrane can be viewed as a delicate semipermeable structure which lies underneath the bacterial cell wall. The membrane is responsible for controlling the passage of solutes into and out of the cell as well as being the site of a variety of enzymes and other proteins involved in degradation and biosynthetic functions. Most studies dealing with the effect of heat on the membrane have measured the release of various cell-associated components or the penetration of substances which are normally excluded by the cell. There is little direct evidence that membrane damage is of primary importance in the thermal injury of bacterial cells (Allwood & Russell, 1970). However, loss of the functional integrity of the membrane may be of importance in the inability of the cell to recovery from such injury.

One of the first approaches designed to extend the understanding of the heat stability of cells has been to correlate the thermostability of an organism with the melting point of the cell lipids. Belehradek (1931) developed the concept that cell constituents were inactivated when the cell lipids are liquid. Gaughran (1947) proposed that cell processes are maintained as long as the cell lipids are liquid. The fact that the degree of saturation, length of the fatty acid chains, and the melting point of lipids increase with increasing growth temperature is in accord with both interpretations (Belehradek, 1931). There is considerable evidence that lipids formed by living organisms at high temperatures are more solid than lipids formed by living organisms at lower temperatures (reviewed by Christophersen, 1973).

Marr & Ingraham (1962), however, reported that the growth of *E. coli* at different temperatures did not result in significant changes in the fatty acid composition of the membrane lipids.

Studies on membranes of thermophilic organisms have concentrated mainly on the fatty acids of the membrane lipid. The fatty acid distribution in mesophilic and thermophilic strains of the genus *Bacillus* (Cho & Salton, 1964; Shen *et al.*, 1970) and in psychrophilic, mesophilic and thermophilic strains of the genus *Clostridium* (Chan, Himes & Akagi, 1971) have been determined. The combined data indicate that membranes of thermophiles generally contain a higher content of saturated and branched-chain fatty acids. In obligate (Daron, 1970) and facultative (Chan *et al.*, 1973) strains of *B. stearothermophilus* and extreme thermophiles (Ray, White & Brock, 1971; Weerkamp & MacElroy, 1972), raising the growth temperature produced a shift in the fatty acid composition in the membrane lipid. The shift was to fatty acids which in general had higher melting points. These results establish a definite correlation between the fatty acid content of the membrane and the temperature of growth and are consistent with the theory that the ability of an organism to grow at an elevated temperature is related to the melting temperature of its membrane lipids. It is tempting to speculate that a specific fatty acid composition of the membrane may set the maximum growth temperature.

There is, however, no simple correlation between the chain length of a fatty acid and its melting point (Sober, 1968).

Another characteristic feature of thermophilic organisms is the presence of branched-chain fatty acids in the membrane lipid. Heinen, Klein & Volkman (1970) have shown that membrane lipids of *T. aquaticus* grown at 80 °C contain the branched-chain fatty acids *iso* C_{16} and *iso* C_{17} which account for almost 10 and 50 per cent, respectively, of the total fatty acid fraction. Lipids extracted from *T. aquaticus* grown at 50 °C contain the two branched-chain fatty acids in approximately equal amounts (30 per cent).

Daron (1970) reported that in a thermophilic *Bacillus* species, the branched-chain fatty acids were more abundant than normal fatty acids and increasing the growth temperature from 40 to 60 °C results in a three- to four-fold increase in the ratio of normal to branched-chain hexadecanoic acids. Chan *et al.* (1973) reported that the fatty acid composition of a facultative thermophile grown at 55 °C contains a higher proportion of iso fatty acids while the anti-iso types were predominant in cells grown at 37 °C.

Changes in the membrane lipid composition can also be affected independently of temperature by alteration of the composition of the growth medium and the stage of growth (Bodman & Welker, 1969). Card, Georgi & Militzer (1969) reported that the major phospholipids in *B. stearothermophilus* 2184 are phosphatidyl glycerol (PG), phosphatidylethanolamine (PE) and cardiolipin (CL). Card (1973) reported that the total phospholipid content of this strain of *B. stearothermophilus* was constant during exponential growth, increased during the transition from exponential to stationary phase of growth, and again increased during the stationary phase of growth. The first increase was a result of an increase in PE and the second increase was a result of an increase in CL. A rapid conversion of PG to CL occurs in cells placed under anaerobic conditions or suspended in a non-growth medium.

Daron (1973) reported that the fatty acid composition of a thermophilic *Bacillus* species was altered by the addition of isobutyrate, isovalerate, α-methylbutyrate, leucine, and isoleucine to the growth medium. By altering the composition of the growth medium, membranes of these cells contain lipids in which the majority of the fatty acids had either 15, 16, or 17 carbons and belonged to each of the three groups of branched-chain fatty acids.

These results are similar to those for *B. subtilis* and are consistent with the scheme for branched-chain fatty acid biosynthesis proposed by Kaneda (1966). He proposed that α-keto analogues of the branched-chain amino acids compete in the rate-limiting, oxidative decarboxylation reaction, and the addition of one amino acid would thus increase one group of branched-chain fatty acids while decreasing the other two groups.

Considerable information is available about the chemical composition of membranes of mesophiles (Bodman & Welker, 1969). Suspension of spheroplasts of most mesophilic bacteria will rupture in hypotonic media or upon mechanical manipulation, even in the presence of osmotic stabilizers (reviewed by Bodman & Welker, 1969). Although much work has been done with spheroplasts or protoplasts of mesophilic bacteria, there have been no direct investigations of their heat stability. Ray & Brock (1971) examined the thermostability of protoplasts of *Sarcina lutea* and *Streptococcus faecalis*. Protoplasts of these organisms did not rupture when held at 0, 37, or 50 °C for 60 min. Incubation at 60 and 70 °C resulted in a slow rupture and complete rupture, respectively, of the protoplast suspension in 5 to 10 min. Protoplast membrane damage, measured by the release of [^{14}C]glycine, occurred at a significant rate at 50 °C.

Abram (1965) reported that elongated intact, or nearly intact, protoplasts could be prepared from water suspensions of thermophilic bacilli but not from mesophilic bacteria. She suggested that the cytoplasmic membrane of thermophiles possesses certain structural elements responsible for this unique property. Bodman & Welker (1969) isolated and characterized the membrane from *B. stearothermophilus* 1503–4R. The gross chemical composition of the membrane of this organism does not significantly differ from the values reported for mesophiles (Bodman & Welker, 1969). Protoplasts of this organism, however, are exceptionally stable to osmotic lysis, do not require a stabilizing medium to keep them from rupturing, are not sensitive to mechanical treatments, and are stable for at least 5 h at 55 °C. Wisdom & Welker (1973) reported that the membrane content of the cell and the protein-to-lipid ratio of the membrane of this organism increase as the growth temperature is increased (Table 1). As the growth temperature is increased, the protein content of the membrane increases and the lipid content decreases. Although a detailed analysis of the membrane lipid of this organism has not been done, it is not unreasonable to assume that there is also a shift to fatty acids having higher melting points as the growth temperature is increased. Oo & Lee (1971) reported similar but larger changes in the membranes of a facultative thermophile grown at 37 and 55 °C.

The increase in the amount of protein in the membrane as the growth

Table 1. *Effect of growth temperature on membrane content and composition of* B. stearothermophilus*

Growth temperature (°C)	Per cent membrane†	Protein (μg mg^{-1} membrane)	Lipid (μg mg^{-1} membrane)	Ratio of protein to lipid
55	16.5	650.0	178.2	3.65
60	16.9	664.1	155.5	4.27
65	17.8	680.0	130.3	5.22
70	18.4	690.1	118.2	5.85

* From Bodman & Welker (1969) and Wisdom & Welker (1973).

† Dry weight of the membranes was determined gravimetrically, and the results are expressed as per cent dry wt of the cells from which they were isolated.

temperature is raised may reflect changes in metabolic or biosynthetic pathways of these organisms. Jung *et al.* (1974) recently reported that a facultative thermophile grown at 37 or 55 °C in a complex medium showed identical maximum specific growth rates and yields in cell mass. Cell-free extracts of this bacterium, however, showed remarkable differences in the activity levels of two enzymes. Relatively high activities of alcohol dehydrogenase and glyceraldehyde-3-phosphate dehydrogenase were found in the extracts of the 55 °C cultures. Extracts of cells grown at 37 °C contained no detectable alcohol dehydrogenase activity. They concluded, that during exponential growth at 55 °C, the organism obtains its energy from both aerobic respiration and fermentation, whereas during exponential growth at 37 °C energy is obtained mainly by aerobic respiration. It is clear from these studies that a change in the protein content of membrane can be a result of a shift in metabolic pathways.

Wisdom & Welker (1973) reported that the thermal stability of protoplasts of *B. stearothermophilus* is enhanced as the growth temperature is raised (Table 2). Protoplasts were prepared from cells grown at each of the indicated temperatures, suspended in high-Mg^{2+} buffer and held at various temperatures for 1 h. The results are presented as per cent protoplast rupture. As the growth temperature was increased, the thermostability of the protoplasts was enhanced. Protoplasts suspended in buffer or ethylenediaminetetraacetic acid (EDTA) buffer rupture at all temperatures regardless of the temperature at which the cells were grown. These results indicate that the enhanced thermostability cannot be explained by a divalent cation protection. As the growth temperature is raised, the enhancement of protoplast stability can be correlated with an increase in the membrane content of the cell and the protein content of the membrane (Table 1). The lipid content

Table 2. *Effect of growth temperature on protoplast rupture, and the membrane composition and content of* B. stearothermophilus 1503–4R*

Growth temperature (°C)	Per cent protoplast rupture† when held at the following temperatures (°C)						Per cent membrane‡	Protein‡	Lipid‡	Ratio of protein to lipid
	55	60	65	70	75	80				
55	0	0	90	100	100	100	16.5	650.0	178.2	3.65
60	0	0	0	80	100	100	16.9	664.1	155.5	4.27
65	0	0	0	0	70	100	17.8	680.0	130.3	5.22
70	0	0	0	0	10	100	18.4	690.1	118.2	5.84

* From Wisdom & Welker (1973). Cells were grown in complex medium to the mid-exponential phase of growth. Protoplasts were prepared from cells grown at each temperature, suspended in Tris–Mg^{2+} buffer, and divided into six samples. A sample of each preparation was held at each temperature for 1 h.

† Protoplast rupture was quantified by direct counts with a phase-contrast microscope and by the liberation of 280 nm- and 260 nm-absorbing material.

‡ From Table 1.

of the membrane, however, decreases. Appropriate controls were run which eliminated the presence of endogenous enzymes which would lyse the protoplast membrane.

From the results presented it is tempting to speculate that thermophilic organisms can adapt to changes in temperature by altering the lipid composition of the membrane and that this heat resistance is related to the melting temperature of the lipids. Esser & Souza (1974) provided indirect evidence for the necessity and the reason for such changes. They measured paramagnetic resonance spectra of spin labels partitioned into spheroplast membranes of *B. stearothermophilus* to indicate lateral lipid-phase separations of the lipid in the membranes. Their spin label data showed that the dynamics of the lipid phase are unchanged at the different growth temperatures. In contrast, a temperature-sensitive mutant of this organism, which apparently has a defect in lipid synthesis fails to change its lipid composition above a certain temperature and can survive only up to the higher temperature boundary for lateral separation. Souza, Kostiw & Tyson (1974) reported that, in wild-type cells, a shift in growth temperature from 42 to 65 °C results in an increase from 42 to 69 per cent of fatty acids with melting points above 55 °C. In the temperature-sensitive mutant, the ability to make such changes is severely limited above 52 °C. They proposed that, in thermophilic organisms, the lowest and highest boundary temperature of any possible lipid mixture synthesized by the organism may determine the minimal growth temperature and the maximal growth temperature. Chan *et al.* (1973) used electron spin resonance spectroscopy to study the rigidity and viscosity of the membrane of lyophilized cells of a facultative thermophile grown at 37 and 55 °C. They concluded that the membrane of cells grown at the higher temperature is more rigid.

Protein

A considerable amount of work has been done studying the shape of curves obtained on heating bacterial cultures for extended periods of time. Usually the $\log_{10}$ of the number of surviving bacteria gives a straight line when plotted against the period of treatment (Allwood & Russell, 1970). Frequently non-linear survivor curves are observed indicating that either a specific amount of cell damage has to take place before the cell loses viability or that cells of a population have an unequal heat resistance. In these cases, cell inactivation is due to multiple events or cumulative damage to the cell. Generally, however, it is assumed that exponential death of heated bacteria is due to a single event and this is the heat denaturation of cell proteins, because heat denaturation studies with pure proteins have been shown to follow first-order kinetics.

It would seem at first sight that enzyme inactivation in mesophiles would be an obvious cause of heat injury and cell death. It is difficult, however, to see how this concept could result in exponential death. Very few experiments have been carried out which would specifically assess the role of enzyme inactivation in bacterial death (Allwood & Russell, 1970). Rosenberg *et al.* (1971) found a numerical correlation between the thermodynamic parameters of protein denaturation and the observed death rates of various organisms, and suggested that protein denaturation was a likely cause of cell death in mesophilic organisms. Some enzymes will be more thermolabile than others and this will vary from one strain to another. The location of an enzyme in the cell will play an important role in determining its inactivation by heat (Brandts, 1967).

A majority of the studies on the effects of temperature on microbial cell constituents have been done with cell extracts or, when possible, with purified preparations of the constituent. It is likely, however, that temperature has quite different effects on cell constituents *in vitro* as compared with *in vivo*. So far, very few studies have been reported which might throw light on possible differences in the temperature sensitivity of biological molecules *in vivo* and *in vitro*. It has been reported that substrates, and coenzymes (Burton, 1951) can protect certain enzymes against thermal denaturation. Boyer and co-workers (Boyer *et al.*, 1946; Boyer, Ballou & Luck, 1947) reported that short- and medium-chain fatty acids can protect some proteins against thermal denaturation.

Indirect evidence that the temperature sensitivity of some enzymes may be different *in vivo* as compared with *in vitro* has come from a number of different sources. Eidlic & Neidhardt (1965) isolated a temperature-sensitive mutant of *E. coli* which grew normally at 30 °C but failed to grow at 37 °C. The parental strain grows optimally at 37 °C. The temperature-sensitive mutant possesses an altered valyl-tRNA synthetase which was active *in vivo* at 30 °C but not at 37 °C. In cell free extracts however, the temperature-sensitive enzyme was not active at the lower temperature.

Bubela & Holdsworth (1966) compared the thermostability of the amino acid-activating system of membranes of *B. stearothermophilus* with a soluble activating system prepared from the membranes. Membranes lost only 20 per cent of their activity for the incorporation of amino acids into proteins after exposure to 63 °C for 10 min whereas the detached system lost 50 per cent of its activity after 10 min.

Ameluxen & Lins (1968) compared the thermostability of 11 different enzymes in crude extracts of a thermophile and a mesophile. With the exception of pyruvate kinase and glutamic-oxaloacetic transaminase the thermophile enzymes showed a much greater thermostability than those of the mesophile. Above 60 °C, pyruvate kinase and the glutamic-oxaloacetic transaminase from the thermophile are inactivated to the same extent as the corresponding enzymes from the mesophile. Although the two thermophile enzymes are inactivated *in vitro* at 70 °C, the thermophile is capable of growth and division at these temperatures. Studies with mixtures of thermophilic and mesophilic protein indicated the absence of stabilizing factors in the cell extracts of thermophiles.

These combined results, although of a preliminary nature, indicate that the nature of the association of the enzyme with the membrane may contribute to the thermostability of the enzyme. One still can argue for the presence of specific or non-specific protective components in the crude extracts used in most of these experiments.

The protective association of an enzyme with the membrane is a fundamental property of protein–lipid, protein–protein, and protein–substrate interactions and is probably not unique to thermophilic organisms.

A general approach which has proved to be extremely useful in studies involving the effects of temperature on biological systems was first used by Crozier and collaborators in the early 1920s (Ingraham,

Table 3. *Fatty acid composition of* A. laidlawii *membrane lipids of cells grown in media supplemented with various fatty acids**

Fatty acid supplemented in the growth medium	Fatty acid composition of the membrane lipids (mole per cent)										
	12:0	13:0	14:0	15:0	16:0	17:0	18:0	18:1c	18:1t	18:2	Unknown
18:1t	4.9	0.2	3.0	0.9	14.0	—	1.8	—	69.8	1.1	4.3
18:0	20.1	2.2	21.2	1.4	16.4	0.4	30.9	1.0	—	1.1	5.3
16:0	7.6	1.2	20.2	0.9	62.7	—	1.7	2.8	—	1.5	1.5
18:1c	5.8	1.1	6.7	1.5	26.4	0.3	6.5	45.1	—	1.4	5.2
18:2	3.5	Trace	7.9	Trace	45.7	Trace	1.9	3.6	—	33.2	4.2

* Table taken from De Kruyff *et al.* (1973).

1962). They studied the effect of temperature on a large number of varied biological processes. They clearly established that for most biological processes there is a considerable range of temperature over which the temperature–activity relation closely follows the Arrhenius equation. This general approach has been used successfully to examine the temperature-dependent activity of a function of the membrane (membrane-bound enzymes) in response to changes in the lipid composition of the membrane phospholipid. The utility of this approach depends upon the controlled manipulation of the composition of the membrane components.

Acholeplasma laidlawii is particularly suited for these studies because the fatty acid composition of the membrane lipids can be manipulated by varying the fatty acid supplement added to the growth medium (McElhaney & Tourtellotte, 1970). The absence of a cell wall also makes this mesophilic organism suitable for studies on lipid-phase transition using differential scanning calorimetry (Steim *et al.*, 1969; Melchior, Morowitz, Sturtevant & Tsong, 1970; De Kruyff, Demel & Van Deenen, 1972) and X-ray diffraction (Engelman, 1970). These studies established that the fatty acids of membrane phospholipids undergo a transition from an ordered crystalline or gel state to a more disordered liquid-crystalline state between 20 and 30 °C.

De Kruyff *et al.* (1973) investigated the membrane-bound NADH oxidase system, *p*-nitrophenylphosphatase and Mg^{2+}-dependent ATPase activity in *A. laidlawii* membranes as a function of temperature and fatty acid composition of the membrane phospholipids. The fatty acid supplement added to the growth medium dramatically increases its

Table 4. *Correlation between the endothermic lipid-phase transition of* A. laidlawii *membrane and the temperature of the break in the Arrhenius plot of ATPase activity**

Fatty acid supplement in growth medium	Mole per cent in membrane lipids	Temperature at which phase transition begins (°C)	Break in the Arrhenius plot of ATPase activity (°C)
Elaidic (18:1t)	69.8	13	15
Stearic (18:0)	30.9	20	18.5
Palmitic (16:0)	62.7	25	18

* Data from De Kruyff *et al.* (1973). (Welker, 1975. Reproduced with permission from Academic Press.)

incorporation into the membrane as well as causing minor fluctuations in the concentration of the other membrane lipid fatty acids (Table 3). Arrhenius plots of the NADH oxidase system and *p*-nitrophenyl-phosphatase activities showed no discontinuities although the membranes underwent a phase transition within the temperature range 5 to 35 °C. Membranes enriched in oleic (18:1) or linoleic (18:2) show no phase transition within this temperature range. In contrast, Arrhenius plots of the ATPase activity in membrane enriched with elaidic, stearic, or palmitic acids show breaks which roughly coincide with the temperature at which the membrane lipid phase transition begins (Table 4). They concluded that a membrane lipid-phase transition induces a change in the activation energy of membrane ATPase. NADH oxidase and *p*-nitrophenylphosphatase although membrane bound are not as sensitive to changes in the fatty acid composition of the membrane lipid.

A number of other studies using *A. laidlawii* or fatty acid auxotrophic mutants of *E. coli* support the theory that the liquid-crystalline state of the membrane lipids is necessary to support growth (Overath, Schairer & Stoffel, 1970: McElhaney, 1974), membrane transport (Overath, Schairer & Stoffel, 1970; Wilson & Fox, 1971; Machtiger & Fox, 1973), respiration (Overath *et al.*, 1970), and the activity of membrane-associated enzymes (Kimelberg & Papajadjapoulos, 1972).

Arrhenius plots for most of these processes were biphasic, the slopes extrapolating to intersections at unique transition temperatures. The transition temperatures for most of these biological processes varied with the degree of unsaturation of the predominant fatty acid in the membrane phospholipid.

In most studies of membrane-bound enzymes the temperature range was not extended beyond 40 or 42 °C. It would have been interesting if data were available indicating what effect the fatty acid composition of the membrane lipids had on the high temperature inactivation (thermal denaturation) of the various enzymes. No data were presented as to the membrane content of the cell or the protein content of the various membrane preparations.

Studies such as those with the fatty acid auxotrophic mutants of *E. coli* or *A. laidlawii* have not been done with thermophilic organisms. This is not surprising since fatty acids of the membrane lipids of thermophiles cannot easily be manipulated in the same manner. Changes in the membrane content of the cell or quantitative and qualitative changes in membrane lipid and proteins of thermophiles can be accomplished, however, by raising the temperature of growth.

Since we had demonstrated that protoplasts of a thermophilic bacillus had an enhanced thermostability as the growth temperature was raised, we were interested in determining whether specific intracellular constituents had also gained an enhanced thermostability. Protoplasts are well suited to study differences in the temperature sensitivity of biological molecules *in vivo* and *in vitro*. Thus, protoplasts prepared from cells grown at various temperatures can be used to examine the role of the membrane in protecting cell constituents and membrane-bound enzymes against thermal inactivation.

Two enzymes were selected on the basis of their cell localization. The enzymes were alkaline phosphatase, an internal or a loosely bound enzyme and the membrane-bound NADH oxidase of the terminal respiratory system. EDTA lysis of protoplasts releases all of the alkaline phosphatase whereas the NADH oxidase remains associated with the protoplast ghosts (Wisdom & Welker, 1973).

Data relating to the thermostability of these enzymes are shown in Table 5. NADH oxidase was less sensitive to thermal inactivation as the growth temperature was increased. The same results were obtained when this experiment was run with protoplast ghosts. The apparent thermostability of NADH oxidase is not dependent on the intactness of the protoplast or the presence of divalent cations.

Free alkaline phosphatase was inactivated to the same extent regardless of the growth temperature. The extent of the inactivation was the same in the presence of excess Mg^{2+}. There is a good correlation between the extent of protoplast rupture (see Table 2) and the thermal inactivation of alkaline phosphatase. Alkaline phosphatase is not inactivated in the intact protoplast. Once released from the protoplast, however, alkaline phosphatase is inactivated at 65 and 70 °C. Some

Table 5. *Thermostability of free and internal alkaline phosphatase and NADH oxidase**

Growth temperature (°C)	Per cent inactivation of alkaline phosphatase free at: 55 °C	60 °C	65 °C	70 °C	internal at: 55 °C	60 °C	65 °C	70 °C	Per cent inactivation of NADH oxidase at: 75 °C	80 °C	85 °C
55	11	15	100	100	0	0	87	100	100	100	100
60	12	14	100	100	0	0	0	92	90	100	100
65	11	15	100	100	0	0	0	20	75	84	100
70	12	16	100	100	0	0	0	5	0	10	87

* Table taken from Wisdom & Welker (1973). Protoplasts were prepared from cells grown at various temperatures. Each protoplast suspension (55 °C-grown protoplasts, etc.) in Mg^{2+} buffer was divided into 13 portions. Protoplasts in five of the samples were ruptured by the addition of EDTA. The membranes were collected by centrifugation, the supernatant fluids were adjusted to an equal protein concentration, and designated 'free alkaline phosphatase'. Five of the remaining samples were designated 'internal alkaline phosphatase' and three designated as 'NADH oxidase'.

A sample of the 'free' and 'internal alkaline phosphatase' were incubated for 1 h at 6 °C (control), 55, 60, 65, or 70 °C. After incubation, 'internal alkaline phosphatase' was released by the addition of EDTA. The 'NADH oxidase' samples were incubated for 1 h at 75, 80 and 85 °C followed by the addition of EDTA. Enzyme activity in each of the samples was measured and the per cent inactivation calculated.

inactivation of alkaline phosphatase occurs in intact 65 °C-grown and 70 °C-grown protoplasts held at 70 °C.

These results indicate that in the intact protoplast, alkaline phosphatase is protected from heat inactivation. The apparent thermostability of the NADH oxidase system increases as the growth temperature is raised. It is difficult, however, to differentiate between inherent protein stability and a protective association of this enzyme system with the protoplast membrane.

It is possible to reconstitute functional NADH oxidase in aggregated membrane after dialysis against Mg^{2+}-buffer of non-ionic-detergent solubilized membranes (Welker, unpublished data). It would now be possible to mix solubilized membranes prepared from cells grown at 55 °C (55 °C-membranes) or 70 °C (70 °C-membranes) in varying proportions, and then determine the thermostability of NADH oxidase in the various aggregated membrane preparations.

The 55 °C-membranes and 70 °C-membranes were differentiated by labelling each membrane preparation with a different isotope. The protein-to-lipid ratio in aggregated membrane was similar to that of the intact protoplast membrane prepared from cells grown at the various temperatures. The effectiveness of the aggregation procedure was demonstrated by the relatively high recovery of protein (92–97 per cent)

Table 6. *Thermostability of reconstituted functional NADH oxidase in aggregated membrane**

Per cent solubilized membrane in mixture		Per cent recovery in aggregated membrane		Per cent inactivation† of NADH oxidase at:		
^{3}H-labelled held at 70 °C	^{14}C-labelled held at 55 °C	[^{3}H]lipid	[^{14}C]lipid	75 °C	80 °C	85 °C
0	100	0	97.5	100	100	100
25	75	23	74	30	72	100
50	50	46	48	7	22	95
75	25	73	23	4	17	90
100	0	92	0	—	—	—
100‡	0	95	0	5 (0)	18 (10)	94 (87)

* Membranes were prepared from cells grown at 55 or 70 °C. The 55 °C-membranes were labelled with ^{14}C and the 70 °C-membranes were labelled with ^{3}H. Membranes were solubilized with Triton X-100, and aggregation of solubilized membranes was accomplished by dialysis against a Mg^{2+}-buffer. The recovery of the 55 °C-membrane and 70 °C-membrane in the aggregated material was determined. The recovery of NADH oxidase activity was determined after exposure at each temperature for 1 h.

† Detergent-solubilized ^{3}H-labelled 70 °C-membranes were heated at 90 °C for 1 h to inactivate NADH oxidase.

‡ 'Unheated' control: detergent-solubilized ^{3}H-labelled 70 °C-membranes were held at 37 °C for 1 h; the percentage NADH oxidase inactivation in non-solubilized membranes at 75, 80 and 85 °C is in parentheses.

(From Welker, 1975. Reproduced with permission from Academic Press.)

and lipid (96–98 per cent). The extent of NADH oxidase recovery varied from 50 to 70 per cent.

The thermostability of reconstituted NADH oxidase in aggregated membrane is shown in Table 6. The heating procedure did not significantly affect the membrane aggregation. However, the NADH oxidase in aggregated membrane was somewhat more sensitive to thermal inactivation when compared to the inactivation of NADH oxidase in the intact protoplast membrane.

In mixtures containing 50 or 75 per cent 70 °C-membrane material, the extent of NADH oxidase inactivation was similar to that observed in the 'unheated' 70 °C-membrane control sample. The thermostability of NADH oxidase is still considerably enhanced when the mixture contains only 25 per cent 70 °C-membrane material. These results suggest that 70 °C-membrane material provides some degree of protection against thermal inactivation of the NADH oxidase present in 55 °C-membranes.

Arrhenius plots of the NADH oxidase activity in protoplast membranes prepared from cells grown at different temperatures are shown in Fig. 1. The Arrhenius plots have virtually identical slope until the high temperature denaturation occurs.

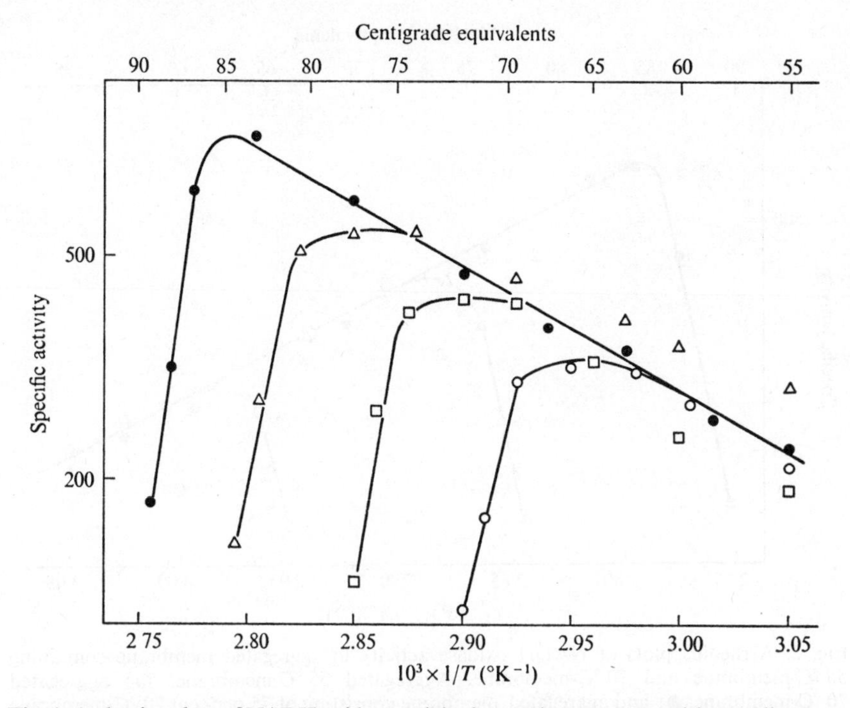

Fig. 1. Arrhenius plots of NADH oxidase activity in membranes prepared from cells grown at 55 (○), 60 (□), 65 (△) and 70 °C (●) (Wisdom & Welker, 1973). Membrane suspensions were adjusted to an equal protein concentration, each suspension was held at the indicated temperature for 1 h, and the residual NADH oxidase activity measured at 55 °C. (Welker, 1975. Reproduced with permission from Academic Press.)

The transition temperature for the high temperature denaturation of NADH oxidase was dependent for the most part on the growth temperature of the cells from which the membranes were obtained. Inactivation of the membrane-bound NADH oxidase was observed between 64 and 66 °C in the 55 °C-membranes; 70 and 75 °C in the 60 °C-membranes; 76 and 79 °C in the 65 °C-membranes; and 84 and 86 °C in the 70 °C-membranes. Temperatures below 55 °C were not used so a low temperature denaturation was not observed (Brandts, 1967).

Arrhenius plots of the NADH oxidase activity in aggregated membranes are shown in Fig. 2. Only the aggregated 55 °C- and 70 °C-membrane samples and an aggregated sample containing 75 per cent 70 °C-membrane (NADH oxidase inactivated by heating at 90 °C for 1 h) and 25 per cent 55 °C-membrane are shown. These three Arrhenius plots have virtually identical slopes with no change in the activation energy until the high temperature denaturation occurs. Inactivation of

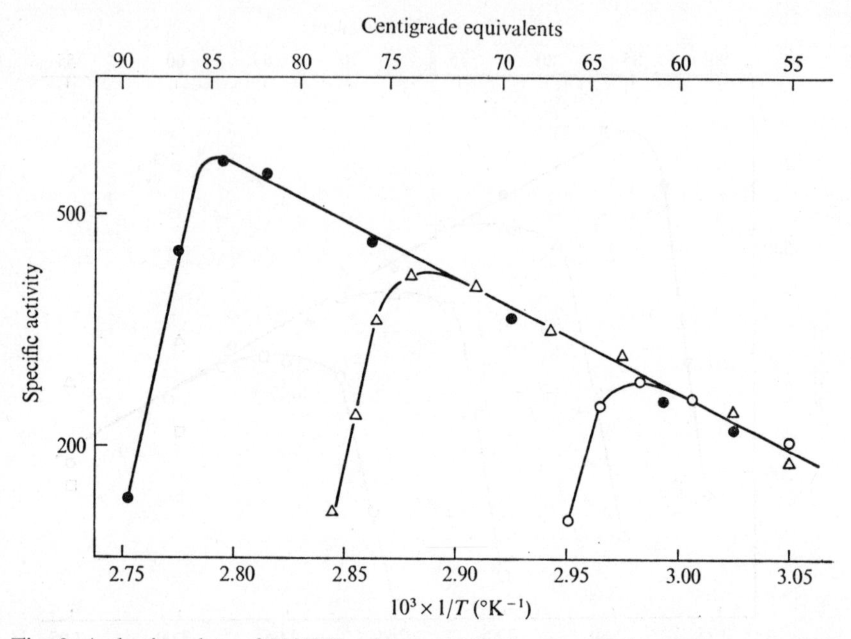

Fig. 2. Arrhenius plots of NADH oxidase activity in aggregated membrane containing 55 °C-membrane and 70 °C-membrane. Aggregated 55 °C-membrane, ○; aggregated 70 °C-membrane, ●; and aggregated membrane consisting of 75 per cent 70 °C-membrane (NADH oxidase was inactivated by heating at 90 °C for 1 h) and 25 per cent 55 °C-membrane, △. (Welker, 1975. Reproduced with permission from Academic Press.)

the NADH oxidase in the aggregated membrane mixture was observed between 73 and 74 °C. These results provide additional evidence that 70 °C-membrane protects the 55 °C-membrane NADH oxidase from heat inactivation.

We conclude from these studies that the thermostability of alkaline phosphatase of *B. stearothermophilus* was not affected by the growth temperature. This enzyme is protected from heat inactivation in the intact protoplast. The extent of this protection appears to be correlated with the composition of the protoplast membranes. It is possible that the membrane acts as an insulator against the transfer of heat from the environment so that alkaline phosphatase and other soluble enzymes are protected against thermal denaturation.

The thermostability of the membrane-bound NADH oxidase system also increases as the temperature of growth is raised but is not dependent on an intact protoplast. It is still difficult, however, to differentiate between inherent protein stability and a protective association of this enzyme system with the protoplast membrane.

Bausum & Matney (1965) reported that facultative thermophiles growing in the mesophilic range require a brief period of adaptation at intermediate temperatures before gaining the capability to grow at temperatures suitable for thermophiles. They concluded that the temperature boundary (adaptation temperature) between bacterial mesophily and thermophily is between 44 and 52 °C. We have isolated from soil a number of facultative thermophilic bacilli which are capable of growing over a relatively broad temperature range. Cells of one of these isolates (strain T1) were inactivated (99 per cent loss of viability) when exposed to 59 °C (challenge temperature) for 5 min after a period of growth at 30 °C. Cells grown at the latter temperature are less sensitive to inactivation (40 per cent) when held at the challenge temperature for 3 h if they are first grown for 25–30 min at a temperature between 53 and 56 °C (adaptation temperature). Since adaptation occurs in the whole population in a relatively short period of time, these results cannot be explained by the presence of a small mutant population. Adapted cells held at the challenge temperature for 1–2 h will again be inactivated at 59 °C if they are grown at 30 °C for at least 30 min (de-adaptation). Only exponential phase cells can be adapted to survive at the challenge temperature. Although the rate of growth doubles when the cells are shifted from 30 °C to the challenge temperature (59 °C), DNA synthesis is inhibited for a period of 30 min and protein and RNA synthesis immediately adjust to a rate characteristic of cells growing at 55 °C. If NaCl is present in the growth medium during the adaptation period, DNA synthesis is not inhibited, and the survival of cells at the challenge temperature is increased markedly.

This same strain was shown to undergo significant changes in the cell wall composition (Novitsky *et al.*, 1974) and rigidity of the membrane (Chan *et al.*, 1973) when grown at 55 °C as compared to when the cells are grown at 37 °C.

The above studies suggest that biochemical changes occur in all the cells of a culture at the adaptation temperature which effectively protects the critical components of the cell from heat inactivation and allows the cell to grow and divide at an elevated temperature. Evidence that changes in the thermostability of specific enzymes occur in facultative thermophiles when grown at low or high temperatures has been presented by several investigators. Campbell (1955) purified α-amylase from facultative strains of *Bacillus coagulans* and *B. stearothermophilus* following growth at 35 or 55 °C. A comparison of the thermostability at 90 °C for 1 h revealed that the 55 °C preparation showed only a 6–10 per cent inactivation, whereas the 35 °C preparations were inactivated by 90–92 per cent. L. L. Campbell (personal communication) examined the thermostability of the α-amylase of the same strains at graded temperatures between 37 and 55 °C and found that the enzyme produced below 46 °C was heat labile whereas above 50 °C was heat stable. Brown, Militzer & Georgi (1957) reported that the thermostability of pyrophosphatase of a thermophilic bacillus increases with an increase of growth temperature. Isono (1970) reported that a partially purified α-amylase from cultures of *B. stearothermophilus* grown at 55 °C was relatively more resistant to heat inactivation than the same enzyme isolated from cultures grown at 37 °C.

Haberstich & Zuber (1974) reported that a thermophilic strain of *B. stearothermophilus* can only be adapted to grow at either 37 or 55 °C provided the culture is first grown for a period at 46 °C. The thermostability of glucokinase, glucose-6-phosphate isomerase, glucose-6-phosphate dehydrogenase, glyceraldehyde-3-phosphate dehydrogenase, and isocitrate dehydrogenase from 37 and 55 °C cultures were determined by heating crude extracts for 30 min at 50, 60, 70, 80 and 90 °C. The enzymes of glycolysis and of the citric acid cycle from the 55 °C culture are more thermostable than those from the 37 °C culture. At the intermediate growth temperature of 46 °C, the thermostability curves of these enzymes were intermediate to those exhibited by these same enzymes of the 37 and 55 °C cultures. They postulated that the 46 °C cultures either contained a mixture of heat-labile and heat-stable enzyme or that these cells contained enzyme of intermediate stability. The adaptation of the extreme thermophile, *Bacillus caldotenex*, was also examined in this investigation. Cultures of *B. caldotenex* grown at 37 °C will grow at 37, 45, 50, 55, 60, 65 and 70 °C. Growth at 30–50 °C started within 1–2 h, whereas at temperatures above 55 °C a lag period of 6–11 h was observed. As in the case of *B. stearothermophilus*, thermolabile enzymes were present in 37 °C cultures, and thermostable enzymes were present in 70 °C cultures. There appears to be a relationship between the appearance of the lag period in the temperature range above 55 °C and the formation of thermostable enzymes. Between 30–50 °C there was only heat-labile glucose-6-phosphate isomerase; at 55 °C mostly the heat-stable form of this enzyme was present; and at temperatures above 60 °C only the latter form was detected. The presence of 'stabilizing factors' in these studies was considered unlikely because neither crude nor partially purified preparations of glucokinase and glucose-6-phosphate isomerase differed significantly from purified enzymes in thermostability. They proposed that these thermophilic bacilli contain a special mechanism for the thermo-adaptation of some enzymes. The exact mechanisms of this adaptation are unknown but they suggested that either (i) these strains contain two genes which code for each form of the different enzymes. The transcription of these genes would be regulated by temperature. (ii) One gene for each enzyme and thermo-adaptation occurs during translation. (iii) The catalytically mature enzyme is modified in some manner. The conversion from thermophilic to mesophilic metabolism and vice versa, however, was only successful in a complex medium.

It appears that the degree of thermo-adaptation depends to a large extent on the culture conditions, especially the composition of the growth medium. Weerkamp & MacElroy (1972) reported that a partially purified lactate dehydrogenase of an extremely thermophilic bacillus was more thermostable when isolated from cells grown in a complex medium than when isolated from cells grown in a minimal medium. Campbell & Williams (1953), using strains of *B. stearothermophilus*, *B. coagulans*, and *Bacillus globigi*, found that all strains gave good growth in a chemically defined medium. However, the cultures could be placed into three groups according to the manner in which the growth temperature affected their nutritional growth requirements. One group showed no differences in growth requirements regardless of the incubation temperature. The second group had additional requirements as the temperature of incubation was increased. The third group required additional requirements as the temperature of incubation was lowered.

These combined results suggest that the ability of a thermophile or facultative thermophile to grow at low or high temperatures depends on whether they are cultivated (or pre-cultivated) in the proper growth medium.

Crabb, Murdock & Amelunxen (1975) have recently shown that cell protein in extracts of cells of a facultative thermophilic strain of *B. coagulans* grown at 55 °C are markedly more thermostable than the proteins in an extract of cells grown at 37 °C. Partially purified glyceraldehyde-3-phosphate dehydrogenase from cells grown at 37 or 55 °C is extensively inactivated at low ionic strength after 24 h at 4 °C, but little inactivation occurs in this enzyme prepared from cells grown at either temperature when the ionic strength was increased by a factor of 1.8. These investigators concluded that this enzyme, produced at either temperature, possesses considerable inherent thermostability but is subject to heat inactivation in an environment of low ionic strength.

The survival of facultative thermophiles at elevated temperatures may be due to a combination of factors. Certain cell components may be stabilized by ions or other molecules or by their association with the cytoplasmic membrane. These organisms also could contain proteins having an inherent thermostability.

A number of attempts, with somewhat controversial results, have been made to adapt mesophilic micro-organisms to grow at temperatures higher than their optimum growth temperature but these have usually failed. Dowben & Weidenmuller (1968), using a genetically marked strain of *B. subtilis* which was capable of some growth at 53 to 55 °C, reported that they could adapt this strain to grow at temperatures as high as 72 °C by slowly increasing the temperature of incubation. When bacteria adapted to growth at elevated temperatures were subsequently grown at 37 °C, the ability to grow at high temperatures was lost and could not be regained unless the adaptation regime was again used. The phenomenon appears to be an adaptation rather than a selection of a mutant population. We have made a number of unsuccessful attempts to repeat these experiments with several mesophilic bacilli, including the strain used by Dowben & Weidenmuller. The molecular mechanisms of this physiological adaptation should be of considerable interest; however, no further studies on this strain of *B. subtilis* have been published.

There have been a considerable number of reports concerning the thermal stability of enzymes and other proteins isolated from thermophiles. If we accept the concept that thermophilic and mesophilic micro-organisms have a common origin, we can also postulate that the various cell constituents must also have evolved from a common source. It follows from this, therefore, that an enzyme from a mesophilic or thermophilic species of the same genus would have similar structural and functional properties but the thermophilic enzyme would be relatively more heat stable than the mesophilic enzyme. There is now considerable evidence that enzymes and other proteins of most thermophiles are relatively more heat stable than the corresponding proteins of mesophiles or psychrophiles. These studies, however, do not reveal the molecular basis for the enhanced thermostability since it is difficult to rule out the presence of protective or stabilizing factors. A number of enzymes and other proteins from thermophiles have now been purified and in some cases their properties have been compared with their counterparts isolated from mesophiles or psychrophiles.

The properties of a number of these enzymes were discussed in a recent review by Singleton & Amelunxen (1973). Ljungdahl & Sherod (1975, see acknowledgments) recently presented a complete and comprehensive review on this topic at a symposium on 'Extreme Environments: Mechanisms of Microbial Adaptation' which was held at the NASA/Ames Research Center, Moffett Field, California, 26–28 June 1974. Since the literature concerning thermophilic proteins is extensive and that a second review of this field would be redundant, I will not discuss any of these proteins in detail, but will briefly summarize the results of these studies.

In all cases the presence of non-protein protective factors have been eliminated. All the proteins from thermophiles, with the exception of one, are remarkably similar in physical properties such as molecular weight, number of subunits, Stokes radius, specific volume, sedimentation constant, etc. One can conclude therefore, that in physical properties proteins from thermophilic bacteria do not differ much from their mesophilic counterparts. In general, the amino acid composition of the various thermophilic proteins does not significantly vary from that of mesophilic organisms.

The only thermostable protein found so far which differs grossly from its mesophilic counterpart is an α-amylase from *B. stearothermophilus* (Manning & Campbell, 1961). Studies of the physical properties of this enzyme (Manning, Campbell & Foster, 1961) revealed that it had a molecular weight of 15600, contained four cysteine residues, and had a large negative optical rotation, indicating a semi-random or randomly coiled conformation. They postulated that the enzyme exists, under physiological conditions in what might be considered as a 'denatured state', and heating which unfolds the α-helix structure of most larger enzyme molecules has little or no effect on this enzyme. Pfueller & Elliott (1969) isolated an α-amylase from the same strain used by Manning & Campbell and reported that their α-amylase has a molecular weight of 53000, does not contain cysteine, and does not possess any unusual thermostability. The α-amylase studied by Endo (see Farrell & Campbell, 1969), which was produced by a different strain of *B. stearothermophilus*, has some properties similar to those reported by Pfueller & Elliott (1969) but was more thermostable. Ogasahara, Imanishi & Isemura (1970*a*, *b*) examined the properties of an α-amylase from a strain of *B. stearothermophilus* different from those used by either of the above investigators, and reported a molecular weight of 48000, one cysteine residue and a thermostability similar to that of the α-amylase studied by Endo. As pointed out by Ljungdahl & Sherod (1975), the α-amylase isolated by Manning & Campbell (1961) was one of the first isolated from a thermophilic organism free from contaminating protein, and it was subsequently assumed that enzymes from thermophiles should be small and exist in semi-random coil conformations.

Almost all of the evidence so far accumulated leads to the conclusion that thermophilic proteins have inherent thermostability, which depends on their amino acid composition and the sequence of the amino acids. As a result of unique sequences of amino acids, thermophilic proteins have additional bonds which stabilize the secondary and tertiary structure, compared to proteins of mesophilic organisms. These bonds include disulphide bridges, hydrogen bonds, ionized group interactions and hydrophobic bonds.

It has been proposed that disulphide bonds may stabilize proteins and, in ribonuclease, it appears that the four disulphide bonds play a significant role in the thermostability of this enzyme. The data from thermophilic proteins do show that these bonds are more numerous in proteins from thermophilic organisms. In fact, in some proteins there are fewer disulphide bonds in proteins from thermophiles as compared with proteins from mesophiles.

Hydrophobic bonds are very important in the stability of proteins. Since the thermostability of these bonds increases with an increase in temperature up to about 65 °C, these bonds may contribute significantly to the thermostability of proteins of thermophiles (Brandts, 1967). Proteins having more amino acids with hydrophobic side chains may possess a greater thermostability. Hydrophobicity has been calculated for a number of proteins from both mesophiles and thermophilic organisms (Singleton & Amelunxen, 1973). Although in some cases thermostable proteins have a slightly higher hydrophobicity than comparable proteins from mesophiles, most thermostable proteins are not more hydrophobic than their mesophilic counterparts.

The amino acid sequences of several proteins isolated from thermophiles and mesophiles have been reported. On the basis of nearly all the comparisons made there appears to be some subtle differences between thermophilic and mesophilic proteins. It is difficult, however, to relate these differences directly to a higher or lower thermostability.

CONCLUSION

The biochemical changes occurring in mesophilic bacteria induced by heat stress are of necessity related to viability. It is extremely difficult, however, to demonstrate a quantitative relationship between loss of viability and the biochemical changes occurring in the cell. It is obvious that heat inactivation of mesophilic bacteria, and their ability to recover, cannot be defined in simple biochemical terms.

It seems reasonable to propose that quantitative data relating to the effect of heat stress on cell viability can be obtained from studies with facultative thermophilic and thermophilic micro-organisms. The validity of this proposal assumes that thermophilic and mesophilic micro-organisms have evolved from common ancestors. If thermophiles and mesophiles have a common origin, it follows that their cell constituents have a common source. Facultative thermophiles may represent an intermediate evolutionary stage between mesophiles and thermophiles. This assumption is most valid when one selects representative species of the same genus. Although the fundamental question as to how thermophilic micro-organisms evolved still remains to be answered, detailed, comparative studies on the properties of cell constituents isolated from mesophiles and thermophiles, or facultative thermophiles grown at low and high temperatures should provide insights into those biochemical properties which are most necessary for a cell's thermophilic existence.

From the evidence presented it seems justified to conclude that no one factor is responsible for the heat inactivation of a bacterial cell. Many proteins from thermophiles are more thermostable when compared to their counterparts from mesophiles. Although there are subtle differences in some of the physical properties, and in amino acid composition and sequence, specific explanations for the difference in thermostability between homologous proteins of thermophiles and mesophiles requires a determination of the amino acid sequence and their three-dimensional structure. In contrast, other proteins from facultative thermophiles and thermophiles do not appear to possess any unusual thermostability and are probably protected from heat denaturation by protective or stabilizing factors or by their association with the membrane.

The membrane lipid of thermophiles differs quantitatively and qualitatively from the membrane lipid of mesophiles. Membranes of thermophiles also have an enhanced thermostability, and the membrane fatty acids from thermophiles have a higher melting point. Similar changes are observed in the membrane fatty acids of facultative thermophiles grown at the higher temperatures. Since the membrane plays a critical role in cell metabolism, changes in the composition of the membrane would be expected to occur in order to ensure the functional integrity of this structure.

It is generally assumed that nucleic acids from thermophiles do not possess any unique thermostability. It is possible, however, that nucleic acids of thermophiles possess a unique structure (methylated nucleotides, specific nucleotide sequences, etc.) which has not been detected by the analytical procedures currently used. In which case, the nucleic acids of thermophiles may also play a significant role in the cells thermophilic existence.

It is clear that heat-induced death in vegetative, non-sporulating bacteria cannot be explained in simple terms. A similar conclusion is reached when one considers the studies on the biochemical basis for thermophily. It is obvious that future investigations should be concerned with the more subtle changes in cell constituents. These studies will not only extend our understanding of the cause of heat inactivation of vegetative bacteria but will lead to a better understanding of the structural and functional properties of proteins, membranes and other cell components.

I am especially grateful to Drs Ljungdahl and Sherod for providing me with a copy of their manuscript prior to its publication. Thanks are also due to Mrs Trudy Pollack for her invaluable assistance in the preparation of this manuscript.

The original work cited from the author's laboratory was supported by Public Health Service research grant AI-06382 from the National Institute of Allergy and Infectious Diseases.

REFERENCES

Abram, D. (1965). Electron microscope observation on intact cells, protoplasts, and the cytoplasmic membrane of *Bacillus stearothermophilus*. *Journal of Bacteriology*, **89**, 855–873.

Algranati, I. D. & Lengyel, P. (1966). Polynucleotide-dependent incorporation of amino acids in a cell-free system from thermophilic bacteria. *Journal of Biological Chemistry*, **241**, 1778–1783.

Allen, M. B. (1953). The thermophilic aerobic spore-forming bacteria. *Bacteriological Reviews*, **17**, 125–173.

Allwood, M. C. & Russell, A. D. (1969). Thermally induced changes in the physical properties of *Staphylococcus aureus*. *Journal of Applied Bacteriology*, **32**, 68–78.

Allwood, M. C. & Russell, A. D. (1970). Mechanisms of thermal injury in non-sporulating bacteria. *Advances in Applied Microbiology*, **12**, 89–119.

Altenburg, L. C. & Saunders, G. F. (1971). Properties of hybrid ribosomes formed from the subunits of mesophilic and thermophilic bacteria. *Journal of Molecular Biology*, **55**, 487–502.

Ameluxen, R. & Lins, M. (1968). Comparative thermostability of enzymes from *Bacillus stearothermophilus* and *Bacillus cereus*. *Archives of Biochemistry and Biophysics*, **125**, 765–769.

Arca, M., Calvori, C., Frontali, L. & Tecce, G. (1964). The enzymic synthesis of aminoacyl derivatives of soluble ribonucleic acid from *Bacillus stearothermophilus*. *Biochemica et Biophysica Acta*, **87**, 440–448.

Bausum, H. T. & Matney, T. S. (1965). Boundary between bacterial mesophilism and thermophilism. *Journal of Bacteriology*, **90**, 50–53.

Belehradek, J. (1931). Le méchanisme physico-chimique de l'adaptation thermique. *Protoplasma*, **12**, 406–434.

Bodman, H. & Welker, N. E. (1969). Isolation of spheroplast membranes and stability of spheroplasts of *Bacillus stearothermophilus*. *Journal of Bacteriology*, **97**, 924–935.

BOYER, P. D., BALLOU, G. A. & LUCK, J. M. (1947). The combination of fatty acids and related compounds with serum albumin. *Journal of Biological Chemistry*, **167**, 407–424.

BOYER, P. D., LUM, F. G., BALLOU, G. A., LUCK, J. M. & RICE, R. G. (1946). The combination of fatty acids and related compounds with serum albumin. I. Stabilization against heat denaturation. *Journal of Biological Chemistry*, **162**, 181–198.

BRANDTS, J. F. (1967). Heat effects on proteins and enzymes. In *Thermobiology*, ed. A. H. Rose, pp. 25–72. New York: Academic Press.

BRIDGES, B. A., ASHWOOD-SMITH, M. J. & MUNSON, R. J. (1969). Correlation of bacterial sensitivities to ionizing radiation and mild heating. *Journal of General Microbiology*, **58**, 115–124.

BROCK, T. D. (1967). Life at high temperatures. *Science, Washington*, **158**, 1012–1019.

BROCK, T. D. & FREEZE, H. (1969). *Thermus aquaticus* gen. n. and sp. n., a non-sporulating extreme thermophile. *Journal of Bacteriology*, **98**, 289–297.

BROWN, D. K., MILITZER, W. & GEORGI, C. E. (1957). The effect of growth temperature on the heat stability of bacterial pyrophosphatase. *Archives of Biochemistry and Biophysics*, **70**, 248–256.

BUBELA, B. & HOLDSWORTH, E. S. (1966). Amino acid uptake, protein and nucleic acid synthesis and turnover in *Bacillus stearothermophilus*. *Biochemica et Biophysica Acta*, **123**, 364–375.

BURTON, K. (1951). The stabilization of D-amino acid oxidase by flavin dinucleotide, substrates and competitive inhibitors. *Biochemical Journal*, **48**, 458–467.

CAMPBELL, L. L. JR (1955). Purification and properties of an α-amylase from facultative thermophilic bacteria. *Archives of Biochemistry and Biophysics*, **54**, 154–161.

CAMPBELL, L. L. & PACE, B. (1968). Physiology of growth at high temperatures. *Journal of Applied Bacteriology*, **31**, 24–35.

CAMPBELL, L. L. JR & WILLIAMS, O. B. (1953). The effect of temperature on the nutritional requirements of facultative and obligate thermophilic bacteria. *Journal of Bacteriology*, **65**, 141–145.

CARD, G. L. (1973). Metabolism of phosphatidylglycerol, phosphatidylethanolamine, and cardiolipin of *Bacillus stearothermophilus*. *Journal of Bacteriology*, **114**, 1125–1137.

CARD, G. L., GEORGI, C. E. & MILITZER, W. E. (1969). Phospholipids from *Bacillus stearothermophilus*. *Journal of Bacteriology*, **97**, 186–192.

CHAN, M., HIMES, R. H. & AKAGI, J. M. (1971). Fatty acid composition of thermophilic, mesophilic, and psychrophilic clostridia. *Journal of Bacteriology*, **106**, 876–881.

CHAN, M., VIRMANI, Y. P., HIMES, R. H. & AKAGI, J. M. (1973). Spin-labeling studies on the membrane of a facultative thermopholic bacillus. *Journal of Bacteriology*, **113**, 322–328.

CHO, K. Y. & SALTON, M. R. J. (1964). Fatty acid composition of the lipids of membranes of gram-positive bacteria and 'walls' of gram-negative bacteria. *Biochemica et Biophysica Acta*, **84**, 773–775.

CHRISTOPHERSEN, J. (1973). Microorganisms. I. Basic aspects of temperature action on microorganism. In *Temperature and Life*, ed. J. Christophersen, H. Hensel & W. Larcher, pp. 3–85. New York: Springer-Verlag.

COONEY, D. G. & EMERSON, R. (1964). *Thermophilic Fungi*, 188 pp. San Francisco: W. H. Freeman & Son.

CRABB, J. W., MURDOCK, A. L. & AMELUNXEN, R. E. (1975). A proposed mechanism of thermophily in facultative thermophiles. *Biochemical and Biophysical Research Communications*, **62**, 627–633.

CROSS, T. (1968). Thermophilic actinomycetes. *Journal of Applied Bacteriology*, **31**, 36–53.

DARON, H. H. (1970). Fatty acid composition of lipid extracts of a thermophilic *Bacillus* species. *Journal of Bacteriology*, **101**, 145–151.

DARON, H. H. (1973). Nutritional alteration of the fatty acid composition of a thermophilic *Bacillus* species. *Journal of Bacteriology*, **116**, 1096–1099.

DE KRUYFF, B., DEMEL, R. A. & VAN DEENEN, L. L. M. (1972). The effect of cholestrol and epicholestrol incorporation on the permeability and on the phase transition of intact *Acholeplasma laidlawii* cell membrane and derived liposomes. *Biochemica et Biophysica Acta*, **255**, 331–347.

DE KRUYFF, B., VAN DIJCK, P. W. M., GOLDBACH, R. W., DEMEL, R. A. & VAN DEENEN, L. L. M. (1973). Influence of fatty acid and sterol composition on the lipid phase transition and activity of membrane-bound enzymes in *Acholeplasma laidlawii*. *Biochemica et Biophysica Acta*, **330**, 269–282.

DOTY, P., BOEDTKER, H., FRESCO, J. R., HASELKORN, R. & LITT, M. (1959). Secondary structure in ribonucleic acids. *Proceedings of the National Academy of Sciences, U.S.A.* **45**, 482–499.

DOWBEN, R. M. & WEIDENMULLER, R. (1968). Adaptation of mesophilic bacteria to growth at elevated temperatures. *Biochemica et Biophysica Acta*, **158**, 255–261.

EIDLIC, L. & NEIDHARDT, F. C. (1965). Protein and nucleic acid synthesis in two mutants of *Escherichia coli* with temperature-sensitive aminoacyl ribonucleic acid synthetases. *Journal of Bacteriology*, **89**, 706–711.

ENGLELMAN, D. M. (1970). X-ray diffraction studies of phase transitions in the membrane of *Mycoplasma laidlawii*. *Journal of Molecular Biology*, **47**, 115–117.

ESSER, A. F. & SOUZA, K. A. (1974). Correlation between thermal death and membrane fluidity in *Bacillus stearothermophilus*. *Proceedings of the National Academy of Sciences, U.S.A.* **71**, 4111–4115.

FARRELL, J. & CAMPBELL, L. L. (1969). Thermophilic bacteria and bacteriophages. *Advances in Microbial Physiology*, **3**, 83–109.

FORRESTER, I. T. & WICKEN, A. J. (1966). The chemical composition of the cell walls of some thermophilic bacilli. *Journal of General Microbiology*, **42**, 147–154.

FRIEDMAN, S. M. (1968). Protein-synthesizing machinery of thermophilic bacteria. *Bacteriological Reviews*, **32**, 27–38.

FRIEDMAN, S. M. & WEINSTEIN, I. B. (1966). Protein synthesis in a subcellular system from *Bacillus stearothermophilus*. *Biochemica et Biophysica Acta*, **114**, 593–605.

GAUGHRAN, E. R. L. (1947). Thermophilic microorganisms. *Bacteriological Reviews*, **11**, 189–225.

GOMEZ, R. F. & SINSKEY, A. J. (1973). Deoxyribonucleic acid breaks in heated *Salmonella typhimurium* LT-2 after exposure to nutritionally complex media. *Journal of Bacteriology*, **115**, 522–528.

GOMEZ, F. R. & SINSKEY, A. J. (1975). Effect of aeration on minimal medium recovery of heated *Salmonella typhimurium*. *Journal of Bacteriology*, **122**, 106–109.

GORDON, R. E. & SMITH, N. R. (1949). Aerobic spore-forming bacteria capable of growth at high temperatures. *Journal of Bacteriology*, **58**, 327–341.

HABERSTICH, H. U. & ZUBER, H. (1974). Thermoadaptation of enzymes in thermophilic and mesophilic cultures of *Bacillus stearothermophilus* and *Bacillus caldotenax*. *Archives of Microbiology*, **98**, 275–287.

HEINEN, W., KLEIN, H. P. & VOLKMAN, C. M. (1970). Fatty acid composition of *Thermus aquaticus* at different growth temperatures. *Archiv für Mikrobiologie*, **72**, 199–202.

Ingraham, J. L. (1962). Temperature relationships. In *The Bacteria*, vol. I, ed. I. C. Gunsalus & R. Y. Stanier, pp. 265–296. New York: Academic Press.

Irwin, C. C., Akagi, J. M. & Himes, R. H. (1973). Ribosomes, polyribosomes, and deoxyribonucleic acid from thermophilic, mesophilic, and psychrophilic *Clostridia*. *Journal of Bacteriology*, **113**, 252–262.

Isono, K. (1970). Enzymological differences of α-amylase from *Bacillus stearothermophilus* grown at 37 °C and 55 °C. *Biochemical and Biophysical Research Communications*, **41**, 852–857.

Jung, L., Jost, R., Stoll, E. & Zuber, H. (1974). Metabolic differences in *Bacillus stearothermophilus* grown at 55 °C and 37 °C. *Archives of Microbiology*, **95**, 125–138.

Kaneda, T. (1966). Biosynthesis of branched-chain fatty acids. IV. Factors affecting relative abundance of fatty acids produced by *Bacillus subtilis*. *Canadian Journal of Microbiology*, **12**, 501–514.

Kimelberg, H. K. & Papajadjapoulos, D. (1972). Phospholipid requirements for ($Na^+ + K^+$)-ATPase activity: Head-group specificity and fatty acid fluidity. *Biochemica et Biophysica Acta*, **282**, 277–292.

Lennox, E. S. (1960). Antigenic analysis of cell structure. In *The Bacteria*, vol. I, ed. I. C. Gunsalus & R. Y. Stanier, pp. 415–441. New York: Academic Press.

Ljungdahl, L. G. & Sherod, D. (1975). Proteins from thermophilic microorganisms. In *Extreme Environments: Mechanisms of Microbial Adaptation*, ed. M. R. Heinrich. New York: Academic Press. (In press.)

Machtiger, N. A. & Fox, C. F. (1973). Biochemistry of Bacterial Membranes. *Annual Reviews of Biochemistry*, **42**, 575–600.

Mangiantini, M. T., Tecce, G., Toschi, G. & Trentalance, A. (1965). A study of ribosomes and of ribonucleic acid from a thermophilic organism. *Biochemica et Biophysica Acta*, **103**, 252–274.

Manning, G. B. & Campbell, L. L. (1961). Thermostable α-amylase of *Bacillus stearothermophilus*. I. Crystallization and some general properties. *Journal of Biological Chemistry*, **236**, 2952–2957.

Manning, G. B., Campbell, L. L. & Foster, R. J. (1961). Thermostable α-amylase of *Bacillus stearothermophilus*. II. Physical properties and molecular weight. *Journal of Biological Chemistry*, **236**, 2958–2961.

Marmur, J. (1960). Thermal denaturation of dexoyribose-nucleic acid isolated from a thermophile. *Biochemica et Biophysica Acta*, **38**, 342–343.

Marmur, J. & Doty, P. (1959). Heterogeneity in deoxyribonucleic acids. *Nature, London*, **183**, 1427–1429.

Marr, A. G. & Ingraham, J. L. (1962). Effect of temperature on the composition of fatty acids in *Escherichia coli*. *Journal of Bacteriology*, **84**, 1260–1267.

Marsh, L. L. & Larsen, D. H. (1953). Characterization of some thermophilic bacteria from the hot springs of Yellowstone National Park. *Journal of Bacteriology*, **65**, 193–197.

McElhaney, R. N. (1974). The effect of alterations in the physical state of the membrane lipids on the ability of *Acholeplasma laidlawii* B to grow at various temperatures. *Journal of Molecular Biology*, **84**, 145–157.

McElhaney, R. N. & Tourtellotte, M. E. (1970). The relationship between fatty acid structure and the positional distribution of esterified fatty acids in the phosphotidyl glycerol from *Mycoplasma laidlawii* B. *Biochemica et Biophysica Acta*, **202**, 120–128.

Melchior, D. L., Morowitz, H. L., Sturtevant, J. M. & Tsong, T. Y. (1970). Characterization of the plasma membrane of *Mycoplasma laidlawii*. VII. Phase transitions of membrane lipids. *Biochemica et Biophysica Acta*, **219**, 114–122.

MIQUEL, P. (1888). Monographie d'un bacille vivant au dela de 70 centrigrades. *Annales de Micrographie*, **1**, 3–10.

NOVITSKY, T. J., CHAN, M., HIMES, R. H. & AKAGI, J. M. (1974). Effect of temperature on the growth and cell wall chemistry of a facultative thermophilic *Bacillus*. *Journal of Bacteriology*, **117**, 858–865.

OGASAHARA, K., IMANISHI, A. & ISEMURA, T. (1970*a*). Studies on thermophilic α-amylase from *Bacillus stearothermophilus*. I. Some general and physicochemical properties of thermophilic α-amylase. *Journal of Biochemistry*, **67**, 65–75.

OGASAHARA, K., IMANISHI, A. & ISEMURA, T. (1970*b*). Studies on thermophilic α-amylase. II. Thermal stability of thermophilic α-amylase. *Journal of Biochemistry*, **67**, 77–82.

OO, K. C. & LEE, K. L. (1971). The lipid content of *Bacillus stearothermophilus* at 37° and at 55°. *Journal of General Microbiology*, **69**, 287–289.

OVERATH, P., SCHAIRER, H. C. & STOFFEL, W. (1970). Correlation of *in vivo* and *in vitro* phase transition of membrane lipids in *Escherichia coli*. *Proceedings of the National Academy of Sciences, U.S.A.* **67**, 606–612.

PACE, B. & CAMPBELL, L. L. (1967). Correlation of maximal growth temperature and ribosome heat stability. *Proceedings of the National Academy of Sciences, U.S.A.* **57**, 1110–1116.

PAULING, C. & BECK, L. A. (1975). Role of DNA ligase in the repair of single strand breaks induced in DNA by mild heating of *Escherichia coli*. *Journal of General Microbiology*, **87**, 181–184.

PFUELLER, S. L. & ELLIOTT, W. K. (1969). The extracellular α-amylase of *Bacillus stearothermophilus*. *Journal of Biological Chemistry*, **244**, 48–54.

RAY, P. H. & BROCK, T. D. (1971). Thermal lysis of bacterial membranes and its prevention by polyamines. *Journal of General Microbiology*, **66**, 133–135.

RAY, P. H., WHITE, D. C. & BROCK, T. D. (1971). Effect of temperature of the fatty acid composition of *Thermus aquaticus*. *Journal of Bacteriology*, **106**, 25–30.

ROSENBERG, B., KEMENY, G., SWITZER, R. C. & HAMILTON, T. C. (1971). Quantitative evidence for protein denaturation as the cause of thermal death. *Nature, London*, **232**, 471–473.

ROSENTHAL, L. J. & IANDOLO, J. J. (1970). Thermally induced intracellular alteration of ribosomal ribonucleic acid. *Journal of Bacteriology*, **103**, 833–835.

SALTON, M. R. J. & PAVLIK, J. G. (1960). Studies of the bacterial cell wall. VI. Wall composition and sensitivity to lysozyme. *Biochemica et Biophysica Acta*, **39**, 398–407.

SAUNDERS, G. F. & CAMPBELL, L. L. (1966). Ribonucleic acid and ribosomes of *Bacillus stearothermophilus*. *Journal of Bacteriology*, **91**, 332–339.

SCHEIE, P. & EHRENSPECK, S. (1973). Large surface blebs on *Escherichia coli* heated to inactivating temperatures. *Journal of Bacteriology*, **114**, 814–818.

SEDGWICK, S. G. & BRIDGES, B. A. (1972). Evidence for indirect production of DNA strand scissions during mild heating of *Escherichia coli*. *Journal of General Microbiology*, **71**, 191–193.

SHEN, P. Y., COLES, E., FOOTE, J. L. & STENESH, J. (1970). Fatty acid distribution in mesophilic and thermophilic strains of the genus *Bacillus*. *Journal of Bacteriology*, **103**, 479–481.

SINGLETON, R. JR & AMELUNXEN, R. E. (1973). Proteins from thermophilic microorganisms. *Bacteriological Reviews*, **37**, 320–342.

SOBER, H. A. (1968). *Handbook of Biochemistry*. Cleveland, Ohio: The Chemical Rubber Co.

Souza, K. A., Kostiw, L. L. & Tyson, B. J. (1974). Alterations in normal fatty acid composition in a temperature-sensitive mutant of a thermophilic bacillus. *Archives of Microbiology*, **97**, 89–102.

Steim, J. M., Tourtelotte, M. E., Reinert, J. C., McElhaney, R. N. & Rader, R. L. (1969). Calorimetric evidence for the liquid-crystalline state of lipids in biomembranes. *Proceedings of the National Academy of Sciences, U.S.A.* **63**, 104–109.

Stenesh, J. & Holazo, A. A. (1967). Studies of the ribosomal ribonucleic acid from mesophilic and thermophilic bacteria. *Biochemica et Biophysica Acta*, **138**, 286–295.

Stenesh, J., Roe, B. A. & Snyder, T. L. (1968). Studies of the deoxyribonucleic acid from mesophilic and thermophilic bacteria. *Biochemica et Biophysica Acta*, **161**, 442–454.

Strange, R. E. & Shon, M. (1964). Effects of thermal stress on viability and ribonucleic acid of *Aerobacter aerogenes* in aqueous suspension. *Journal of General Microbiology*, **34**, 99–114.

Sutow, A. B. & Welker, N. E. (1967). Chemical composition of the cell walls of *Bacillus stearothermophilus*. *Journal of Bacteriology*, **93**, 1452–1457.

Watanabe, K., Oshima, T., Saneyoshi, M. & Nishimura, S. (1974). Replacement of ribothymidine by 5-methyl-2-thiouridine in sequence GTΨC in tRNA of an extreme thermophile. *FEBS Letters*, **43**, 59–63.

Weerkamp, A. & MacElroy, R. D. (1972). Lactate dehydrogenase from an extremely thermophilic bacillus. *Archiv für Mikrobiologie*, **85**, 113–122.

Welker, N. E. (1971). Structure of the cell wall of *Bacillus stearothermophilus*: Mode of action of a thermophilic bacteriophage lytic enzyme. *Journal of Bacteriology*, **107**, 697–703.

Welker, N. E. (1975). Effect of temperature on membrane proteins. In *Extreme Environments: Mechanisms of Microbial Adaptation*, ed. M. R. Heinrich. New York: Academic Press. (In press.)

Welker, N. E. & Campbell, L. L. (1965). Induction and properties of a temperate bacteriophage from *Bacillus stearothermophilus*. *Journal of Bacteriology*, **89**, 175–184.

Wilson, G. & Fox, C. F. (1971). Biogenesis of microbial transport systems: Evidence for coupled incorporation of newly synthesized lipids and proteins into membrane. *Journal of Molecular Biology*, **55**, 49–60.

Wisdom, C. & Welker, N. E. (1973). Membranes of *Bacillus stearothermophilus*: Factors affecting protoplast stability and thermostability of alkaline phosphatase and reduced nicotinamide adenine dinucleotide oxidase. *Journal of Bacteriology*, **114**, 1336–1345.

Woodcock, E. & Grigg, G. W. (1972). Repair of thermally induced DNA breakage in *Escherichia coli*. *Nature New Biology*, **237**, 76–79.

Zeikus, J. G., Taylor, M. W. & Brock, T. D. (1970). Thermal stability of ribosomes and RNA from *Thermus aquaticus*. *Biochemica et Biophysica Acta*, **204**, 512–520.

SURVIVAL OF BACTERIA IN COLD AND MODERATE HYDROSTATIC PRESSURE ENVIRONMENTS WITH SPECIAL REFERENCE TO PSYCHROPHILIC AND BAROPHILIC BACTERIA*

R. Y. MORITA

Department of Microbiology and School of Oceanography
Oregon State University, Corvallis, Oregon, 97331, *USA*

INTRODUCTION

Most of the biosphere is cold. Approximately 14 per cent of the earth's surface is in the polar regions and 90 per cent of the marine environment is 5 °C or less. Any aquatic organism living below the surface of a water mass is subjected also to hydrostatic pressure which in the marine environment may go as high as 1100 atm (1.11×10^5 k Pa) which is approximately equal to the pressure at the bottom of the Challenger Deep, the deepest known part of the oceans.

The true cold-loving bacteria were first reported by Tsiklinsky (1908) but were not rediscovered until 1964 (Hagen, Kushner & Gibbons, 1964; Morita & Haight, 1964; Sieburth, 1964). Only a few investigators are currently working on various problems related to psychrophilic or barophilic bacteria and hence, there are insufficient data on which to base any definitive conclusions.

This paper will deal with the effects of changes in the natural environment on the survival of barophilic and psychrophilic bacteria. Unfortunately, the study of survival of marine bacteria in the environment, especially the psychrophiles and barophiles, has been neglected and so many of my statements will be speculative.

Psychrophiles are defined, for the purpose of this paper, as microorganisms having an optimal temperature for growth at about 15 °C or lower, a maximal temperature for growth at about 20 °C or lower, and a minimal temperature for growth at 0 °C or below (Morita, 1975). The term psychrotroph will be used for organisms that do not meet the above definition but have the ability to grow at lower temperature (Eddy, 1960). Since many organisms discussed in the literature do not

* Published as technical paper no. 4040, Oregon Agricultural Experiment Station.

meet the above criteria for psychrophiles, the literature on which I can draw is limited.

Barophiles are defined as organisms that grow well at elevated pressure (ZoBell & Johnson, 1949). Eurybaric organisms are those that are capable of growth between 1 and 600 atm (ZoBell, 1968). The terms barophobic and baroduric are also used for describing various bacteria living in the oceans.

THE NATURAL ENVIRONMENT FOR PSYCHROPHILES AND BAROPHILES

Psychrophiles

Permanently cold environments are the natural habitats for psychrophilic bacteria since any significant rise in temperature will bring about death. All cold environments do not have a psychrophilic microflora since it appears that the evolution of psychrophilic bacteria from their mesophilic counterpart requires considerable time (Brock, Passman & Yoder, 1973). In natural, cold environments, either psychrotrophic or a combination of psychrotrophic and psychrophilic bacteria exist. Although mesophilic and thermophilic bacteria may be present (Bartholomew & Rittenberg, 1949; McBee & McBee, 1956; Bartholomew & Paik, 1966; Trüper, 1969), they are non-functional and only serve as a source of organic matter for other biological forms. The distribution of psychrophiles is reviewed by Morita (1975). The marine biosphere is the principle environment for psychrophiles but freshwater psychrophiles are known to occur in arctic waters.

The temperature range found in the ocean is from −3 °C to about +42 °C (Sverdrup, Johnson & Fleming, 1942). The water mass above the thermocline is characterized by variations in temperature, depending upon its geographical location, while the water below the thermocline is uniformly cold (−1.5 °C to +4.5 °C). Convergence (sinking of a water mass), divergence (upwelling of a water mass), tides and currents in the ocean may expose psychrophiles to changes in temperature, salinity and/or pressure and, as a result of these natural variations in the environment, survival of psychrophiles becomes a matter of interest. In addition, the amount of organic matter (energy) in any given water mass will greatly influence the survival of the psychrophilic bacteria.

Barophiles

The natural habitat for barophilic bacteria is the deep sea. This environment is characterized by a low temperature (less than 5 °C), increased hydrostatic pressure (up to *c.* 1100 atm), and low organic matter content. For example, the oxygen concentration at 10000 m in the Philippine Trench is approximately 3 ml l^{-1} of sea water (sufficient for aerobic forms), a salinity of *c.* 3.47 per cent and a temperature of *c.* 2.5 °C. These deep layers are characterized by great horizontal and vertical uniformity and slow movement. There is no significant influx of new water masses except for that coming in with the Antarctic circumpolar current (Bruun, Greve, Mielche & Spärck, 1956).

Concentrations of particulate and dissolved organic carbon at all depths of the ocean below 200 or 300 m is between 3 and 10 μg C l^{-1} and 0.35 and 0.7 mg C l^{-1} respectively (Menzel & Ryther, 1970). The organic matter in the deep sea is largely composed of nitrogenous compounds but free monomers, e.g. amino acids, fatty acids and sugars, amount to less than 10 per cent of the total organic matter (Degens, 1970). Bases and nucleic acids are found only in traces at concentrations less than 1 to 5 μg l^{-1} (Degens, 1970). Simple calculations show that in a water column 5000 m deep, the dissolved organic matter must be of considerable age (Menzel & Ryther, 1970).

The occurrence of bacteria in the deeper parts of the oceans was demonstrated during the Galathea Deep Sea Expedition (1950–2) by ZoBell (1952) and ZoBell & Morita (1957, 1959) when barophilic and baroduric bacteria were found. However, due to the sampling techniques used, bacteria that could not tolerate a decrease in hydrostatic pressure and increase in temperature of the water above the thermocline during retrieval (stenobarophiles) were not isolated. It. s known that both decompression and increases in temperature during sampling have an adverse effect on the growth (and survival?) of bacteria (Seki & Robinson, 1969; ZoBell, 1970).

SURVIVAL OF PSYCHROPHILES AND BAROPHILES IN THE NATURAL ENVIRONMENT

Psychrophiles

When studying bacteria from the natural environment, most of the data concern those bacteria which can reproduce on agar or in liquid media, Unfortunately, available cultural methods are only effective in detecting a fraction of the organisms from nature and even for those organisms that can be cultured, we cannot detect any stages between life and death because we use reproductive ability as a criterion of life. Moribund and resting cells cannot be detected by our methods of analysis. In all probability, there are cells in the natural environment that are capable of metabolizing but are unable to reproduce. Direct counts of bacteria in sea water show the presence of 13 to 9700 times as many organisms as determined by cultural methods (Jannasch & Jones, 1959). Unfortunately, none of the cultural methods employed by Jannasch and Jones (1959) were used to detect psychrophiles.

When the temperature of cold oceanic water is raised as a result of divergence or when a salt wedge (salt water intrusion) occurs in a bay, the temperatures may be high enough to inactivate psychrophilic bacteria. Some of the psychrophiles that have been studied have maximum temperatures for growth of 10 °C to 15 °C (Morita, 1975; Christian & Wiebe, 1974) and are killed, therefore, in some natural marine environments. In bays and estuaries, a salt wedge can be diluted as a result of a freshwater intrusion so that salinity can also affect the microbial cells.

Temperature elevation in water masses. Where a water mass with the indigenous psychrophilic microflora reaches temperatures above the maximum for growth, death may ensue. Since water has a high heat capacity, temperature changes may be slow unless mixing with another water mass of higher or colder temperature occurs. Since water masses may be only a few degrees above the maximum for growth of psychrophiles, temperature increases of 5 to 10 °C may be an unnatural occurrence in the ocean. However, the period of exposure of psychrophiles to high temperatures must be taken into consideration as well as a temperature increase of just a few degrees above the organism's maximum for growth.

Although a water temperature of 20 °C is rarely attained in nature, a temperature of 15 °C is common. Some of the psychrophiles mentioned earlier have maximum growth temperatures between 10 to 15 °C and so most of the data on survival result from work with mesophiles.

The growth phase of a population governs its response to the effects of temperature (Kenis & Morita, 1968). In the aquatic environment, it is difficult, if not impossible, to determine the growth phase of the organisms because of the lack of suitable techniques. However, when a psychrophilic vibrio, *Vibrio marinus* MP-1, was exposed to 20.5 °C (the organism's maximal temperature for growth) for up to 30 min in the absence of nutrients, it survived (Robison & Morita, 1966). Exposure for longer than 30 min brought about a reduction in endogenous oxygen uptake and death, thereby indicating a thermal lesion had occurred in certain metabolic pathways. However, in the presence of nutrients, survival was much more pronounced at temperatures above the organism's maximum growth temperature (Haight & Morita, 1966).

Langridge & Morita (1966) found that exposure of the same strain to a temperature above its optimal for growth (*c.* 15 °C), caused thermal denaturation of malate dehydrogenase. Approximately 25 per cent of malate dehydrogenase was lost after 1 h exposure to 20 °C. Inactivation of the enzyme also occurred when it had been partially purified, thereby indicating that the enzyme is abnormally thermolabile. Other enzymes shown to be abnormally thermolabile in psychrophilic bacteria include hexokinase, lactate dehydrogenase, aldolase and the phosphofructokinase/glyceraldehyde 3-phosphate dehydrogenase complex (Mathemeier, 1966).

Thermally induced leakage has been reported in psychrophilic bacteria, even at the maximum temperature for growth (Robison & Morita, 1966; Haight & Morita, 1966; Kenis & Morita, 1968; Harder & Veldkamp, 1967; Geesey & Morita, 1975). The question of how much intracellular material must leak out of the cells at temperatures above its maximum for growth before death occurs is unanswered. The ability to maintan intracellular material depends upon the generation of energy. At temperatures close to the maximum for growth, psychrophiles respire at a faster rate (Harder & Veldkamp, 1967; Christian & Wiebe, 1974; Geesey & Morita, 1975). According to Geesey & Morita (1975), there appears to be an inhibition of glucose uptake at the maximum growth temperature and this coupled with fast respiration results in starvation. As stated by Christian & Wiebe (1974), Harder & Veldkamp (1967), and Brown (1957) the portion of energy capable of being produced via respiration which is utilized for reproduction decreases as temperature increases.

Thermally induced lysis also occurs in psychrophiles. Autolytic enzymes, which are inactive at low temperature, may be activated at higher temperature and cannot be inactivated by returning the culture to a cold environment (Hagen *et al.*, 1964; Madeley *et al.*, 1967). Lipid phosphorus occurs in the menstruum after the onset of lysis (Hagen *et al.*, 1964; Korngold & Kushner, 1968). Temperatures above the organism's maximum cause loss of control of the membrane functions of the cell and as a result cells die. Membrane permeability has also been shown to be salt-dependent in marine bacteria (Drapeau & MacLeod, 1963; Drapeau, Matula & MacLeod, 1966; MacLeod, 1971).

Salinity changes. In the colder latitudes, bays and estuaries are inundated with freshwater which results in salinity changes. These salinity changes can bring about changes in the response of psychrophiles to temperature changes. Stanley & Morita (1967) demonstrated that *V. marinus* MP-1 can change its maximum growth temperature by 10.5 deg C, depending upon the salt concentration of the medium. The lower the salinity, the lower the maximum growth temperature. Although all salt-requiring micro-organisms we have tested behave in this way, the extent of the change varies from species to species. *Vibrio psychroerythrus* requires both low temperature and salt for growth (D'Aoust & Kushner, 1971; Korngold & Kushner, 1968). Lysis can occur at low ionic strength (Madeley *et al.*, 1967).

From the above, it appears that survival of marine organisms is dependent upon the salinity–temperature regime. Our preliminary data concern the survival of *V. anguillarum*. When the organism is placed in a starvation medium, there is a rapid initial decrease in the number of viable organisms, independent of salinity. Prolonged starvation leads to the formation of small spherical forms and survival of these depends on salinity. Low salinity favours survival more than high salinity. Bays and estuaries in the colder regions are inundated with freshwater mainly in the winter months because of rainfall but whether or not our preliminary data are relevant to such changes remains to be seen.

Organic matter. In areas of divergence the water is rich in nutrients and as a result phytoplankton growth occurs readily and this results in organic-rich waters. Also, where cold water from a salt wedge mixes with surface water in bays and estuaries, the nutrient level is quite high. If, however, cold water due to circulation patterns ends up in the open sea, nutrients are depleted and this may result in starvation of organisms.

The concentration of dissolved organic matter in sea water is approximately 1/5000 that in ordinary garden soil (ZoBell, 1946). When sea water is stored only 10 to 15 per cent of the organic matter is utilized by bacteria (Keys, Christensen & Krogh, 1935), although Waksman & Carey (1935) indicate that 50 per cent of the total organic content is readily decomposed by bacteria. The resistance of organic matter to decomposition is one of the main factors which limits the multiplication of bacteria in sea water (Waksman & Carey, 1935). ZoBell & Grant (1943) found that most marine bacteria multiply in mineral media containing 0.1 mg of glucose or peptone per litre and solid surfaces promote utilization of the organic matter. Barber (1968) demonstrated that bacteria are capable of decomposing dissolved organic matter except in deep sea water, where the concentration of dissolved organic matter is too low for further decomposition.

In surface waters the rate at which the organic matter is supplied to the environment will depend upon the photosynthetic activities of the phytoplankton which, in turn, are dependent upon the availability of nutrients. If the rate of supply is sufficient, then organic matter is not critical for the survival of most bacteria in the sea. For most of the ocean there is a low rate of organic matter supply. Below the photic zone the supply rate will depend upon many factors such as the productivity above a specific water mass, depth of the water column, movement of water masses, presence or absence of a thermocline, etc.

The question that must be answered in the future is: what is the threshold amount of organic matter necessary for the survival of marine organisms? Threshold levels of substrates were determined in a chemostat but they are not directly applicable to sea water (Jannasch, 1967).

A vibrio isolated from Antarctic waters (designated as Ant-300) underwent a two- to nine-fold increase in cell numbers when starved, depending upon the age of the culture prior to starvation. The best survival of this organism occurred at 5 °C at a salt concentration of 4 per cent Rila Marine Mix (equivalent to 35 parts per thousand salinity). The cells became very small and 50 per cent passed through a 0.4 μm filter. During the first 48 h starvation, endogenous respiration decreased by 80 per cent and after 75 h of starvation, over 75 per cent of the DNA had been degraded or leaked from the cells. Between 75 and 250 h, the DNA content remained unchanged. The addition of an amino acid mixture (23 $\mu g\ l^{-1}$) to the starvation mixture had little effect on their survival.

Pressure–temperature relationships. Brown, Johnson & Marsland (1942) and Strehler & Johnson (1954) studied the variation in luminescence by bacteria under various pressures and temperatures. At low temperature, increasing pressures decreased the amount of luminescence. At temperatures close to and above the optimum for growth, increased pressures increased the luminescence. These results were interpreted as follows: at low temperatures the major effect of pressure is to limit the enzyme reaction and at higher tempratures the amount of reversibly denatured enzyme becomes increasingly more important as a limiting factor in the overall rate. In the latter situation, the equilibrium shifts, under pressure, in favour of the undenatured state, thereby increasing the amount of active enzyme.

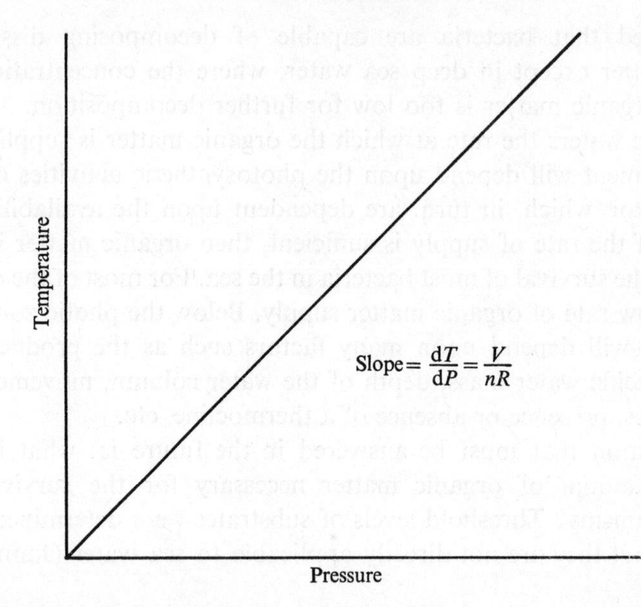

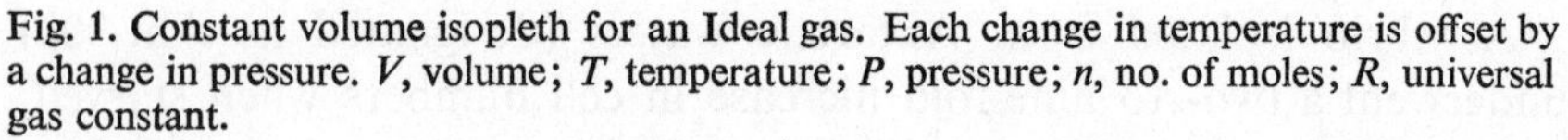

Fig. 1. Constant volume isopleth for an Ideal gas. Each change in temperature is offset by a change in pressure. V, volume; T, temperature; P, pressure; n, no. of moles; R, universal gas constant.

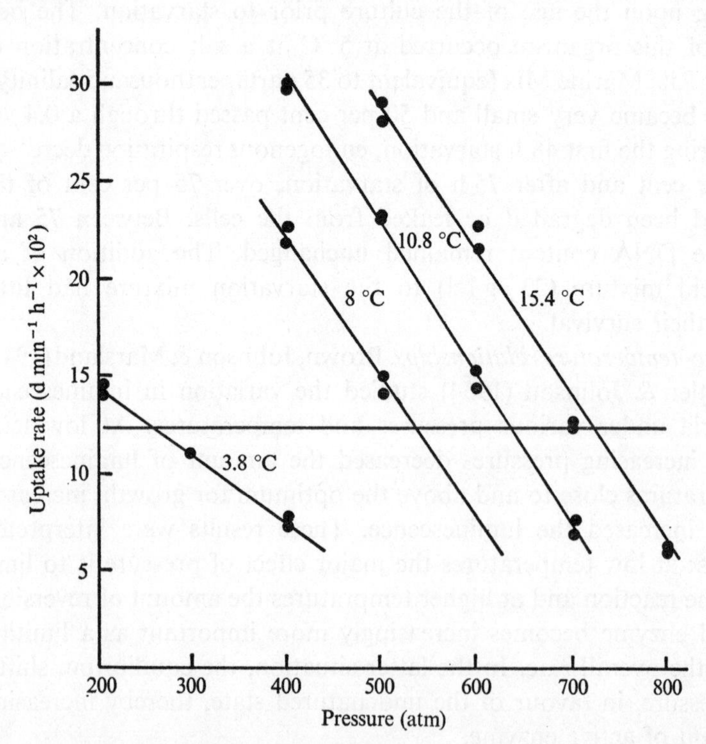

Fig. 2. Uptake rate of [U-^{14}C]glutamic acid by the antarctic psychrophile Ant-300 at various temperatures and pressures. It is possible to choose combinations of pressure and temperature that will allow an uptake rate of 1250 d min^{-1} h^{-1}.

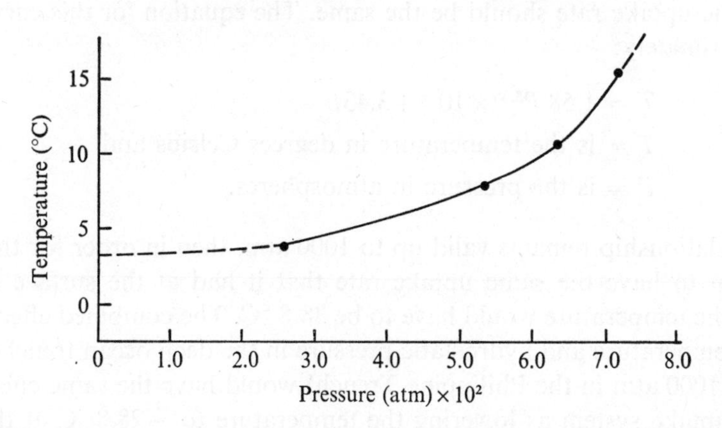

Fig. 3. Constant uptake rate isopleth for [U-^{14}C]glutamic acid by Ant-300. At any combination of temperature and pressure on this line the uptake rate should be 1250 d min^{-1} h^{-1}.

The effect of increased temperature upon other enzyme systems can be offset by increasing the pressure. Thus, Morita & Haight (1962) and Morita & Mathemeier (1964) were able to demonstrate the activity of malate dehydrogenase at 101 °C and inorganic pyrophosphatase at 105 °C under pressure. This same principle can be applied to the growth of bacteria under pressure. ZoBell & Johnson (1949) noted that many marine and terrestrial bacteria grew better at elevated temperatures under pressure than at 1 atm.

Morita (1972) stressed the fact that increasing hydrostatic pressure with a concomitant decrease in temperature results in an additive effect on various systems mainly because both changes result in a decrease in the molecular volume of the system. If some physiological parameter (such as molecular volume, enzyme reaction rate, growth, respiration rate or transport) were measured under a given set of temperature–pressure conditions and found to have a certain magnitude, it should be possible to maintain this value by offsetting each change in temperature with a concomitant change in pressure (Fig. 1). This hypothesis was tested by F. J. Hanus & J. A. Baross (personal communication). [U-^{14}C]glutamic acid uptake by a psychrophile, Ant-300 (a vibrio species) was measured under a wide range of temperature–pressure conditions (Fig. 2). The uptake rate was 1250 d min^{-1} h^{-1} at 3.8 °C and 260 atm, at 8 °C and 535 atm, at 10.8 °C and 635 atm, and 15.4 °C at 720 atm. A temperature–pressure isopleth was constructed for the 1250 d min^{-1} h^{-1} uptake rate (Fig. 3). Under any set of temperature–pressure conditions on this curve, the uptake rate should be the same. The equation for this curve is approximately:

$$T = 1.58\,P^{3.35} \times 10^{-8} + 3.43,$$

where T = is the temperature in degrees Celsius and

P = is the pressure in atmospheres.

If this relationship remains valid up to 1000 atm, then in order for this organism to have the same uptake rate that it had at the surface at 3.4 °C, the temperature would have to be 38.8 °C. The combined effects of low temperature and hydrostatic pressure in the deep ocean trenches (2.5 °C, 1000 atm in the Philippine Trench) would have the same effect on the uptake system as lowering the temperature to −28.8 °C at the surface.

The calculated figure of −28.8 °C would metabolically 'freeze' many of the bacteria sedimenting to the deeper parts of the ocean. Due to the decreased molecular volume of cellular components, resulting from a decrease in temperature and an increase in pressure, as well as a low supply of organic matter, it is not surprising that Jannasch, Eimhjellen, Wirsen & Farmanfarmaian (1971) found low rates of metabolic activity in their deep sea samples. From a survival viewpoint, low metabolic rates are of direct benefit to an organism in the deep sea since the rate of energy supply is low. This metabolic 'frozen state' would also prevent enzymatic lysis of the cells.

The activity of the indigenous microbial flora of the waters in the Antarctic convergence at temperatures near 0 °C has been measured employing the heterotrophic activity method. This method (Parsons & Strickland, 1962; Wright & Hobbie, 1965; Hobbie & Crawford, 1969) is a kinetic one where the maximum velocity of the uptake (V_{max}), the turnover time (T_t), and the sum of a 'transport constant' and the natural substrate concentration ($K_t + S_n$) can be calculated. It is based on the principles of the Michaelis–Menten theory and employs various concentrations of uniformly labelled ^{14}C-substrates. The following formula is employed to obtain the necessary data to make a modified Lineweaver–Burk plot:

$$\frac{C\mu t}{c} = \frac{1}{V_{max}}(A) + \frac{(K_t + S_n)}{V_{max}},$$

where c is the radioactivity of cells recovered by filtration plus $^{14}CO_2$ respired by the organisms in disintegrations per min (d.p.m.); S_n is the natural substrate concentration in $\mu g\,l^{-1}$; A is the added quantity of

Table 1. *Heterotrophic activity on uniformly labelled [^{14}C]glutamic acid by the indigenous microflora in water samples taken from various depths at Station* 15 *during Cruise* 51 *of the USNS Eltanin*

Depth (m)	V_{max} (μg l^{-1} h^{-1}) $\times 10^{-3}$	T_t (h)	$K_t + S_n$ (μg l^{-1})	Per cent respired
0	11.7	366	4.3	38
22	9.7	534	4.9	42
50	13.1	535	7.0	41
80	13.7	322	4.4	39
150	123.0	116	14.3	25
300	1.2	3330	4.1	51

Water temperature was 0 °C. Percent respiration = $^{14}CO_2 \times 100/^{14}$C-uptake + $^{14}CO_2$.

Table 2. *Survival of mesophilic human enteric bacteria and marine psychrophilic bacteria in sea water at various pressures at* 4 °C

Organism (temp. of growth)	Exposure time (h)	No. of viable organisms per ml pressure (atm)				
		1	250	500	800	1000
Escherichia coli	Initial	3×10^7	3×10^7	3×10^7	—	3×10^7
(8 to 45 °C)	5	3×10^7	3×10^7	3×10^7	—	3×10^7
	100	1×10^6	5×10^6	3×10^7	—	2×10^5
	350	1×10^4	1×10^6	3×10^7	—	2×10^1
Streptococcus faecalis	Initial	1×10^8	1×10^8	1×10^8	—	1×10^8
(8 to 45 °C)	5	1×10^8	1×10^8	1×10^8	—	1×10^8
	100	5×10^7	9×10^7	1×10^8	—	5×10^7
	350	4×10^6	5×10^7	9×10^7	—	3×10^6
Clostridium perfringens	Initial	2×10^6	2×10^6	2×10^6	—	2×10^6
	5	2×10^6	2×10^6	1×10^6	—	1×10^6
(15 to 45 °C)	100	8×10^5	2×10^4	1×10^2	—	< 10
	350	1×10^2	< 10	< 10	—	< 10
Vibrio MP-1	Initial	5×10^6	5×10^6	5×10^6	5×10^6	—
(< 0 to 20 °C)	5	5×10^6	1×10^5	6×10^6	< 10	—
	100	5×10^6	< 10	< 10	—	—
	350	—	—	—	—	—
Vibrio Ant-300	Initial	5×10^6	5×10^6	5×10^6	5×10^6	—
(< 0 to 13 °C)	5	5×10^6	5×10^6	2×10^6	< 10	—
	100	5×10^6	5×10^6	1×10^5	—	—
	350	—	—	< 10	—	—

substrate in μg l^{-1}; $C = 2.22\times10^6$ d.p.m. for one microcurie of ^{14}C-labelled substrate; μ is the quantity of ^{14}C added in microcuries per sample bottle; and t is incubation time in hours. The heterotrophic activity method is performed immediately after the water sample is taken and at the same temperature as the environment. The data (Table 1) obtained indicate that the rate of microbial activity is high (Morita, Griffiths & Hayasaka, 1975).

The psychrophilic bacteria found in abundance in the surface waters of the Antarctic convergence sink with the water mass. Although the nutrient level at the surface may be quite high, the water mass that sinks is not replenished with organic nutrients and it should be noted that very little heterotrophic activity occurs below 500 m in the Antarctic region (Morita *et al.*, 1975). It has been calculated by Li (personal communication) that 1100 years is needed for water to converge and subsequently diverge. During that time survival of the psychrophilic and psychrotrophic microflora must be taken into consideration. Table 2 compares the action of hydrostatic pressure at 4 °C on two psychrophilic bacteria (*V. marinus* MP-1 and Ant-300) and three mesophiles; survival of the mesophiles is much better than the psychrophiles.

Barophiles

Occurrence and activities of deep sea organisms. The term 'deep sea' is ill defined and therefore I have included the depth measurement where applicable. The average depth of the oceans is 3800 m and the deepest part is the Challenger Deep (11034 m). Bacteria have been found in bottom sediments (water depths from 1707 to 5952 m) between San Diego, California and Hawaii as well as in the mid-Pacific mountain range during the Mid-Pacific Expedition of 1950 (Morita & ZoBell, 1955). Patchiness in distribution of microbes in waters as deep as 1100 m has been noted by Sieburth (1971).

Many of the bacteria isolated from the deeper portion of the oceans have the ability to grow at 1 atm at both temperatures *in situ* (near 5 °C) and 20 °C (Morita & ZoBell, 1955; ZoBell & Morita, 1957, 1959; Quigley & Colwell, 1968; Sieburth, 1971). Others have various threshold tolerances to pressure (Oppenheimer & ZoBell, 1952) and the eurybaric organisms grow better at 1 atm than at elevated pressures (ZoBell, 1968). *Pseudomonas* 8113 is able to produce 20 per cent more biomass at 200 to 400 atm than at 1 atm (Mitskevich & Kriss, 1966). Details on the effect of hydrostatic pressure on micro-organisms are given by ZoBell (1964), Morita (1967, 1972), Zimmerman (1970), Brauer (1972), Sleigh & Macdonald (1972) and Colwell & Morita (1974). A sulphate-reducing bacterium has been obtained in an enrichment culture from the sediments of the Weber Deep (7250 m) and grown in the laboratory for many years. The formation of a black precipitate did not occur unless the material was incubated for at least 10 months at 3–5 °C at 700 atm. Controls held at 3–5 °C and 1 atm were negative (ZoBell & Morita, 1959). Subsequent observations for five years have confirmed these findings.

Kriss (1962) demonstrated that a barotolerant organism isolated from the north polar region was capable of growth at both 1 atm and elevated pressures. He considered barotolerance to be a stable feature, even after prolonged cultivation at 1 atm. He found that many bacteria isolated from garden soil were capable of growth within 24 h at 336 or 504 atm and 26 °C. On the other hand, where pressures exceed 800 atm in the oceanic trenches, microbial forms capable of growth under pressure do not always outnumber those which can reproduce at 1 atm. However, Kriss (1962) observed that the number of bacteria developing at elevated pressures drops off considerably when deep sea mud is stored at atmospheric pressure. He suggested that barotolerant and barophilic bacteria were being replaced by other forms that reproduced during storage.

After the discovery of barophilic and baroduric bacteria during the Danish Galathea Deep Sea Expedition, other investigators have studied the activities of the microbes in the deep sea. Jannasch *et al.* (1971) reported that the rates of organic matter degradation were 10 to 100 times slower in the deep sea (1540 m) compared to controls at comparable temperatures. Foods recovered from the sunken ship *Alvin* were found to be well-preserved probably because they were enclosed in lunch boxes. However, Sieburth & Dietz (1974) demonstrated that decomposition of food material in the sea does occur within 10 weeks at 1–3 °C at 5200 m. Jannasch & Wirsen (1973) inoculated deep sea material (1830 m) into a sterile organic medium *in situ* and found an extremely slow conversion rate of the organic matter within one year. Seki (1972) grew deep sea bacteria in sterile media placed 50 cm above the ocean floor (5200 m). He found that bacterial carbon increased by 150 mg l^{-1} and the organic nitrogen by 45 mg l^{-1}. The generation time was calculated to be 6 h. In another experiment there was an increase in bacterial carbon of 21 mg l^{-1} and 5.7 mg l^{-1} of organic nitrogen and the generation time of the microbial population was four days.

Laboratory experiments performed with *Pseudomonas bathycetes*, an organism isolated from the Challenger Deep, demonstrated that the organism had a very long lag period of approximately four months at 1000 atm and 3 °C. The culture reached the mid-logarithmic phase after eight months and the stationary phase after one year. The generation time was calculated to be approximately 33 days. There were also low rates of incorporation of labelled substrate into protein, RNA and DNA (Schwarz & Colwell, 1975*a*).

The indigenous microflora in sediments collected from the Puerto Rico Trench (7750 and 8130 m) were incubated in the presence of

^{14}C-labelled amino acids at in-situ pressure and temperature (3 °C) and at 1 atm (Schwarz & Colwell, 1975*b*). The atmospheric controls at 3 °C took up five times more ^{14}C-labelled amino acids, produced 45 times more carbon dioxide through respiration, and assimilated 69 times more of the radioactive label. On the other hand, the intestinal microflora of a deep sea amphipod taken from the Aleutian Trench (7050 m), when examined under simulated natural conditions, had a generation time of 1 h and a substrate degradation rate (starch, urea, and *N*-acetyl-D-glucosamine) approximately equal to, or greater than, atmospheric controls during short incubation times (Schwarz, Yayanos & Colwell, 1975). Clearly, the microflora of the intestinal tract is not inhibited by the hydrostatic pressure that occurs in the deeper portions of the oceans. However, the extrapolation of data from one or several locations to the entire deep sea system is unwarranted since the ocean is not a homogeneous system.

Organic matter. When marine chemists analyse sea water for its dissolved organic content, the organic matter in microbes is not measured separately. Filtration of sea water would not help to measure this fraction since many cells would lyse. Lysis can take place on membrane filters and if the membrane filter has a pore size greater than 0.4 μm, it could not remove the small cells formed during starvation. With this in mind, one wonders what percentage of the 'dissolved' organic matter in the deep sea comes from microbial cells.

Both the amount of organic matter and the bacterial cell count are high in the sediment of the various trenches and deeps but not in the sediments of the open ocean. The presence of recalcitrant material in the deep sea may restrict the metabolism of the microbial population and this may be a factor promoting survival of organisms.

SOME SPECULATIONS CONCERNING THE MICROBIOLOGY OF THE DEEP SEA

Low temperatures, elevated hydrostatic pressures and a lack of an energy source will limit rates of metabolism, macromolecular synthesis, growth and enzyme action and cause molecular volume changes (ZoBell, 1964; Morita, 1967, 1972; Zimmerman, 1970; Brauer, 1972; Sleigh & MacDonald, 1972; Colwell & Morita, 1974). This may be a blessing in disguise because otherwise the sources of energy would be exhausted very rapidly. This depletion of energy sources would be extremely detrimental to all forms of life in the deep sea.

Psychrophilic bacteria cannot survive elevated hydrostatic pressure at 4 °C although some mesophiles are preserved at low temperature and increased hydrostatic pressure (Kriss, 1962; Baross, Hanus & Morita, 1975). It is not surprising, therefore, that I was unable to isolate true psychrophiles from sediment taken from the Challenger Deep (10 482 m), but only organisms capable of withstanding a wide range of environmental factors (eurybiotic forms), e.g. *Pseudomonas bathycetes*. The organisms that we isolated from the crab and the organism isolated from the deep sea amphipod (Schwarz & Colwell, 1975*b*) also have the ability to grow at room temperature as well as low temperature and elevated pressure. Some of the soil forms that Kriss (1962) refers to are probably eurybiotic bacteria and Menzies & Selvakumaran (1974) found that the eurybiotic higher forms have a greater pressure tolerance. In other words, stenobiotic microbes will not tolerate high pressure whereas eurybiotic ones will. As stated by Baross *et al.* (1975):

Microorganisms can tolerate higher pressures at temperatures below their minimum for growth than at temperatures within their growth range. It is possible that enzymatic activity, macromolecular synthesis, or substrate transport mechanisms are pressure sensitive within the range of activity for these systems, whereas at temperatures below the minimum for activity, no damage is incurred. It is apparent, therefore, that the primary physical parameter affecting microorganisms in the deep ocean is temperature, and as the pressure is increased the resulting physical effect on organisms is analogous to a further decrease in temperature within the limits whereby pressure is not physically inactivating macromolecules or their synthetic systems.

Menzies, George & Rowe (1973) discussed various mechanisms by which organic matter reaches the deeper parts of the oceans. Approximately 90 per cent of the organic matter produced in the photic zone is mineralized and the remaining 10 per cent is lost at an unknown rate in the water column. Deeper waters and oceanic sediments have a low organic content and I think that organic matter reaches the deeper parts of the ocean as microbial cells. These cells are probably very small (as shown by our preliminary starvation studies) and cannot be readily separated from the water by filtration. Microbes in the deep sea could be present as dead cells that do not lyse (lytic enzymes are inactive due to decreased temperature and the added effect of pressure), dormant cells (metabolic mechanisms are 'frozen'), starving cells, slowly metabolizing cells, eurybiotic cells and stenobarophilic cells. Cultural analysis of the cells is impossible since many may have lost their ability to multiply. Many would lyse when being brought to the surface for examination due to the decrease in pressure and increase in temperature as the material passes through the thermocline into warmer water. Since the process of sinking to the bottom is a continuous one taking place over aeons of time, one does not have to consider the sinking rate of these cells. They are not removed while sinking since the water is sparse in other forms of life and many of these forms cannot filter out bacterial cells directly unless they are attached to detritus particles. When they sediment to the bottom, many of the bottom feeders ingest the bacteria with sediment particles. Their indigenous intestinal microflora can utilize these microbial cells as a source of energy and both the ingested and intestinal microflora become food for the higher organism. Efficient use of the organic matter in sediment by higher forms is essential and so most of the abyssal benthic fauna is small, but generally their intestinal tracts are very long or large. Our starvation studies indicate that DNA decreases upon starvation of the cells, thereby leaving the cells sedimenting to the bottom mainly as protein, amino acids and lipid material.

The amount of pertinent data concerning aquatic bacteria (especially psychrophiles and barophiles) is insufficient to form any clear concepts on their survival. The number of samples taken for microbiological analysis in relation to the vast area and volume of the aquatic environment is extremely small. This situation is further complicated by other environmental factors (other biological forms, organic matter, pressure, patchiness, temperature, etc.). Nevertheless, some insight into the problems involved in the survival of bacteria in the aquatic environment is being achieved.

I wish to thank my colleagues (Drs J. A. Baross, R. P. Griffiths, S. S. Hayasaka L.-A. Meyer-Reil and Messrs F. J. Hanus and J. A. Novitsky) for the use of unpublished data obtained in my laboratory. The preparation of this manuscript was supported by the National Science Foundation grant DES 73-06611 A02.

REFERENCES

Barber, R. T. (1968). Dissolved organic carbon from deep waters resists microbial oxidation. *Nature, London*, **220**, 274–275.

Bartholomew, J. W. & Paik, G. (1966). Isolation and identification of obligate thermophilic sporeforming bacilli from ocean basin cores. *Journal of Bacteriology*, **92**, 635–638.

Bartholomew, J. W. & Rittenberg, S. C. (1949). Thermophilic bacteria from deep ocean bottom cores. *Journal of Bacteriology*, **57**, 659.

Baross, J. A., Hanus, F. J. & Morita, R. Y. (1975). The survival of human enteric and other sewage microorganisms under simulated deep sea conditions. *Applied Microbiology*, **21**, 309–318.

Brauer, R. W. (1972). *Barobiology and the Experimental Biology of the Deep Sea.* Chapel Hill: North Carolina Sea Grant Program.

Brock, T. D., Passman, F. & Yoder, I. (1973). Absence of obligately psychrophilic bacteria in constantly cold springs associated with caves in southern Indiana. *American Midland Naturalist*, **90**, 240–246.

BROWN, A. D. (1957). Some general properties of a psychrophilic pseudomonad: The effects of temperature on some of these properties and the utilization of glucose by this organism and *Pseudomonas aeruginosa*. *Journal of General Microbiology*, **17**, 640–648.

BROWN, D. D., JOHNSON, F. H. & MARSLAND, D. A. (1942). The pressure-temperature relations of bacterial luminescence. *Journal of Cellular and Comparative Physiology*, **20**, 151–168.

BRUUN, A. F., GREVE, S. V., MIELCHE, H. & SPÄRCK, R. (1956). *The Galathea Deep Sea Expedition* 1950–1952. London: George Allen & Unwin.

CHRISTIAN, R. R. & WIEBE, W. J. (1974). The effects of temperature upon the reproduction and respiration of a marine obligate psychrophile. *Canadian Journal of Microbiology*, **29**, 1341–1345.

COLWELL, R. R. & MORITA, R. Y. (1974). *Effect of the Ocean Environment on Microbial Activities*. Baltimore: University Park Press.

D'AOUST, J. Y. & KUSHNER, D. J. (1972). *Vibrio psychroerythrus* sp. n.: classification of the psychrophilic marine bacterium, NRC 1004. *Journal of Bacteriology*, **111**, 340–342.

DEGENS, E. T. (1970). Molecular nature of nitrogenous compounds in sea water and recent marine sediments. In *Organic Matter in Natural Waters*, ed. D. W. Hood, pp. 77–106. Institute of Marine Science Occasional Publication No. 1. University of Alaska.

DRAPEAU, G. R. & MACLEOD, R. A. (1963). Na^+ dependent active transport of α-aminoisobutyric acid into cells of a marine pseudomonad. *Biochemical and Biophysical Research Communications*, **12**, 111–115.

DRAPEAU, G. R., MATULA, T. I. & MACLEOD, R. A. (1966). Nutrition and metabolism of marine bacteria. XV. Relation of Na^+-activated transport of the Na^+ requirement of a marine pseudomonad for growth. *Journal of Bacteriology*, **92**, 63–71.

EDDY, B. P. (1960). The use and meaning of the term 'psychrophilic'. *Journal of Applied Bacteriology*, **23**, 189–190.

GEESEY, G. G. & MORITA, R. Y. (1975). Some physiological effects of near maximum growth temperatures on an obligately psychrophilic marine bacterium. *Canadian Journal of Microbiology*, **21**, 811–818.

HAGEN, P. D., KUSHNER, D. J., & GIBBONS N. E. (1964). Temperature induced death and lysis in a psychrophilic bacterium. *Canadian Journal of Microbiology*, **10**, 813–823.

HAIGHT, R. D. & MORITA, R. Y. (1962). Interaction between the parameters of hydrostatic pressure and temperature on aspartase of *Escherichia coli*. *Journal of Bacteriology*, **83**, 112–120.

HAIGHT, R. D. & MORITA, R. Y. (1966). Thermally induced leakage from *Vibrio marinus*, an obligately psychrophilic bacterium. *Journal of Bacteriology*, **92**, 1388–1393.

HARDER, W. & VELDKAMP, H. (1967). A continuous culture study of an obligately psychrophilic marine *Pseudomonas* species. *Archiv für Mikrobiologie*, **59**, 123–130.

HOBBIE, J. E. & CRAWFORD, C. C. (1969). Respiration corrections for bacterial uptake of dissolved organic compounds in natural waters. *Limnology and Oceanography*, **14**, 528–532.

JANNASCH, H. W. (1967). Growth of marine bacteria at limiting concentrations of organic carbon in seawater. *Limnology and Oceanography*, **12**, 264–271.

JANNASCH, H. W. & JONES, G. E. (1959). Bacterial populations in sea water as determined by different methods of enumeration. *Limnology and Oceanography*, **4**, 128–139.

JANNASCH, H. W., EIMHJELLEN, K., WIRSEN, C. O. & FARMANFARMAIAN, A. (1971). Microbial degradation of organic matter in the deep sea. *Science, Washington*, **171**, 672–675.

JANNASCH, H. W. & WIRSEN, C. O. (1973). Deep-sea microorganisms: *in situ* response to nutrient enrichment. *Science, Washington*, **180**, 641–643.

KENIS, P. R. & MORITA, R. Y. (1968). Thermally induced leakage of cellular material and viability of *Vibrio marinus*, a psychrophilic marine bacterium. *Canadian Journal of Microbiology*, **14**, 1239–1244.

KEYS, A., CHRISTENSEN, E. H. & KROGH, A. (1935). The organic metabolism of sea-water with special reference to the ultimate food cycle in the sea. *Journal of the Marine Biological Association of the United Kingdom*, **20**, 181–196.

KORNGOLD, R. R. & KUSHNER, D. J. (1968). Responses of a psychrophilic marine bacterium to changes in its ionic environment. *Canadian Journal of Microbiology*, **14**, 253–263.

KRISS, A. E. (1962). *Marine Microbiology* (*Deep Sea*). Translated by J. M. Shewan & Z. Kabata. Edinburgh & London: Oliver & Boyd.

LANGRIDGE, P. & MORITA, R. Y. (1966). Thermolability of malic dehydrogenase from the obligate psychrophile, *Vibrio marinus*. *Journal of Bacteriology*, **92**, 418–423.

MCBEE, R. H. & MCBEE, V. H. (1956). The incidence of thermophilic bacteria in Arctic soils and waters. *Journal of Bacteriology*, **71**, 182–185.

MACLEOD, R. A. (1971). Salinity – bacteria, fungi, and blue-green algae. In *Marine Ecology*, vol. I, part 2, ed. O. Kinne, pp. 689–703. New York: Wiley–Interscience.

MADELEY, J. R., KORNGOLD, R. R., KUSHNER, D. J. & GIBBONS, N. E. (1967). The lysis of a psychrophilic marine bacterium as studied by micro-electrophoresis. *Canadian Journal of Microbiology*, **13**, 45–55.

MATHEMEIER, P. F. (1966). Thermal inactivation of some enzymes from *Vibrio marinus*, an obligately psychrophilic marine bacterium. PhD Thesis. Oregon State University.

MENZEL, D. W. & RYTHER, J. H. (1970). Distribution and cycling of organic matter in the oceans. In *Organic Matter in Natural Waters*, ed. D. W. Hood, pp. 31–54. Institute of Marine Science Occasional Publication No. 1, University of Alsaka.

MENZIES, R. J., GEORGE, R. Y. & ROWE, G. T. (1973). *Abyssal Environment and the Ecology of the World Oceans*. New York: John Wiley & Sons.

MENZIES, R. J. & SELVAKUMARAN, M. (1974). The effect of hydrostatic pressure on living aquatic organisms. V. Eurybiotic environmental capacity as a factor in high pressure tolerance. *Internationale Revue der gesamten Hydrobiologie*, **59**, 207–212.

MITSKEVICH, I. N. & KRISS, A. E. (1966). High-pressure tolerance of *Pseudomonas* sp., strain 8113, isolated from the bottom of a deep-water basin of the Black Sea. *Doklady Akademii nauk SSSR, Biological Science Section*, **171**, 822–824.

MORITA, R. Y. (1967). Effect of hydrostatic pressure on marine microorganisms. *Annual Review of Oceanography and Marine Biology*, **5**, 187–203.

MORITA, R. Y. (1972). Pressure–bacteria, fungi and blue green algae. In *Marine Ecology*, vol. I, part 3, ed. O. Kinne, pp. 1361–1388. New York: Wiley–Interscience.

MORITA, R. Y. (1975). Psychrophilic bacteria. *Bacteriological Reviews*, **39**, 144–167.

MORITA, R. Y., GRIFFITHS, R. P. & HAYASAKA, S. S. (1975). Heterotrophic potential of microorganisms in Antarctic waters. In *Third SCAR/IUBS Symposium on Antarctic Biology*, ed. G. R. Llano. Washington: National Academy of Science (in press).

MORITA, R. Y. & HAIGHT, R. D. (1962). Malic dehydrogenase activity at 101 °C under hydrostatic pressure. *Journal of Bacteriology*, **83**, 1341–1346.

MORITA, R. Y. & HAIGHT, R. D. (1964). Temperature effects on the growth of an obligate psychrophilic marine bacterium. *Limnology and Oceanography*, **9**, 102–106.

MORITA, R. Y. & MATHEMEIER, P. F. (1964). Temperature–hydrostatic pressure studies on partially purified inorganic pyrophosphatase activity. *Journal of Bacteriology*, **88**, 1667–1671.

MORITA, R. Y. & ZOBELL, C. E. (1955). Occurrence of bacteria in pelagic sediments collected during the Mid-Pacific Expedition. *Deep-Sea Research*, **3**, 66–72.

OPPENHEIMER, C. H. & ZOBELL, C. E. (1952). The growth and viability of sixty-three species of marine bacteria as influenced by hydrostatic pressure. *Journal of Marine Research*, **11**, 10–18.

PARSONS, T. R. & STRICKLAND, J. D. H. (1962). On the production of particulate organic carbon by heterotrophic processes in sea water. *Deep-Sea Research*, **8**, 211–222.

QUIGLEY, M. M. & COLWELL, R. R. (1968). Properties of bacteria isolated from deep-sea sediments. *Journal of Bacteriology*, **95**, 211–230.

ROBISON, S. M. & MORITA, R. Y. (1966). The effect of moderate temperature on the respiration and viability of *Vibrio marinus*. *Zeitschrift für allgemeine Mikrobiologie*, **6**, 181–187.

SCHWARZ, J. R. & COLWELL, R. R. (1975*a*). *Pseudomonas bathycetes*: macromolecular biosynthesis at elevated pressure. *Journal of Bacteriology*, (in press).

SCHWARZ, J. R. & COLWELL, R. R. (1975*b*). Metabolic activities of deep-sea bacteria. *Applied Microbiology*, (in press).

SCHWARZ, J. R., YAYANOS, A. A. & COLWELL, R. R. (1975). Metabolic activities of the intestinal flora of a deep-sea invertebrate. *Applied Microbiology*, (in press).

SEKI, H. (1972). The role of microorganisms in the marine food chain with reference to organic aggregates. In *Detritus and Its Role in Aquatic Ecosystems*, ed. U. Melchiorri-Santolini & J. W. Hopton. Proceeding of an IBP–UNESCO Symposium. *Memorie dell'Instituto di Idrobiologia*, **29**, Supplement, 245–259.

SEKI, H. & ROBINSON, D. B. (1969). Effect of decompression on activity of microorganisms in sea water. *International Revue der gesamten Hydrobiologie*, **54**, 201–205.

SIEBURTH, J. MCN. (1964). Polymorphism of a marine bacterium (*Arthrobacter*) as a function of multiple temperature optima and nutrition. *Symposium on Experimental Marine Ecology, University of Rhode Island Occasional Publications*, **2**, 11–16.

SIEBURTH, J. MCN. (1971). Distribution and activity of oceanic bacteria. *Deep-Sea Research*, **18**, 1111–1121.

SIEBURTH, J. MCN. & DIETZ, A. A. (1974). Biodeterioration in the sea and its inhibition. In *Effect of the Ocean Environment on Microbial Activities*, ed. R. R. Colwell & R. Y. Morita, pp. 318–326. Baltimore: University Park Press.

SLEIGH, M. A. & MACDONALD, A. G. (1972). *The effects of Pressure on Organisms. Symposia of the Society for Experimental Biology*, **26**. London: Cambridge University Press.

STANLEY, S. O. & MORITA, R. Y. (1968). Salinity effect on the maximum growth temperature of some bacteria isolated from marine environments. *Journal of Bacteriology*, **95**, 169–173.

STREHLER, B. L. & JOHNSON, F. H. (1954). The temperature–pressure inhibitor relations of bacterial luminescence in vitro. *Proceedings of the National Academy of Sciences, U.S.A.* **40**, 606–617.

SVERDRUP, H. U., JOHNSON, M. W. & FLEMING, R. H. (1942). *The Oceans*. Englewood Cliffs: Prentice-Hall.

TSIKLINSKY, M. (1908). *Expedition Antarctique Francaise*, 1903–1905. Paris: Masson et Cie.

TRÜPER, H. G. (1969). Bacterial sulfate reduction in the Red Sea hot brines. In *Hot Brines and Recent Heavy Metal Deposits in the Red Sea*, ed. E. T. Degens & D. A. Ross, pp. 263–273. Berlin: Springer-Verlag.

WAKSMAN, S. A. & CAREY, C. L. (1935). Decomposition of organic matter in sea water by bacteria. II. Influence of addition of organic substances upon bacterial activities. *Journal of Bacteriology*, **29**, 545–561.

WRIGHT, R. T. & HOBBIE, J. E. (1965). The uptake of organic solutes in lake water. *Limnology and Oceanography*, **10**, 22–28.

ZIMMERMAN, A. M. (1970). *High Pressure Effects on Cellular Processes*. New York and London: Academic Press.

ZOBELL, C. E. (1946). *Marine Microbiology*. Watham: Chronica Botanica.

ZOBELL, C. E. (1952). Bacterial life at the bottom of the Philippine Trench. *Science, Pennsylvania*, **115**, 507–508.

ZOBELL, C. E. (1964). Hydrostatic pressure as a factor affecting the activities of marine microbes. In *Recent Researches in the Fields of Hydrosphere, Atmosphere, and Nuclear Geochemistry*, ed. Y. Miyake & T. Koyama, pp. 83–116. Tokyo: Maruzen Co. Ltd.

ZOBELL, C. E. (1968). Bacterial life in the deep sea. *Bulletin of the Misaki Marine Biological Institute, Kyoto University*, **12**, 77–96.

ZOBELL, C. E. (1970). Pressure effects on morphology and life processes. In *High Pressure Effects on Cellular Processes*, ed. A. Zimmerman, pp. 85–130. New York and London: Academic Press.

ZOBELL, C. E. & GRANT, C. W. (1943). Bacterial utilization of low concentrations of organic matter. *Journal of Bacteriology*, **45**, 555–564.

ZOBELL, C. E. & JOHNSON, F. H. (1949). The influence of hydrostatic pressure on the growth and viability of terrestrial and marine bacteria. *Journal of Bacteriology*, **57**, 179–189.

ZOBELL, C. E. & MORITA, R. Y. (1957). Barophilic bacteria in some deep sea sediments. *Journal of Bacteriology*, **73**, 563–568.

ZOBELL, C. E. & MORITA, R. Y. (1959). Deep-sea bacteria. *Galathea Reports, Copenhagen*, **1**, 139–154.

SURVIVAL OF VEGETATIVE BACTERIA IN ANIMALS

H. SMITH

Department of Microbiology, University of Birmingham
PO Box 363, *Birmingham B*15 2*TT, West Midlands*

Survival is the avoidance of death threatened by adverse circumstances. These circumstances can vary widely in their nature and effect on microbial activity. Most of the articles in this volume deal with the survival of micro-organisms under physical or chemical conditions that are themselves inimical to microbial growth and multiplication, such as extreme temperatures and lack of water or nutrients. And those contributions dealing with microbial survival against the lethal action of light, radiations and disinfectants, involve in the main micro-organisms in nutritional environments which would not normally support rapid growth and multiplication. In contrast, this paper deals with the survival of micro-organisms in a physico-chemical environment – that of animal tissues and fluids – which would normally support active proliferation of microbes were it not for the restrictive influence of the numerous anti-microbial mechanisms that can be mounted by a living animal. Sometimes the conditions within an animal host are unsuitable for the growth and multiplication of relevant microbes. Thus some nutritionally demanding mutants of *Salmonella typhi* were unable to grow in animal tissues because of the lack of required nutrients (Burrows, 1960) and Brazilian strains of *Pasteurella pestis* were unable to infect guinea pigs because the latter lacked asparagine, a required nutrient for these strains (Burrows & Gillett, 1971). Anaerobic pathogens such as those causing tetanus and gas gangrene cannot multiply in undamaged oxygenated tissues. Redox potential (E_h) levels also affect the flora in different parts of the alimentary tract. A few potential pathogens like *Fusiformis necrophorus* of sheep foot-rot have a nutritional dependence on other microbes in the same lesion (Roberts, 1967). Also, competition for nutrients may be a factor involved in destruction or survival of microbes on mucous surfaces. Yet, despite these contrary examples, it appears that most vegetative pathogens and commensals find the physico-chemical and nutritional conditions within the animal host more than adequate for growth and multiplication. This is evident from the following facts. Luxuriant microbial flora occur in some tissues *in vivo*, for example in the alimentary tract. Tissue extracts, blood and serum have been used over many years as media for growing most pathogens and commensals *in vitro*. Pathogens like *Bacillus anthracis* and *Streptococcus pneumoniae* multiply rapidly *in vivo* in the terminal stages of infection when host defence mechanisms have been overcome (Smith, 1958). *Brucella abortus* not only proliferates abundantly in certain foetal tissues due to the presence there of a preferred nutrient, erythritol, but also multiplies in the maternal udder using other carbohydrate sources and is shed into milk over long periods. The emergence of fatal aspergillosis and candidiasis in immuno-suppressed patients following transplant surgery or cancer treatment, shows that human tissues can support rapid fungal growth when host defences have been reduced. Finally, the rapid microbial proliferation that occurs within the tissues *post mortem* also testifies to the nutrients available to microbes in the animal host although in life physical barriers, as well as host defence mechanisms, may prevent access to some of them.

Hence, with notable exceptions such as the anaerobes, survival of vegetative micro-organisms in the animal host depends more on prevention of destruction by host defence mechanisms than on finding the minimum physico-chemical conditions for maintenance or growth. Furthermore, because most vegetative pathogens can multiply in the tissue environment if unmolested by defence mechanisms, the survival of microbial populations *in vivo* can be achieved by an equilibrium between destruction of those microbes which cannot resist host defences and multiplication of those that can. This dynamic process probably operates for survival of microbial populations in animals more often than persistence of a proportion of the original organisms without multiplication. The equilibrium can be short lived as during the critical primary lodgement period of infection (Miles, Miles & Burke, 1957; Polk & Miles, 1973) or long lived as in chronic infections and carrier states. At any time it can break down, leading either to increased destruction or multiplication of microbes with consequent elimination or progression of disease.

Until recently the fact that microbial destruction and multiplication could occur side by side in animal tissues was suspected but not proved because a method for measuring true microbial division rates *in vivo* was not available. The numbers of viable microbes in the tissues of an infected animal can be measured at any time after inoculation but they are the resultants of multiplication and destruction and, for microbial populations which have outside access such as those of the alimentary tract, mechanical removal. Over the past decade a method for measuring division rates of bacteria *in vivo* has been devised and, just as important for this paper, the method allows the calculation of death rates by relating the division rates to the overall increase or decrease in viable counts (Meynell, 1959; Meynell & Subbaiah, 1963; Maw & Meynell, 1968; Polk & Miles, 1973). The method relies on introducing into a pathogenic bacterium a genetic marker which neither integrates into the chromosome nor replicates in the cytoplasm. On division the marker is distributed to only one of the two daughter cells in each succeeding generation. If the proportion of a bacterial population containing such non-replicating markers is determined at inoculation and at the end of a known period, the number of generations occurring during the period can be calculated. The reasonable assumption is made that bacteria containing the marker multiply or die at the same rate as those without it. Two methods have been used for introducing the marker, superinfecting bacteriophage (Meynell, 1959; Maw & Meynell, 1968; Polk & Miles, 1973) and abortive transduction (Meynell & Subbaiah, 1963). In the first, the pathogenic bacteria are lysogenized by a temperate phage and then a related phage is introduced; this superinfecting phage does not lyse the bacteria and at their division goes into only one of the daughter cells. Ultraviolet irradiation releases both the temperature and superinfecting phage. Plating out irradiated homogenates of the infected tissue on a bacterial strain susceptible to both phages and counting the plaques gives the total viable count in the homogenate; and plating on a strain susceptible to only the superinfecting phage gives the number of bacteria in the homogenate containing that phage. In abortive transduction a genome fragment (carrying a suitable biochemical marker) of a donor bacterium is transferred by a bacteriophage to the recipient pathogenic bacterium but, unlike normal transduction, the fragment is not incorporated into the recipient genome and on division it behaves as described above.

The method is confined to those bacteria – until now *Escherichia coli* and *Salmonella typhimurium* – on which the required genetic manipulation is possible and unfortunately it has been used in few investigations. However, these show clearly that bacterial multiplication and destruction can occur side by side sometimes in near equilibrium, and that an increase in numbers ensues immediately host defence mechanisms are impaired. In the spleens of mice, *S. typhimurium* multiplied and died slowly (Maw & Meynell, 1968); the doubling time was about 5 h, equivalent to five generations in 24 h, but the 50 per cent death time was about 8 h, equivalent to three generations in 24 h. Thus the net rate of increase of the population was only about two generations or four-fold in 24 h; the influence of host defence mechanisms on the system is indicated by the fact that *S. typhimurium* can increase four-fold every hour in media prepared from animal tissues. In the muscle of mice, *E. coli* was killed without multiplication during the first 4 h after inoculation until only 5 per cent of the inoculum remained. Then, during the next 3 h, slight multiplication (about two generations) took place so that the decline in viable counts was arrested and near equilibrium conditions prevailed before a final decline to complete disappearance of the viable counts in 72 h (Polk & Miles, 1973). The nutritional completeness of muscle for multiplication of *E. coli* was indicated by parallel experiments using recently killed mice, when immediate multiplication occurred, two to three generations being accomplished in 4 h and seven to eight in 7 h. Also, when local adrenalin or ferric iron was given to live mice with the inoculum, the muscle bactericidal mechanisms were impaired within 1–2 h and the viable count rose, due to the reduced destruction and some multiplication (up to three generations). In studies on the survival and growth of a streptomycin-resistant strain of *S. typhimurium* in the alimentary tract of mice (Meynell & Subbaiah, 1963), the bacterial excretion rate was estimated (by comparing, in relation to their viable counts, the dry weights of caecal and colonic contents with that of voided faeces) as well as the total viable counts in the intestinal contents and the doubling times. The death rate was then obtained by appropriate calculation. In normal mice, the bacteria multiplied slowly, making one division in the first 6 h. However, in the same time the viable counts in the gut contents decreased 10- to 100-fold because of a weak bactericidal activity and mechanical excretion. The bactericidal activity appeared to be due to volatile fatty acids produced by commensal anaerobes and a low E_h. If the commensals were removed by treating the mice with streptomycin, the viable counts of the streptomycin-resistant *S. typhimurium* increased due to prevention of killing and a rapid multiplication rate which, in the guts of some mice, approached that in a broth medium.

To reiterate, in most cases the crux of microbial survival in the animal host is non-stimulation or impairment of host defence mechanisms. The remainder of this paper describes how this is accomplished by vegetative bacteria on mucous surfaces and within tissues and fluids. It is restricted to bacteria because viruses and spores are excluded by the title of the Symposium, and little is known of the survival mechanisms of fungi, protozoa and mycoplasmas. Lack of space precludes full descriptions of the numerous anti-microbial mechanisms possessed by an animal host (Coombs & Smith, 1968, 1975) but the most important mechanisms are summarized at appropriate points as preludes to descriptions of bacterial resistance to them. It will become obvious that although host defence mechanisms and bacterial factors counteracting them have been recognized and even identified chemically, their respective modes of action in either destroying or preventing destruction of the bacteria are usually not clear.

SURVIVAL OF BACTERIA ON MUCOUS SURFACES

Two categories of bacteria exist, with fungi and other microbes, on the mucous surfaces of the alimentary, respiratory and urogenital tracts. There are vast numbers of indigenous organisms or commensals that are not usually associated with disease. Then, from time to time, there are relatively few members of a pathogenic species beginning the first phase of their disease process. This process may develop solely on the mucous surface as in cholera or whooping cough but usually it progresses by penetrating either into the surface epithelium or through it into the underlying tissues as in dysentery and streptococcal infections respectively. Both commensals and pathogens must have the following properties to survive and grow on mucous surfaces. First, they need ability to adhere to the mucous surface and thus to withstand the flushing action of moving lumen contents or mucociliary action. Second, they must be unaffected by the pH and the bacteriostatic or bactericidal action of host secretions at various surface sites. Third, they should be able to compete with their neighbours for space on the host surface and possibly for nutrients as well as being unaffected by any changes in E_h or potentially inhibitory metabolites produced by them (Savage, 1972). The competing activities of the indigenous bacteria can be counted as a third host defence mechanism that must be resisted by pathogens if disease is to occur. And pathogens must be well endowed with these properties of resistance if relatively few of them are to survive in the face of a massive established indigenous population.

Commensals and pathogens not only adhere to mucous surfaces but do so selectively (Savage, 1972). Thus, oral *Streptococcus salivarius* adheres well to cheek and tongue cells but not to teeth, whereas *S. mutans* adheres to teeth and not to cheek and tongue cells. In mice and rats, lactobacilli stick to the non-secreting stomach epithelium, short rods and cocci to the ileal epithelial cells and anaerobic fusiform bacteria to caecum and colon epithelium; if removed by antibiotic treatment they all recolonize selectively when treatment ceases. In human cholera and *E. coli* infections of swine and calves the pathogen adheres only to epithelial cells of the small intestine and, for *E. coli*, more to cells near the villous tips than at the bases (Staley, Jones & Corley, 1969; Drees & Wexler, 1970; Savage, 1972). The sophistication of some adherence mechanisms is indicated by recent observations that a filamentous segmented bacteria indigenous to mouse intestinal tract has a nipple-like structure on the segment nearest the epithelium which is inserted into the epithelial cells (Davis & Savage, 1974). Obviously these adherences are due to specific interactions between surface components of the bacteria and those of the host cells but the nature of these components is known only in a few instances and only for the bacteria. A specific mucopolysaccharide appears to be involved in attaching indigenous lactobacilli to the keratinized epithelium of mouse stomach (Savage, 1972). *Streptococcus mutans*, the causative organism for dental caries, produces from sucrose *in vivo* and *in vitro* an insoluble dextran-levan by which the bacteria stick to teeth and is thus important in dental plaque formation (Gibbons & van Houte, 1971; Mukasa & Slade, 1973, 1974); mutants lacking the ability to produce the dextran had less ability to produce plaque in pathogen-free and gnotobiotic rats (Tanzer *et al.*, 1974). Enteropathogenic strains of *E. coli* attach to the brush border of the small intestine of piglets because the bacteria possess K88 protein antigens (Smith & Linggood, 1971; Bertschinger, Moon & Whipp, 1972*a*, *b*; Jones & Rutter, 1972; Rutter & Jones, 1973), although not all piglets seem to have the corresponding host receptor (Sellwood *et al.*, 1974). On the other hand the strains of *E. coli* infecting calves and humans do not have K88 antigens and whether they adhere strongly to the epithelium and by what means is still an open question (Wilson & Hohmann, 1974). *Streptococcus pyogenes* must adhere to cells of the upper alimentary and respiratory tract to produce sore throats and tonsillitis and this adhesion appears to be due to the M protein of virulent strains (Ellen & Gibbons, 1972). Gonococci adhere strongly to epithelial cells of the urogenital tract (Ward & Watt, 1972) but former suggestions that pili play the major role in this attachment (Ward, Watt & Robertson, 1974; Punsalang & Sawyer, 1973) do not appear to be true at least for female tissues since non-pilated types stick to human fallopian tube and inner cervix almost as well as pilated types (Taylor-Robinson *et al.*, 1974; G. M. Tebbutt, D. R. Veale,

J. G. P. Hutchinson & H. Smith; unpublished observations). Also there are few pili resembling those produced *in vitro* on gonococci grown *in vivo* (Novotny, Short & Walker, 1975) and other larger appendages of doubtful function may be present (Grimble & Armitage, 1974).

Host secretions from mucous surfaces contain non-specific and specific (e.g. the immunoglobulin IgA) anti-bacterial substances. Some can prevent adherence, for example the action of parotid fluid on streptococci (Ørstavik, Kraus & Henshaw, 1974) and K88 antibody on *E. coli* (Rutter & Jones, 1973). Also they can be bacteriostatic or bactericidal and not the least consideration in this respect is the pH of the secretions, alkaline in some sites and acid in others. Thus signs and symptoms of cholera were produced in human volunteers by a few orally administered *Vibrio cholerae* vibrios if the acid content of the stomach was first neutralized by sodium bicarbonate (Cash *et al.*, 1974). Finally, large numbers of phagocytes are extruded from the mucous surfaces (Hirsch, 1972) and this can be mediated by immune phenomena (Bellamy & Neilsen, 1974). Thus the cellular defences can operate as described later in conjunction with mechanical removal of the phagocytes and their contained bacteria by the moving lumen contents. How bacteria survive these anti-microbial mechanisms of mucous secretions is not clear. Presumably any bacterial aggressins (see below) that are effective later against humoral defences in the blood and tissues as well as those which inhibit the cellular defences can also operate similarly on mucous surfaces. Sheer quantity and lack of accessibility of bacteria in the lumen contents undoubtedly promote their survival since the contents are moving and the anti-bacterial mechanisms not only vary in strength at different sites along the body tracts but take time to act. Thus a few survivors from the killing of a large initial population in one part of the tract may find conditions better further along it. In the human volunteer experiments described above, cholera signs and symptoms were produced in men not receiving bicarbonate provided the dose of vibrios was large.

The indigenous bacterial flora on mucous membranes varies with the animal species, with the site of the membrane and the age, diet and degree of physical stress (Savage, 1972; Davis, McAllister & Savage, 1973; Tannock & Savage, 1974). The competition and antagonism between various members of the flora and especially between them and extraneous pathogens were indicated by changes in the flora and by the fate of introduced pathogens (staphylococci, streptococci, *V. cholerae*, salmonellae, shigellae) when some indigenous bacteria had been removed by antibiotic treatment. This has been confirmed in many instances by mixed culture experiments *in vitro* and in gnotobiotic animals with mixtures of various indigenous and pathogenic bacteria (Savage, 1972; Maier & Hengtes, 1972; Crowe, Sanders & Langley, 1973; Levison, 1973; Tannock & Savage, 1974). Thus, competition and antagonism are proven but the mechanisms involved in destruction or survival are still not clear. The low E_h determining the anaerobic flora in the lower bowel probably occurs because the limited supplies of oxygen are consumed by other members of the population (Hobson, 1974) and it possibly prevents the establishment in that site of some pathogens which require at least microaerophillic conditions. Other possible mechanisms are nutrient deprivation, bacteriocin production, preferential adherence and production of inhibitory metabolites (Savage, 1972) but only for the latter is there any real evidence and then only in one case. The intestinal anaerobic fusiforms produce from mucus volatile fatty acids which, in a reducing environment, are weakly bactericidal to salmonellae and shigellae (Meynell & Subbaiah, 1963; Hentges & Maier, 1970; Baskett & Hentges, 1973; Levison, 1973). How relatively small numbers of such pathogens as salmonellae and shigellae survive the antagonistic activities of the indigenous bacteria during the initial stages of diseases like typhoid or dysentry is unknown. A recent paper (Ducluzeau & Raibaud, 1974) indicates that at least *Shigella flexneri* can mount a resistance mechanism *in vivo*. When *S. flexneri* was grown in gnotobiotic mice for only a short period (1 day) before introducing *E. coli* it was eliminated by the latter competing species in eight days. This was not so, however, if the shigellae were present alone in the mice for three months; resistance to elimination by *E. coli* mounted stepwise over this period. The nature of the resistance is not known. Paradoxically, and underlining the difference in microbial behaviour *in vivo* and *in vitro*, the shigellae which were resistant *in vivo* were still eliminated by *E. coli* in mixed culture *in vitro*. In epidemics of dysentry, organisms grown *in vivo* are transferred directly by faecal contamination to a new host and it is likely that any characteristics of resistance to the indigenous flora would not be lost in the few generations which might occur in the faeces during transference.

SURVIVAL OF BACTERIA WITHIN ANIMAL TISSUES

Only relatively few bacterial species – the pathogens – can survive within animal tissues. These species form surface or extracellular products which interfere with host defence mechanisms that destroy or remove other bacteria including commensals if they penetrate into the tissues. Such products have been called aggressins, auxillary pathogenic factors (Coombs & Smith, 1968, 1975) or impedins (Glynn, 1972) and they determine the survival of most pathogens in animals.

There can be two important survival periods in the life of pathogens within the host. The first is short term and always occurs after any pathogen enters the host tissues and before the significant increase in population which results in disease. The second is long term and occurs only in a few instances after the subsidence of the acute phase of the disease. The two survival periods correspond with the timing of action of the major host defence mechanisms first non-specific and then specific.

During their first few hours within the host, pathogens must cope with powerful bactericidal mechanisms already present (Miles, 1962) and phagocytes mobilized by inflammation soon after the tissues are irritated. In this primary lodgement period of infection (Miles *et al.*, 1957) the anti-bacterial mechanisms of the host are heavily weighted against the few invading bacteria and many infections fail to survive this period. If survival is achieved at the primary site any spread of infection is immediately opposed by phagocytes fixed in the lymph-nodes, spleen and liver. Successful survival against the activities of these phagocytes results in acute disease which sometimes kills the host but usually subsides after a time for the following reason.

All the initial defence mechanisms described in the previous paragraph are non-specific. However, a few days after infection there is a progressive rise in the immune reactions of the host. These reactions can not only increase considerably the efficiency of the phagocytic defence but provide antibodies capable of direct neutralization of those bacterial products which are important in the disease process (Coombs & Smith, 1968, 1975). Eventually, most infections are eliminated by immune reactions and often this is hastened by therapeutic treatment started at the onset of the disease signs. Only in chronic disease and carrier states is long-term survival achieved against such formidable anti-bacterial defences.

Before describing the bacterial aggressins that may be operative in short-term and long-term survival, the defence mechanisms of the host are summarized.

Anti-bacterial mechanisms of the host

Non-specific humoral factors

These are soluble materials present in the tissues and body fluids such as blood, lymph, milk, tears, saliva and mucus.

Lysozyme. This is found in many body fluids. By splitting cell-wall peptide it lyses some Gram-positive bacteria and enhances the action of complement on some Gram-negative ones (Muschel, 1963).

β-Lysins. Heat-stable substances found in sera of most animals; they lyse Gram-positive bacteria and their amounts increase in infection (Myrvik & Leake, 1960). Recent electron microscopy on *Bacillus subtilis* and *Listeria monocytogenes* indicates that β-lysins act primarily on the plasma and mesosomal membranes rather than on the cell wall (Matheson & Donaldson, 1970; Shultz & Wilder, 1973).

Basic polypeptides. These kill Gram-positive and Gram-negative bacteria and, at least for staphylococci, cause leakage of protoplasmic contents after binding to the cell wall and damaging both it and the plasma membrane (Hirsch, 1958; Hibbitt & Benians, 1971).

Iron-binding transferrin and lactoferrin. These are present in serum, milk and other secretions. Alone and sometimes in conjunction with antibody and complement, they prevent the growth of many pathogens (*Clostridium welchii, Pasteurella septica, P. pestis, E. coli, Pseudomonas aeruginosa*) by denying them essential iron which may result in some cases in an inhibition of RNA synthesis (Weinberg, 1971; Bullen, Rogers & Griffiths, 1972; Bullen, Ward & Wallis, 1974).

Complement. Complement is a multicomponent system with many biological properties. Some bacteria can by lysed by it after being sensitized by non-immune mechanisms (Coombs & Smith, 1968, 1975). However, the mechanisms of its action on bacteria is not clear although a primary attack on the plasma membrane seems likely.

Non-specific cellular factors: the phagocytic defence system

Phagocytes vary in origin, morphology, constituents and bactericidal function (Coombs & Smith, 1968, 1975; Hirsch, 1972). There are two main types each having two subdivisions; polymorphonuclear (neutrophils and eosinophils) and mononuclear (blood monocytes and tissue macrophages) phagocytes. Polymorphonuclear phagocytes are cells with a short life derived from different stem cells from the long-lived mononuclear phagocytes. The blood, tissues and inflammatory exudates contain all types of phagocytes but only macrophages are found in the fixed phagocytic system in the lymph nodes, spleen and liver.

Ingestion occurs by invagination of the phagocytic membrane around the bacterium and killing by discharge of cytoplasmic granules (lysosomes) into the vacuole (Hirsch, 1972). The anti-bacterial factors of the vacuoles with their discharged granules are many; low pH, lactic acid, hydrogen peroxide, hydrogen peroxide acting with myeloperoxidase and halide, basic proteins, alkaline phosphatase, lactoferrin, lysozyme, peroxidase and others yet unknown especially for macrophages (Hirsch, 1972). In any particular situation it is difficult to determine which bactericidal factor of a phagocyte is mainly responsible for killing. There have been only a few investigations of their effects on bacteria and, as might be expected, most of them indicate an attack on the cell wall. Cohn (1963) showed that isotopically labelled bacteria were digested rapidly to give small molecules, which moved from the vacuole to the cytoplasm, and indigestible particles which remained. The myeloperoxidase–hydrogen peroxide–chloride system may work by forming chloramines with amino compounds in the bacterial cell wall; these chloramines could then degrade to form aldehyde groups with a bactericidal action (Sbarra *et al.*, 1972). A mixture of lysosomal enzymes degraded the cell walls of brucellae and tubercle bacilli (Willett & Thacore, 1967; McGhee & Freeman, 1970*a*). *β*-*N*-acetylglucosaminidase from leucocytes degraded the C-polysaccharide of Group A streptococci (Ayoub & McCarty, 1968). Lysates of human leucocytes which contained acid hydrolases readily degraded the cell walls of *Staphylococcus albus* and to a lesser extent those of *S. aureus*; in contrast the cell walls of a group A *Streptococcus* were stable to the lysates, a fact which fits with the persistence of streptococcal cell wall components in animal tissues for long periods (Lahav *et al.*, 1974). Rapid killing of *E. coli* by intact or disrupted rabbit phagocytes occurred without major structural disorganization and stoppage of macromolecular synthesis; it appeared to be connected with a very rapid (within minutes) increase in permeability of the envelope detected by entry of materials like actinomycin D which are normally excluded by *E. coli* (Beckerdite *et al.*, 1974). A strain of *Salmonella typhimurium* was also killed rapidly with the increase in permeability but strains of *Serratia marcescens* and *P. aeruginosa* were resistant to both effects.

Specific humoral and cellular defences

These have been reviewed (Coombs & Smith, 1968, 1975) and are so diverse that they cannot be described adequately in the space available. For this article, however, it is sufficient to say that antibodies can: (1) neutralize bacterial aggressins – the compounds that inhibit host defences – directly; (2) sensitize bacteria to lysis by complement; and (3) possibly block surface components such as permeases or iron-chelating molecules which are vital to bacterial survival. Furthermore, the efficiency of the phagocytic defences is considerably increased: (*a*) by antibodies opsonizing bacteria so that they are more easily ingested and destroyed by all types of phagocytes, and (*b*) by stimulating the activities of mononuclear phagocytes through a variety of mechanisms.

Interference with host defence mechanisms in the early stages of disease

Pathogenic bacteria produce aggressins which interfere with the host defence mechanisms that have been described. Somewhat arbitrarily I am dividing these aggressins into those which seem important in the short-term survival period immediately following bacterial intrusion into the tissues, and those which, although they may also influence initial invasion, appear to be important in long-term survival. The former – those interfering with non-specific humoral defences, with ingestion by polymorphonuclear and mononuclear phagocytes and with intracellular killing by the short-lived polymorphonuclear cells – are dealt with here. The latter – those contributing to inhibition of immune defences and intracellular survival in mononuclear phagocytes – will be described later.

Interference with humoral defences

Resistance to humoral bactericidins has been shown by virulent strains of *B. anthracis* (Keppie, Harris-Smith & Smith, 1963), *Staphylococcus aureus* (Cybulska & Jeljaszewicz, 1966), *Leptospira* spp. (Shultz & Wilder, 1971; Stalheim, 1971), Enterobacteriaceae (Braun & Siva Sankar, 1960) and *Brucella abortus* (Ellwood, Keppie & Smith, 1967). The resistance can vary with growth phase and environmental conditions. Stationary-phase *Streptococcus faecalis* was more resistant than logarithmic-phase organisms to rat and rabbit bactericidins (Guze, Hubert & Kalmanson, 1970). Gonococci from urethral pus were more resistant to human bactericidins than they were after subculture on laboratory media unless the latter contained extract of human prostate (Ward, Watt & Glynn, 1970; Watt, Glynn & Ward, 1972). Interference with the various humoral bactericidins is thus well established for virulent strains of many species but the aggressins responsible for the survival against such defences are known only in a few instances.

Capsular poly-D-glutamic acid and the three-component toxin of *Bacillus anthracis* interfered with the bactericidins (probably β-lysins and basic proteins) of horse serum (Keppie *et al.*, 1963). The acidic polysaccharide, K antigens of *E. coli* infecting the urinary tract, interfere with antibody and complement (Glynn, 1972; Hanson, 1973; Kaijser, 1973). A cell wall component from *Brucella abortus* containing protein, carbohydrate, formyl residues and about 40 per cent of lipid interferes with the bactericidins in bovine serum (Ellwood *et al.*, 1967). The elastase of *P. aeruginosa* interferes with complement (Schultz & Miller, 1975). Transferrin and lactoferrin activity against bacteria may be counteracted by the production of surface compounds which chelate iron such as enterochelin – a cyclic trimer of 2,3-dihydroxybenzoyl serine – by *Salmonella typhimurium*, *E. coli* and *Klebsiella aerogenes*, or mycobactins by tubercle bacilli (Bullen *et al.*, 1972).

Interference with the initial action of polymorphonuclear and mononuclear phagocytes

Bacteria can inhibit the bactericidal activities of phagocytes by preventing one or more of the following processes: mobilization (inflammation), contact, ingestion and intracellular killing. Only the first three processes are considered here.

Inflammation. In mice, virulent staphylococci multiply more rapidly and produce more severe lesions by suppressing the inflammatory response (Agarwal, 1967). A cell wall mucopeptide which appears to prevent the release of kinins is responsible for this anti-inflammatory activity (Hill, 1968, 1969; Easmon, Hamilton & Glynn, 1973). The significance of this mucopeptide in staphylococcal survival in human infections is not known (Glynn, 1972).

Contact. Phagocytes contact bacteria by random hits, by trapping in filtration systems like the spleen and by chemotaxis. Chemotaxis was prevented *in vitro* by fractions from tubercle bacilli (Dubos & Hirsch, 1965) and a cell wall product from staphylococci which was different from the anti-inflammatory mucopeptide (Weksler & Hill, 1969; Glynn, 1972). Whether these potential aggressins act in infection is not known.

Ingestion. Once engulfed, many bacteria (salmonellae, pneumococci, anthrax and plague bacilli) are usually destroyed and digested (Dubos & Hirsch, 1965; MacKaness, 1960). Resistance to ingestion, thus avoiding intracellular destruction is the main survival mechanism of these bacteria. There are two types of aggressins: surface and capsular products, which do not appear to harm the phagocytes, and toxic materials which produce direct damage. Examples of the first are the capsular polysaccharides of the pneumococci (Dubos & Hirsch, 1965) and meningococci (Gotschlich, Goldschneider & Artenstein, 1969; Roberts, 1970), the capsular hyaluronic acid and M protein of streptococci (Fox, 1974), the O somatic antigens of the Enterobacteriaceae (Friedberg & Shilo, 1970; Luderitz, Staub & Westphal, 1966; Medearis, Camitta & Heath, 1968; Nakano & Saito, 1968, 1969; Valtonen & Makela, 1971), the capsular polyglutamic acid of *Bacillus anthracis* (Keppie *et al.*, 1963), the Vi antigen (poly-*N*-acetyl-D-galacto-samino-uronic acid) of *S. typhi* (Clark, McLaughlin & Webster, 1958), the acid polysaccharide K antigens of *E. coli* (Glynn, 1972), the protein carbohydrate envelope substance of *Pasteurella pestis* (Dubos & Hirsch, 1965), a cell wall mucopeptide from staphylococci (Shayegani, Hitatsune & Mudd, 1970) and a surface slime from *Pseudomonas aeruginosa* (Kobayashi, 1971; Sensakovic & Bartell, 1974). Examples of the second are the leucocidins of the staphylococci (Dubos & Hirsch, 1965) and the anthrax toxic complex (Keppie *et al.*, 1963).

Knowledge of the chemical structure of the O antigens of mutants of the Enterobacteriaceae has allowed investigations of the relation between structure and antiphagocytic activity. Resistance of *E. coli* to phagocytosis by mouse polymorphonuclear leucocytes appeared to depend on a complete polysaccharide side chain in the cell wall; a mutant lacking colitose in its side chain was significantly more susceptible to phagocytosis (and less virulent) than the wild type and a mutant lacking galactose, glucose, N-acetyl glucosamine and colitose was even more susceptible (Medearis *et al.*, 1968). Similarly, work with mutants and with media which induce phenotypic change has indicated that complete sugar sequences in the core and polysaccharide side chain of the O antigens are necessary for full phagocytosis resistance and virulence of *S. typhimurium* for mice; the tetrasaccharide sequences abesquosyl-mannosyl-rhamnosyl-galactose have been suggested as the determinant group, acetyl and glucosyl groups being less important (Nakano & Saito, 1968, 1969; Friedberg & Shilo, 1970; Valtonen & Makela, 1971; Stendahl & Edebo, 1972). Recently Makela, Valtonen & Valtonen (1973) observed that small changes in O antigen structure of *S. typhimurium* can occur which influence virulence without any apparent effect on phagocytosis resistance.

While we are beginning to learn something of the connection between structure and prevention of ingestion, the mode of action of these aggressins is still not clear. Non-toxic aggressins may interfere with ingestion by purely mechanical means, by inhibition and absorption of serum opsonins as seems to occur for *B. anthracis* (Keppie *et al.*,

1963) and staphylococci (Shayegani *et al.*, 1970); and possibly by rendering the bacterial surface less foreign to the host as might occur for the M protein of streptococci which may share common antigenic determinants with host tissue components (Fox, 1974). Toxic aggressins probably harm phagocytes in the same way as they affect ordinary cells; for example interference with membrane function by the leucocidin α toxin of the staphylococci.

Survival within the short-lived polymorphonuclear phagocytes

Polymorphonuclear phagocytes live only for a few days *in vivo* (Hirsch, 1972). Ability to resist their bactericidins can therefore contribute to bacterial survival in the early phases of disease; at death of the phagocytes, the intracellular bacteria will be liberated and can infect surrounding tissues which may, due to movement of the phagocytes, be distant from the original site of ingestion. Thus infection can be spread by this process which seems to occur in brucellosis (Smith & FitzGeorge, 1964), in plague (Jansen & Surgalla, 1969), in staphylococcal infections (Adlam, Pearce & Smith, 1970), in gonorrhoea (Witt, Veale & Smith, 1975; Veale, Smith & Witt, 1975; Veale, Witt & Smith, 1975) and probably in other infections. However, in only two cases have the mechanisms of this intracellular survival been probed.

Virulent strains of *Brucella abortus* survived and grew in the phagocytes (mainly polymorphonuclear with some mononuclear) of bovine 'buffy coat' whereas attenuated strains were gradually destroyed (Smith & FitzGeorge, 1964). This was not due either to a greater ability of the virulent strains to use the nutritional conditions within the phagocyte or to their higher catalase content (which could afford greater protection against phagocyte hydrogen peroxide) (Smith, 1968). The ability of the virulent strains to survive intracellularly relies on the production, under the growth conditions *in vivo* as well as simulant ones *in vitro*, of a cell wall substance which interferes with the bactericidal mechanisms of the phagocytes. Virulent brucellae obtained from infected bovine placental tissue (Smith & FitzGeorge, 1964), or from cultures in laboratory media supplemented by bovine placental extracts or foetal fluids (FitzGeorge & Smith, 1966), had an increased ability to survive intracellularly compared with the same strain grown in unsupplemented laboratory media. Cell wall preparations of the organisms from infected bovine placenta and from the supplemented media inhibited the intracellular destruction of an avirulent strain of *B. abortus* (Smith & FitzGeorge, 1964; FitzGeorge & Smith, 1966). Finally, from *B. abortus* grown in supplemented medium, the antigen which appeared to be responsible for the inhibition of the phagocyte bactericidin was removed by washing with an ether–water mixture (Frost *et al.*, 1972). This material, which prevented intracellular destruction of *B. abortus* was serologically different from the other cell wall material described previously which interfered with the humoral bactericidins of bovine serum (Frost *et al.*, 1972). It is not known whether it prevents phagosome discharge or interferes with the action of the discharged phagosomes (cf. the action of tubercle and leprosy bacilli, see below).

Staphylococci grown in rabbits were more resistant than staphylococci grown *in vitro* to killing by rabbit polymorphonuclear phagocytes and their bactericidal extracts (Adlam *et al.*, 1970; Pearce, Scragg & Kolawole, 1975). This resistance to intracellular killing of the organisms grown *in vivo* appeared to be due to a surface layer of host protein, which may result to some extent from the action of free and bound coagulase (Kolawole, Scragg & Pearce, 1975). Gladstone, Walton & Kay (1974) confirmed that the organisms grown *in vivo* by Pearce and his colleagues were resistant to killing by the cationic proteins of polymorphonuclear phagocytes.

Mechanisms which may contribute to long-term survival in animal tissues

Here I emphasize that in this Symposium we are interested in the mechanisms of survival of vegetative bacteria in chronic disease and carrier states. Some chronic diseases such as *E. coli*, nephritis and sequelae of streptococcal infections may be caused by immunopathological phenomena following the persistence of antigens alone after the death of the bacteria (Miller, Smith & Sanford, 1971; Lahav *et al.*, 1974).

Suppression of immune responses

Although there is less experimental evidence in bacteriology compared with virology, the immune defences seem to be suppressed in some bacterial diseases and by some bacterial compounds. Although certain cell wall preparations containing endotoxins have been considered to have an adjuvant effect on vaccines, in some cases endotoxins have decreased antibody response (Finger, Fresenius & Angerer, 1971). Antibody response was also decreased by fractions from Group A streptococci and *E. coli* (Malakian & Schwab, 1967; Kirpatovski & Stanislavski, 1971) and L-asparaginase from *E. coli* (Friedman & Chakrabarty, 1971). However, sheep red blood cells were the antigenic preparation in most of these experiments and not a relevant bacterial infection or product. Cell-mediated immunity (which affects macrophage function) also appeared to be depressed in tuberculosis (Fanconi & Wallgren, 1952) and in leprosy (Turk, 1970). Whether or not such depressions of immune responses could be significant in long-term bacterial survival is difficult to judge. I incline to the view that they are not, since antibody and delayed hypersensitivity persist in the chronic and carrier states of most of the diseases that have been mentioned. It is hard to believe that the humoral and cellular immune defences cannot cope with the relatively small number of bacteria persisting unless other protective mechanisms are involved.

Survival in long-lived cells

Within animal cells bacteria can find sanctuary from specific and non-specific humoral bactericidins and often from injected antibacterial drugs. Ingestion by cells can therefore contribute substantially to bacterial survival within the host, provided the bacteria can withstand the intracellular environment and the cells themselves are not short lived (as are the polymorphonuclear phagocytes) or shed to the outside as are some epithelial cells.

Phagocytes are not the only animal cells that can be penetrated by bacteria; for example, dysentery bacilli and gonococci enter epithelial cells of the alimentary and urogenital tracts respectively and brucellae enter the cells of ungulate placentae. Nevertheless the phagocytes usually pick up more bacteria than other cells because they have well-developed powers of ingestion; and if a bacterial population can survive within phagocytes despite the many intracellular bactericidal mechanisms, they should survive also if they penetrate other types of cells that are less well endowed with bactericidins. In the past few years we have learned something of the mechanisms whereby the typical intracellular pathogens brucellae, tubercle bacilli and leprosy bacilli survive and grow within long-lived macrophages.

Brucellae survived and grew in bovine macrophages separated from the mixed cell population of bovine 'buffy coat' by prolonged attachment to glass (FitzGeorge, Solotorovsky & Smith, 1967; Solotorovsky & Tewari, 1974). Although not certain, it is probable that the bactericidal mechanisms of these macrophages were inhibited by the same cell wall material which has been shown to inhibit the bactericidins of the mixed polymorphonuclear and mononuclear phagocytes of 'buffy coat' (see above).

Mycobacterium tuberculosis and *M. microti* (causal agent of vole tuberculosis) appear to resist intracellular bactericidins by a mechanism different from that of *M. lepraemurium*. In mouse peritoneal macrophages infected with virulent *M. tuberculosis* and *M. microti* the lysosomes do not discharge into the phagosomes containing intact bacteria (Armstrong & D'Arcy Hart, 1971; D'Arcy Hart *et al.*, 1972; D'Arcy Hart & Armstrong, 1974). On the other hand, *M. lepraemurium* in mouse peritoneal macrophages and in rat fibroblasts survives and grows despite lysosomal discharge into the phagosomes (Brown & Draper, 1970; D'Arcy Hart *et al.*, 1972). The aggressins determining the different types of resistance seem to be connected with the electron transparent layers that surround these organisms *in vivo* (Imaeda, 1965; Imaeda *et al.*, 1969; Draper & Rees, 1970). Recently the material from *M. lepraemurium* was isolated from infected mouse liver and spleen; it was a mycoside (a peptido-glycolipid) of type C known to occur in some species of mycobacteria (Draper & Rees, 1973). It is possible that *M. microti* prevents discharge of lysosomes by producing in the vacuoles cyclic adenyl monophosphate (Lowrie, Jackett & Ratcliffe, 1975).

In addition to the typical intracellular pathogens which are well known to cause long-term chronic disease, other bacteria can survive within mononuclear phagocytes; salmonellae (Roantree, 1967; Baskerville *et al.*, 1972; Solotorovsky & Tewari, 1974), shigellae (Yee & Buffenmyer, 1970), plague bacilli (Jansen & Surgalla, 1969) and gonococci (Veale, Smith & Witt, 1975; Veale, Witt & Smith, 1975). At present we do not know the mechanisms operating in such cases of survival but presumably they are similar to those described for brucellae, tubercle bacilli and leprosy bacilli. Although these other bacteria tend to produce acute disease and perhaps rely on resistance to ingestion more than resistance to intracellular killing for survival during the early stages, their capacity to survive in the long-lived mononuclear phagocytes may play a role in the chronic and carrier states that occasionally occur as sequelae to the acute disease.

Now, in this discussion I should stress what was said at the beginning of the article – survival *in vivo* is a dynamic equilibrium between bacterial destruction and multiplication so there is neither a gross reduction nor a large increase in the overall bacterial population. Indications as to how it might be achieved intracellularly emerge from the cell culture studies of brucellae, tubercle bacilli and leprosy bacilli described above (Smith & FitzGeorge, 1964; FitzGeorge & Smith, 1966; Brown & Draper, 1970; Armstrong & D'Arcy Hart, 1971; D'Arcy Hart *et al.*, 1972; Frost *et al.*, 1972; D'Arcy Hart & Armstrong, 1974). First, intracellular destruction of avirulent strains and multiplication of virulent strains was slow; days rather than hours being needed for two- to four-fold decrease or increase in populations. Second, there was ample evidence for a range of ability to survive and multiply within a single bacterial population and between strains of differing virulence including examples of near equilibrium between destruction and growth. Thus some members of the virulent 544 strain of *B. abortus* were killed after 5 h within bovine phagocytes and between 5 and 18 h near-equilibrium conditions prevailed before an increase of the survivors occurred; this applied to organisms grown both *in vitro* and *in vivo* (Smith & FitzGeorge, 1964). Several other less virulent strains (27, C102, 28) behaved like 544 during the first 18 h in bovine phagocytes and then remained almost stationary for a further 24 h (Smith & FitzGeorge, 1964). Furthermore, when the killing mechanisms of bovine phagocytes were inhibited by cell wall preparations from the virulent 544 strain (see above), an avirulent 54/0 strain was less easily destroyed but showed no signs of intracellular growth, whereas an attenuated strain (58/20) which normally was also destroyed by phagocytes increased slightly within the cell wall treated phagocytes (Smith & FitzGeorge, 1964). Similarly, a few members of an avirulent strain of *M. tuberculosis* seemed to prevent lysosomal discharge into the vacuoles of macrophages as did the majority of members of a virulent population (D'Arcy Hart & Armstrong, 1974). It would be interesting to measure actual multiplication and death rates in these cell culture studies by Meynell's technique (see above).

If this range of ability for intracellular survival within bacterial populations is coupled with the known heterogeneity of phagocyte function within phagocyte populations of the same type (neutrophils or macrophages) and between phagocyte populations of different types (Coombs & Smith, 1968, 1975; Smith, 1968), there is ample scope for bacterial multiplication almost equalling destruction in some individuals on certain occasions. What is not clear, is the mechanism of the occasional disturbance of the equilibrium which leads to increased shedding of organisms or return of the signs and symptoms of acute disease.

Sequestration in areas of diminished host defence

Bacteria are more likely to persist in those tissues where the host defences are lowered. The kidney is the best example. It is prone to persistent infection by many Gram-negative organisms and staphylococci; this is due to some extent to the hypertonicity of the tissue, which reduces phagocyte mobilization and action, and to the presence of ammonia ion which may prevent the action of complement (Smith, 1968, 1972). Other examples may be the udder and brain.

DEFICIENT CELL WALLS AND MEMBRANES OF BACTERIA *IN VIVO*

The nutritional conditions within host tissues are different from those normally provided for growth of bacteria *in vitro*. Consequently cell wall and capsular components of bacteria grown *in vivo* may not be the same in quality or quantity as those of bacteria grown *in vitro* (Smith, 1964, 1972). Some of the components produced *in vivo* are the aggressins (see above), which contribute to inhibition of host defence and survival within the host. In some instances, however, another change takes place: the cell wall and membrane become more permeable and the former may largely disappear to give the so called L-phase forms. *Bacillus anthracis* purified from infected guinea pig blood by differential centrifugation in saline, became swollen on washing with water and dissolved on adding a trace of ammonium carbonate, in contrast to *B. anthracis* grown in four different media *in vitro* (Smith, Keppie & Stanley, 1953). Hanks (1968) regarded the lack of metabolic activity of *M. tuberculosis* isolated from mice as due to loss of essential constituents through the 'leaky' cell wall during the isolation processes. Moulder (1962) considered 'leakiness' as one property of an intracellular parasite; large chlamydia which developed from small infective chlamydia in the vacuoles of mouse fibroblasts lacked cell wall peptidoglycan and were more permeable (Moulder, 1969). Finally there are the cell wall-deficient L-phase variants of brucellae, mycobacteria, streptococci, gonococci and urinary tract bacteria which are associated with chronicity and latency in many cases and which have been produced in long-term tissue culture (Mattman, 1974; McGhee & Freeman, 1970*b*; Hatten & Sulkin, 1966).

It is tempting to believe that increased permeability and cell wall deficiency contribute to survival within the animal host by allowing freer entry of nutrients. Indeed, this seems to occur for the large chlamydial cells described by Moulder (1969). L-phase variants of bacteria should be able to survive under treatment with antibiotics that act by interfering with cell wall synthesis and many of them are in fact produced by such treatment. Nevertheless, there is no reason to believe that cell wall-deficient bacteria are more able than normal bacteria, with their cell wall aggressins intact, to resist the many and varied host defence mechanisms that have been described. It is possible, for example, that some deficiencies in the cell walls have been produced by incomplete action of host lytic agencies. To sum up, some bacteria *in vivo* are more permeable than those grown *in vitro* and even lack a complete cell wall but the contribution of these phenomena to bacterial survival within animal tissues is not known.

SUMMARY

Most bacterial species found in animals could multiply freely in the physico-chemical environment of the tissues if it were not for the restrictive influence of the many anti-bacterial mechanisms that can be mounted by living animal hosts. Hence, with notable exceptions such as for the anaerobes, the crux of bacterial survival *in vivo* is non-stimulation or impairment of host defence mechanisms rather than finding the minimum physico-chemical conditions for maintenance and growth. A dynamic equilibrium between multiplication of those bacteria which can resist host defences and destruction of those which cannot probably operates for survival of bacterial populations *in vivo*, more often than persistence of members of an original population without multiplication. At any time the steady state can break down, leading to either an increase or decrease of bacterial numbers with consequent elimination or progression of the disease.

Bacterial survival *in vivo* is important during three overlapping stages of disease; first, at initiation of infection on the mucous surfaces; second, immediately following the entry of the tissues by the bacteria; and third, in chronic and carrier states. We have some knowledge of the mechanisms whereby bacteria inhibit host defence mechanisms in the relatively short-term survival needed during the first two stages of acute disease. But the processes operating for long-term survival of vegetative bacteria in chronic and carrier states are still obscure.

REFERENCES

ADLAM, C., PEARCE, J. H. & SMITH, H. (1970). The interaction of staphylococci grown *in vivo* and *in vitro* with polymorphonuclear leucocytes. *Journal of Medical Microbiology*, **3**, 157–163.

AGARWAL, D. S. (1967). Subcutaneous staphylococcal infection in mice. II. The inflammatory response to different strains of staphylococci and micrococci. *British Journal of Experimental Pathology*, **48**, 468–482.

ARMSTRONG, J. A. & D'ARCY HART, P. (1971). Response of cultured macrophages to *Myobacterium tuberculosis*, with observations on fusion of lysosomes with phagosomes. *Journal of Experimental Medicine*, **134**, 713–740.

AYOUB, E. M. & MCCARTY, M. (1968). Intraphagocytic β-N-acetylglucosaminidase. Properties of the enzyme and its activity on group A streptococcal carbohydrate in comparison with a soil bacillus enzyme. *Journal of Experimental Medicine*, **127**, 833–851.

BASKERVILLE, A., DOW, C., CURRAN, W. L. & HANNA, J. (1972). Ultrastructure of phagocytosis of *Salmonella cholerae-suis* by pulmonary macrophages *in vivo*. *British Journal of Experimental Pathology*, **53**, 641–647.

BASKETT, R. C. & HENTGES, D. J. (1973). *Shigella flexneri* inhibition by acetic acid. *Infection and Immunity*, **8**, 91–97.

BECKERDITE, S., MOONEY, C., WEISS, J., FRANSON, R. & ELSBACH, P. (1974). Early and discrete changes in permeability of *Escherichia coli* and certain other Gram-negative bacteria during killing by granulocytes. *Journal of Experimental Medicine*, **140**, 396–409.

BELLAMY, J. E. C. & NEILSEN, N. O. (1974). Immune-mediated emigration of neutrophils into the lumen of the small intestine. *Infection and Immunity*, **9**, 615–619.

BERTSCHINGER, H. U., MOON, H. W. & WHIPP, S. C. (1972*a*). Association of *Escherichia coli* with the small intestinal epithelium. I. Comparison of enteropathogenic and non enteropathogenic porcine strains in pigs. *Infection and Immunity*, **5**, 595–605.

BERTSCHINGER, H. U., MOON, H. W. & WHIPP, S. C. (1972*b*). Association of *Escherichia coli* with the small intestinal epithelium. II. Variation in association index and relationship between association index and enterosorption in pigs. *Infection and Immunity*, **5**, 606–611.

BRAUN, W. & SIVA SANKAR, D. V. (1960). Biochemical aspects of microbial pathogenicity. *Annals of the New York Academy of Sciences*, **88**, 1021–1318.

BROWN, C. A. & DRAPER, P. (1970). An electron microscope study of rat fibroblasts infected with *Mycobacterium lepraemurium*. *Journal of Pathology*, **102**, 21–26.

BULLEN, J. J., ROGERS, H. J. & GRIFFITHS, E. (1972). Iron binding proteins and infection. *British Journal of Haematology*, **23**, 389–92.

BULLEN, J. J., WARD, C. G. & WALLIS, S. N. (1974). Virulence and the role of iron in *Pseudomonas aeruginosa* infection. *Infection and Immunity*, **10**, 443–450.

BURROWS, T. W. (1960). Biochemical properties of virulent and avirulent strains of bacteria: *Salmonella typhosa* and *Pasteurella pestis*. *Annals of the New York Academy of Sciences*, **88**, 1125–1135.

BURROWS, T. W. & GILLETT, W. A. (1971). Host specificity of Brazilian strains of *Pasteurella pestis*. *Nature, London*, **229**, 51–52.

CASH, R. A., MUSIC, S. I., LIBONATI, J. P., SNYDER, M. J., WENZEL, R. P. & HORNICK, R. B. (1974). Response of man to infection with *Vibrio cholerae*. I. Clinical, serologic, and bacteriologic responses to a known inoculum. *Journal of Infectious Diseases*, **129**, 45–52.

CLARK, W. R., MCLAUGHLIN, J. & WEBSTER, M. E. (1958). An amino-hexuronic acid as the principal hydrolytic component of the VI antigen. *Journal of Biological Chemistry*, **230**, 81–89.

COHN, Z. A. (1963). The fate of bacteria within phagocyte cells. I. The degradation of isotopically labelled bacteria by polymorphonuclear leucocytes and macrophages. *Journal of Experimental Medicine*, **117**, 27–42.

COOMBS, R. R. A. & SMITH, H. (1968, 1975). The allergic response and immunity. In *Clinical Aspects of Immunology*, ed. P. G. H. Gell & R. R. A. Coombs, 2nd edn, pp. 423–456, 3rd edn, pp. 473–505. Oxford: Blackwell

CROWE, C. C., SANDERS, W. E. & LANGLEY, S. (1973). Bacterial interference. II. Role of the normal throat flora in prevention of colonization by Group A streptococcus. *Journal of Infectious Diseases*, **128**, 527–532.

CYBULSKA, J. & JELJASZEWICZ, J. (1966). Bacteriostatic activity of serum against staphylococci. *Journal of Bacteriology*, **91**, 953–962.

D'ARCY HART, P. & ARMSTRONG, J. A. (1974). Strain, virulence and the lysosomal response in macrophages infected with *Mycobacterium tuberculosis*. *Infection and Immunity*, **10**, 742–746.

D'ARCY HART, P., ARMSTRONG, J. A., BROWN, C. A. & DRAPER, P. (1972). Ultrastructural study of the behaviour of macrophages toward parasitic mycobacteria. *Infection and Immunity*, **5**, 803–807.

DAVIS, C. P., MCALLISTER, J. S. & SAVAGE, D. C. (1973). Microbial colonization of the intestinal epithelium in suckling mice. *Infection and Immunity*, **7**, 666–672.

DAVIS, C. P. & SAVAGE, D. C. (1974). Habitat, succession, attachment and morphology of segmented, filamentous microbes indigenous to the murine gastrointestinal tract. *Infection and Immunity*, **10**, 948–956.

DRAPER, P. & REES, R. J. W. (1970). Electron-transparent zone of mycobacteria may be a defence system. *Nature, London*, **228**, 860–861.

DRAPER, P. & REES, R. J. W. (1973). The nature of the electron-transparent zone that surrounds *Myobacterium lepraemurium* inside host cells. *Journal of General Microbiology*, **77**, 79–87.

DREES, D. T. & WEXLER, G. L. (1970). Enteric colibacillosis in gnotobiotic swine: a fluorescence microscopic study. *American Journal of Veterinary Research*, **31**, 1147–57.

DUBOS, R. J. & HIRSCH, J. G. (1965). *Bacterial and Mycotic Infections of Man*, 4th edn. Philadelphia: Lippincott.

DUCLUZEAU, R. & RAIBAUD, P. (1974). Interaction between *Escherichia coli* and *Shigella flexneri* in the digestive tract of 'gnotobiotic' mice *Infection and Immunity*, **9**, 730–733.

EASMON, C. S. F., HAMILTON, I. & GLYNN, A. A. (1973). Mode of action of a staphylococcal anti-inflammation factor. *British Journal of Experimental Pathology*, **54**, 638–645.

ELLEN, R. P. & GIBBONS, R. J. (1972). M protein-associated adherence of *Streptococcus pyogenes* to epithelial surfaces; prerequisite for virulence. *Infection and Immunity*, **5**, 826–830.

ELLWOOD, D. C., KEPPIE, J. & SMITH, H. (1967). The chemical basis of the virulence of *Brucella abortus*. VIII. The identity of purified immunogenic material from culture filtrate and from the cell-wall of *Brucella abortus* grown *in vitro*. *British Journal of Experimental Pathology*, **48**, 28–39.

FANCONI, G. & WALLGREN, A. (1952). In *Textbook of Paediatrics*, p. 445. London: Heinemann.

FINGER, H., FRESENIUS, H. & ANGERER, M. (1971). Bacterial endotoxins as immunosuppressive agents. *Experientia*, **27**, 456–458.

FITZGEORGE, R. B. & SMITH, H. (1966). The chemical basis of the virulence of *Brucella abortus*. VII. The production *in vitro* of organisms with an enhanced capacity to survive intracellularly in bovine phagocytes. *British Journal of Experimental Pathology*, **47**, 558–562.

FITZGEORGE, R. B., SOLOTOROVSKY, M. & SMITH, H. (1967). The behaviour of *Brucella abortus* within macrophages separated from the blood of normal and immune cattle by adherence to glass. *British Journal of Experimental Pathology*, **48**, 522–28.

FOX, E. N. (1974). M proteins of group A streptococci. *Bacteriological Reviews*, **38**, 57–86.

FRIEDBERG, D. & SHILO, M. (1970). Role of cell wall structure of *Salmonella* in the interaction with phagocytes. *Infection and Immunity*, **2**, 279–285.

FRIEDMAN, H. & CHAKRABARTY, A. K. (1971). L-asparaginase induced inhibition of transplantation immunity; prolongation of skin allografts in enzyme-treated mice. *Transplantation Proceedings*, **3**, 826–830.

Frost, A. J., Smith, H., Witt, K. & Keppie, J. (1972). The chemical basis of the virulence of *Brucella abortus*. X. A surface virulence factor which facilitates intracellular growth of *Brucella abortus* in bovine phagocytes. *British Journal of Experimental Pathology*, **53**, 587–596.

Gibbons, R. J. & van Houte, J. (1971). Selective bacterial adherence to oral epithelial surfaces and its role as an ecological determinant. *Infection and Immunity*, **3**, 567–573.

Gladstone, G. P., Walton, E. & Kay, U. (1974). The effect of cultural conditions on the susceptibility of staphylococci to killing by the cationic proteins from rabbit polymorphonuclear leucocytes. *British Journal of Experimental Pathology*, **55**, 427–447.

Glynn, A. A. (1972). Bacterial factors inhibiting host defence mechanisms. *Symposia of the Society for General Microbiology*, **22**, 75–112.

Gotschlich, E. C., Goldschneider, I. & Artenstein, M. S. (1969). Human immunity to the meningococcus. IV. Immunogenicity of group A and group C meningococcal polysaccharides in human volunteers. *Journal of Experimental Medicine*, **129**, 1367–1384.

Grimble, A. & Armitage, L. R. G. (1974). Surface structures of the gonococcus. *British Journal of Venereal Diseases*, **50**, 354–359.

Guze, L. B., Hubert, E. G. & Kalmanson, G. M. (1970). Pyelonephritis. XI. Effect of growth phase of *Streptococcus faecalis* on serum susceptibility and virulence. *Infection and Immunity*, **1**, 532–537.

Hanks, J. H. (1968). Metabolism of *in vivo* grown mycobacteria. *Annals of the New York Academy of Sciences*, **154**, 68–78.

Hanson, L. A. (1973). Host–parasite relationships in urinary tract infections. *Journal of Infectious Diseases*, **127**, 726–730.

Hatten, B. A. & Sulkin, S. E. (1966). Intracellular production of *Brucella* L forms. I. Recovery of L-forms from tissue culture cells infected with *Brucella abortus*. *Journal of Bacteriology*, **91**, 285–296.

Hentges, D. J. & Maier, B. R. (1970). Inhibition of *Shigella flexneri* by the normal intestinal flora. III. Interactions with *Bacteroides fragilis* strains *in vitro*. *Infection and Immunity*, **2**, 364–370.

Hibbitt, K. G. & Benians, M. (1971). Some effects *in vitro* of the teat canal and effects *in vitro* of cationic proteins on staphylococci. *Journal of General Microbiology*, **68**, 123–128.

Hill, M. J. (1968). A staphylococcal aggressin. *Journal of Medical Microbiology*, **1**, 33–43.

Hill, M. J. (1969). Protection of mice against infection by *Staphylococcus aureus*. *Journal of Medical Microbiology*, **2**, 1–7.

Hirsch, J. G. (1958). Bactericidal action of histone. *Journal of Experimental Medicine*, **108**, 925–944.

Hirsch, J. G. (1972). The phagocytic defence system. *Symposia of the Society for General Microbiology*, **22**, 59–74.

Hobson, P. N. (1974). Microorganisms in anaerobic environments. *Proceedings of the Society for General Microbiology*, **2**, 1.

Imaeda, T. (1965). Electron microscopy. Approach to leprosy research. *International Journal of Leprosy*, **33**, 669–688.

Imaeda, T., Kanetsuna, F., Rieber, M., Galindo, B. & Cesari, I. M. (1969). Ultrastructural characteristics of mycobacterial growth. *Journal of Medical Microbiology*, **2**, 181–186.

Jansen, W. A. & Surgalla, M. J. (1969). Plague bacillus: survival within host phagocytes. *Science, Washington*, **163**, 950–952.

Jones, G. W. & Rutter, J. M. (1972). Role of the K88 antigen in the pathogenesis of neonatal diarrhea caused by *Escherichia coli* in piglets. *Infection and Immunity*, **6**, 918–927.

KAIJSER, B. (1973). Immunology of *Escherichia coli*: K antigen and its relation to urinary-tract infection. *Journal of Infectious Diseases*, **127**, 670–677.

KEPPIE, J., HARRIS-SMITH, P. W. & SMITH, H. (1963). The chemical basis of the virulence of *Bacillus anthracis*. IX. Its aggressins and their mode of action. *British Journal of Experimental Pathology*, **44**, 446–453.

KIRPATOVSKI, I. D. & STANISLAVSKI, E. S. (1971). Immunosuppressive effects of cell-free extracts from *Escherichia coli*. *Transplantation Proceedings*, **3**, 831–834.

KOBAYASHI, F. (1971). Experimental infection with *Pseudomonas aeruginosa* in mice. II. The fate of highly and low virulent strains in the peritoneal cavity and organs of mice. *Japanese Journal of Microbiology*, **15** (4), 301–307.

KOLAWOLE, D. O., SCRAGG, M. A. & PEARCE, J. H. (1975). Mechanisms of resistance of staphylococci grown *in vivo* to polymorph bactericidins. *Proceedings of the Society for General Microbiology*, **2**, 73.

LAHAV, M., NE'EMAN, H., ADLER, E. & GINSBURG, I. (1974). Effect of leukocyte hydrolases on bacteria. I. Degradation of ^{14}C-labelled *Streptococcus* and *Staphylococcus* by leucocyte lysates *in vitro*. *Journal of Infectious Diseases*, **1129**, 528–537.

LEVISON, M. E. (1973). Effect of colon flora and short-chain fatty acids on growth *in vitro* of *Pseudomonas aeruginosa* and Enterobacteriaceae. *Infection and Immunity*, **8**, 30–35.

LOWRIE, D. B., JACKETT, P. S. & RATCLIFFE, N. A. (1975). *Myobacterium microti* may protect itself from intracellular destruction cyclic AMP into phagosomes. *Nature, London*, **254**, 600–602.

LUDERITZ, O., STAUB, A. M. & WESTPHAL, O. (1966). Immuno-chemistry of O and R antigens of *Salmonella* and related Enterobacteriaceae. *Bacteriological Reviews*, **30**, 192–255.

MACKANESS, G. B. (1960). The phagocytosis and inactivation of staphylococci by macrophages of normal rabbits. *Journal of Experimental Medicine*, **112**, 35–53.

MAIER, B. R. & HENGTES, D. J. (1972). Experimental *Shigella* infections in laboratory animals. I. Antagonism by human normal flora components in gnotobiotic mice. *Infection and Immunity*, **6**, 168–173.

MAKELA, P. A., VALTONEN, V. V. & VALTONEN, M. (1973). Factors in the virulence of Salmonella. *Journal of Infectious Diseases*, **128**, 581–586.

MALAKIAN, A. & SCHWAB, J. (1967). Immunosuppressant from Group A streptococci. *Science, Washington*, **159**, 880–888.

MATHESON, A. & DONALDSON, D. M. (1970). Effect of β-lysin on isolated cell walls and protoplasts of *Bacillus subtilis*. *Journal of Bacteriology*, **101**, 314–317.

MATTMAN, L. (1974). *Cell-Wall Deficient Forms*. Cleveland, Ohio: CRC Press.

MAW, J. & MEYNELL, G. G. (1968). The true division and death rates of *Salmonella typhimurium* in the mouse spleen determined with superinfected phage P_{22}. *British Journal of Experimental Pathology*, **49**, 597–613.

MCGHEE, J. R. & FREEMAN, B. A. (1970*a*). Effect of lysosomal enzymes on *Brucella*. *Journal of the Reticuloendothelial Society*, **8**, 208–219.

MCGHEE, J. R. & FREEMAN, B. A. (1970*b*). Osmotically sensitive *Brucella* in infected normal and immune macrophages. *Infection and Immunity*, **1**, 146–150.

MEDEARIS, D. N. JR, CAMITTA, B. M. & HEATH, E. C. (1968). Cell wall composition and virulence in *Escherichia coli*. *Journal of Experimental Medicine*, **128**, 399–414.

MEYNELL, G. G. (1959). Use of superinfected phage for estimating the division rate of lysogenic bacteria in infected animals. *Journal of General Microbiology*, **21**, 421–437.

MEYNELL, G. G. & SUBBAIAH, T. V. (1963). Antibacterial mechanisms of the mouse gut. I. Kinetics of infection by *Salmonella typhimurium* in normal and streptomycin-treated mice studied with abortive transductants. *British Journal of Experimental Pathology*, **44**, 197–208.

MILES, A. A. (1962). Mechanisms of immunity in bacterial infections. *Recent Progress in Microbiology*, **7**, 399–407.

MILES, A. A., MILES, E. M. & BURKE, J. (1957). The value and duration of defence reactions of the skin to the primary lodgement of bacteria. *British Journal of Experimental Pathology*, **38**, 79–96.

MILLER, T. E., SMITH, J. W. & SANFORD, J. P. (1971). Antibody synthesis in kidney, spleen and lymph nodes in acute and healed focal pyelonephritis. *British Journal of Experimental Pathology*, **52**, 678–683.

MOULDER, J. W. (1962). *The Biochemistry of Intracellular Parasitism*. Chicago: University of Chicago Press.

MOULDER, J. W. (1969). A model for studying the biology of parasitism. *Chlamydia psittaci* and mouse fibroblasts (L cells). *BioScience*, **19**, 875–881.

MUKASA, H. & SLADE, H. D. (1973). Mechanisms of adherence of *Streptococcus mutans* to smooth surfaces. I. Roles of insoluble dextran-levan synthetase enzymes and cell wall polysaccharide antigen in plaque formation. *Infection and Immunity*, **8**, 555–562.

MUKASA, H. & SLADE, H. D. (1974). Mechanisms of adherence of *Streptococcus mutans* to smooth surfaces. II. Nature of the binding site and the absorption of the dextran-lavan synthetase enzymes on the cell-wall surface of the Streptococcus. *Infection and Immunity*, **9**, 419–429.

MUSCHEL, L. H. (1963). Activity of the antibody-complement system and lysozyme against Gram-negative organisms. *Proceedings of the Federation of American Societies of Experimental Biology*, **22**, 673–678.

MYRVIK, Q. N. & LEAKE, E. S. (1960). Studies on antibacterial factors in mammalian tissues and fluids. IV. Demonstration of two non-dialyzable components in the serum bactericidin system for *Bacillus subtilis*. *Journal of Immunology*, **84**, 247–250.

NAKANO, M. & SAITO, K. (1968). The chemical compositions in the cell wall of *Salmonella typhimurium* affecting the clearance rate in mouse. *Japanese Journal of Microbiology*, **12**, 471–478.

NAKANO, M. & SAITO, K. (1969). Chemical components in the cell wall of *Salmonella typhimurium* affecting its virulence and immunogenicity in mice. *Nature, London*, **222**, 1085–1086.

NOVOTNY, P., SHORT, J. A. & WALKER, P. D. (1975). An electron microscopy comparison between cultured and naturally occurring gonococci and host cell gonococcal interactions in urethral pus. *Journal of Medical Microbiology*, **8**, 413–427.

ØRSTAVIK, D., KRAUS, F. W. & HENSHAW, L. C. (1974). *In vitro* attachment of streptococci to the tooth surface. *Infection and Immunity*, **9**, 794–800.

PEARCE, J. H., SCRAGG, M. A. & KOLAWOLE, D. O. (1975). Uptake and phagocytic killing of staphylococci grown *in vivo* and *in vitro*. *Proceedings of the Society for General Microbiology*, **2**, 73.

POLK, H. C. & MILES, A. A. (1973). The decisive period in the primary infection of muscle by *Escherichia coli*. *British Journal of Experimental Pathology*, **54**, 99–109.

PUNSALANG, A. P. JR & SAWYER, W. D. (1973). Role of pili in the virulence of *Neisseria gonorrheae*. *Infection and Immunity*, **8**, 255–263.

ROANTREE, R. J. (1967). Salmonella O antigens and virulence. *Annual Reviews of Microbiology*, **21**, 443–466.

ROBERTS, D. S. (1967). The pathogenic synergy of *Fusiformis necrophorus* and *Corynebacterium pyogenes*. II. The response of *F. necrophorus* to a filterable product of *C. pyogenes*. *British Journal of Experimental Pathology*, **48**, 674–679.

ROBERTS, R. B. (1970). The relationship between group A and group C meningococcal polysaccharides and serum opsonins in man. *Journal of Experimental Medicine*, **131**, 499–513.

RUTTER, J. M. & JONES, G. W. (1973). Protection against enteric disease caused by *Escherichia coli* – a model for vaccination with a virulence determinant? *Nature, London*, **242**, 531–532.

SAVAGE, D. C. (1972). Survival on mucosal epithelia, epithelial penetration and growth in tissues of pathogenic bacteria. *Symposia of the Society for General Microbiology*, **22**, 25–57.

SBARRA, A. J., PAUL, B. B., JACOBS, A. A., STRAUSS, R. R. & MITCHELL, G. W. JR (1972). Role of the phagocyte in host–parasite interactions. XXXVIII. Metabolic activities of the phagocyte as related to antimicrobial action. *Journal of Reticuloendothelial Society*, **12**, 109–126.

SCHULTZ, D. R. & MILLER, K. D. (1974). Elastase of *Pseudomonas aeruginosa*. Inactivation of complement components and complement-derived chemotactic and phagocytic factors. *Infection and Immunity*, **10**, 128–135.

SELLWOOD, R., GIBBONS, R. A., JONES, G. W. & RUTTER, J. M. (1974). A possible basis for the breeding of pigs relatively resistant to neonatal diarrhoea. *The Veterinary Record*, **95**, 574–575.

SENSAKOVIC, J. W. & BARTELL, P. F. (1974). The slime of *Pseudomonas aeruginosa*. Biological characterization and possible role in experimental infection. *Journal of Infectious Diseases*, **129**, 101–109.

SHAYEGANI, M., HITATSUNE, K. & MUDD, S. (1970). Cell wall component which affects the ability of serum to promote phagocytosis and killing of *Staphylococcus aureus*. *Infection and Immunity*, **2**, 750–756.

SHULTZ, L. D. & WILDER, M. S. (1971). Cytotoxicity of rabbit blood for *Listeria monocytogenes*. *Infection and Immunity*, **4**, 703–708.

SHULTZ, L. D. & WILDER, M. S. (1973). Fate of *Listeria monocytogenes* in normal rabbit serum. *Infection and Immunity*, **7**, 289–297.

SMITH, H. (1958). The use of bacteria *in vivo* for studies on the basis of their pathogenicity. *Annual Reviews of Microbiology*, **12**, 77–102.

SMITH, H. (1964). Microbial behaviour in natural and artificial environments. *Symposia of the Society for General Microbiology*, **14**, 1–29.

SMITH, H. (1968). Biochemical challenge of microbial pathogenicity. *Bacteriological Reviews*, **32**, 164–184.

SMITH, H. (1972). The little known determinants of microbial pathogenicity. *Symposia of the Society for General Microbiology*, **22**, 1–24.

SMITH, H. & FITZGEORGE, R. B. (1964). The chemical basis of the virulence of *Brucella abortus*. V. The basis of intracellular survival and growth in bovine phagocytes. *British Journal of Experimental Pathology*, **45**, 174–186.

SMITH, H., KEPPIE, J. & STANLEY, J. L. (1953). The chemical basis of the virulence of *Bacillus anthracis*. I. Properties of bacteria grown *in vivo* and preparation of extracts. *British Journal of Experimental Pathology*, **34**, 477–485.

SMITH, H. W. & LINGGOOD, M. A. (1971). Observations on the pathogenic properties of the K88, HLy and Ent plasmids of *Escherichia coli* with particular reference to porcine diarrhoea. *Journal of Medical Microbiology*, **4**, 467–485.

SOLOTOROVSKY, M. & TEWARI, R. P. (1973). Interaction of macrophages with facultative intracellular parasites – *Brucella* and *Salmonella*. In *Activation of Macrophages*, Workshop Conferences Hoechst, vol. II, ed. W. H. Wagner & H. Hahn, pp. 238–248. Amsterdam: Excerpta Medica.

STALEY, T. E., JONES, E. W. & CORLEY, L. D. (1969). Attachment and penetration of *Escherichia coli* into intestinal epithelium of the ileum in new-born pigs. *The American Journal of Pathology*, **56**, 371–392.

STALHEIM, O. H. V. (1971). Virulent and avirulent leptospires: biochemical activities and survival in blood. *American Journal of Veterinary Research*, **32**, 843–849.

STENDAHL, O. & EDEBO, L. (1972). Phagocytosis of mutants of *Salmonella typhimurium* by rabbit polymorphonuclear cells. *Acta pathologica microbiologica scandinavia*, **80**B, 481–485.

TANNOCK, G. W. & SAVAGE, D. C. (1974). Influences of dietary and environmental stress on microbial populations in the murine gastrointestinal tract. *Infection and Immunity*, **9**, 591–598.

TANZER, J. M., FREEDMAN, M. L., FITZGERALD, R. J. & LARSON, R. H. (1974). Diminished virulence of glucan synthesis-defective mutants of *Streptococcus mutans*. *Infection and Immunity*, **10**, 197–203.

TAYLOR-ROBINSON, D., WHYTOCK, S., GREEN, C. J. & CARNEY, F. E. JR (1974). Effect of *Neisseria gonorrhoeae* on human and rabbits oviducts. *British Journal of Venereal Diseases*, **50**, 279–288.

TURK, J. L. (1970). Immunological aspects of clinical leprosy. *Proceedings of the Royal Society of Medicine*, **63**, 1053–1056.

VALTONEN, V. V. & MAKELA, P. H. (1971). The effect of lipopolysaccharide modifications – antigenic factors 1, 5, 12_2 and 27 on the virulence of *Salmonella* strains for mice. *Journal of General Microbiology*, **69**, 107–115.

VEALE, D. R., SMITH, H. & WITT, K. (1975). Penetration of penicillin into human phagocytes containing gonococci. *Lancet*, i, 306.

VEALE, D. R., WITT, K. & SMITH, H. (1975). Penetration of penicillin into human phagocytes containing gonococci. Intracellular survival at minimal concentrations of antibiotic. *Proceedings of the Society for General Microbiology*, **2**, 75.

WARD, M. E. & WATT, P. J. (1972). Adherence of *Neisseria gonorrhoeae* to urethral mucosal cells; an electron-microscopic study of human gonorrhoeae. *Journal of Infectious Diseases*, **126**, 601–605.

WARD, M. E., WATT, P. J. & GLYNN, A. A. (1970). Gonococci in urethral exudates possess a virulence factor lost on subculture. *Nature, London*, **227**, 382–384.

WARD, M. E., WATT, P. J. & ROBERTSON, J. N. (1974). The human fallopian tube; a laboratory model for gonococcal infection. *Journal of Infectious Diseases*, **129**, 650–659.

WATT, P. J., GLYNN, A. A. & WARD, M. E. (1972). Maintenance of virulent gonococci in laboratory culture. *Nature, London*, **236**, 186.

WEINBERG, E. D. (1971). Roles of iron in host–parasite interactions. *Journal of Infectious Diseases*, **124**, 401–410.

WEKSLER, B. B. & HILL, M. J. (1969). Inhibition of leukocyte migration by a staphylococcal factor. *Journal of Bacteriology*, **98**, 1030–1035.

WILLET, H. P. & THACORE, H. (1967). Formation of spheroplasts of *Mycobacterium tuberculosis* by lysozome in combination with certain enzymes of rabbit peritoneal monocytes. *Canadian Journal of Microbiology*, **13**, 481–488.

WILSON, M. R. & HOHMANN, A. W. (1974). Immunity to *Escherichia coli* in pigs: adhesion of enteropathogenic *Escherichia coli* to isolated intestinal epithelial cells. *Infection and Immunity*, **10**, 776–782.

WITT, K., VEALE, D. R. & SMITH, H. (1975). Resistance to ingestion and digestion of *Neisseria gonorrhoea* by phagocytes of human buffy coat. *Journal of Medical Microbiology* (in press).

YEE, R. B. & BUFFENMYER, C. L. (1970). Infection of cultured mouse macrophages with *Shigella flexneri*. *Infection and Immunity*, **1**, 459–463.

SURVIVAL OF VEGETATIVE MICROBES IN SOIL

T. R. G. GRAY
Hartley Botanical Laboratories, University of Liverpool,
PO Box 147, *Liverpool L*69 3*BX*, *Merseyside*

INTRODUCTION

Micro-organisms have developed a range of survival mechanisms enabling them to persist in soil. Fungi, actinomycetes and a few bacteria produce resistant spores, but there is a large number of organisms which do not form spores and yet can be isolated from environments which are demonstrably unfavourable for growth. These organisms persist in soil as vegetative cells, often in a state of lowered metabolic activity, using up their energy reserves slowly. If this were not the case, then the occurrence of such organisms would be transient and resemble that of some animal and plant pathogens which disappear rapidly in soil.

The range of resting structures formed by soil micro-organisms has been reviewed in an earlier symposium of this society (Gray & Williams, 1971), together with evidence suggesting that micro-organisms grow rather slowly in soil. In this paper I shall concentrate on the nature of the environmental stresses which have to be withstood and the nature of survival mechanisms. Two general types of survival will be considered: firstly, the maintenance of a growing population of organisms in the presence of organisms better adapted for growth, as exemplified by the concept of competitive saprophytic ability (Garrett, 1950); and secondly, the production of vegetative organisms able to survive during periods of no growth. Finally, I shall consider the difficulty associated with determining which survival mechanism is operating and the difficulty of measuring the viability of a natural population, both of which prevent a proper assessment of survival.

NATURE OF STRESSES IN THE NATURAL ENVIRONMENT

Distribution of stress

The variability of the soil environment has been enshrined in the use of the terms micro-habitat and molecular environment. In general, micro-habitats represent small volumes of soil which are favourable for growth, the size of the volume depending on the sensitivity of the microbe to environmental change, the size of the organism and the steepness of the gradient of environmental conditions. By implication, the remaining volume of the soil is unfavourable for growth. Unfortunately, sufficiently sensitive probes for measuring environmental gradients in soil have not been developed and even if such probes were available, it is likely that their insertion into soil would alter the gradient.

The major environmental stresses that occur in soil include unavailability and low rate of supply of nutrients, unavailability of water or oxygen, development of acidity and toxic compounds. The occurrence of these stresses will now be considered.

Unavailability and low rate of supply of nutrients

The amount of dry matter produced by plants is approximately 1 kg m^{-2} p.a., and of this about 93 per cent is decomposed by microbes in estuarine muds, 54 per cent in grazed meadows, 30 per cent in temperate beech woods and 33 per cent in tropical rain forests (Macfadyen, 1970). The remaining material is utilized by grazing herbivores or soil animals. Recent estimates of the partition of this material between fungi and bacteria show that about one-fifth of it is used by bacteria and four-fifths by fungi in an oak forest soil (Anderson & Domsch, 1975). If this figure is of general application and the other assumptions made by Gray, Hissett & Duxbury (1973) are correct, the mean generation time for bacteria in a deciduous woodland soil would be about 20 days. Shields *et al.* (1973) suggested that the mean generation time for micro-organisms in prairie soils is also low and they calculated that the maintenance coefficient for soil bacteria growing *in situ* might be about 0.002 h^{-1}.

The implication of these calculations is that either the growth of soil organisms is continuous and very slow, or that it occurs in short but rapid bursts, interspersed with long periods of inactivity. The former situation can give rise to partly dead but nonetheless surviving populations (see Dawes, this volume). The latter situation, however, is the more likely and is supported by work on the growth of *Bacillus subtilis* on microscope slides buried in soil (Siala & Gray, 1974). Growth, as judged by increases in population numbers, occurred for only about one week and during this time the generation time was about 9.5 days. In the subsequent eight weeks, the population declined slowly and about half the vegetative propagules were transformed into spores while the remainder disappeared. Siala & Gray (1974) showed also that

Table 1. *Survival of* Arthrobacter globiformis *in acidic buffer at pH* 4.1. (*Data of B. M. Luscombe*)

Limiting nutrient	Cultural conditions before transfer to 0.075 M citrate-phosphate buffer: Temperature (°C)	Dilution rate (h^{-1})	Morphology	Percentage of survivors after different time intervals (min): 60	90	120	150	180	240
Carbon	25	0.05	Coccus	60	32	21	12	6.3	0.83
	25	0.3	Rod	30	14	6.1	2.2	0.81	0.044
	10	0.01	Coccus	38	16	5.1	2.3	0.71	0.19
	10	0.05	Rod	34	25	13	4.3	1.4	0.59
Nitrogen	25	0.05	Coccus	53	46	34	20	14	8.2
	25	0.3	Rod	35	20	8.2	6.9	2	0.94
	10	0.01	Coccus	40	29	15	9.0	6.1	2.5
	10	0.05	Rod	31	27	14	6.9	1.6	0.58

growth only commenced in an acid soil when the environment had been altered by the growth of fungi. This dependence upon fungi has been noted for other soil organisms which do not form spores, e.g. arthrobacters (Lowe & Gray, 1973*a*), and it seems likely, therefore, that the following sequence of events takes place in soil.

1. Addition of plant material, mostly in polymerized form and often in an acidic environment.
2. Priming of plant material by soil animals to produce large surface area.
3. Breakdown of plant material initiated by fungi and a few bacteria.
4. Breakdown of fungal products and some plant material by bacteria.
5. Period of inactivity.

The period of rapid growth may last only a few weeks whereas the period of inactivity or slow growth, which may be long, will be determined by the survival capacity of the organism in starvation conditions. This capacity will be influenced by other unfavourable conditions for it has been shown that, while 15 per cent of a population of *Arthrobacter globiformis* could survive in neutral, nutrient-free solutions for between 50 and 70 days, only 0.1–10 per cent survived for 4 h in a similar solution at pH 4.1 (Table 1).

So far, I have assumed that most of the organic matter reaching soil is readily available to micro-organisms. However, Clark & Paul (1970) have recognized three fractions of the soil organic matter based on availability. These fractions consist of relatively readily available material, e.g. decomposing plant residues which turn over at least once every few years, relatively stable products of microbial metabolism with a half-life of 5–25 years and very resistant organic matter, mainly humic in nature, ranging in age from 250–2500 years. This last material is synthesized from degradation products of lignin, cellulose and microbial metabolites and among the reasons put forward to explain its resistance are its polyphenolic structure, the adsorption of the carbonaceous part of the molecule by silicates, the formation of complexes with toxic trivalent cations and the irregular cross-linking within the molecule by a variety of covalent bonds. Furthermore, the stability of humic components may influence the stability of other more inherently decomposable materials such as soil polysaccharides which may complex with them (Cheshire, Greaves & Mundie, 1974). The presence of this recalcitrant material imposes a further stress on those organisms unable to utilize it and, despite the large number of investigations that have been made on the degradation of humus, it is surprising how few soil organisms are able to use it *in vitro*.

Unavailability of water

As a soil dries out, many microbes are killed by desiccation or survive as resting propagules. These effects can be interpreted in terms of the water potential of a soil, the sum of the osmotic and matric potentials (usually expressed in bars where 1 bar = 10^6 dyne cm^{-2} = 1022 cm water = pF 3.0) (Griffin, 1972). The water potential of the soil determines the degree of difficulty a microbe will experience in extracting water from the soil. Figure 1 shows the relationship between water potential and soil moisture content for several soil types, and indicates that this relationship is different for soils undergoing drying or wetting. Griffin (1972) has summarized the main effects changes in water potential are likely to have on microbial activity. At potentials of less than -145 bar, fungi such as aspergilli and penicillia predominate, but the majority of fungi are most frequently isolated when the potential is greater than -40 bar. The lower limit for bacterial activity in soil is probably about -80 bar but activity may cease at higher water potentials, even at -5 bar (Cook & Papendick, 1970) because of other effects such as decreased mobility of the organisms. Vegetative streptomycetes have a lower limit for activity of about -55 bar, but they can survive as spores in much drier soils. Therefore bacteria are likely to suffer from desiccation effects more than fungi and need to be able to survive periods of drought. Bacteria will almost certainly be inactive when a clay soil contains less than 12 per cent, a loam 7 per cent and

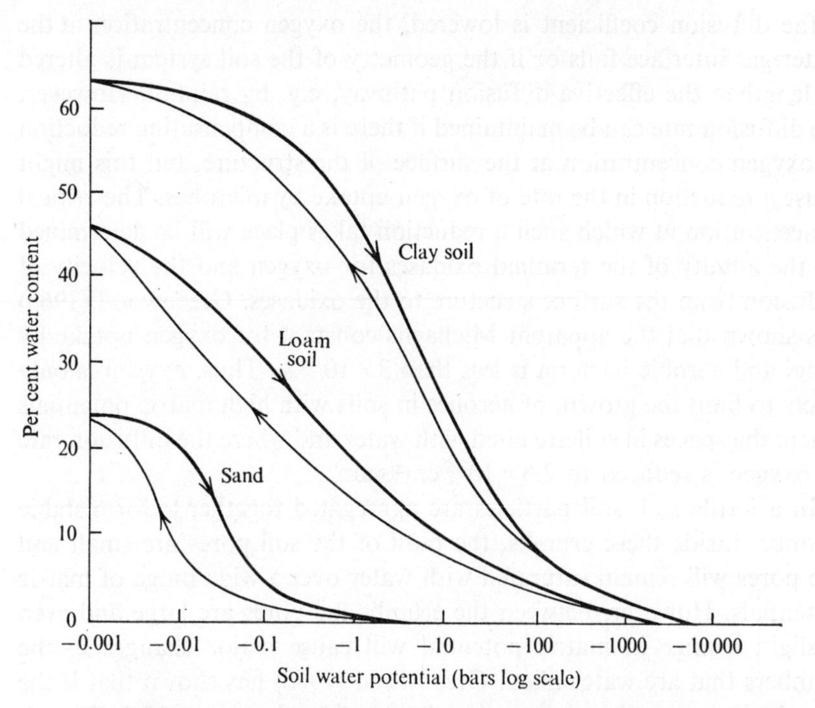

Fig. 1. Moisture characteristic curves for three soils. The drying (upper) and wetting (lower) boundary curves are both shown. (After Griffin, 1972.)

sandy soil 1 per cent water, and may even cease activity at 23 per cent, 12 per cent and 2 per cent, respectively, due to lack of mobility and decreased rates of solute diffusion. Fungal activity will almost certainly have ceased when the water content drops to 10 per cent, 5 per cent and 8 per cent, respectively.

The differential effect that water potential exerts on soil organisms can cause complications when examining their survival. Thus Bruehl & Lai (1968) showed that hyphae of *Cephalosporium gramineum* at 15 °C survived best at −270 bar and worst at −145 and −200 bar. This was explained by postulating that, at the last two potentials, penicillia were still active and decreased the viability of the test fungus. Thus the results of experiments performed on the effect of water potential on microbes in pure culture should be applied cautiously to natural environments.

Variation in oxygen concentration

Griffin (1972) has also reviewed the factors affecting oxygen concentrations in soil and the distribution of conditions inimical to microbial growth. He points out that the rate of oxygen diffusion in soil will fall if the diffusion coefficient is lowered, the oxygen concentration at the water/gas interface falls or if the geometry of the soil system is altered to lengthen the effective diffusion pathway, e.g. by rainfall. However, the diffusion rate can be maintained if there is a compensating reduction in oxygen concentration at the surface of the structure, but this might cause a reduction in the rate of oxygen uptake by microbes. The critical concentration at which such a reduction takes place will be determined by the affinity of the terminal oxidases for oxygen and the velocity of diffusion from the surface structure to the oxidases. Greenwood (1968) has shown that the apparent Michaelis constant for oxygen uptake by fungi and aerobic bacteria is less than 3×10^{-6} M. Thus, oxygen is only likely to limit the growth of aerobes in soils with high matric potentials where the spaces in soil are filled with water and where the diffusion rate of oxygen is reduced to 2.6×10^{-5} cm^2 sec^{-1}.

In a fertile soil, soil particles are aggregated together to form stable crumbs. Inside these crumbs, the radii of the soil pores are small and the pores will remain saturated with water over a wide range of matric potentials. However, between the crumbs the pores are large and even a slight change in matric potential will cause major changes in the numbers that are water filled. Greenwood (1968) has shown that if the crumbs have a radius greater than 3 mm, then the centre of the crumb will be anaerobic, and in a large crumb, both aerobic and anaerobic zones will develop. Since the pores are small, it has been argued that microbes found within the crumbs must have been incorporated at the time of crumb formation (Allison, 1968). If these organisms are to survive they must be able to withstand low oxygen tensions and low substrate concentrations.

Conversley, active growth of microbes will take place at the surface of the soil crumbs and between them. The rates of diffusion of oxygen in the inter-crumb pores are sensitive to changes in gas-filled porosity for, if water fills the pores, the continuity of the gas phase will be broken and this will reduce the respiration rates of the microbes in the soil pores (Greenwood & Goodman, 1965). Rixon & Bridge (1968) also showed that when the gas-filled pore space changed from 10–20 per cent, the respiratory quotient changed from two to unity, suggesting the onset of aerobic respiration. Thus anaerobic organisms will be under stress within the inter-crumb pores and near the surface of soil crumbs.

Acidity and alkalinity

Experiments in pure culture have shown that many of the bacteria isolated from soil are sensitive to acidity and that growth is very slow or else ceases below pH 5.0. On the other hand, fungi are much more tolerant of changes in pH and continue to grow down to pH 3.0. Gross measurements of pH in soil suggest that the majority of soils, except in arid regions, are acid, due to the continual leaching of bases by rain-water. Acidity may vary from horizon to horizon in a soil profile, and the boundaries between horizons can be very sharp in some cases. Nye (1972) has shown that in acid soils the H_3O^+-H_2O acid–base pair is the most important in determining pH changes while in alkaline soils, the H_2CO_3–HCO_3^- pair is dominant. In neutral and slightly acid soils, soluble organic matter and the $H_2PO_4^-$–HPO_4^{2-} may also contribute. He derived a soil acidity diffusion coefficient which he defined as $v_l f_l / b_{HS} \Sigma b_{HB} D_{lHB}$, where v_l is the volume fraction of the soil solution, f_l the impedance factor for the liquid diffusion pathway, b_{HS} the pH buffer capacity of the soil, b_{HB} the pH buffer capacity of each mobile acid–base pair and D_{lHB} the diffusion coefficient of each mobile acid–base pair in free solution, and the sum is taken over all mobile acid–base pairs. The soil acidity diffusion coefficient is high in acid and alkaline soils and at a minimum in slightly acid soil. Consequently, in a block of soil in a steady state in which one end of the block is at pH 4.0 and the other at pH 8.0 and the mobile acid–base pairs are H_3O^+–H_2O and H_2CO_3–HCO_3^-, a steep gradient of pH will occur at pH 6.1 where the diffusion coefficient is minimal (Fig. 2*a*). Precisely this situation can occur in natural soil profiles and in Fig. 2*b* are shown the pH values found at different depths in a sand dune soil where surface acidity is maintained by the decomposition of pine needles and alkalinity in the deeper parent sand by the solution of calcium from mollusc shell fragments.

Within acid soils, the pH also varies from point to point and is greatly influenced by presence of negatively charged colloidal clay particles. McLaren & Skujins (1968) have shown that the ideas developed by Hartley & Roe (1940) on the pH at a charged interface and its relationship to pH in the adjacent bulk phase can be applied to soil. Thus $pH_s = pH_b + 0.325\,\mu$, where pH_s is the pH at the charged interface, pH_b the pH of the adjacent bulk phase and μ the mobility of the charged particle in micrometres sec^{-1} $volt^{-1}$ cm^{-1}.

This implies a pH gradient in soil perhaps extending from the clay particle surface into the bulk phase with the pH at the surface being more acid than in the bulk phase. Many extracellular enzymes and the organisms that produced them are adsorbed onto soil particles and, consequently, growth may not be taking place in soil even when measurements of bulk pH suggest that it should be. Weiss (1963) has also shown

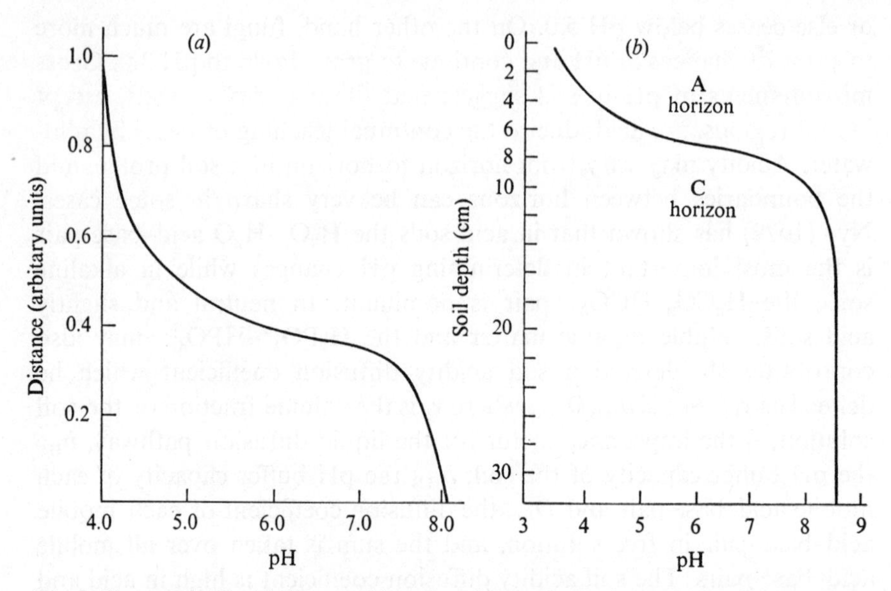

Fig. 2. Change of pH with distance (*a*) across a block of soil whose ends are kept at pH 4.0 and 8.0 (after Nye, 1972) and (*b*) in a pine forest soil where the surface mineral horizon (A) is at pH 3.5 and the lower horizon (C) is at pH 8.3.

that the behaviour of penicillinase from *Bacillus subtilis* when bound to the outside of the cells is like that found in an environment of lower pH than the bulk phase, and thus we must conclude that the pH in the immediate environment of both cells and soil particles is more acid. Changes in pH also affect the solubility of other ions, so that as soils become more acid, iron, manganese and aluminium become more soluble while calcium, magnesium and potassium become less soluble (Foth & Turk, 1972). This prevents a clear analysis of which factors inhibit microbial growth and necessitate survival.

Another important ion which influences pH in soil is the ammonium ion. Thus Williams & Mayfield (1971) have shown that ammonia, released from the decomposition of nitrogenous materials added to soil, often accumulates around organic matter causing the pH near such particles to change from 4.0 to between 6.0 and 7.0. This may lead to the survival of soil micro-organisms in acid soils, through the continuation of growth. Mayfield & Williams demonstrated the survival of non-acid-tolerant streptomycetes in this way and Lowe & Gray (1973*a*) showed that arthrobacter populations behaved in a similar manner. In certain instances ammonia may become toxic to microbes as Court, Stephens & Waid (1964) have shown on studies on the decomposition of urea fertilizers.

It is important to note that the reaction of leaf litter also changes markedly during decomposition. Initially most leaf litter is acid, but this becomes less marked during the first month of contact with soil due to the leaching of organic acids from the leaves (Nykvist, 1959). However, this trend is soon reversed and most decomposing litter becomes very acid with a pH below 4.5 (Plice, 1934). Decomposition is restricted to a small range of tolerant forms, the remaining species becoming dormant.

Toxic substances

Substances toxic to microbes may be found as solids, liquids or gases in the soil. However, little work has been done on their effects on survival since most interest has centred on their role in controlling activity. This has been complicated by problems of measuring effective concentrations of these substances, for most assays have been done on large samples without reference to localized concentrations in micro-environments. Among the toxic substances known to be present in soil are antibiotics, polyphenolic compounds, aluminium ions, gases such as ethylene and carbon dioxide, and some alkaloids, e.g. chlorogenic acids which affect nitrifying bacteria. A discussion of antibiotics, ethylene and carbon dioxide in soil will serve to illustrate the difficulties involved in quantifying their effects.

Soulides (1964) has shown that antibiotics can be found in soil at low concentrations. He extracted soils with buffer, concentrated the extract by evaporation and solvent extraction and determined the concentration of antibiotic by bioassay and found minimum detectable quantities of aureomycin (0.02–0.08 μg g^{-1} soil), terramycin (0.08–0.2 μg g^{-1}), carbomycin (0.3–90 μg g^{-1}) and streptomycin (5–100 μg g^{-1}). He also inoculated sterilized soil with *Streptomyces rimosus* and *S. aureofaciens* and showed that production of terramycin and aureomycin was directly related to the concentration of added organic matter (Soulides, 1965). Wright (1956) found this effect when he located gliotoxin only in wheat straws buried in soils. Baker (1968) has pointed out that antibiotics can be inactivated in soils by an increase in instability in acid conditions, adsorption on clay colloids and humus particles and microbiological degradation. Not surprisingly a good deal of work on antibiotics in soil has been directed at the control of soil-borne plant pathogens but as the evidence for control rests largely on correlation between disease suppression and increases in numbers of antibiotic-producing organisms no proof of their importance can be obtained.

Ethylene has been proposed as a regulator of biological activity in soil. Smith (1973) showed that, if a saturated air stream containing ethylene at 1 ppm (v/v) was passed over soil containing sclerotia of *Sclerotium rolfsii*, their germination was inhibited. Further, mould growth on the soil surface only occurred in control tubes in which less than 0.01 ppm (v/v) of ethylene was introduced. He also suggested (Smith, 1975) that ethylene was produced primarily in the anaerobic regions of the soil and that as it diffused into the aerobic zones it depressed the growth of aerobes. In turn, the decreased oxygen uptake by aerobes caused the anaerobic zones to contract in size, leading to less ethylene production and the re-growth of aerobes. He suggested that this might be a mechanism for fungistasis in soils.

The concentrations of ethylene present in some soils have been assessed by Dowdell *et al.* (1972). They found that the ethylene concentration in soil varied with rainfall. The highest concentrations (1–5 ppm) at three depths (30, 60 and 90 cm) were found after heavy rain in January and February. However, there were enormous and rapid spatial and temporal variations in concentrations. The recorded concentrations of ethylene are high enough to have inhibitory effects on root growth but the fluctuations are such that conclusions on its effect on plant and microbial growth in soil would be unwarranted. Recently Lynch & Harper (1974*a*) have shown that ethylene is produced by several aerobic fungi and that oxygen promotes ethylene production, casting doubt on Smith's theory of regulation of activity of aerobes. They suggest that the explanation for ethylene production in natural soils being favoured by low oxygen tensions (Smith & Restall, 1971) is that substrates necessary for its production are mobilized under anaerobic conditions. Lynch & Harper (1974*b*) also showed that the rate of ethylene production per unit mass of *Mucor hiemalis* was highest at low specific growth rates, suggesting that more ethylene would be produced in soil than in laboratory batch culture where growth rates are high. Thus at present, the role of ethylene as an inhibitor of growth in soil is not proven.

Carbon dioxide has been suggested as a differential inhibitor of heterotrophs in soil (Alexander, 1964). Thus Papavizas & Davey (1962) showed that growth of *Rhizoctonia solani* in soil was reduced by 10–20 per cent carbon dioxide concentrations while at 30 per cent it was drastically inhibited. Burges & Fenton (1953) also showed that a concentration of 5 per cent carbon dioxide inhibited the growth of *Penicillium nigricans*, a fungus restricted to the top 5 cm of the soil. They showed that concentrations of carbon dioxide could reach 9.2 per cent shortly after rain, even in a coarse-grained, sandy soil. Rommell (1922) has recorded even higher maximum values for ricefields (20.9 per cent), woodland (15.2 per cent) and grassland soil at 40 cm depth (14.3 per cent).

NATURE OF SURVIVAL IN THE NATURAL ENVIRONMENT

Indigenous and alien populations of micro-organisms in soil may be subjected to the stresses outlined in the previous section. Three alternative responses are possible: the formation of dormant spores or vegetative cells, the continuation of growth (even if at a low rate) enabling the population to survive, or death. The properties that an organism require to form dormant vegetative cells or to continue growing in unfavourable conditions are different and are considered separately below.

Factors contributing to survival of growing populations

Growth patterns. The surface area/volume ratio of a hypha will be less favourable for growth than that of an equivalent amount of unicells and yet the predominant organisms in the soil environment are often hyphal, e.g. fungi and actinomycetes. The importance of the hyphal pattern must lie elsewhere and one of the likely explanations is connected with survival. Hyphae extend for considerable distances as shown by the length of mycelium detectable in a gram of soil, which averages about 5 m. Burges (1960) has shown that some of this mycelium is not associated with particulate substrates but grows through the soil. It is likely in view of the information available on the size of micro-environments in the soil, that such hyphae will never lie completely in unfavourable areas for growth and so survival will be achieved by local outgrowth of new mycelium. The importance of position of a propagule in relation to favourable conditions has also been cited by Garrett (1970) as an important factor in relation to the colonization of roots by soil-borne fungi and he suggested that it could compensate for possession of a relatively slow growth rate in competitive situations.

A further advantage is gained by those fungi whose hyphae are aggregated together to form simple strands or complex rhizomorphs. These structures are often differentiated so that the outer hyphae serve a protective function for the internal hyphae. If the fungus grows from a sufficiently large food base, e.g. a rotting tree stump, rhizomorphs may wander through soil containing no substrate for very long distances until a further tree stump is encountered. Griffin (1972) has summarized the information on the survival of these structures in *Armillariella elegans* and *A. mellea* in relation to oxygenation of the environment. The rhizomorphs of these fungi have a meristematic apex which produces a hollow tube with walls made up of closely packed hyphae. The central tube is open to the atmosphere at its origin which is important since the rate of growth of the rhizomorph into columns of agar has been shown to be sensitive to changes in partial pressure of oxygen over a certain range at its origin. The production of melanin in the rhizomorph is governed by the action of laccase which is sensitive to oxygen concentration throughout the range 0–20 per cent. In a water film, the rate of melanization is halved compared with the rate in air and so the tips of the rhizomorphs stay white, but in air they rapidly go brown and growth stops. Thus, optimal growth is achieved if there is a high partial pressure of oxygen within the apex but a low partial pressure outside it. As rhizomorphs cross gas-filled pores, abortive side branches are produced which allow oxygen to enter the rhizomorph along its length. Field data confirm these findings and show that rhizomorph activity is reduced in very dry soils but not in wet soils since the rhizomorphs on the tree stumps are probably still in contact with the atmosphere. The role of melanin in survival of dormant structures is discussed later.

Competitive saprophytic ability and saprophytic survival. The greatest body of work on the survival of growing populations of microbes in face of competition from other microbes concerns the ability of plant pathogenic fungi which attack roots to survive, during the absence of a host, by growing saprophytically on dead organic matter on or in soil. Garrett (1950, 1956) coined the term competitive saprophytic ability for root-infecting fungi to describe 'the summation of physiological characteristics that make for success in competitive colonization of dead organic substrates'. Later, Garrett (1970) pointed out that the kinds of competitive saprophytic ability were as diverse as the types of substrate available for colonization.

Those fungi which possess a high competitive saprophytic ability are able to survive for long periods in the absence of a host and much work has since been directed towards determining which physiological characteristics are most important in giving high competitive saprophytic ability. A brief description of one method measuring competitive saprophytic ability will illustrate the way in which mycologists have approached this problem.

The fungus to be tested is grown in a sand–maize meal mixture and, when the fungus is fully grown, it is mixed in differing proportions (0–100 per cent) with soil. Pieces of dead plant tissue (often wheat straws for investigation of cereal foot-rot fungi) are placed in the mixture and incubated for four weeks. The straws are examined for colonization by the fungus, in the face of competition from the whole soil microflora. Fungi with a high competitive saprophytic ability are those able to colonize wheat straws, even when they are present in small quantities relative to the mass of the soil microflora. On the basis of results from such experiments, Garrett (1950) put forward the following properties which would influence competitive saprophytic ability.

(*a*) Rapid response of dormant fungal propagules and rapid growth rate on exposure to soluble nutrients diffusing from a substrate.

(*b*) A wide range of enzymes, especially those able to attack resistant residues in plant tissues, e.g. cellulases and ligninolytic enzymes.

(*c*) Production of antibiotics and other compounds affecting the soil microbes.

(*d*) Tolerance of these compounds produced by the soil microbes.

Subsequently other determining factors and refinements of the above factors have been suggested. Thus, Macer (1961) showed that one important factor was the ability of the fungus to penetrate the straws and this was shown to be unrelated to growth rate over agar. Burgess & Griffin (1967) have found that an ability to withstand low temperatures of about 10 °C is important in allowing wheat straw colonization, and suggested that wheat stubble should be ploughed into the soil in summer when temperatures are high enough to allow the soil microflora to suppress saprophytic colonization. Finally, Byther (1965) showed that fungi with high competitive saprophytic ability were able to utilize inorganic nitrogen faster and more efficiently in pure culture when low levels of carbon were present in the medium or when the cultures were grown at a high pH or low temperature. Thus the dominance of *Fusarium roseum* over *F. solani* in competitive situations was enhanced when nitrogen was limiting growth (Lindsey, 1965).

The time of saprophytic survival of root pathogens in soil free from living hosts is only partly dependent on their competitive saprophytic ability and so, to assess this properly, it must be certain that spores of the fungus are not present. Thus, fungi which are not known to produce spores during the saprophytic period of growth in soil are likely to give the most reliable results, e.g. *Ophiobolus graminis*. If spores are produced, e.g. *Cochliobolus sativus*, steps to ensure their removal must be taken. Usually, this is achieved by removal of leaf sheaths around the straws and vigorous washing of the remaining tissue. These precautions may fail, however, if the fungus can produce spores inside the tissue, especially if, as in *Fusarium* species, chlamydospores form within lengths of mycelium. A proper assessment of survival also depends on

Table 2. *Cellulolysis adequacy indices (CAI) for five cereal foot-rot fungi (Garrett, 1970)*

	Per cent loss in filter paper dry wt after 7 weeks at 25 °C		Linear growth rate of fungus (mm day^{-1}) at 22.5 °C	Mean value for CAI
	Range	Mean		
Ophiobolus graminis	2.1–3.6	2.8	7.3	0.38
Fusarium roseum f. *cerealis*	6.5–8.5	7.5	11.0	0.68
Cercosporella herpetrichoides	0.8–1.1	0.9	1.3	0.69
Curvularia ramosa	—	6.6	5.3	1.25
Cochliobolus sativus	5.4–8.4	7.5	3.6	2.10

a good method of recovery of the fungus from soil. This is achieved either by incubating straws in damp chambers containing sand when identifiable spores are produced, or by planting seeds of susceptible plants in the lumen of a straw inserted into damp sand and watching for the development of characteristic disease symptoms.

Garrett (1936) found that *O. graminis* survived poorly in straws when soil was at normal temperatures, an equable moisture content and was well aerated, i.e. when soil microbial activity was high. On the other hand, survival was good at low temperatures and in either air-dried or waterlogged soil. This was confirmed by Fellows (1941) who showed that the same conditions governed the disappearance of *O. graminis* from naturally infested soils, maintained under controlled conditions in caves and greenhouses. Survival is also affected by the nitrogen-content of the soil and plant tissue, since cellulose decomposition stops when the fungus ceases to be in direct contact with its substrate following dissolution, and new mycelial extension is required to maintain cellulose breakdown (Garrett, 1970). Thus, survival of *O. graminis* was best when nitrogen was not limiting growth (Garrett, 1940; Chambers & Flentje, 1969) and when microbial activity was high. The effect of nitrogen is not uniform, however, and *Cochliobolus sativus* shows the opposite response while *Curvularia ramosa* is indifferent. Garrett (1966) has now shown that the different effects are related to the cellulolysis adequacy index (CAI), i.e. the cellulolysis rate (per cent loss in dry weight of filter paper over seven weeks at 22.5 °C) divided by linear growth rate (mm day^{-1} at 22.5 °C). This is an index of the adequacy of cellulolysis to supply carbon for maintaining growth. Those fungi which had a positive survival response to nitrogen had a CAI of less than 1 and those with a negative response an index greater than 1. Garrett postulated that the latter fungi survived less well because extra nitrogen would promote

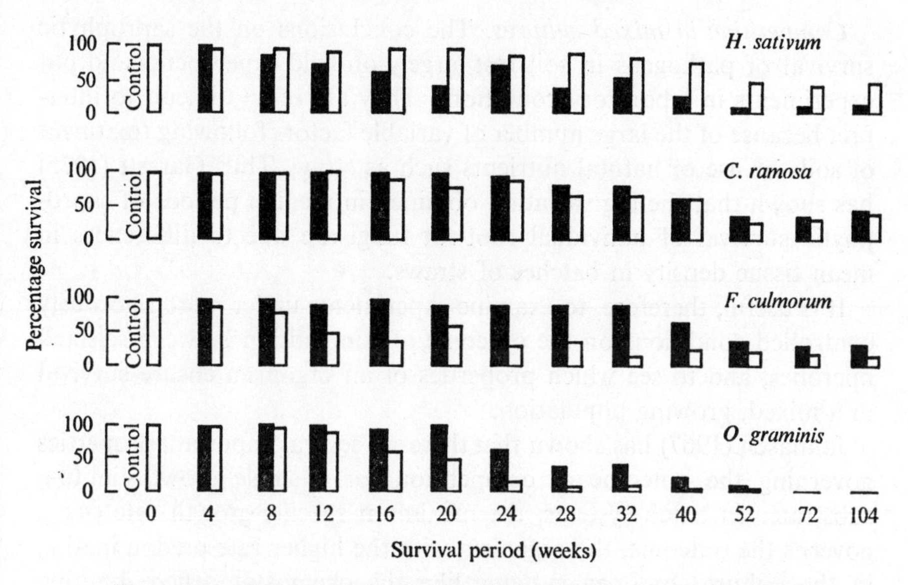

Fig. 3. Saprophytic survival of four cereal foot-rot fungi in wheat straw (■ in soil of high fertility; □ in soil of low fertility). (After Butler, 1959.)

an uneconomic rate of cellulose consumption resulting in rapid exhaustion of the substrate. On the other hand *O. graminis* required nitrogen to promote cellulose decomposition to a level which could maintain growth (Table 2). More recently Garrett (1975) has suggested that high nitrogen content may cause the premature displacement of fungi such as *Cochliobolus* by stimulating the development of a high microbial population.

Despite these variations, saprophytic survival times have been obtained for several fungi. *O. graminis* can survive for up to one year, but for only about 32 weeks in soils low in nitrogen. By contrast over 50 per cent of *Cochliobolus sativus* (*H. sativum*) survive for one year in soils low in nitrogen but under 10 per cent survive in nitrogen-rich soils (Fig. 3).

The reasons for the relatively prolonged saprophytic survival phase is not clear, especially for *Ophiobolus*. This fungus does not produce dormant propagules and is very sensitive to fungistatic products of soil microbes. Garrett (1970) suggests that it establishes itself in almost pure culture in living host tissue but how it then prevents invasion of other organisms on the death of the host is unknown. It does not produce fungistatic substances and even if it did, they would probably be produced only during periods of rapid growth rather than during the saprophytic survival phase.

Competition in mixed cultures. The conclusions on the saprophytic survival of pathogens in soil rest largely on field experiments and pot experiments in laboratory conditions. They are often difficult to interpret because of the large number of variable factors following treatment of soils or use of natural nutrients such as straw. Thus Garrett (1975) has shown that the big variation obtained in median periods of saprophytic survival of individual root-rot fungi are due to differences in mean tissue density in batches of straws.

It is useful, therefore, to examine experiments under more rigorously controlled conditions on the outcome of competition between selected microbes, and to see which properties of an organism ensure survival in a mixed, growing population.

Jannasch (1967) has shown that there are several important properties governing the outcome of competition for a single growth-limiting substrate. In batch systems, the maximum specific growth rate (μ_{max}) governs the outcome, the organism with the higher rate predominating in the culture. In open systems like the chemostat, where limiting substrate concentrations may be low, then the substrate affinity as measured by the saturation coefficient (K_s) may also be critical. Thus in competition experiments between a spirillum and a pseudomonad, he showed that spirilla had a higher μ_{max} than pseudomonads at low organic substrate concentrations while the reverse was true at higher substrate concentrations (Fig. 4). Harder & Veldkamp (1971) pointed out that temperature could affect the outcome of such experiments and that at low temperature (2 °C), pseudomonads could still outgrow spirilla at any concentration of the growth-limiting substrate because their μ_{max} was higher than that of spirilla. At higher temperatures (4 and10°C) the outcome depended upon the concentration of the growth-limiting substrate, while at 16 °C, the spirilla always predominated because their μ_{max} was always higher than that of *Pseudomonas*. Veldkamp & Kuenen (1973) have now shown that the same considerations apply to low concentrations of inorganic substrate, e.g. phosphate, although here the K_s values are very much lower than those for organic substrates. In general, K_s values for organic carbon compounds are of the order of a few mg l^{-1}, for organic nitrogen compounds $\mu g\ l^{-1}$, for inorganic substrates about 3.5×10^{-8} M. Substrate affinities of non-sporing organisms from soil have not been examined extensively, though Gray *et al.* (1975) have found that the K_s values of a strain of *Arthrobacter globiformis* for glucose and sucrose are 3 mg l^{-1} and 1.3 mg l^{-1} respectively. These values are lower than those obtained for organisms such as *Escherichia coli* and S. J. Chapman (unpublished results) has shown

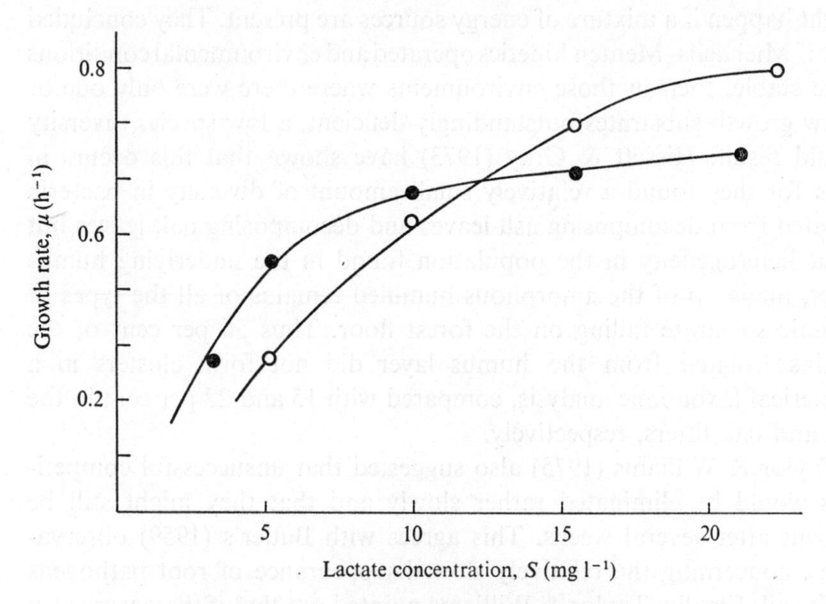

Fig. 4. Saturation curves for *Pseudomonas* (○) and *Spirillum* (●) in lactate-limited culture. (After Jannasch, 1967.)

that *A. globiformis* can displace this organism in competition for glucose at 25 °C.

These experiments suggest that observations on the saprophytic survival of root-infecting fungi could be interpreted, not only in terms of growth rate, but also substrate affinities. It would be interesting to determine the saturation coefficients for fungi with high and low competitive saprophytic abilities, especially in respect of the temperature effects reported by Burgess & Griffin (1967) for *O. graminis*. The experiments on the effects of pH and temperature on nitrogen assimilation reported by Byther (1965) might also be interpreted in this way (see p. 339). A fungus with high competitive saprophytic ability might have a relatively high maximum specific growth rate under soil conditions and a low saturation coefficient for the limiting nutrient. The different behaviour of *O. graminis* and *Cochliobolus sativus* in nitrogen-limited and non-nitrogen-limited conditions might be explained by the relationship of the saturation coefficients of the surviving fungi for nitrogen and other limiting nutrients to those of the rest of the microflora.

The situations described so far differ from those encountered in nature because only a single energy source has been supplied to the competing organisms. Taylor & Williams (1975) have considered what might happen if a mixture of energy sources are present. They concluded that if Michaelis–Menten kinetics operated and environmental conditions were stable, then in those environments where there were only one or a few growth substrates outstandingly deficient, a low species diversity would result. Hissett & Gray (1973) have shown that this occurs in soils for they found a relatively small amount of diversity in bacteria isolated from decomposing ash leaves and decomposing oak leaves but great heterogeneity in the population found in the underlying humus layer, made up of the amorphous humified remains of all the types of organic substrate falling on the forest floor. Thus 56 per cent of the strains isolated from the humus layer did not form clusters in a numerical taxonomic analysis, compared with 15 and 22 per cent in the ash and oak litters, respectively.

Taylor & Williams (1975) also suggested that unsuccessful competitors would be eliminated rather slowly and that they might still be present after several weeks. This agrees with Butler's (1959) observations, concerning the relatively slow disappearance of root pathogens from soil. Finally, Taylor & Williams pointed out that if there were also negative interactions between the competitors, e.g. toxin production, then the stability of the system would be decreased, whereas positive interactions, e.g. growth factor production, would lead to increased stability. In the event of environmental changes occurring, in favour of the previously weaker competitor, then an oscillation between the competitors would ensue.

Meers & Tempest (1968) have shown that there may be significant departures from Michaelis–Menten type kinetics when microbial products which influence growth are formed. Thus in a mixed culture of *Bacillus megaterium* and *Torula utilis* or Gram-negative bacteria, limited by magnesium, it was shown that the outcome depended upon the relative amounts of each species in the culture immediately after mixing the two populations. If the bacilli were present in large relative numbers, they then outgrew the yeast, but if present in low numbers, they failed to establish themselves. They showed that this effect was not due to production of inhibitors by the bacilli but to the secretion of substances which facilitated magnesium uptake by bacilli more than the yeast. These substances could be obtained in cell-free extracts and were found to survive heating at 100 °C for 30 min, diffuse through a dialysis sac and separate on a Sephadex column (G10) with material of molecular weight greater than 700. It was not clear whether this 'population-product factor' affected μ_{max}, K_s or both. Garrett (1970) suggested that the most important factor in survival of root-infecting fungi is their presence in almost pure culture in living root tissues and postulated that this gave them an advantage when the roots died. He showed that this advantage was usually not exploited by the production of inhibitory substances to keep out the rest of the microflora and concluded that 'Possession is nine points of the law'. This could be explained by an extension of the theory of population-product factors to cover the uptake of substances other than magnesium. Such factors should also be taken into account during the measurement of competitive saprophytic ability described earlier (p. 338).

FACTORS CONTRIBUTING TO THE SURVIVAL OF DORMANT VEGETATIVE MICROBES

Energy reserves

Dawes (see p. 37) has reviewed the evidence of the importance of energy reserves in prolonging the life of dormant vegetative microbes and has pointed out the very long survival times reported for some soil bacteria, including *Arthrobacter* and *Nocardia.*

The role of energy reserves in promoting the survival of *Arthrobacter* has been investigated by several workers. Boylen & Ensign (1970) showed that the utilization of carbohydrate reserves of rods and cocci of these organisms were different, but examination of their methods suggests that the differences lay not between the morphological types but between the cultural histories of the organisms. Rods were produced by growing the organism on peptone yeast extract while cocci were formed on a glucose–mineral salts medium. The latter medium, with a carbon/nitrogen ratio of 9.4, would probably have been nitrogen-limited and might be expected to promote the accumulation of polysaccharide by increasing the supply of glucose-1-phosphate, the precursor of glycogen. Such an effect has been demonstrated by Mulder & Zevenhuizen (1967) who found that phosphohexoisomerase was less active in nitrogen-deficient cultures, allowing glucose-6-phosphate to be converted to glucose-1-phosphate in increasing amounts, through the action of phosphoglucomutase. The peptone–yeast extract medium used to produce rods would not have been nitrogen deficient and so the accumulation of glycogen would not have been promoted. Mulder & Zevenhuizen (1967) showed that while the intracellular polysaccharide concentration of an arthrobacter was 27 per cent in media with a carbon/nitrogen ratio of 6.1, it increased to 60 per cent if the carbon/nitrogen ratio was 18.2:1. It is not surprising that Boylen & Ensign (1970) found that

Table 3. *Composition, morphology and viability of cells of* A. globiformis *grown under carbon-limiting conditions at varying dilution rates. Amounts expressed as percentages of total dry weight.* (*Data of S. J. Chapman*)

Dilution rate (h^{-1})...	0.34	0.29	0.085	0.052	0.028	0.015	0.0096
Morphology	Rod	Rod	Coccus	Coccus	Coccus	Coccus	Coccus
Viability (per cent)	96.3	93.0	91.1	88.4	—	87.6	92.5
Carbohydrate*	34.5	23.0	22.2	20.0	20.9	18.2	16.2
Protein	30.0	34.3	39.5	38.5	37.1	35.9	39.5
RNA	15.5	16.1	15.2	—	—	—	14.2
DNA	4.1	3.1	5.7	4.1	3.7	3.9	3.0
Lipid	9.9	5.7	—	—	—	—	—
Yield (mg l^{-1})	—	111.3	104.1	101.1	91.0	88.3	91.6

* Wall carbohydrate constitutes about 12 per cent of the dry weight of rods and 18 per cent of cocci.

40 per cent of the dry weight of cocci consisted of polysaccharide while rods contained only 10 per cent, especially as Zevenhuizen (1966) also found that the use of amino acids as sole sources of carbon and nitrogen gave good growth but a low carbohydrate content.

Luscombe & Gray (1974) have shown that the main factor controlling morphogenesis in several arthrobacters is growth rate and the production of rods and cocci in Boylen & Ensign's experiments can be attributed to such differences, caused by the relative inability of *A. crystallopoietes* to utilize glucose as a carbon source. It is difficult, therefore, to be certain that the differences observed in carbohydrate utilization between rods and cocci are related to growth rate, morphological type or cultural history. Recent studies by S. J. Chapman, reported below, have contributed some new information on the accumulation of carbohydrates in *A. globiformis* NCIB10683 in more carefully defined conditions, showing that the amounts of storage material produced are related to both growth rate and the nature of the growth-limiting nutrient.

In common with other arthrobacters, *A. globiformis* NCIB10683, originally isolated from a sand dune soil (Lowe & Gray, 1972) produces glycogen as its main carbohydrate reserve and has a polymer of glucose, galactose and rhamnose, making up about 60 per cent of the cell wall (Gray *et al.* 1975). This latter polymer represents 12 per cent of the weight of a rod and 18 per cent of the weight of a coccus. If the bacteria are grown at different growth rates in a chemostat under either carbon-limiting or nitrogen-limiting conditions, the amount of carbohydrate varies. Table 3 shows that as the growth rate of carbon-limited organisms decreases, so does the carbohydrate content. At a dilution rate of 0.01h^{-1}

Table 4. *Composition, morphology and viability of cells of* A. globiformis *grown under nitrogen-limiting conditions at varying dilution rates. Amounts expressed as percentages of total dry weight.* (*Data of S. J. Chapman*)

Dilution rate (h^{-1}) ...	0.34	0.27	0.092	0.025	0.01	0.007
Morphology	Rod	Rod	Coccus	Coccus	Coccus	Coccus
Viability (per cent)	98.5	—	97.5	93.7	—	91.1
Carbohydrate	41.6	41.1	50.5	52.1	56.7	64.7
Protein	28.8	23.6	28.1	14.5	11.9	12.0
RNA	—	—	11.5	5.6	5.4	—
DNA	—	—	4.0	2.6	2.3	—
Lipid	—	—	4.6	—	3.4	—
Yield (mg l^{-1})	136.9	136.6	149.1	207.7	300.2	—

the carbohydrate content is so low that most of it can probably be accounted for by wall material. If the organism grew at a low rate under natural conditions, which is not unlikely in view of the rate of supply of substrate (Hissett & Gray, 1976), then it would probably not produce very much reserve material. Table 3 also shows that, at low growth rates, other major compounds, e.g. protein, RNA and DNA, do not increase markedly; no measurements of mucopeptide material were made.

When the experiment was repeated under nitrogen-limiting growth conditions, very different results were obtained (Table 4). The levels of carbohydrate were higher at all growth rates and actually increased with decreasing growth rate. At a dilution rate of 0.01 h^{-1}, 56.7 per cent of the cell weight consisted of carbohydrate. This increase was largely at the expense of the protein content of the cells which fell from 28.8 per cent at a dilution rate of 0.34 h^{-1} to 11.9 per cent at 0.01 h^{-1}. In all cases where viability was determined, values of over 87 per cent were found.

These high levels of carbohydrate cause the cells to increase in size and weight, so much so that they resemble the cystites reported from batch cultures in which excessive amounts of nitrogen are present (Mulder & Zevenhuizen, 1967). Electron micrographs of cystites show the very large amounts of reserve material in the cell and the virtual absence of ribosomes from many thin sections (Duxbury, 1973). The synthesis of the reserve material causes the cell to expand in size, rendering the wall liable to fracture and distortion during the sectioning process. However, many cystites are capable of 'germination' on restoration to a normal medium.

These experiments suggest that in *A. globiformis*, if the amount of carbohydrate reserve material is important in controlling longevity, nitrogen-limited populations should survive for longer periods than carbon-limited populations during starvation. Experiments to determine the rate of loss of reserve materials and viability in nutrient-free buffer solutions have shown that the rate of carbohydrate loss from nitrogen-limited cells occurs at a faster rate than from carbon-limited organisms grown at a dilution rate of 0.01 h^{-1}. After about 20 days, the percentage losses of carbohydrate from starved carbon and nitrogen-limited organisms were 27.3 and 40.2 per cent respectively. The viability of the carbon-limited cells was still high (89.6 per cent) but at the time of writing, data were not available on the viability of the starved nitrogen-limited cells. However, it is clear that the carbon-limited cultures retained their viability for long periods when the amount of carbohydrate in the organism was low and probably mostly in the wall fraction. Some other factor influencing survival is indicated, especially since viability was still high after 53 days (73.1 per cent) and carbohydrate had dropped only to 65.6 per cent of the amount present at the start.

Fungi also contain large amounts of storage reserve material in sclerotia. These reserves may be polysaccharides such as glycogen (Ergle, 1947), or even soluble carbohydrates like trehalose and mannitol (Cooke & Mitchell, 1969). Glycerol and fatty acids have been reported in some forms (Howell & Fergus, 1964) and in others large amounts of lipid have been found (Mitchell & Cooke, 1968). *Phymatotrichum omnivorum* sclerotia contain up to 35 per cent of their dry weight as glycogen but this is used rapidly during germination and about half of it disappears during the first 12 days of germination (Ergle, 1948). *Claviceps purpurea* sclerotia have about 50 per cent of their dry weight as lipid and this falls to below 40 per cent during germination. The amounts of reserve material in some sclerotia have also been found to change during dormancy. Thus, Gordee & Porter (1961) found that one month old sclerotia of *Verticillium dahliae* contained 80 per cent carbohydrate and 6 per cent lipid but after a further eight months, these values had changed to 59 per cent and 22 per cent respectively. The significance of this change is not clear.

Endogenous metabolism

If an organism is to survive a long period of starvation, it is important that its energy reserves are used at a slow rate. Dawes (this volume) has pointed out that, if this is not the case, the energy runs to waste and the bacteria die relatively rapidly. Zevenhuizen (1966) has suggested that the rate of glycogen degradation in arthrobacter is limited by the slow action of the α-1,6-glucosidase enzyme responsible for debranching of the molecule. However, Chen & Alexander (1972) found that endogenous respiration rates of some soil bacteria apparently resistant to

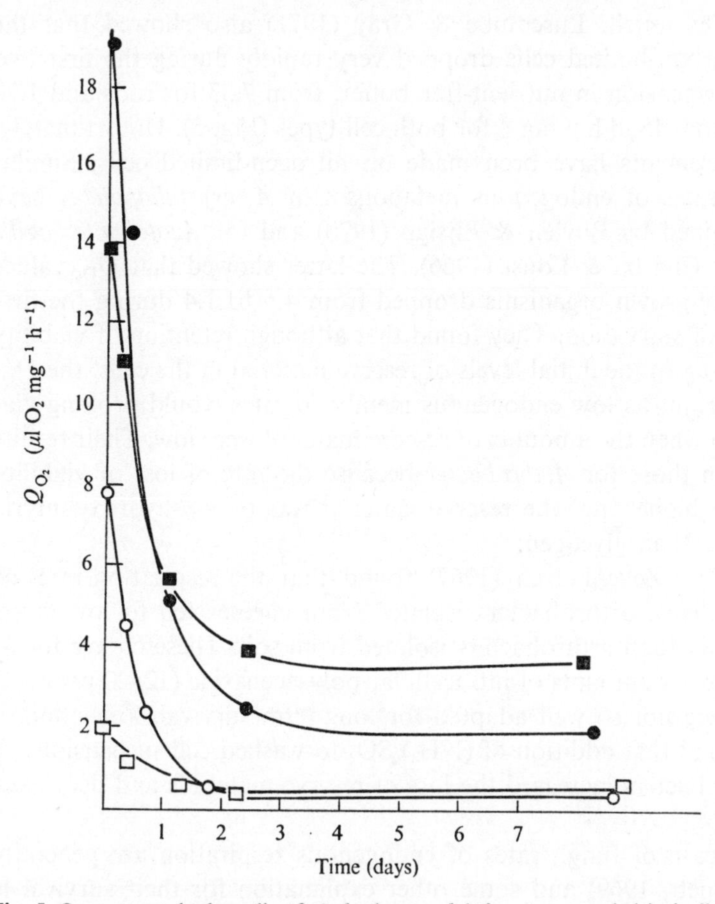

Fig. 5. Oxygen uptake by cells of *Arthrobacter globiformis* suspended in buffer at pH 7.0, estimated by respirometry (○, rods; □, cocci), and polarography (●, rods; ■, cocci). (After Luscombe & Gray, 1974.)

starvation were relatively high. They also found that some starvation-resistant cells were rich in poly-β-hydroxybutyric acid (34–301 mg g^{-1} cells). Their resistant bacteria did not show very high viability upon starvation, however, as less than 10 per cent survived a few days of starvation.

In our own experiments on *Arthrobacter*, the rate of utilization of carbohydrate by nitrogen-limited cultures is relatively high and so the rate of endogenous metabolism is probably greater than that reported for carbon-limited organisms (Q_{O_2} of 0.45 after two days in buffer) by Luscombe & Gray (1974). In absolute terms, 0.035 μg carbohydrate were used by 10^6 carbon-limited organisms (dilution = 0.01 h^{-1}) during 20 days starvation while nitrogen-limited ones used 0.16 μg, nearly

12 times as much. Luscombe & Gray (1971) also showed that the Q_{O_2} of carbon-limited cells dropped very rapidly during the first two days of suspension in nutrient-free buffer, from 7.33 for rods and 1.74 for cocci to 0.45 μl h^{-1} mg^{-1} for both cell types (Fig. 5). Unfortunately, no measurements have been made on nitrogen-limited cells. Similar data for rates of endogenous metabolism in *A. crystallopoietes* have been obtained by Boylen & Ensign (1970) and for *Azotobacter agilis* by Sobek, Charba & Foust (1966). The latter showed that Q_{O_2} values for glucose-grown organisms dropped from 4.6 to 1.4 during the first two days of starvation. They found that although retention of viability depended upon the initial levels of reserve material in the cells, the Q_{O_2} was important as low endogenous metabolic rates would prolong viability even when the amounts of reserve material were low. Their results differ from those for *Arthrobacter* because the rate of loss of viability was much higher and the reserve material was poly-β-hydroxybutyric acid rather than glycogen.

Mulder & Zevenhuizen (1967) found that the respiration rates of washed cells of arthrobacters isolated from cheeses fell to low levels more slowly than arthrobacters isolated from soil. These cheese forms also had lower amounts of intracellular polysaccharide (12–20 per cent) and so were not so well adapted for long-term survival. Zevenhuizen (1966) found that addition of $(NH_4)_2SO_4$ to washed cell-suspensions of soil arthrobacters increased the loss of reserve materials and decreased viability.

In the case of fungi, rates of endogenous respiration are generally high (Burnett, 1969) and some other explanation for their survival is needed. However, Mitchell & Cooke (1968) have found that the Q_{O_2} of sclerotia of *Claviceps purpurea* is about 0.4 μl h^{-1} mg^{-1}, rising to 2.4 μl h^{-1} mg^{-1} upon germination, because of the metabolism of lipids and their partial conversion to trehalose via the glyoxylate cycle.

Resistance to lysis

An organism which can survive for long periods of time because it has a low endogenous rate of metabolism or a plentiful supply of energy reserve material needs to have a structure that is resistant to the action of lytic enzymes. Alexander (1971) has pointed out that in many natural environments, those species which are dominant possess compounds, often associated with the cell wall, that prevent rapid lysis. Resistance to both autolysis and heterolysis may occur and in this latter connection the deposition of melanins and heteropolysaccharides in walls has been suggested as important.

Arthrobacter globiformis has a wall polysaccharide made of glucose, galactose and rhamnose, resembling one found by Zevenhuizen (1966) in *A. simplex* which contained additionally mannose and glucosamine. Duxbury (1973) suggested that this material protects the peptidoglycan fraction of the wall from the action of lytic enzymes. This might be achieved by the layering of the wall with a polymer resistant to lysis on the outside of the cell or the linking of the monomers in the polymer to the peptidoglycan in a variety of ways. Pengra, Cole & Alexander (1969) have suggested such a mechanism for the resistance of the walls of the fungus *Mortierella parvispora* to lysis by glucanase-chitinase mixtures and heterogeneous microbial populations in soil. In this fungus, the walls possess a fucose-containing heteropolysaccharide and the authors suggested that several enzymes produced by different organisms might be needed to hydrolyse the different linkages. The probability that all these organisms and their enzymes would be present simultaneously at the same point is low. Ballesta & Alexander (1968) have also detected a fucose-uronic acid fraction in the walls of four *Zygorhyncus* species which could not be hydrolysed enzymatically by a variety of microbes and lytic enzyme preparations.

Other fungi have melanin-like pigments in the walls and melanins are also synthesized by many streptomycetes. These may protect the organisms against photo-inactivation but they are also important in dark environments. Alexander (1971) has reviewed the evidence for the importance of this substance in preventing lysis. Thus, the hyphae of *Rhizoctonia solani* which possess melanin at the surface cannot be lysed (Potgieter & Alexander, 1966). In the fungus, *Sclerotium rolfsii*, only the sclerotia contain melanin and these are the only structures resistant to lysis in this fungus. Removal of melanin spicules from *Aspergillus phoenicis* conidia renders them susceptible to lysis (Bloomfield & Alexander, 1967), while in *A. nidulans*, the heterolysis of hyphae is inversely proportional to their melanin content. The mechanism of the protective effect could be due to any of the factors considered in relation to polysaccharides, though Bull (1970) has shown that melanin can inhibit polysaccharases. The occurrence of melanin in rhizomorphs also contributes to their survival in zones unfavourable for growth (see p. 338).

Fungal sclerotia are of diverse structure but can be divided into two groups, those that possess rind tissue and those that do not. In the rindless types, cell walls may or may not be thickened according to the type of fungus. In those with both thickened and unthickened walls such as *Verticillium dahliae*, the cells with thickened walls are often heavily melanized and contain a complete set of organelles, while the thin-walled cells are unpigmented and lack organized membrane systems and nuclei (Brown & Wyllie, 1970). In those types with a rind, the rind cells may be protected by an outer skin and may enclose thin-walled cortical cells and filamentous hyphae, e.g. *Sclerotium rolfsii.* The outer skin consists of cell wall residues and the rind of melanin-rich, empty, thick-walled cells. The function of these layers is not definitely established but there is a strong presumption that they are protective, even though no correlation between presence of a rind and longevity has been established (Coley-Smith & Cooke, 1971).

Lytic enzymes are necessary to micro-organisms since they allow the softening of cell walls and the insertion of new wall material. If the relative amounts of lytic and synthetic enzymes are disturbed, lysis may result, as shown by the effect of penicillin on growing bacterial cells. It is likely that those organisms which have a high degree of control on the production of lytic enzymes will be able to survive longer. Gray *et al.* (1975) investigated the budding growth of *Arthrobacter globiformis* and showed that the production of buds from cells occurred sequentially. Cocci generally elongated to form a rod and then a cross-wall was laid down at the site of the original coccus wall to produce a two-celled structure made up of a rod and a coccus. The coccus then produced a second bud which was cut off as a rod. Although this process of sequential budding occurred several times, the outline of the original coccus wall was still discernible after 10 h and sometimes remained visible as a very phase-dark region for over a day. Separation of the buds from the parent took place by a slow bending process caused by the differential growth of a further bud which ruptured part of the outer wall layer (Fig. 6). In this way, potentially lethal enzyme activity was restricted to one small area of the wall at a time. Examination of thin sections of budding arthrobacters has shown that the mother cell contains less ribosomes than the daughter cell (Gray *et al.*, 1975), suggesting that a process of ageing analogous to that discussed by Postgate (this volume) for *Rhodomicrobium vannielli* is occurring. However, in view of the high viability of populations of these organisms in continuous flow culture, even at low growth rates (Luscombe & Gray, 1974), it does not seem likely that this is an important factor causing death.

A further factor that might promote longevity is the structure of the peptidoglycan or other skeletal material in the wall. However, investigation of the peptidoglycans of a range of arthrobacters have not produced conclusive evidence on this point. They do not possess any unique amino acids in the interpeptide bridges, and the composition of the

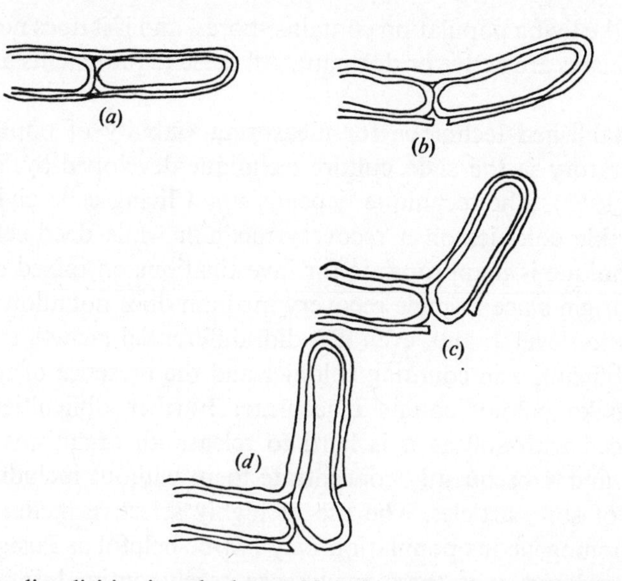

Fig. 6. Bending division in arthrobacter. Note the rupture in outer wall layer, caused by the outgrowth of a new bud. (After Gray *et al.*, 1975.)

bridges does not correlate wih susceptibility to lysis (Duxbury, 1973). However, organisms which have many amino acids in the interpeptide bridge are often more susceptible to lysis. *Staphylococcus aureus* with five or six residues and *S. afermentans* with four residues are less stable than those with three residues or less, which include many of the arthrobacters (Schleifer & Kandler, 1972). Krulwich *et al.* (1967) showed that the length of the interpeptide bridges of *Arthrobacter crystallopoietes* increased during the transition from rod to coccus and suggested this might lead to a weakening of the wall. These suggestions are difficult to test since so many other factors such as occurrence of polysaccharide in the wall may be involved.

Microbes may also produce toxins inhibiting soil predators and Coler, Gunner & Zuckerman (1968) have shown that some arthrobacters produce water-soluble metabolites which deter protozoa and are toxic for some aquatic vertebrates and invertebrates.

MEASUREMENT OF SURVIVAL IN SOIL

Measurement of survival of vegetative microbes in soil depends upon the use of an accurate technique to estimate the numbers and proportions of living and dead organisms in soil at different times. If useful information on survival is to be obtained, it is also necessary to know whether the living population contains spores, and if it does not, whether the organisms are active or dormant. All these requirements are difficult to fulfil.

An established technique for measuring viability of populations in the laboratory is the slide culture technique developed by Postgate & Hunter (1962). The technique depends upon living cells giving rise to recognizable colonies on a recovery medium while dead cells do not. This technique is inappropriate for investigations on mixed cultures of natural origin since a single recovery medium does not allow all microbial cells to develop and, even if it did, differential growth rates would lead to difficulties in counting colonies and the presence of motile cells would make colony counts inaccurate. Further difficulties may be experienced with soil as it is hard to release all organisms from soil particles and subsequently concentrate them without including a large number of soil particles. The use of highly selective media to obtain a more homogeneous population may not be helpful as Postgate (1969) has pointed out that these media are rarely optimal for the types selected. Many of these considerations also apply to the use of viable counting techniques which are all likely to underestimate the numbers of organisms present. Specific problems connected with the use of these techniques for soil organisms have been reviewed by Parkinson, Gray & Williams (1971) and Jensen (1968).

Many soil microbiologists have used stains that are taken up only by dead cells to determine the number of viable bacteria, following work of Strugger (1948) with acridine orange. Postgate (1967) has suggested that in situations where mild stresses, e.g. starvation, are applied these techniques are unreliable since the permeability barriers of the cell may persist after death, thus preventing uptake of dye. Anderson & Westmoreland (1971) proposed the use of a mixture of europium chelate and a fluorescent brightener. Recently, Anderson & Slinger (1975) have claimed that only living (or recently dead) cells containing nucleic acids pick up the europium chelate and stain red while all cells pick up the green fluorescent brightener. When both stains are taken up, the red dye always masks or displaces the green dye. The assumption that nucleic acids disappear from dead cells quite quickly seems to be unsupported and the occurrence of 'viabilities' of 68 per cent in a sandy clay-loam would be high, considering the stresses the population may be subjected to. Despite this, direct counts of this type seem to offer the only prospect of estimating the viability of mixed populations, although viability of specific organisms in a mixed population can still be achieved only by determining numbers of viable organisms on selective media.

Data obtained by using viable counting techniques have to be interpreted with care for if the organism produces spores, it will not be clear whether the survivors are vegetative. Some evidence on this point has been obtained by Warcup (1955) for fungi. He prepared dilution plates and after a short incubation period searched for young fungal colonies and determined the nature of the propagule giving rise to the colony. The procedure was repeated twice daily until no further colonies developed and in this way he showed that almost all the colonies arose from spores. Thus, unless it is clear that spores are not produced, measurements of the viability and survival of populations of vegetative fungi are difficult, with the possible exception of the survival of structures such as sclerotia. Vegetative spore-forming bacteria can be counted on dilution plates, identifying the colonies with fluorescent antisera, and repeating the experiment after heating duplicate soil samples (Siala, Hill & Gray, 1974).

If it is shown that the survivors in a population are vegetative, then it is also necessary to determine whether they are actively growing in restricted micro-environments and thus surviving saprophytically, *sensu* Garrett (1950), or they are dormant. The approach to this problem has usually been indirect and has been considered earlier, but in view of these difficulties, it is not surprising that the data on survival of dormant vegetative organisms in soil is difficult to interpret. However, some of the better-documented cases will now be considered, concerning the survival of alien species such as plant pathogens and indigenous organisms such as rhizobia and arthrobacters.

Plant pathogenic bacteria

Buddenhagen (1965) has suggested that it is unrealistic to expect plant pathogens to have a large soil-saprophytic population and that it is more likely they behave like other zymogenous forms and decline to low and undetectable levels in the absence of their normal substrate. Survival of plant pathogens takes place principally in the host material and so there have been practically no studies on the fundamental problems of why these bacteria are poor survivors in soil. In a few cases, plant pathogenic bacteria have been shown to survive in soil and in some cases may even multiply in the absence of the host. An outstanding example is *Agrobacterium tumefaciens* which can survive in many soils for long periods, providing that they remain at equable temperatures, do not dry out and remain close to neutrality. Thus, Dickey (1961) showed that, in unsterilized soil at pH 7.2, cells declined in number from 1.3×10^7 g^{-1} soil to 1.93×10^5 g^{-1} soil after 90 days and 2.55×10^4 g^{-1} soil after 146 days. In contrast, addition of sulphur to the soil, which lowered the pH to 4.2, caused the population to drop to 2.0×10^2 g^{-1} soil after 146 days.

Lee (1920) showed that *Xanthomonas citri* survived for less than nine days in unsterile soil, but for more than 14 days in sterile soil, kept in the laboratory. Under field conditions, survival was rarely for more than three days. This suggests that the soil microflora plays an active role in removing these bacteria from soil, especially as this same bacterium can survive in air-dried soil for 166 days (Fulton, 1920). It is difficult to interpret these survival experiments since the pathogen was detected by induction of disease symptom in test plants, the object of such experiments being to demonstrate whether the organism could over-winter in soil and infect plants in the next growing season. It is possible that the pathogen remains viable for much longer periods, albeit in reduced numbers.

Corynebacterium insidiosum behaves in a similar manner (Nelson & Semeniuk, 1963), for high populations of between 10^6 and 10^9 g^{-1} soil disappear from unsterile soil in four to seven days although they may survive for more than 39 days in sterile soil. Temperature seems to be another important factor affecting the survival of this organism, for in experiments carried out near freezing point the bacterium persisted for more than 66 days, even in unsterile soil.

As mentioned earlier, many plant pathogens survive for longer periods on host debris and *Xanthomonas malvacearum* is a case in point. The bacteria survived for six to seven months in plant debris on the soil surface, but declined to low levels in 40–107 days when the debris was buried in the soil. The decline was hastened at temperatures optimal for decomposition of the debris (Brinkerhoff & Fink, 1964).

Rhizobia

More quantitative information is available concerning the survival of rhizobia in soil in the absence of the host plant, principally because of the interest in preparing commercial inoculants for application to soils. Vincent (1958) expressed the loss of viability of rhizobia with time as k, the average logarithmic decline per week, its reciprocal giving the time in weeks required for a 10-fold reduction in the numbers of viable cells. He found that two death-rate categories were applicable to rhizobia: (*a*) where $k < 0.2$ such as survival in agar and peat cultures, low temperature storage, and freeze-drying; and (*b*) where k was between 0.5 and 100 such as in thin films on seed and in contact with fertilizers. Temperature coefficients for these death rates were all between 1.3 and

1.9. The suitability of soil as a medium for survival of rhizobia evidently depends upon several factors of which pH seems to be the most important. Thus in an experiment in which 1.1×10^4 clover rhizobia were added to each gram of soil, no viable cells could be recovered from soil with a pH of 4.8–5.0 after 24 h. If the soil pH was adjusted to 5.2 or 8.7, although there was a rapid fall in numbers, some survivors remained and multiplied so that a population larger than the inoculum had built up after 10 days. No initial loss of viability occurred when the pH was between 5.7 and 7.1. Similar data were obtained for field soils where introduced rhizobia failed to persist after lucerne had been harvested if the soil pH was between 5.3 and 5.7.

Survival in peat culture is also well documented. Roughley & Vincent (1967) have calculated that a commercial culture with an initial bacterial number of $2 \times 10^8\ g^{-1}$ of moist unsterilized peat could be expected to contain $1 \times 10^6\ g^{-1}$ after a period ranging from 18–115 weeks. They found that the most important factors prolonging survival were the sterilization of peat (especially for a cowpea strain), an effect most marked with γ irradiation, and a 40–50 per cent moisture content (wet wt basis) in the finished culture. Survival was adversely affected by the generation of heat of wetting as broth was added to dried peat and the generation of inhibitory compounds during drying of the peat. Roughley (1968) has also shown that restriction of aeration caused a rapid drop in the number of survivors, during the first week after preparation of the peat culture. This effect became more marked as the culture aged so that during the 8–12 week period, the weekly logarithmic death rate (k) was 0.047–0.074 in cotton wool plugged tubes but 0.25–1.2 in cans and mylarpolythene laminate containers. Steinborn & Roughley (1974) showed that some coastal peats which contain sodium chloride adversely affected survival but leaching of the peats reduced the weekly logarithmic death rate from 2.13 to 0.04.

Arthrobacter

A considerable part of this review has concerned the longevity of arthrobacters but much of this information comes from experiments in culture which may or may not be relevant to natural conditions. Surprisingly little is known about the survival of this organism in soil, apart from its resistance to desiccation. Robinson, Salonius & Chase (1965) compared the ability of *Arthrobacter* species and *Pseudomonas* species to survive air-drying and concluded that, following an initial fall, numbers of arthrobacters hardly changed after one month of drying, while there was a further 100-fold reduction in the numbers of pseudomonads. Boylen (1973) confirmed this and showed that when either rods or cocci of *A. crystallopoietes* were added to sand and air-dried, about half the population died quickly while the remainder survived for at least six months. He calculated that desiccated cocci converted 0.0005 per cent h^{-1} of their carbon to carbon dioxide, implying a half-life of about 12 years for self consumption.

On the other hand Lowe & Gray (1973*a*) have shown that *A. globiformis* is sensitive to acidity in soil. When this organism was reintroduced into a forest soil of pH 4.3 from which it had originally been isolated, bacterial populations of 10^6 g^{-1} soil had fallen to less than 10^4 g^{-1} within 24 h and no oxygen uptake could be detected. They found that populations declined in this way unless the pH was amended to a level above 5.5 and that amendment of the pH to 7.5 allowed growth and the formation of populations in excess of 10^8 g^{-1}.

The ability of arthrobacters to survive in mixed cultures in soil has also been investigated by Salonius, Robinson & Chase (1970) who found that the outcome of competition experiments between *A. globiformis* and *Pseudomonas fluorescens* in γ-irradiated soil depended upon temperature and moisture content: high temperatures and lower water contents favoured arthrobacter. Lowe & Gray (1973*b*) found that arthrobacters competed with varying success when grown with staphylococci or micrococci in soil. In some cases, arthrobacters and cocci grew without affecting the total yield of the competitors while in other cases, severe reductions in yield of both competitors were found. No correlation between the ability of the organisms to compete and their overall similarity to one another was found, as might have been expected, but this could have been due to the spatial separation of potential competitors on the surfaces of the soil particles.

Sclerotia

Measurements of the longevity of the vegetative stages of fungi in soil are more difficult to interpret because of the possibility of the presence of spores. However, Coley-Smith & Cooke (1971) have summarized the data concerning survival of sclerotia, including some records of the viability of sclerotia after varying periods of time. Extremely long survival was recorded, whether sclerotia were kept in dry soil, moist soil or field conditions. Thus, after keeping sclerotia of *Phymatotrichum omnivorum* in moist soil in the laboratory, 8 per cent of sclerotia were still viable after seven to nine years, while 100 per cent of sclerotia of *Rhizoctonia tuliparum* were viable after two years in the field, and 70 per cent of sclerotia of *Botrytis convoluta* after one year in dry storage. One problem in interpreting these figures is that some sclerotia can germinate to produce secondary sclerotia so that those recorded at the end of the experiment may not be the original ones. Coley-Smith & Cooke (1971) concluded that the most important environmental factors determining longevity of sclerotia were moisture, temperature, aeration and soil organic matter. In general, sclerotia are extremely resistant to desiccation and often survive less well in moist soils, probably because of the increased rate of decay of the sclerotia caused by other soil organisms. Fluctuating temperatures may cause sclerotia to germinate and consume large amounts of reserve material, leading to death; likewise, oxygen may cause germination of the sclerotium, so that sclerotia remain intact longest below the soil surface. However, Griffin & Nair (1968) showed that the failure of *Sclerotium rolfsii* sclerotia to germinate was not due to low oxygen or high carbon dioxide tensions, which suggests that other volatile inhibitors may be important (Coley-Smith & Cooke, 1971).

Most of the work on the effect of environment on sclerotia has thus been concerned with germination rather than survival, but the two phenomena are closely related since germination depletes the energy reserve of the sclerotium each time it occurs. This depletion may be caused either by germination itself or by the activities of the rest of the soil microflora following germination.

I wish to thank Mr S. J. Chapman, Dr T. Duxbury and Dr B. M. Luscombe for permission to use some of their data, and the Science Research Council for a grant enabling part of the work on *Arthrobacter* to be carried out.

REFERENCES

ALEXANDER, M. (1964). Biochemical ecology of soil micro-organisms *Annual Review of Microbiology*, **18**, 217–252.

ALEXANDER, M. (1971). Biochemical ecology of micro-organisms. *Annual Review of Microbiology*, **25**, 361–392.

ALLISON, F. E. (1968). Soil aggregation – some facts and fallacies as seen by a microbiologist. *Soil Science*, **106**, 136–143.

ANDERSON, J. P. E. & DOMSCH, K. H. (1975). Measurement of bacterial and fungal contributions to respiration of selected agricultural and forest soils. *Canadian Journal of Microbiology*, **21**, 314–322.

ANDERSON, J. R. & SLINGER, J. M. (1975). Europium chelate and fluorescent brightener staining of soil propagules and their photomicrographic counting. II. Efficiency. *Soil Biology and Biochemistry*, **7**, 211–215.

ANDERSON, J. R. & WESTMORELAND, D. (1971). Direct counts of soil organisms using a fluorescent brightener and a europium chelate. *Soil Biology and Biochemistry*, **3**, 85–87.

BAKER, R. (1968). Mechanisms of biological control of soil-borne pathogens. *Annual Review of Phytopathology*, **6**, 263–294.

BALLESTA, J.-P. G. & ALEXANDER, M. (1968). Cell walls and lysis of *Zygorhyncus* species. *Bacteriological Proceedings*, 1968, 39.

BLOOMFIELD, B. J. & ALEXANDER, M. (1967). Melanins and resistance of fungi to lysis. *Journal of Bacteriology*, **93**, 1276–1280.

BOYLEN, C. W. (1973). Survival of *Arthrobacter crystallopoietes* during prolonged periods of extreme desiccation. *Journal of Bacteriology*, **113**, 33–57.

BOYLEN, C. W. & ENSIGN, J. C. (1970). Intracellular substrates for endogenous metabolism during long-term starvation of rod and spherical cells of *Arthrobacter crystallopoietes. Journal of Bacteriology*, **103**, 578–587.

BRINKERHOFF, L. A. & FINK, G. B. (1964). Survival and infectivity of *Xanthomonas malvacearum* in cotton plant debris and soil. *Phytopathology*, **54**, 1198–1201.

BROWN, M. F. & WYLLIE, T. D. (1970). Ultrastructure of micro-sclerotia of *Verticillium albo-atrum. Phytopathology*, **60**, 538–542.

BRUEHL, G. W. & LAI, P. (1968). Influence of soil pH and humidity on survival of *Cephalosporium gramineum* in infested wheat straw. *Canadian Journal of Plant Sciences*, **48**, 245–252.

BUDDENHAGEN, I. W. (1965). The relation of plant-pathogenic bacteria to the soil. In *The Ecology of Soil-borne Plant Pathogens*, ed. K. F. Baker & W. C. Snyder, pp. 269–284. Berkeley: University of California Press.

BULL, A. T. (1970). Inhibition of polysaccharases by melanin: enzyme inhibition in relation to mycolysis. *Archives of Biochemistry and Biophysics*, **137**, 345–356.

BURGES, A. (1960). Dynamic equilibria in the soil. In *The Ecology of Soil Fungi*, ed. D. Parkinson & J. S. Waid, pp. 185–191. Liverpool: Liverpool University Press.

BURGES, A. & FENTON, E. (1953). The effect of carbon dioxide on certain soil fungi. *Transactions of the British Mycological Society*, **36**, 104–108.

BURGESS, L. W. & GRIFFIN, D. M. (1967). Competitive saprophytic colonization of wheat straw. *Annals of Applied Biology*, **60**, 137–142.

BURNETT, J. H. (1969). *Fundamentals of Mycology*. London: Edward Arnold.

BUTLER, F. C. (1959). Saprophytic behaviour of some cereal root-rot fungi. IV. Saprophytic survival in soils of high and low fertility. *Annals of Applied Biology*, **47**, 28–36.

BYTHER, R. (1965). Ecology of plant pathogens in soil. V. Inorganic nitrogen utilization as a factor of competitive saprophytic ability of *Fusarium roseum* and *F. solani. Phytopathology*, **55**, 852–858.

CHAMBERS, S. C. & FLENTJE, N. T. (1969). Relative effects of soil nitrogen and soil organisms on survival of *Ophiobolus graminis. Australian Journal of Biological Sciences*, **22**, 275–278.

CHEN, M. & ALEXANDER, M. (1972). Resistance of soil microorganisms to starvation. *Soil Biology and Biochemistry*, **4**, 283–288.

CHESHIRE, M. V., GREAVES, M. P. & MUNDIE, C. M. (1974). Decomposition of soil polysaccharide. *Journal of Soil Science*, **25**, 483–498.

CLARK, F. E. & PAUL, E. A. (1970). The microflora of grassland. *Advances in Agronomy*, **22**, 375–435.

COLER, R. A., GUNNER, H. B. & ZUCKERMAN, B. M. (1968). Model mechanisms for soil population selection. *Bacteriological Proceedings*, 1968, 3.

COLEY-SMITH, J. R. & COOKE, R. C. (1971). Survival and germination of fungal sclerotia. *Annual Review of Phytopathology*, **9**, 65–92.

COOK, R. J. & PAPENDICK, R. I. (1970). Soil water potential as a factor in the ecology of *Fusarium roseum f.* sp. *cerealis* 'culmorum'. *Plant and Soil*, **32**, 131–145.

COOKE, R. C. & MITCHELL, D. T. (1969). Sugars and polyols in sclerotia of *Claviceps purpurea, C. nigricans* and *Sclerotinia curreyana* during germination. *Transactions of the British Mycological Society*, **52**, 365–372.

COURT, M. N., STEPHENS, R. C. & WAID, J. S. (1964). Toxicity as a cause of the inefficiency of urea as a fertilizer. II. Experimental. *Journal of Soil Science*, **15**, 49–65.

DICKEY, R. S. (1961). Relation of some edaphic factors to *Agrobacterium tumefaciens*. *Phytopathology*, **51**, 607–614.

DOWDELL, R. J., SMITH, K. A., CREIS, R. & RESTALL, S. W. F. (1972). Field studies of ethylene in the soil atmosphere – equipment and preliminary results. *Soil Biology and Biochemistry*, **4**, 325–331.

DUXBURY, T. (1973). The cell wall as a factor in the survival of *Arthrobacter*. Ph.D. Thesis, University of Liverpool.

ERGLE, D. R. (1947). The glycogen content of *Phymatotrichum* sclerotia. *Journal of the American Chemical Society*, **69**, 2061–2062.

ERGLE, D. R. (1948). The carbohydrate metabolism of germinating *Phymatotrichum* sclerotia with special reference to glycogen. *Phytopathology*, **38**, 142–151.

FELLOWS, H. (1941). Effect of certain environmental conditions on the prevalence of *Ophiobolus graminis* in the soil. *Journal of Agricultural Research*, **63**, 715–726.

FOTH, H. D. & TURK, L. M. (1972). *Fundamentals of Soil Science*, 5th edn. New York: John Wiley.

FULTON, H. R. (1920). Decline of *Pseudomonas citri* in the soil. *Journal of Agricultural Research*, **19**, 207–223.

GARRETT, S. D. (1936). Soil conditions and the take-all disease of wheat. *Annals of Applied Biology*, **23**, 667–699.

GARRETT, S. D. (1940). Soil conditions and the take-all disease of wheat. V. Further experiments on the survival of *Ophiobolus graminis* in infected wheat stubble buried in the soil. *Annals of Applied Biology*, **27**, 199–204.

GARRETT, S. D. (1950). Ecology of the root-inhabiting fungi. *Biological Reviews*, **25**, 220–254.

GARRETT, S. D. (1956). *Biology of Root-infecting Fungi*. London: Cambridge University Press.

GARRETT, S. D. (1966). Cellulose-decomposing ability of some cereal foot-rot fungi in relation to their saprophytic survival. *Transactions of the British Mycological Society*, **50**, 519–524.

GARRETT, S. D. (1970). *Pathogenic Root-infecting Fungi*. London: Cambridge University Press.

GARRETT, S. D. (1975). Sources of variability in determinations of the median period of saprophytic survival by *Cochliobolus sativus*. *Transactions of the British Mycological Society*, **64**, 351–355.

GORDEE, R. S. & PORTER, C. L. (1961). Structure, germination and physiology of micro-sclerotia of *Verticillium albo-atrum*. *Mycologia*. **53**, 171–182.

GRAY, T. R. G., DUXBURY, T., LUSCOMBE, B. M. & CHAPMAN, S. J. (1975). *Arthrobacter globiformis*, a successful soil bacterium. In *Proceedings of the 1st Intersectional Congress of the International Association of Microbiological Societies*, **2**, 262–276.

GRAY, T. R. G., HISSETT, R. & DUXBURY, T. (1973). Bacterial populations of litter and soil in a deciduous woodland. II. Numbers, biomass and growth rates. *Revue d'Ecologie et Biologie du Sol*, **11**, 15–26.

GRAY, T. R. G. & WILLIAMS, S. T. (1971). Microbial productivity in soil. *Symposia of the Society for General Microbiology*, **21**, 255–286.

GREENWOOD, D. J. (1968). Measurement of microbial metabolism in soil. In *The Ecology of Soil Bacteria*, ed. T. R. G. Gray & D. Parkinson, pp. 138–157. Liverpool: Liverpool University Press.

GREENWOOD, D. J. & GOODMAN, D. (1965). Oxygen diffusion and aerobic respiration in columns of fine crumbs. *Journal of Science of Food and Agriculture*, **16**, 152–160.

GRIFFIN, D. M. (1972). *Ecology of Soil Fungi*. London: Chapman & Hall.

GRIFFIN, D. M. & NAIR, N. G. (1968). Growth of *Sclerotium rolfsii* in different concentrations of oxygen and carbon dioxide. *Journal of Experimental Botany*, **19**, 812–816.

HARDER, W. & VELDKAMP, H. (1971). Competition of marine psychrophilic bacteria at low temperatutes. *Antonie van Leeuwenhoek*, **37**, 51–63.

HARTLEY, G. S. & ROE, J. W. (1940). Ionic concentrations at surfaces. *Transactions of the Faraday Society*, **36**, 101–109.

HISSETT, R. & GRAY, T. R. G. (1973). Bacterial populations of litter and soil in a deciduous woodland. 1. Qualitative studies. *Revue d'Ecologie et Biologie du Sol*, **10**, 495–508.

HISSETT, R. & GRAY, T. R. G. (1976). Microsites and time changes in soil microbe ecology. In *The Role of Aquatic and Terrestrial Organisms in Decomposition Processes*, ed. J. Anderson. London: British Ecological Society, (in press).

HOWELL, D. M. & FERGUS, C. L. (1964). The component fatty acids found in sclerotia of *Sclerotium rolfsii*. *Canadian Journal of Microbiology*, **10**, 616–618.

JANNASCH, H. W. (1967). Enrichments of aquatic bacteria in continuous culture. *Archiv für Mikrobiologie*, **59**, 165–173.

JENSEN, V. (1968). The plate count technique. In *The Ecology of Soil Bacteria*, ed. T. R. G. Gray & D. Parkinson, pp. 158–170. Liverpool: Liverpool University Press.

KRULWICH, T. A., ENSIGN, J. C., TIPPER, D. J. & STROMINGER, J. L. (1967). Sphere–rod morphogenesis in *Arthrobacter crystallopoietes*. II. Peptides of the cell wall peptidoglycan. *Journal of Bacteriology*, **94**, 741–750.

LEE, H. A. (1920). Behaviour of the citrus canker organisms in the soil. *Journal of Agricultural Research*, **19**, 189–206.

LINDSEY, D. L. (1965). Ecology of plant pathogens in soil. III. Competition between soil fungi. *Phytopathology*, **55**, 104–110.

LOWE, W. E. & GRAY, T. R. G. (1972). Ecological studies on coccoid bacteria in a pine forest soil I. Classification. *Soil Biology and Biochemistry*, **4**, 459–467.

LOWE, W. E. & GRAY, T. R. G. (1973*a*). Ecological studies on coccoid bacteria in a pine forest soil. II. Growth of bacteria introduced into soil. *Soil Biology and Biochemistry*, **5**, 449–462.

LOWE, W. E. & GRAY, T. R. G. (1973*b*). Ecological studies on coccoid bacteria in a pine forest soil. III. Competitive interactions between bacterial strains in soil. *Soil Biology and Biochemistry*, **5**, 463–472.

LUSCOMBE, B. M. & GRAY, T. R. G. (1974). Characteristics of Arthrobacter grown in continuous culture. *Journal of General Microbiology*, **82**, 213–222.

LYNCH, J. M. & HARPER, S. H. T. (1974*a*). Ethylene formation by a soil fungus. *Journal of General Microbiology*, **80**, 187–195.

LYNCH, J. M. & HARPER, S. H. T. (1974*b*). Fungal growth rate and the formation of ethylene in soil. *Journal of General Microbiology*, **85**, 91–96.

MACER, R. C. F. (1961). Saprophytic colonization of wheat straw by *Cercosporella herpotrichoides* Fron and other fungi. *Annals of Applied Biology*, **49**, 239–256.

MACFADYEN, A. (1970). Soil metabolism in relation to ecosystem energy flow and to primary and secondary production. In *Methods for the Study of Production and Energy Flow in Soil Communities*, ed. J. Phillipson, pp. 167–172. Paris: Unesco.

McLaren, A. D. & Skujins, J. (1968). The physical environment of micro-organisms in soil. In *The Ecology of Soil Bacteria*, ed. T. R. G. Gray & D. Parkinson, pp. 3–24. Liverpool: Liverpool University Press.

Meers, J. L. & Tempest, D. W. (1968). The influence of extracellular products on the behaviour of mixed microbial populations in magnesium-limited chemostat cultures. *Journal of General Microbiology*, **52**, 309–317.

Mitchell, D. T. & Cooke, R. C. (1968). Water uptake, respiration pattern and lipid utilization in sclerotia of *Claviceps purpurea* during dormancy and germination. *Transactions of the British Mycological Society*, **51**, 7331–7336.

Mulder, E. G. & Zevenhuizen, L. P. T. M. (1967). Coryneform bacteria of the *Arthrobacter* type and their reserve material. *Archiv für Mikrobiologie*, **59**, 345–354.

Nelson, G. A. & Semeniuk, G. (1963). Persistence of *Corynebacterium insidiosum* in the soil. *Phytopathology*, **53**, 1167–1169.

Nye, P. H. (1972). The measurement and mechanism of ion diffusion in soils. VIII. A theory for the propagation of changes of pH in soils. *Journal of Soil Science*, **23**, 82–92.

Nykvist, N. (1959). Leaching and decomposition of litter. I. Experiments on leaf litter of *Fraxinus excelsior*. *Oikos*, **10**, 190–211.

Papavizas, G. C. & Davey, C. B. (1962). Activity of *Rhizoctonia* in soil as affected by carbon dioxide. *Phytopathology*, **52**, 759–766.

Parkinson, D., Gray, T. R. G. & Williams, S. T. (1971). *Methods for Studying the Ecology of Soil Microorganisms*. IBP Handbook no. 18. Oxford: Blackwell.

Pengra, R. M., Cole, M. A. & Alexander, M. (1969). Cell walls and lysis of *Mortierella parvispora* hyphae. *Journal of Bacteriology*. **37**, 1056–1061.

Plice, M. J. (1934). Acidity, antacid buffering, and nutrient content of forest litter in relation to humus and soil. *Memoirs of the Cornell University Agricultural Experiment Station*, **166**, 1–32.

Postgate, J. R. (1967). Viability measurements and the survival of microbes under minimum stress. *Advances in Microbial Physiology*, **1**, 1–23.

Postgate, J. R. (1969). Viable counts and viability. In *Methods in Microbiology*, **1**, ed. J. R. Norris & D. W. Ribbons, pp. 611–628. London: Academic Press.

Postgate, J. R. & Hunter, J. R. (1962). The survival of starved bacteria. *Journal of General Microbiology*, **29**, 233–263.

Potgieter, H. J. & Alexander, M. (1966). Susceptibility and resistance of several fungi to microbial lysis. *Journal of Bacteriology*, **91**, 1526–1532.

Rixon, A. J. & Bridge, B. J. (1968). Respiration quotient arising from microbial activity in relation to matric suction and air-filled pore space of soil. *Nature, London*, **218**, 961–962.

Robinson, J. B., Salonius, P. O. & Chase, F. E. (1965). A note on the differential response of *Arthrobacter* spp. and *Pseudomonas* spp. to drying soil *Canadian Journal of Microbiology*, **11**, 746–748

Rommell, L. G. (1922). Luft vaxlingen i marken som ekologisk faktor. (Die bodenventilation als okolozischer Faktor). *Meddelser om Skogsforsoksvaesen*, **19**.

Roughley, R. J. (1968). Some factors influencing the growth and survival of root-nodule bacteria in peat culture. *Journal of Applied Bacteriology*, **31**, 259–265.

Roughley, R. J. & Vincent, J. M. (1967). Growth and survival of *Rhizobium* spp. in peat culture. *Journal of Applied Bacteriology*, **30**, 362–376.

Salonius, P. O., Robinson, J. B. & Chase, F. E. (1970). The mutual growth of *Arthrobacter globiformis* and *Pseudomonas fluorescens* in gamma sterilised soil. *Plant and Soil*, **32**, 216–326.

Schleifer, K. H. & Kandler, O. (1972). Peptidoglycan types of bacterial cell walls and their taxonomic implications. *Bacteriological Reviews*, **36**, 407–477.

SHIELDS, J. A., LOWE, W. E., PAUL, E. A. & PARKINSON, D. (1973). Turnover of microbial tissue under field conditions. *Soil Biology and Biochemistry*, **5**, 753–764.

SIALA, A. H. & GRAY, T. R. G. (1974). Growth of *Bacillus subtilis* and spore germination in soil observed by a fluorescent antibody technique. *Journal of General Microbiology*, **81**, 191–198.

SIALA, A., HILL, I. R. & GRAY, T. R. G. (1974). Populations of spore-forming bacteria in an acid forest soil, with special reference to *Bacillus subtilis*. *Journal of General Microbiology*, **81**, 183–190.

SMITH, A. M. (1973). Ethylene as a cause of soil fungistasis. *Nature, London*, **246**, 311–313.

SMITH, A. M. (1975). Ethylene as a critical regulator of microbial activity in soil. *Proceedings of the 1st Intersectional Congress of the International Association of Microbiological Societies*, **2**, 463–473.

SMITH, K. A. & RESTALL, S. W. F. (1971). The occurrence of ethylene in anaerobic soil. *Journal of Soil Science*, **22**, 430–443.

SOBEK, J. M., CHARBA, J. F. & FOUST, W. N. (1966). Endogenous metabolism of *Azotobacter agilis*. *Journal of Bacteriology*, **92**, 687–695.

SOULIDES, D. A. (1964). Antibiotics in soil. VI. Determination of micro-quantities of antibiotics in soil. *Soil Science*, **97**, 286–289.

SOULIDES, D. A. (1965). Antibiotics in soil. VII. Production of streptomycin and tetracyclines in soil. *Soil Science*, **100**, 200–206.

STEINBORN, J. & ROUGHLEY, R. J. (1974). Sodium chloride as a cause of low numbers of *Rhizobium* in legume inoculants. *Journal of Applied Bacteriology*, **37**, 93–99.

STRUGGER, S. (1948). Fluorescence microscope examination of bacteria in soil. *Canadian Journal of Research B*, **26**, 188–193.

TAYLOR, J. & WILLIAMS, P. (1975). Theoretical studies on the coexistence of competing species under continuous flow conditions. *Canadian Journal of Microbiology*, **21**, 90–98.

VELDKAMP, H. & KUENEN, J. G. (1973). The chemostat as a model system for ecological studies. *Bulletin from the Ecological Research Committee*, **17**, ed. T. Rosswall, pp. 347–355. Stockholm: Swedish Natural Science Research Council.

VINCENT, J. M. (1958). Survival of the root-nodule bacteria. In *Nutrition of the Legumes*, ed. E. G. Hallsworth, pp. 108–123. London: Butterworth.

WARCUP, J. H. (1955). On the origin of colonies of fungi developing on soil dilution plates. *Transactions of the British Mycological Society*, **38**, 298–301.

WEISS, L. (1963). The pH value at the surface of *Bacillus subtilis*. *Journal of General Microbiology*, **32**, 331–340.

WILLIAMS, S. T. & MAYFIELD, C. I. (1971). Studies on the ecology of actinomycetes in soil. III. The behaviour of neutrophilic streptomycetes in acid soils. *Soil Biology and Biochemistry*, **3**, 197–208.

WRIGHT, J. M. (1956). The production of antibiotics in soil. III. Production of gliotoxin in wheat straw buried in soil. *Annals of Applied Biology*, **44**, 461–466.

ZEVENHUIZEN, L. P. T. M. (1966). Function, structure and metabolism of the intracellular polysaccharide of *Arthrobacter*. *Mededelingen Landbouwhogeschool, Wageningen*, **66–10**, 1–80.

THE CONSEQUENCES OF THYMINE STARVATION

K. A. STACEY

Biological Laboratory, University of Kent at Canterbury, Canterbury CT2 7NJ, Kent

INTRODUCTION

Thymine occupies a special place in cellular metabolism and starvation for thymine has special consequences. This is true for all dividing cells but I shall be concerned here only with thymine starvation in bacteria and in *Escherichia coli* particularly. Thymine is the base that distinguishes DNA from RNA and DNA cannot be replicated in the absence of deoxythymidine triphosphate, the actual metabolite that is lacking when we speak of thymine starvation. It was curiosity about the effects of stopping DNA synthesis specifically – while (presumably) allowing all other metabolic processes to continue normally – that led Seymour Cohen (Cohen & Barner, 1954) to the discovery of thymineless death – the loss of colony-forming ability that follows a prolonged period of growth of a thymine auxotroph in the absence of thymine. Starvation for a growth requirement is usually simply bacteriostatic and most of the exceptions can be easily explained. Thus the death that follows starvation for diaminopimelic acid is due to the failure to make the rigid cell wall necessary to protect bacterial cells from the osmotic pressure difference between the cell and the surrounding medium (Meadow & Work, 1956). An explanation for thymineless death has proved much more elusive but the phenomenon has exerted a curious attraction for microbiologists in the twenty years since the first report from Cohen's laboratory. Thymine starvation has obvious parallels with the effect of ultra-violet light (UV) irradiation; it, too, is mutagenic and recombinogenic and, in some lysogenic strains, leads to prophage induction. A number of hypotheses to account for these effects have been advanced but, even now, which of these is correct is uncertain.

BIOSYNTHESIS OF THYMIDINE TRIPHOSPHATE

Thymine does not exist as such in the cell; thymidylate is synthesized by the methylation of deoxyuridylate and this is further phosphorylated to give thymidine triphosphate (TTP) (Fig. 1; Blakely, 1969). The

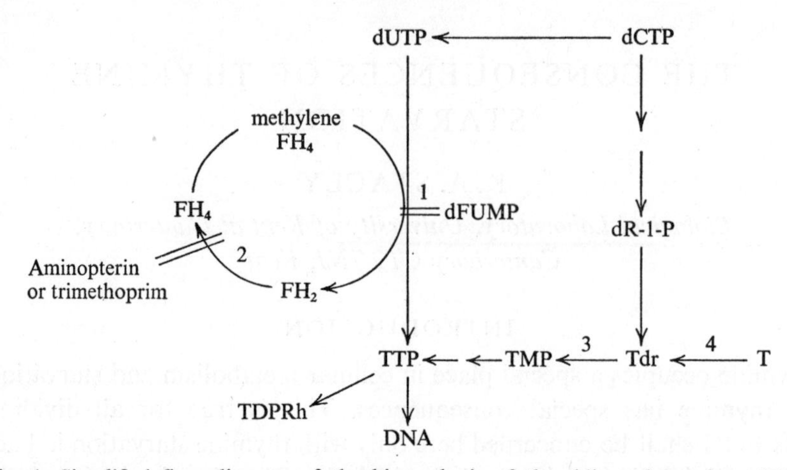

Fig. 1. Simplified flow diagram of the biosynthesis of thymidine triphosphate (TTP). Enzymes referred to in the text: 1, thymidylate synthetase; 2, dihydrofolate reductase; 3, thymidine kinase; 4, thymidine phosphorylase. dCTP, deoxycytidine triphosphate; dUTP, deoxyuridine triphosphate; dFUMP, 5-fluorodeoxyuridylate; dR-1-P, deoxyribose-1-phosphate; T, thymine; Tdr, thymidine; TMP, thymidylate; FH_2, dihydrofolate; FH_4, tetrahydrofolate; methylene FH_4, N^5, N^{10}-methylene tetrahydrofolate; TDPRh, thymidine diphosphate rhamnose.

thymidine residues found in transfer RNA (tRNA) result from the methylation of uridine at specific points in the tRNA precursors by a quite different enzyme, one that has *S*-adenosylmethionine for the methyl donor. They are formed only after the synthesis of the polynucleotide, presumably as part of the maturation process that confers the great specificity required of them in their role in protein synthesis. Although unproven, it is very unlikely that these methylations suffer any interference in thymine auxotrophs starving for thymine. The TTP pool is also drawn upon in the synthesis of thymidine diphosphate rhamnose (TDPRh) which is widely found in bacteria as a donor of the sugar residue for the synthesis of certain polysaccharides. It is unlikely that failure to add this sugar to those surface polymers that contain it has any significant effect on the course of the development of the lethal effects of thymine starvation; rhamnose appears to be one of the sugars that give individuality to the termini of the cell wall polysaccharides and is not involved in the backbone polymers responsible for rigidity. Certain mutant strains lack TDPRh, e.g. *E. coli* K12, C600, but the effect of thymine starvation in *thy*⁻ derivatives of this strain are in no way different from those observed in other strains of *E. coli* K12 (Stacey, unpublished observations).

It will be clear from Fig. 1 that the supply of TTP can be interrupted at a number of points and in a variety of ways. In *E. coli*, thymine auxotrophy is the result of the loss of function of thymidylate synthetase which is encoded by the *thy* gene. In *Bacillus subtilis* the genetics are more complex (Wilson, Farmer & Rothman, 1966).

Starvation for thymidylate can also be brought about by metabolic inhibition. 5-Fluorodeoxyuridylate is a powerful inhibitor of thymidylate synthetase and it is the conversion of fluorouracil to this compound by the same enzymes that act on thymine (see Fig. 1) that accounts for the major part of the toxicity of this compound (Heidelberger, 1965). Cells exposed to fluorodeoxyuridine (FUDR) (1 μg ml^{-1}) die much as if they were starved of thymine; the kinetics of the loss of viability are very similar and the treatments that prevent thymineless death protect equally against the action of FUDR (Cohen, Flaks, Barner & Lichtenstein, 1958). High concentrations of cytosine arabinoside (1 mg ml^{-1}) have a similar effect (Atkinson & Stacey, 1968) but in this case it is probable that it is the supply of deoxyuridylate which is limited. It is thought that the high concentrations are required because cytosine arabinoside is rapidly de-aminated to the inactive uracil derivative but it is not clear at these concentrations what the precise mechanism of its inhibitory action is on the supply of thymidylate. Again, conditions which limit thymineless death reduce the toxic action of cytosine arabinoside. Adenine arabinoside is far more toxic but the cause of death is even more obscure (Cohen, 1967).

The supply of the other substrate for thymidylate synthetase, i.e. N^5,N^{10}-methylene tetrahydrofolate, can also be restricted. Thymidylate synthetase is alone among the enzymes of 1C-metabolism that have tetrahydrofolate co-factors, in that the co-factor is oxidized to dihydrofolate in the course of the enzymic reaction. Thymidylate synthesis therefore requires a stoichiometric equivalent reduction of dihydrofolate (Fig. 1). Agents which block dihydrofolate reductase will, in rich media, produce the symptoms and the effects of thymine starvation. In minimal medium the lethal effect is far less striking because the depletion of the tetrahydrofolate pool brings to a halt the biosynthesis of purines and methionine, the formation of *N*-formyl-methionyl-tRNA, and the interconversion of glycine and serine (Blakely, 1969) and as we shall see, cells that cannot grow are protected against the lethal effects of thymine starvation. Although in rich media the bacteriocidal effect of agents, such as aminopterin and trimethoprim, which block dihydrofolate reductase is almost certainly due to thymineless death while in minimal media their effects are more nearly bacteriostatic, the selection of thymine auxotrophs is better in minimal media (Okada, Homma & Sonohara, 1962; Stacey & Simson, 1965). The continued action of thymidylate synthetase depletes the pool of tetrahydrofolate which is not replenished while the reductase is inhibited: cells in which the synthetase is inactivated by mutation escape this physiological impasse, and can synthesize all their growth requirements (Bertino & Stacey, 1966).

Thymine is mobilized by the sequential action of thymidine phosphorylase and thymidine kinase. Mutants lacking the phosphorylase will only grow if given thymidine; for those lacking the kinase, a mutation in the *thy* gene is lethal. Bacteria which lack a thymine-salvage pathway cannot be made thymine-dependent (Øvrebø & Kleppe, 1973). The pool of deoxyribose-1-phosphate (dR-1-P) derived from deoxycytidine triphosphate (dCTP) (Neuhard & Thomassen, 1971; Jensen, Leer & Nygaard, 1973), is increased if the enzymes for the pathway responsible for deoxyribose catabolism are inactivated by mutation, but even strains able to grow on low concentrations of thymine are unable to maintain as high an intracellular concentration of TTP as wild-type cells. Nevertheless, they grow with virtually the same doubling time as the wild-type under the same conditions by increasing the number of chromosome growing points to compensate for the lowered rate of chromosome replication (Meacock & Pritchard, 1975).

Although bromodeoxyuridine can be incorporated into DNA in place of thymidine, high levels of substitution lead to non-viability: the complete replacement in *thy* mutants by substituting bromouracil for the thymine in the growth medium leads to loss of viability with the same kinetics as thymineless death (Strelzoff, 1962) except that the lag phase is lengthened by a period approximately equivalent to one doubling time (Stacey, unpublished data).

THE EFFECT OF THYMINE STARVATION

The kinetics of the effect of thymine deprivation are usually studied by transferring a culture of a thymine auxotroph in the exponential phase of growth to the same medium lacking thymine. The other growth processes continue unabated and only DNA synthesis is immediately affected; the intracellular pool is usually exhausted in two or three minutes. Cell division falls off rapidly as, presumably, only those cells already triggered can divide; the continued growth is taken up by cell elongation and prolonged thymine starvation produces long filaments.

If the number of cells able to form colonies is assayed at regular intervals the viability after cell division stays constant for a period approximately equivalent to one doubling time in that medium (Fig. 2).

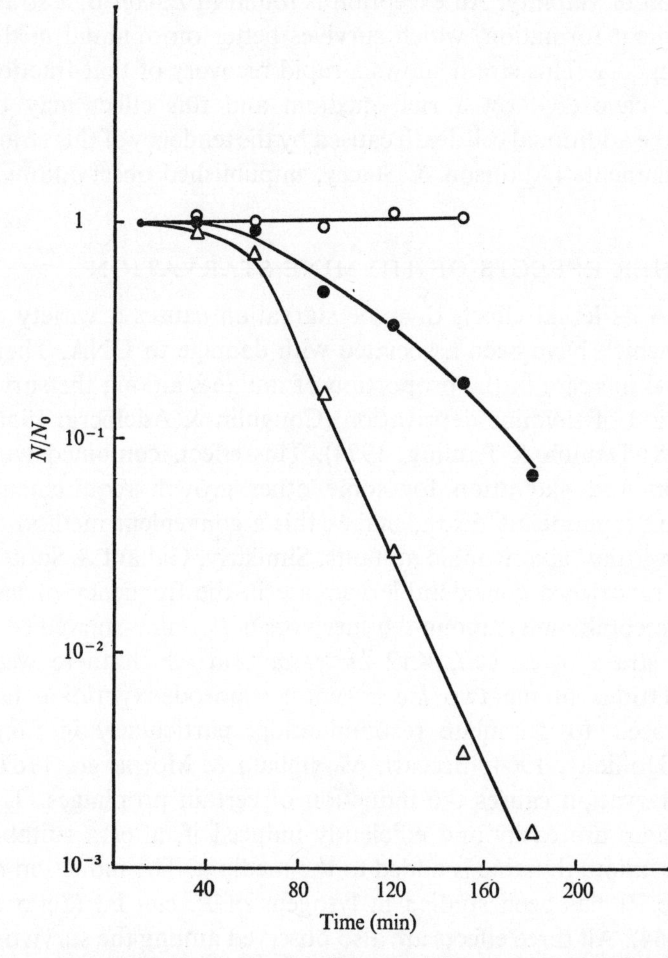

Fig. 2. Thymineless death of a thymine- and arginine-requiring strain of *E. coli* (B3003) under various conditions. (*a*) In M9 medium with arginine (△—△); (*b*) in the same medium lacking arginine (○—○); (*c*) the same medium lacking arginine but with 200 μg ml^{-1} chloramphenicol. (Atkinson & Stacey, unpublished data.) These curves show the lag phase, in this strain about the same as the doubling period (40 min in this medium) and the exponential phase. They show too that the rate of death is affected by conditions that affect growth. N/N_0 is the number of survivors divided by the original number of cells present.

The viability then falls steadily, sometimes until the surviving fraction (N/N_0) is as low as 10^{-4}, at a rate which reflects the growth rate.

A variety of methods of assaying viability show that the loss of colony formation is not caused by the transfer to a solid medium (Nakayama & Couch, 1973). The re-addition of thymine to the starvation medium immediately stops the decline in viability and although chromosomal single-strand breaks are rapidly repaired (see below) there is not, usually, any restoration of viability. An exception is found in *E. coli* B, a strain prone to filament formation, which survives better on minimal media than on rich media. This strain shows a rapid recovery of that fraction which would have died on a rich medium and this effect may be attributed to the additional cell death caused by the tendency of this strain to produce filaments (Atkinson & Stacey, unpublished observations).

OTHER EFFECTS OF THYMINE STARVATION

In addition to its lethal effect, thymine starvation causes a variety of other effects which have been associated with damage to DNA. There is a substantial increase in the proportion of mutants among the survivors of a period of thymine deprivation (Coughlin & Adelberg, 1956; Kanazir, 1958; Deutch & Pauling, 1974). This effect, combined with the protection that starvation for some other growth requirements affords against thymineless death, makes this a convenient method of selecting 'non-leaky' auxotrophic mutants. Similarly, Gallant & Spottswood (1965) reported a considerable increase in the frequency of lactose-positive recombinants among the survivors of thymine-starved cells of a diploid strain of *E. coli*, K12 $lac^-/Flac^-$, in which there were different mutations in the two *lac z* genes. Fluorodeoxyuridine has often been used to stimulate recombination, particularly in fungi (Esposito & Holliday, 1964; Beccari, Modigliani & Morpurgo, 1967).

Thymine starvation causes the induction of certain prophages. The lambdoid phages are easily and efficiently induced if, after a suitable period of starvation, thymine is added to the medium. The induction of bacteriophage P1 has been studied in lysogens of *E. coli* B3 (Seno & Melechen, 1964). All three effects are also observed among the survivors of UV irradiation and are thought to be consequences of DNA damage or side-effects of its repair. Consistent with this idea is the observation that thymine starvation causes a striking sensitization to the effects of UV light (Gallant & Suskind, 1961). It is the shoulder rather than the slope of the survival curve which is changed and the effect is not nearly as marked in these radiation sensitive mutants with small shoulders, such as those with defects in excision repair (uvr^-). In the very radiation-resistant *Micrococcus radiodurans* the reduction in the shoulder by a period of thymine starvation before UV irradiation results in a dramatic sensitization (Little & Hanawalt, 1973). Conversely UV irradiated cells have a shorter lag period before the onset of the lethal phase of thymine starvation.

Finally, thymine starvation can sometimes be used as a means of obtaining strains 'cured' of a plasmid; that is the survivors in a starved culture are more likely to have 'lost' the plasmid than the cells in a culture growing normally. Clowes, Moody & Pritchard (1965) reported this as a way of obtaining cured derivatives of strains carrying Col factors not easily obtained otherwise. The method is useful for curing some plasmids but not others and Pinney, Bremer & Smith (1974) have suggested the difference between the two kinds lies in whether or not the plasmid carries a gene for a nuclease which is induced under conditions of thymine deprivation.

An alternative view is that repair events are random and if the repair has been more efficient for the chromosome than for the plasmid the cell will survive but without the plasmid. If damage and its repair are distributed statistically then the difference between one plasmid and another might be determined by other factors, such as extent to which they are transcribed. However, the F factor sustains far more breaks per unit time than does an equivalent length of chromosomal DNA (Hill & Fangman, 1973).

FACTORS THAT AFFECT THYMINELESS DEATH

If, at any time during the period of thymine starvation, the ATP pool is reduced abruptly, e.g. by bubbling nitrogen through the culture, the fraction of viable cells stays constant from that moment until metabolism is renewed by returning the culture to aerobic conditions. Similarly, starvation in the absence of an energy source, e.g. glucose, does not cause any loss of viability (Freifelder & Maaløe, 1964). If, in addition to depriving a culture of thymine, a multiple auxotroph is simultaneously starved of another growth requirement the loss of viability is either not seen or is very much slower, both in onset and in the rate of loss once started (Fig. 2, curve *b*).

Equally, metabolic inhibitors reduce or abolish the lethal effects of thymine starvation. Chloramphenicol, 5-methyltryptophan (Cummings & Kusy, 1969), puromycin and rifampicin (Nakayama & Hanawalt, 1975) at levels sufficient to stop protein synthesis, greatly reduce the loss of viability. Although the results vary with the strain studied, it seems a good generalization that the rate of thymineless death in a given medium is closely correlated with the rate of growth observed in that medium given an adequate supply of thymine. However, in high concentrations of chloramphenicol, which paradoxically do permit more RNA turnover than is observed in low concentrations, thymineless death does occur (Atkinson & Stacey, unpublished observations) (Fig. 2).

BACTERIAL MUTATIONS WHICH AFFECT THYMINELESS DEATH

It is important to distinguish between the two stages of thymine starvation; the lag phase and the exponential phase. Certain mutants appear more sensitive only because the lag phase is shorter. The sensitivity of *pol*A mutants is due to this change (Berg & O'Neill, 1973) and the increased sensitivity of strains prone to filament formation, *E. coli* B, and *lon*$^-$ mutants of *E. coli* K12 as compared with the more UV resistant *E. coli* B/r and *lon*$^+$ strains, arises for the same reason. Treatments that suppress filament formation, e.g. the addition of pantoyl lactone to the plating medium, abolish the difference by restoring the lag phase to its normal duration. But the short lag phase for *pol*A mutants and the marked sensitivity of a mutant with a temperature-sensitive DNA ligase to thymine starvation (Pauling & Hamm, 1968) imply that both these functions are involved in the resistance of the wild-type strain. One bacterial mutant that is much more resistant to thymine starvation has been studied by Bailone (1974). It was isolated by Devoret & Blanco (1970) from the survivors of repeated cycles of thymine starvation of a K12(λ) strain of *E. coli*. Many of these proved to be *rec*A mutants, in which λ induction is prevented, and some contained mutants of the prophage, but among the others there were bacterial mutants which were given the symbol *itd*$^-$ (*i*nducibility by *t*hymine *d*eprivation). One of these, *itd*-2, proved to be a *gro*E mutant. The mutation has been transferred to other strains; these have been shown to be more resistant to thymine starvation and this effect is not simply due to a slower growth rate. The nature of the defect in *gro*E mutants is unknown but it has been suggested that it affects a site on the membrane (Sternberg, 1973) which may be involved in the morphogenesis of the bacterium as well as the phage.

THEORIES TO EXPLAIN THYMINELESS DEATH

Since its discovery a variety of suggestions have been made to account for the lethal and other effects of thymine starvation. The first came from Cohen & Barner (1954) who suggested that death was due to 'unbalanced growth', but they were also quick to point out the similarities between thymineless death and the effects of UV irradiation which have been a major theme of research in this area ever since. At the time of its discovery thymine starvation was the only means available of stopping DNA synthesis selectively that did not, at first sight at least, involve a treatment known to cause damage. A number of inhibitors of DNA synthesis have been found since (Kornberg, 1974), and more importantly a number of mutants have been isolated, in which DNA synthesis is temperature-sensitive. Stopping DNA synthesis with these inhibitors or maintaining these mutants at non-permissive temperatures is also lethal. The kinetics of the loss of viability of the temperature-sensitive mutants vary with the particular mutation and the gene which it makes defective. However, *dna*B mutants, in particular, die with simple kinetics and at a rate comparable to thymineless death. It is useful to bear these other methods of inhibiting DNA in mind when reviewing the hypotheses suggested to explain the effects of thymine starvation. Although they overlap to some extent, it is convenient to group these theories under four headings; unbalanced growth, prophage induction, DNA damage and DNA damage due to transcription.

Unbalanced growth

The explanation for the lethal effect of thymine starvation favoured by Cohen in his earlier papers was that it was due to 'unbalanced growth'; cells that are able to grow in all but DNA content enlarge, become filamentous and eventually enter a condition from which they cannot recover. Starvation for other growth requirements stops growth and so simultaneous starvation for thymine and, say, an amino acid has much less effect.

This hypothesis explains much but it fell out of favour as death due to other agents that cause filament formation, such as UV irradiation and mitomycin treatment, was attributed directly to DNA damage and the unbalanced growth regarded as a secondary consequence. It remains, however, the best explanation of the cause of death for mutants with defective regulation of RNA synthesis transferred from a rich to a poor medium (Alföldi, Stent, Hoogs & Hill, 1963). Mutants with the 'relaxed' phenotype continue to accumulate ribosomal RNA when deprived of a required amino acid while in strains with the 'stringent' phenotype ribosomal RNA synthesis stops immediately. It is likely that relaxed strains subject to 'shift-down' accumulate so much ribosomal RNA that they are driven into a physiological impasse from which they cannot escape.

Prophage induction

As we have seen, thymine starvation triggers prophage induction: it also induces colicin synthesis. It has been suggested that it is this induction that is the cause of death (Mennigmann, 1964) and, by extension, strains that do not produce phage particles are killed by defective phages. This theory appeared more attractive when it was demonstrated that the first *thy*$^-$ strain, *E. coli* 15T$^-$, produced particles that resemble phage tails after a period of thymine deprivation (Endo, Ayabe, Amako & Takeya, 1965) and that strains, such as JG 151, derived from it and which were not induced, were far more resistant to thymineless death. However, the rate of thymineless death of JG 151 is much affected by the culture conditions (Reichenbach, Schaiberger & Sallman, 1971). Also, strain JG 151 has all the properties of a *rec*A mutant (P. Barth, personal communication) in which prophage induction is inhibited. Indeed this lack of inducibility of phage λ in *rec*A strains can be exploited as a means of selecting recombination-defective strains by repeated cycles of thymine starvation (Devoret & Blanco, 1970). In the case best studied, λ-lysogeny sensitizes a strain to thymine deprivation only in the sense that the lag period is shorter and rate of loss of colony-forming capacity is slightly faster, but together these two factors are sufficient to allow selection of non-inducible strains.

One such strain of *B. subtilis* has been isolated and, significantly, thymineless death in this mutant was associated with much less DNA degradation, as judged by the co-transformation index of linked markers, than was the case in the inducible parent strain (Ephrati-Elizur, Yosur, Shmueli & Horowitz, 1974).

Thymine starvation causes an imbalance in the deoxynucleotide pools and a breakdown in the regulation of their normal levels: in several K12 strains there is a rapid, five-fold rise in the size of the dATP (deoxyadenosine triphosphate) pool until it is about one-quarter that of the ATP pool and a much larger (ten-fold) rise in the dCTP pool which began only after thirty minutes in thymineless conditions (Neuhard, 1966). Deoxynucleotides have been implicated as effectors of the derepression of phage λ and the addition of adenine to λ derivatives of strains of *E. coli* carrying the *tif* mutation causes induction under conditions when these strains would otherwise be stable (Kirby, Jacob & Goldthwait, 1967). The theory that thymineless death is due to prophage induction is attractive; the ubiquity of defective prophages is not in itself unlikely and most of the observations concerning thymine starvation are consistent with it. Simultaneous starvation for a required amino acid prevents induction by thymine starvation (at least in *E. coli* B and K12 strains).

The principal discrepancy between experimental results and the induction hypothesis is the behaviour of *rec*A mutants. Otherwise isogenic strains differing only in the *rec*A allele die at the same rate (Cummings

& Mondale 1967; Anderson & Barbour, 1973; see, however, Inouye, 1971). Since phage induction is prevented in *rec*A strains along with a number of DNA repair functions, it seems likely that some other mechanism is responsible.

DNA damage

McFall & Magasanik (1962) were the first to suggest explicitly DNA damage as the cause of lethality and Mennigmann & Szybalski (1962) were the first to examine the molecular weight of chromosomal DNA in cells starved for thymine (by exposure to FUDR). They found a considerable reduction in molecular weight, as measured in the ultracentrifuge, and suggested that thymineless death resulted from irreversible damage to the chromosome. However, the strain of *B. subtilis* that they examined was lysogenic and this damage could well have been a secondary consequence of induction. The observations of Ephrati-Elizur *et al.* (1974) strongly support this conclusion.

Although studies on populations of cells that include non-viable cells are always ambiguous, because it is impossible to tell if the results refer mainly to 'dead' cells, Breitman, Maury & Toal (1972) showed that there was appreciable loss of DNA by breakdown during thymine starvation. They examined a number of strains and found this to be true of all of them, although most striking in strains of *E. coli* 15T$^-$.

The integrity of the DNA of thymine-deprived cells has been examined by a number of authors following the introduction of alkaline sucrose gradient centrifugation, a technique so powerful in the study of UV and X-ray damage of DNA because single-strand breaks can be detected. However, the results have been surprisingly contradictory.

Freifelder (1969) was the first to demonstrate with certainty single-strand 'nicking'. He studied a DNA fraction that could be identified unambiguously; the small chromosome of the F factor. Thymine starvation of a strain carrying F or F*lac* caused single-strand breaks in the plasmid at a constant rate proportional to the size of the plasmid. This process started immediately the thymine was removed but did not take place if the cells were starved simultaneously for glucose or phosphate. However, simultaneous starvation for a required amino acid which also protects against the lethal effect did not affect the rate of accumulation of single-strand breaks. These results suggested to Freifelder that thymine starvation activated a nuclease or interfered with repair of single-strand breaks by DNA ligase and, subsequently, Freifelder & Levine (1972) presented evidence for enhanced nucleolytic activity in thymine-starved cells. As we have seen, Smith and his colleagues have extended this idea to explain the differential effects of thymine starvation as a 'curing' treatment (Tweats, Pinney & Smith, 1974).

Studies of single-strand breaks in the chromosomal DNA of thymine-starved cells have yielded a variety of results not entirely consistent with this simple picture. Walker (1970) and Reichenbach *et al.* (1971) found evidence for single-strand breaks while Baker & Hewitt (1971) and Sedgwick & Bridges (1971) did not. The most recent study (Nakayama & Hanawalt, 1975) showed that, in part, these discrepancies are the result of studying different strains, for they studied a number of different strains and obtained a variety of results depending, presumably, upon the genetic background. Unlike Sedgwick & Bridges they did find single-strand breaks in the DNA obtained from *pol*A$^-$ mutants.

That thymine starvation causes double-strand breaks is less certain but Yoshinaga (1973) using *E. coli* 15 TAU argued that there was a progressive accumulation with time of double-strand scissions, as judged by centrifugation of DNA extracted from thymine-starved cells on neutral sucrose gradients. However, this result could be an effect of induction again rather than the primary cause of cell death.

Hill & Fangman (1973) compared treatment with nalidixic acid or hydroxyurea, thymine starvation and the effects of prolonged maintenance of a strain carrying a *dna*B mutant (Fangman & Novick, 1968) at 42 °C, a non-permissive temperature. Hydroxyurea is an inhibitor DNA synthesis; it restricts the supply of all the deoxynucleotide triphosphates by inhibition of the enzyme ribonucleotide reductase (Sinha & Snustad, 1972). It, too, causes a slow loss of viability at concentrations greater than 0.1 M, but Hill & Fangman used a concentration that was not lethal, nor did it cause any single-strand breaks. Nalidixic acid and thymine starvation caused far more DNA damage than was observed in the *dna*B mutant, some thirty breaks per chromosome at the $1/e$ dose (the dose corresponding to one lethal hit per cell) as compared with three in the *dna*B strain.

DNA damage due to transcription

Thymineless death is a consequence of metabolism: the rate cells die is related to their growth rate and the lethal effects of thymine starvation are reduced by metabolic inhibitors and in *E. coli* B and K12 strains by concurrent starvation for a required amino acid. It is tempting to re-define the concept of unbalanced growth and say that the damage to chromosomal DNA discussed in the previous section is the consequence of continued transcription (Hanawalt, 1963) without the concomitant repair that occurs in the presence of thymine.

This view is argued in detail by Nakayama & Hanawalt (1975). They found the accumulation of strand breaks was suppressed by inhibitors of protein and RNA synthesis, particularly by rifampicin in *pol*A strains. However, the extent of the breakage and of its inhibition by metabolic inhibitors varied both with the strain and the inhibitor. Nevertheless, the authors argued that their results suggest that the strong correlations in their data between the extent of single-strand damage, transcription and thymineless death support the view that transcription may lead to single-strand breaks that are normally repaired substantially by DNA polymerase I and that this repair is less demanding on the pool of thymidylate (created by DNA breakdown) because it closes breaks with shorter patches than alternative forms of repair.

However, as they point out, there are discrepancies; the most striking is the small effect the addition of puromycin has on the accumulation of strand breaks in a *pol*A strain while affording substantial protection from thymineless death. Furthermore, although there is no effect on the viability of the culture, most of these breaks are repaired within ten minutes of the addition of thymine.

One would expect on this hypothesis that rifampicin, an inhibitor of transcription, would protect an organism against thymineless death more efficiently than inhibitors of translation or amino acid starvation in which messenger synthesis continues. However, in cells in which metabolism is restricted under special physiological conditions, e.g. 'competent' cells of *B. subtilis*, the effects of thymine starvation are small, if any (J. L. Farmer, personal communication).

CONCLUSIONS

The principal facts concerning thymine starvation may be summarized as follows.

(*a*) Thymineless death ensues after a period of starvation roughly equivalent to the doubling time of the strain in the presence of thymine, except in cultures that have been sensitized by another treatment, such as UV irradition or exposure to nalidixic acid. The lag phase is also substantially reduced in *pol*A and ligase mutants and slightly reduced in strains lysogenic for easily induced prophages: it is not affected by mutations in the *rec*A gene.

(*b*) The time of onset can be increased and the rate of loss of viability can be reduced or abolished by inhibitors of metabolism or by starvation for energy in all strains and for a required amino acid or purine in *E. coli* B and K12 strains.

(*c*) Thymine starvation causes mutation, enhances recombination, leads to prophage induction and, for some plasmids, is a means of 'curing'.

(*d*) Damage, particularly single-strand breaks, accumulates in both chromosomal and plasmid DNA under many circumstances. The factors that affect thymineless death affect the amount of damage observed. This damage is subject to repair, although it is not clear whether the repairable damage is associated with the cause of the loss of viability.

(*e*) There are striking analogies between thymineless death and the loss of viability that follows other selective methods of halting DNA synthesis, notably incubation of temperature-sensitive DNA synthesis mutants at non-permissive temperatures. UV irradiation and thymine starvation are synergistic.

The analogy with treatments known to cause DNA damage is striking and most of the evidence points to this as the cause of death. If one takes this as a starting point, the principal question is whether this damage is a normal consequence of transcription, and thymineless death ensues when damage has accumulated to the point where repair cannot prevent irreversible changes in the integrity of the chromosome, or whether thymine starvation (and other forms of inhibition of DNA synthesis) induces a nuclease that inflicts this damage. The evidence from metabolic inhibition is ambiguous; both hypotheses require protein synthesis. However, the synthesis of a nuclease fits the data more simply. There is evidence of increased nuclease activity in thymine-starved cells (Freifelder & Levine, 1972) and its synthesis could be affected by inhibitors of protein synthesis more efficiently than transcription and translation generally. It must be argued that this nuclease is induced in *rec*A mutants and is not, therefore, under the control of the *rec*A gene which appears to be responsible for the control of many repair functions (Gudas & Pardee, 1975).

The problem of the effects of thymine starvation remain obscure. For the author, who has followed this field largely from the side-lines, the balance of the evidence now seems to favour the hypothesis that crucial DNA damage is inflicted by induced nucleases, perhaps as an abortive attempt at repair. Many repair systems appear to involve nucleolytic activities.

REFERENCES

ALFÖLDI, L., STENT, G. S., HOOGS, M. & HILL, R. (1963). Physiological effects of the RNA control (RC) gene in *E. coli. Zeitschrift für Vererbungslehre*, **94**, 285–302.

ANDERSON, J. A. & BARBOUR, S. D. (1973). Effect of thymine starvation on deoxyribonucleic acid repair systems of *Escherichia coli. Journal of Bacteriology*, **113**, 114–121.

ATKINSON, C. & STACEY, K. A. (1968). Thymineless death induced by cytosine arabinoside. *Biochimica et Biophysica Acta*, **166**, 705–707.

BAILONE, A. (1974). Phénotype d'un mutant d'*E. coli* résistant à la mort par carence en thymine. *Comptes rendus de l'Academie des Sciences, Paris*, Ser. D, **279**, 2169–2171.

BAKER, M. L. & HEWITT, R. R. (1971). Influence of thymine starvation on the integrity of deoxyribonucleic acid in *Escherichia coli. Journal of Bacteriology*, **105**, 733–738.

BECCARI, E., MODIGLIANI, P. & MORPURGO, G. (1967). Induction of inter- and intragenic mitotic recombination by fluorodeoxyuridine and fluorouracil in *Aspergillus nidulans. Genetics*, **56**, 7–12.

BERG, C. M. & O'NEILL, J. M. (1973). Thymineless death in *pol*A$^+$ and *pol*A$^-$ strains of *Escherichia coli. Journal of Bacteriology*, **115**, 707–708.

BERTINO, J. B. & STACEY, K. A. (1966). A suggested mechanism for the selective procedure for isolating thymine-requiring mutants of *Escherichia coli. Biochemical Journal*, **101**, 32C–33C.

BLAKELY, R. L. (1969). *The Biochemistry of Folic Acid and Related Pteridines.* Amsterdam & London: North-Holland Publishing Co.

BREITMAN, R. T., MAURY, P. B. & TOAL, J. N. (1972). Loss of deoxyribonucleic acid thymine during thymine starvation of *Escherichia coli. Journal of Bacteriology*, **112**, 646–648.

CLOWES, R. C., MOODY, E. M. & PRITCHARD, R. H. (1965). The elimination of extrachromosomal elements in thymineless strains of *Escherichia coli* K12. *Genetical Research*, **6**, 147–152.

COHEN, S. S. (1967). Introduction to the biochemistry of D-arabinosyl nucleosides. In *Progress in Nucleic Acid Research and Molecular Biology*, **5**, ed. J. N. Davidson & W. E. Cohn, pp. 1–88. New York & London: Academic Press.

COHEN, S. S. & BARNER, H. D. (1954). Studies on unbalanced growth in *E. coli. Proceedings of the National Academy of Sciences, U.S.A.* **40**, 885–893.

COHEN, S. S., FLAKS, J. G., BARNER, D. H. & LICHTENSTEIN, J. (1958). The mode of action of 5-fluorouracil and its derivatives. *Proceedings of the National Academy of Sciences, U.S.A.* **44**, 1004–1012.

COUGHLIN, C. A. & ADELBERG, E. A. (1956). Bacterial mutation induced by thymine starvation. *Nature, London*, **178**, 531–532.

CUMMINGS, D. J. & KUSY, A. R. (1969). Thymineless death in *Escherichia coli*: inactivation and recovery. *Journal of Bacteriology*, **99**, 558–566.

CUMMINGS, D. J. & MONDALE, L. (1967). Thymineless death in *Escherichia coli*: strain specificity. *Journal of Bacteriology*, **93**, 1917–1924.

DEUTCH, C. E. & PAULING, C. (1974). Thymineless mutagenesis in *Escherichia coli. Journal of Bacteriology*, **119**, 861–867.

DEVORET, R. & BLANCO, M. (1970). Mutants of *E. coli* K12 (λ) non-inducible by thymine starvation. *Molecular and General Genetics*, **107**, 272–280.

ENDO, H., AYABE, K., AMAKO, K. & TAKEYA, K. (1965). Inducible phage of *E. coli* 15. *Virology*, **25**, 469–471.

EPHRATI-ELIZUR, E., YOSUR, D., SHMUELI, E. & HOROWITZ, A. (1974). Thymineless death in *Bacillus subtilis*: correlation between cell lysis and deoxyribonucleic acid breakdown. *Journal of Bacteriology*, **119**, 36–43.

ESPOSITO, R. E. & HOLLIDAY, R. (1964). The effect of 5-fluorodeoxyuridine on genetic recombination and mitotic crossing over in synchronized cultures of *Ustilago maydis*. *Genetics*, **59**, 1009–1017.

FANGMAN, W. L. & NOVICK, A. (1968). Characterisation of two bacterial mutants with temperature sensitive synthesis of DNA. *Genetics*, **60**, 1–17.

FREIFELDER, D. (1969). Single strand breaks in bacterial DNA associated with thymine starvation. *Journal of Molecular Biology*, **45**, 1–7.

FREIFELDER, D. & LEVINE, E. (1972). Stimulation of nuclease activity by thymine-starvation. *Biochemical and Biophysical Research Communications*, **46**, 1782–1787.

FREIFELDER, D. & MAALØE, Ø. (1964). Energy requirement for thymineless death in cells of *Escherichia coli*. *Journal of Bacteriology*, **88**, 897–990.

GALLANT, J. & SPOTTSWOOD, T. (1965). The recombinogenic effect of thymidylate starvation in *Escherichia coli* merodiploids. *Genetics*, **52**, 107–118.

GALLANT, J. & SUSKIND, S. R. (1961). Relationship between thymineless death and ultra violet inactivation in *Escherichia coli*. *Journal of Bacteriology*, **82**, 187–194.

GUDAS, L. J. & PARDEE, A. B. (1975). Model for regulation of *Escherichia coli* DNA repair functions. *Proceedings of National Academy of Sciences*, **72**, 2330–2334.

HANAWALT, P. C. (1963). Involvement of synthesis of RNA in thymineless death. *Nature, London*, **198**, 286.

HEIDELBERGER, C. (1965). Fluorinated pyrimidines. In *Progress in Nucleic Acid Research and Molecular Biology*, **4**, ed. J. N. Davidson & W. E. Cohen, pp. 2–50. New York & London: Academic Press.

HILL, W. E. & FANGMAN, W. L. (1973). Single strand breaks in deoxyribonucleic acid viability loss during deoxyribonucleic acid synthesis inhibition in *Escherichia coli*. *Journal of Bacteriology*, **116**, 1329–1335.

INOUYE, M. (1971). Pleiotropic effect of the *recA* gene of *Escherichia coli*: uncoupling of cell division from deoxyribonucleic acid replication. *Journal of Bacteriology*, **106**, 539–542.

JENSEN, K. F., LEER, J. C. & NYGAARD, P. (1973). Thymine utilization in *Escherichia coli* K12 on the role of deoxyribose 1-phosphate and thymidine phosphorylase. *European Journal of Biochemistry*, **40**, 345–354.

KANAZIR (1958). The apparent mutagenicity of thymine deficiency. *Biochimica et Biophysica Acta*, **30**, 20–23.

KIRBY, E. P., JACOB, F. & GOLDTHWAIT, D. A. (1967). Prophage induction and filament formation in a mutant strain of *Escherichia coli*. *Proceedings of the National Academy of Sciences, U.S.A.* **58**, 1903–1910.

KORNBERG, A. (1974). *DNA Synthesis*. San Francisco: W. H. Freeman.

LITTLE, J. G. & HANAWALT, P. C. (1973). Thymineless death and ultra violet sensitivity in *Micrococcus radiodurans*. *Journal of Bacteriology*, **113**, 233–240.

MCFALL, E. & MAGASANIK, B. (1962). The effects of thymine deprivation on the synthesis of protein in *Escherichia coli*. *Biochimica et Biophysica Acta*, **55**, 920–928.

MEACOCK, P. A. & PRITCHARD, R. H. (1975). Relationship between chromosome replication and cell division in a thymineless mutant of *Escherichia coli* B/r. *Journal of Bacteriology*, **122**, 931–942.

MEADOW, P. & WORK, E. (1956). Inter-relationships between diaminopimelic acid, lysine and their analogues in mutants of *Escherichia coli*. *Biochemical Journal*, **64**, 11P.

MENNIGMANN, H-D. (1964). Induction in *E. coli* 15 of the colicinogenic factor by thymineless death. *Biochemical and Biophysical Research Commiuncations*, **16**, 373–378.

MENNIGMANN, H-D. & SZYBALSKI, W. (1962). Molecular mechanism of thymineless death. *Biochemical and Biophysical Research Communications*, **9**, 398–404.

NAKAYAMA, H. & COUCH, J. L. (1973). Thymineless death in *Escherichia coli* in various assay systems. *Journal of Bacteriology*, **114**, 228–232.

NAKAYAMA, H. & HANAWALT, P. (1975). Sedimentation analysis of deoxyribonucleic acid from thymine starved *Escherichia coli. Journal of Bacteriology*, **121**, 537–547.

NEUHARD, J. (1966). Studies on the acid soluble nucleotide pool in thymine requiring mutants of *E. coli* during thymine starvation. III. *Biochimica et Biophysica Acta*, **129**, 104–115.

NEUHARD, J. & THOMASSEN, E. (1971). Turnover of the deoxyribonucleoside triphosphates in *Escherichia coli* 15T$^-$ during thymine starvation. *European Journal of Biochemistry*, **20**, 36–43.

OKADA, T., HOMMA, J. & SONOHARA, M. (1962). Improved method for obtaining thymineless mutants of *Escherichia coli* and *Salmonella typhimurium. Journal of Bacteriology*, **84**, 602–603.

ØVREBØ, S. & KLEPPE, K. (1973). Pyrimidine metabolism in *Acinetobacter calcoaceticus. Journal of Bacteriology*, **116**, 331–336.

PAULING, C. & HAMM, L. (1968). Properties of a temperature-sensitive, radiation-sensitive mutant of *Escherichia coli. Proceedings of the National Academy of Sciences, U.S.A.* **60**, 1495–1502.

PINNEY, R. J., BREMER, K. & SMITH, J. T. (1974). R-factor elimination by inhibition of thymidylate synthetase (fluorodeoxyuridine and showdomycin) and the occurrence of SS breaks in plasmid DNA. *Molecular and General Genetics*, **133**, 163–174.

REICHENBACH, D. L., SCHAIBERGER, G. E. & SALLMAN, B. (1971). The effect of thymine starvation on the chromosomal structure of *E. coli* JG 151. *Biochemical and Biophysical Research Communications*, **42**, 23–30.

SEDGWICK, S. G. & BRIDGES, B. A. (1971). Alkaline sucrose gradient sedimentation of chromosomal deoxyribonucleic acid from *Escherichia coli polA*$^+$ and *polA*$^-$ strains during thymine starvation. *Journal of Bacteriology*, **108**, 1422–1423.

SENO, T. & MELECHEN, N. E. (1964). Macromolecular synthesis in the initiation of bacteriophage P1. *Journal of Molecular Biology*, **9**, 340–351.

SINHA, N. K. & SNUSTAD, D. P. (1972). Mechanism of deoxyribonucleic acid synthesis in *Escherichia coli* by hydroxyurea. *Journal of Bacteriology*, **112**, 1321–1334.

STACEY, K. A. & SIMSON, E. (1965). Improved method for the isolation of thymine-requiring mutants of *Escherichia coli. Journal of Bacteriology*, **90**, 554–555.

STERNBERG, N. (1973). Properties of a mutant of *Escherichia coli* defective in bacteriophage λ head formation (*gro* E). II. The propagation of phage λ. *Journal of Molecular Biology*, **76**, 25–44.

STRELZOFF, R. (1962). Substitution of thymine by bromouracil in thymine-dependent *Escherichia coli. Zeitschrift für Vererbungslehre*, **93**, 287–291.

TWEATS, D. J., PINNEY, R. J. & SMITH, J. T. (1974). R-factor mediated nuclease activity involved in thymineless elimination. *Journal of Bacteriology*, **118**, 790–795.

WALKER, J. R. (1970). Thymine starvation and single-strand breaks in chromosomal deoxyribonucleic acid of *Escherichia coli. Journal of Bacteriology*, **104**, 1391–1392.

WILSON, M. C., FARMER, J. L. & ROTHMAN, F. J. (1966). Thymidylate synthesis and aminopterin resistance in *Bacillus subtilis*. *Journal of Bacteriology*, **92**, 186–196.

YOSHINAGA, K. (1973). Double-strand scission of DNA involved in thymineless death of *Escherichia coli* 15 TAU. *Biochimica et Biophysica Acta*, **294**, 204–213.

A GLOSSARY OF THE GENETIC LOCI MENTIONED IN THIS ARTICLE

Gene	Mutant phenotype
*dna*B	Unable to synthesize DNA at non-permissive temperatures.
F*lac*	Increased nucleolytic activity during thymine starvation causing single-strand breaks in the F factor chromosome.
*gro*E	Do not propagate λ unless the phage has a mutation in the phage E gene.
itd	λ-lysogens are not induced by thymine starvation.
lac z	Possess a defective β galactosidase.
lon	Form filaments much more than the parent strain after u.v. irradiation.
*pol*A	Possess a defective DNA Polymerase I.
*rec*A	Defective for genetic recombination; strikingly sensitive to u.v. and X-irradiation.
tif	λ-lysogens are induced by incubation at temperatures > 40 °C.
thy	Possess a defective thymidylate synthetase.
uvr	Lack part of the system responsible for DNA repair by excision.

SURVIVAL OF MICROBES EXPOSED TO CHEMICAL STRESS

W. B. HUGO
Department of Pharmacy, University of Nottingham, Nottingham NG7 2RD, Nottinghamshire

INTRODUCTION

Microbes are exposed to chemical stress by man and also in the ever-changing natural environment. This paper will consider only certain chemical stresses deliberately used by man to further his survival, health and well-being. It is possible to see two such distinct areas of usage. The first includes disinfection, chemical sterilization and preservation and covers such diverse materials as skin, superficial wounds, the undamaged eye, the alimentary tract, wood, textiles, paper, cutting oils, foods, multi-dose containers of injections, contaminated utensils and working surfaces in the food and pharmaceutical industry. The second concerns chemotherapy, pioneered by Paul Ehrlich at the end of the last century, in which chemicals are used to attack the invading agent while leaving the host unscathed. Today antibiotics, sulphonamides, trimethoprim, *p*-aminosalicylic acid, isoniazid and other compounds have revolutionized the treatment of infections. For the sake of historical accuracy, however, it should be recalled that the empirical use of cinchona bark for the treatment of malaria and ipecacuanha for the treatment of amoebiasis, were introduced into Europe from South America in the seventeenth century. Chemotherapy and biochemical studies of antimicrobial drugs were the subjects of the eighth and sixteenth symposia of this Society respectively (Cowan & Rowatt, 1958; Newton & Reynolds, 1966).

Three areas may be recognized in studies of disinfection mechanisms, and resistance and revival of microbes therefrom: (1) studies of the kinetics of disinfection; (2) biochemical and biophysical studies of the basic mechanisms of chemical stress and its reversal; (3) methods of evaluating disinfectants. Only the kinetic and the biochemical and biophysical studies concerning the effects of chemical stress and their reversal will be dealt with here. Interactions of chemicals with whole cells, cell walls, membranes and the cytoplasm will be considered in turn, followed by suggestions on the possible mechanism of survival of chemical stress.

THE KINETIC APPROACH

In an exemplary paper, Kronig & Paul (1897, reprinted in Brock, 1961) laid the foundation for the kinetic approach to chemical sterilization and the fundamental conditions to be observed in chemical studies of the process. Having codified rules which included the notion that comparative toxicity studies should be carried out at equimolecular proportions, with known numbers of bacteria in pure culture at constant temperature and under similar conditions of culture, Kronig & Paul went on to apply the emerging rules of chemical kinetics to the disinfection process, a procedure they said was valid because the process must be a chemical one. They were the first to plot the logarithm of the surviving organisms against time, which they found to give a linear response.

The theme of linear log survival/time was developed by Madsen & Nyman (1907) and survival investigated by many other workers notably Chick (1908), Knaysi (1930), Knaysi & Morris (1930), Jordan & Jacobs (1944), Rahn (1945), Berry & Michaels (1947), Eddy (1953), Jacobs (1960) and Prokop & Humphrey (1970).

Departures from true linearity were often encountered and Madsen & Nyman concluded that the different rates of destruction of bacteria under the influence of lethal agents were determined essentially by variability of resistance among the cells in a population. Rahn & Schroeder (1941) thought that in each bacterial species there was a single vulnerable molecule, the destruction of which was lethal. Such an hypothesis is not supported by our present knowledge of the varied mode of action of lethal chemicals.

More recently Eddy & Hinshelwood (Eddy, 1953) investigated the death rate of *Klebsiella aerogenes* under chemical stress. They used 3,6-diaminoacridine, *m*-cresol and acid. Eddy could find no satisfactory evidence that cells in a given culture possessed variable resistance, and indeed showed that survivors in a disinfection process gave rise to a population no more resistant than the originals. The often-observed lag phase, before decrease in viability proceeded logarithmically, suggests that a number of events must occur before a significant number of cells began to die; the number of these events may be the variable factor.

Eddy believed that at first certain essential metabolic capabilities were impaired. This phase was not considered lethal but gave rise to the initial lag phase. In the light of other evidence it is probable that this represents bacteriostasis, for it is reversible at least if the contact time is not prolonged. As time proceeds the cells die according to the logarithmic law. The slowing up of the death rate after a further lapse of time was taken as indicative of an adaptive (?phenotypic) process, enabling the cells which had survived by chance to adjust to the chemical stress. However, these adapted cells show the same death pattern as in the original experiment when they are subcultured and re-challenged.

A factor which can be calculated from survival data is known as the concentration exponent, n, which measures the effect of dilution on the activity of a stressing agent. It may be calculated from the following expression: (log death time at concentration C_2) − (log death time at concentration C_1)/log C_1 − log C_2. Typical values of n for phenol are 6, for mercuric chloride and formaldehyde 1, and for ethanol 9. There is an exponential relationship between loss of activity and dilution. Thus in the case of phenol, where $n = 6$, a three-fold dilution will mean a decrease in activity by a factor of 3^6 or 729. In practical terms, rapid dilution of a solution of a chemical antimicrobial agent with a high value of n will nullify the stress and the survivor level amongst the remaining population will stabilize. Can we learn anything from the vast array of such data? Very little, it is to be feared, about mechanisms of disinfection, for the curves represent interactions of chemicals having differing cellular targets and modes of action with highly complex micro-organisms at different stages of growth, and with different structures and chemical compositions.

INTERACTIONS WITH THE WHOLE CELL

However organelle- or enzyme-selective a drug may finally turn out to be, its first apparent interaction is with the whole cell and this may be examined as an adsorption process or by micro-electrophoresis.

Adsorption

Herzog & Betzel (1911) realized the importance of adsorption in the disinfection process in their studies with baker's yeast, and since then many others have measured the uptake of drugs by cells. The technique consists essentially of adding a suspension of cells to a solution of the drug and at suitable time intervals removing a sample, centrifuging it, and determining the residual amount of drug in the cell-free supernatant solution. If the cells have removed drug from solution, the concentration of the drug in the supernatant fluid will have diminished. From these data, adsorption isotherms may be plotted and information concerning the rate and total amount of uptake may be computed. Furthermore, by a consideration of the nature of the isotherm, some notion of the adsorptive mechanism may be inferred (Giles, MacEwan, Nakhwa & Smith, 1960). These authors considered four patterns of adsorption which they called S, L, H and C.

The 'S (S-shaped) pattern' is found when the solute molecule is monofunctional, has moderate intermolecular attraction, causing it to orientate vertically, and meets strong competition for substrate sites from molecules of the solvent or by another adsorbed species. Monohydric phenols when adsorbed on a polar substrate from water usually give this pattern.

In the 'L (Langmuir) pattern', as more sites are filled it becomes increasingly difficult for a bombarding solute molecule to find a vacant site. The adsorbed solute molecule is either not orientated vertically or there is strong competition from the solvent. If vertical orientation does occur there is a strong intramolecular attraction between the adsorbed molecules. Amongst the phenols, resorcinol shows this type of behaviour.

The 'H (high affinity) pattern' is obtained when the solute is almost completely adsorbed. Sometimes the process is accompanied by ion exchange, as in many bacteriological staining procedures. It is also shown by the uptake of iodine from an iodophor by yeast (Hugo & Newton, 1964).

The 'C (constant partition) pattern' is obtained when the solutes penetrate more readily into the adsorbate than does the solvent. It has been shown to occur when aqueous solutions of phenols are adsorbed by synthetic polypeptides and it might also be expected to occur when phenols are adsorbed from an aqueous solution by bacteria containing a high proportion of lipid in their cell wall.

More recent studies on the adsorption of antibacterial substances by micro-organisms include the uptake of CTAB* by bacteria (Salton, 1951), the adsorption of hexylresorcinol by *Escherichia coli* (Beckett, Patki & Robinson, 1959), of iodine by *E. coli*, *Staphylococcus aureus* and *Saccharomyces cerevisiae* (Hugo & Newton, 1964), of chlorhexidine by *Staphylococcus aureus*, *E. coli* and *Clostridium perfringens* (Hugo & Longworth, 1964; Hugo & Daltrey, 1974), of basic dyes by fixed yeast cells (Giles & McKay, 1965), of phenols by *E. coli* (Bean & Das, 1966) and *Micrococcus lysodeikticus* (Judis, 1966), of dequalinium by *E. coli* and *S. aureus* (Hugo & Frier, 1969) and of Fentichlor by *E. coli* and *S. aureus* (Hugo & Bloomfield, 1971*a*).

* CTAB or Cetrimide; a commercial, cationic detergent containing mainly cetyltrimethylammonium bromide.

Information on the site of adsorption may be obtained by studying the process at different pH values but it should be borne in mind that the ionization of the disinfectant as well as receptor sites on the cell surface may be affected by changes in pH (Salton, 1957; Hugo & Longworth, 1964).

Adsorption studies of a different nature involving the uptake of drugs by nucleic acids have been used extensively to study actions on this molecule and are dealt with later (p. 398).

The action of such substances as serum and organic debris in reducing the stress caused by some chemical bactericides may be due to their ability to compete with the bacterial cell for some of the active agent. In some cases, drugs may be desorbed from the cell after initial adsorption, thereby decreasing the stress (Bennett, 1959).

Changes in electrophoretic mobility

Bacterial cells are normally negatively charged and, if suspended in water or a suitable electrolyte solution containing electrodes to which a potential has been applied, the cells will migrate to the positively charged electrode. This phenomenon may be placed on a quantitative basis by observing the rate of migration of a single cell to the electrode by timing over a measured distance using a microscope and calibrated eyepiece micrometer. Once the system has been standardized, the effect of drugs on mobility may be studied and from the data so obtained some notion of the drug–cell interaction and the effect of drugs on the charged bacterial cell surface may be deduced. The subject of bacterial cell electrophoresis has been reviewed in detail by Lerch (1953) and James (1965, 1972) and the application of electrophoretic studies to the study of chemical stress has been dealt with in the review by Hugo (1967). In general, it can be said that while providing an exact tool for studying drug–cell interactions, electrophoresis has not provided much insight into the mechanisms of death.

INTERACTIONS WITH THE CELL WALL

Antibiotics which have the cell wall as their target include penicillin, cephalosporin, cycloserine, vancomycin and ristocetin. In addition, Pulvertaft & Lumb (1948) showed that *E. coli*, streptococci and staphylococci lysed almost completely when rapidly growing cultures were exposed to low concentrations of antiseptics such as formalin (0.012 per cent), phenol (0.032 per cent), mercuric chloride (0.0008 per cent), sodium hypochlorite (0.005 per cent) and merthiolate (0.0004 per cent).

They presumed that the autolytic enzymes were not inhibited at the low concentrations of antiseptic used and compared this with the action of penicillin. The involvement of autolytic enzymes in penicillin action has been postulated subsequently by Rogers (1967), Rogers & Forsberg (1971) and Tomasz, Albino & Zanati (1970). Hugo (unpublished data) could not prepare protoplasts by allowing phenol, chlorocresol and sodium hypochlorite (at their lytic concentrations) to act on bacteria in the presence of 0.33 M sucrose as an osmotic stabilizer, possibly because these substances also damaged the membrane.

In further studies, Delpy & Champsey (1949) found that thiomersalate increased the susceptibility of *Bacillus anthracis* to lysis, and Norris (1957) confirmed this observation with *B. cereus*. Schaechter & Santomassino (1962) noted that low concentrations of some mercury compounds, which included mercuric chloride, phenylmercuric acetate and merthiolate, lysed growing cultures of *E. coli*. Isolated cell walls were not affected and so they inferred interference with the formation of disulphide bonds in the cell wall.

Bolle & Kellenberger (1958) found that sodium lauryl sulphate lysed non-respiring (cyanide-treated) cells of *E. coli*. The organisms enlarged into globular forms and then lysed rapidly. Actively metabolizing cells were not susceptible. Shafa & Salton (1960) found that anionic detergents caused disaggregation of walls isolated from six Gram-negative bacteria and suggested that this was due to their action upon the lipid-containing compounds of the wall, rather than to lytic enzymes.

Srivastava & Thompson (1966) showed that 0.5 per cent phenol only caused lysis at the time of separation of the pairs of daughter cells of *E. coli*, i.e. when the cytoplasmic membrane was weak and exposed and its phospholipid content minimal. However, phenol will act on non-dividing cells and Srivastava & Thompson's results merely underline the complexity of the reaction to phenol.

Evidence of cell wall damage may be inferred from the phenomenon of drug-induced, long forms of bacteria (Hughes, 1956). Thus, Braun & Solomon (1918) and Moltke (1927) both reported the induction of long forms in *Proteus vulgaris* by treatment with 1–2 per cent phenol, while Spray & Lodge (1943) described a similar effect of resorcinol and *m*-cresol on *Klebsiella aerogenes*. Ainley-Walker & Murray (1904) and Wahlin & Almaden (1939) induced long forms in *Salmonella typhi* with methyl violet, methyl green, fuchsin and methylene blue.

The involvement of lytic enzymes in long form induction cannot be excluded. Gentian violet was said to prevent the synthesis of mucopeptide in *Staphylococccus aureus* but at an earlier point than that at which the wall-inhibiting antibiotics are known to act (Strominger, 1959). However, a repetition of this experiment with six different samples of gentian violet failed to demonstrate wall inhibition. Ghuysen (1968) has reviewed the involvement of lytic enzymes in some of these phenomena.

INTERACTIONS WITH THE CYTOPLASMIC MEMBRANE

Early work in this area was concerned with drug-induced leakage of material and there is little doubt that this contributes to stasis or death according to the time and intensity of exposure to the chemical stress concerned. More recently, reactions at the molecular level have been revealed and it is these reactions which are currently the most exciting. They include uncoupling of oxidative phosphorylation and inhibition of energy-dependent transport.

Leakage of cell constituents and modification of cell permeability

The phenomenon of haemolysis is well-known and can be induced by surface active agents (Schulman & Rideal, 1937; Pethica & Schulman, 1953). Kuhn & Bielig (1940) made the suggestion that cationic detergents of the quaternary ammonium class (QAC) might act on the bacterial cell membrane by dissociating conjugated proteins and, in a manner analogous to haemolysis, damage it so much that death would ensue. Hotchkiss (1944) proved that membrane damage was occurring, by demonstrating that nitrogen- and phosphorus-containing compounds leaked from staphylococci treated with QAC and the polypeptide antibiotic, tyrocidin.

Gale & Taylor (1947) concluded that the lytic action of the antibacterial compounds they studied, which included CTAB and phenol, was sufficient to explain their disinfectant action. Salton (1950, 1951) demonstrated that the activity of CTAB effecting a 99.99 per cent kill was related to the amounts of purine- or pyrimidine-containing components leaking from the cells in the first 5 min after treatment. Similar observations were made by Beckett, Patki & Robinson (1958, 1959) who found that increasing concentrations of hexylresorcinol increased the amount of these materials which attained a maximum value short of the total pool, i.e. the material released by mechanical disruption of the cell.

Judis (1962) showed that phenol, *p*-chloro-*m*-xylenol, *p*-chloro-*m*-cresol, *p*-chloro-*o*-cresol, 2,4-dichlorophenol, 2,4,6-trichlorophenol and

2,4-dichloro-*m*-xylenol caused release of [^{14}C]glutamate from *E. coli* grown in the presence of the label. Tween 80 protected *E. coli* from the lethal effects of *p*-chloro-*m*-xylenol and prevented, in part, leakage of cell contents. Woodroffe & Wilkinson (1966) demonstrated that tetrachlorosalicylanilide (TCS) released ninhydrin-positive material from *Staphylococcus aureus*, and the bacteriostatic activity of TCS was related to this effect. Addition of horse serum to TCS-treated cells up to 2 h after the initial drug–cell contact resulted in their complete recovery.

Hugo & Longworth (1964) investigated the ability of chlorhexidine to promote leakage of intracellular material from *E. coli* and *S. aureus* and found a diphasic leakage/concentration pattern. They proposed that the first part of the curve represented increasing of leakage with increasing concentration of antiseptic, but at high concentrations the protoplasmic contents or cytoplasmic membrane became gradually coagulated so that the leakage became progressively less. Electron micrographs of thin sections of bacteria taken after suitable dose treatments confirmed this view (Hugo & Longworth, 1965). They also showed that viability decreased at concentrations of the drug which caused very little leakage, suggesting a more subtle effect may be involved. A very similar study was carried out using hexachlorophene and *Bacillus megaterium* by Joswick, Corner, Silvernale & Gerhardt (1971). They found a diphasic leakage/concentration curve and concluded that the release of intracellular solutes was a secondary effect. They later identified the lethal event as an inhibition of a part of the electron transport chain (Frederick, Corner & Gerhardt, 1974). However, Hugo & Bloomfield (1971*a*, *b*, *c*) in their studies on the mode of action of 2,2′-thiobis(4-chlorophenol), Fentichlor, found a close correlation between bactericidal action on *E. coli* and *S. aureus* and ability to promote leakage of material absorbing at 260 nm. Bacteriostatic concentrations of the drug did not cause leakage.

Many solvents including butanol (Pethica, 1958), ethanol (Salton, 1963), toluene (Jackson & De Moss, 1965) and phenylethanol (Silver & Wendt, 1967; Woldringh, 1973) cause the release of intracellular constituents. The phenylethanol/*E. coli* interaction is reversible, suggesting that the structural integrity of the membrane is not seriously impaired (Silver & Wendt, 1967). Ivanov, Markov, Golovinskii & Kharisanova (1964) found 22 per cent of the total lipid from *Pseudomonas aeruginosa* was extractable with light petroleum but viability was unaffected. However, cells were more sensitive to phenol, acetone, ethanol, mercuric chloride, acids, alkalis and some antibiotics. Access of these substances to the cell may be facilitated by the extraction procedure.

Anionic detergents such as sodium dodecyl sulphate (SDS) are much less toxic to bacteria although their target is also the cytoplasmic membrane (Gilby & Few, 1960; Razin & Argaman, 1963). This may be due to the fact that in SDS and other anionic detergents the active ion is negatively charged and may be repelled by the negatively charged bacterial surface.

Non-ionic surface active agents are practically non-toxic to bacteria; indeed some, e.g. nonidet, are useful biochemical tools for preparing enzymically active bacterial membranes. Non-ionic agents have been shown to act synergistically when combined with various other antibacterial compounds (Moore & Hardwick, 1956; Brown & Richards, 1964) but can also protect cells from phenols (Judis, 1962).

Recently, ion-specific electrodes have allowed very accurate determinations of ion efflux from bacterial cells treated with membrane-active antimicrobial agents. Lambert & Hammond (1973) have concluded that the order of release of cell constituents from *E. coli* treated with 0.2 mM cetrimide was K^+ then PO_4^{3-} followed by material absorbing at 260 nm. The release of K^+ was complete in 30 min.

It is clear that solvents and certain antibacterial agents promote leakage of ions, labile nucleic acids and their component purines, pyrimidines, pentoses and inorganic phosphorus, and detection of all these substances is used to determine membrane damage. It is unlikely that this is a rapidly fatal process and this type of damage may sometimes be reparable (Pullman & Reynolds, 1965).

Yet another aspect of cytolysis may be observed in studies on the time course of cytoplasmic leakage. Very often a rapid initial release is followed, after some hours, by a further release of material; this may be due to breakdown of large molecular weight compounds or ribosomal-bound compounds or to further time-dependent damage to a permeability barrier. Both may be due to the involvement of autolytic enzymes and the two phenomena may be distinguished by conducting the experiment at 2 °C and at 36–40 °C. Damage to a permeability barrier will be little different at either temperature but lysis due to enzymic action will be much less at 2 °C than at 36–40 °C.

A novel approach to membrane permeability and the effect of drugs thereon was made by Hamilton (1970), who prepared protoplasts from *B. megaterium* by lysozyme treatment in 0.4 M sucrose. The protoplasts were collected by centrifugation and re-suspended in glycerol and a variety of salt solutions. Immediate lysis occurred in 0.4 M glycerol and 0.5 M ammonium acetate, but 0.4 M NH_4NO_3, NH_4Cl, KNO_3 KCl, $NaNO_3$ and NaCl continued to act as osmotic stabilizers. From this it was concluded that the protoplast membrane of this organism was permeable to glycerol, NH_4^+ and CH_3COO^- and impermeable to K^+, Na^+, NO_3^- and Cl^-. By repeating the experiments in the presence of the stabilizing salts and CTAB, tetrachlorosalicylanilide (TCS) and trichloroacetanilide (TCC), and noting the presence or absence of lysis it was concluded that CTAB caused such a generalized effect on the membrane that it became permeable to all the ions. TCS increased the membrane's permeability to NO_3^- and TCC to Cl^- and NO_3^-. Similar experiments with *Clostridium perfringens* protoplasts and chlorhexidine (Daltrey & Hugo, 1974) suggested that this drug at concentrations of 25 $\mu g\ ml^{-1}$ also caused generalized membrane damage.

Salton (1968) has reviewed the mode of action of detergents on bacteria and he concluded that the sequence of events following exposure of microbes to detergents was: (1) adsorption onto the cell followed by a penetration into the largely porous cell wall; (2) reactions with lipid/protein complexes of the cytoplasmic membrane leading to its disorganization; (3) leakage of low molecular weight components from the cytoplasm; (4) degradation of proteins and nucleic acids; and (5) wall lysis caused by autolytic enzymes. It is very important when considering these propositions and any other interaction of drugs with micro-organisms to bear in mind the difference in structure of the walls of Gram-positive and Gram-negative bacteria and especially the role of the outer layers of the walls of the latter group.

Inhibition of energy processes

Any discussion on this topic must be linked with both uncoupling of oxidative phosphorylation and the chemiosmotic theory of membrane transport (Mitchell, 1968, 1970, 1972). Mitchell proposed that oxidative phosphorylation, ATP synthesis, active transport and the maintenance of intracellular solute levels are powered by a protonmotive force, generated by metabolic oxido-reductions, which is apparent as a chemical and electrical gradient or potential difference across the cytoplasmic membrane. It may be expressed in mathematical terms thus:

$$\Delta p = \Delta\Psi - Z\Delta pH,$$

where Δp is the protonmotive force, $\Delta\Psi$ is the membrane electrical potential in mV and ΔpH is the trans-membrane pH gradient. Z is a factor converting pH values to mV, and at 37 °C has a value of 62. The expression $-Z\Delta pH$, therefore, is a pH difference expressed in mV. Typical reported experimental values of these potentials in bacteria at 37 °C are, -120 mV for $\Delta\Psi$ and $+62$ mV for $Z\Delta pH$. Thus

$$\Delta p = -182\ \text{mV}.$$

Oxidative phosphorylation and its uncoupling. Cross, Taggart, Covo & Green (1949) investigated the relationship between the structures of a series of nitrophenols and other compounds and their ability to inhibit phosphorylation. All the phenols showing activity contained two nitro groups attached to the benzene ring and activity was also shown by 2,4-dinitro-1-naphthol, 2,2′-methylenebis(4-nitrophenol), and bis(3,5,6-trichloro-2-hydroxyphenylmethane). Slight activity was shown by 2,2′-methylenebis(4-chlorophenol) and 2,2′-methylenebis(4-bromophenol). Pentachlorophenol uncouples oxidation from phosphorylation in animal tissue extracts (Weinbach, 1957) and in *Streptococcus faecalis* (Harold & Baarda, 1968).

Uncoupling agents (Mitchell, 1961) act as specific conductors of protons across bacterial, mitochondrial and artificial membranes, thereby modifying part of the protonmotive force across the membrane. The earlier observations on the mode of action of these compounds, reviewed by Simon (1953) and Hugo (1957, 1967) are probably secondary effects resulting from this primary lesion. Harold & Baarda (1968), using the experimental system of Mitchell & Moyle (1967), showed that TCS (tetrachlorosalicylanilide), known to be an uncoupling agent (Williamson & Metcalf, 1967), partially discharged the proton gradient. Harold, Pavlasova & Baarda (1970) also demonstrated a trans-membrane pH gradient in *Streptococcus faecalis* using the reagent dimethyloxazolidinedione. This organism lacks a cytochrome system and relies on glycolysis/substrate level phosphorylation for energy production. They were able to show the dispersal of the pH gradient by TCS and a relationship between proton extrusion and K^+ accumulation.

It will be recalled that K^+ loss is a very early event in general membrane damage and the biochemical importance of K^+ transport in relation to membrane function indicates that a loss of this ion may play a part in the mechanisms of action of membrane-active agents less specific than those known to be uncouplers and dischargers of proton gradients. Mickelson (1974) has shown that uncoupling reagents, which included 2,4-dinitrophenol (DNP) and pentachlorophenol, affected the molar growth yield of *Streptococcus agalactiae* whether it was growing aerobically or anaerobically. He inferred that uncouplers affect both oxidative and substrate level phosphorylation. Bloomfield (1974) has shown a similar proton translocation effect on *Staphylococcus aureus* with the antibacterial agent Fentichlor.

The existence of a proton gradient in obligate anaerobic micro-organisms was demonstrated in *Clostridium perfringens* by Daltrey & Hugo (1974) (Fig. 1*a*). TCS in combination with valinomycin partially

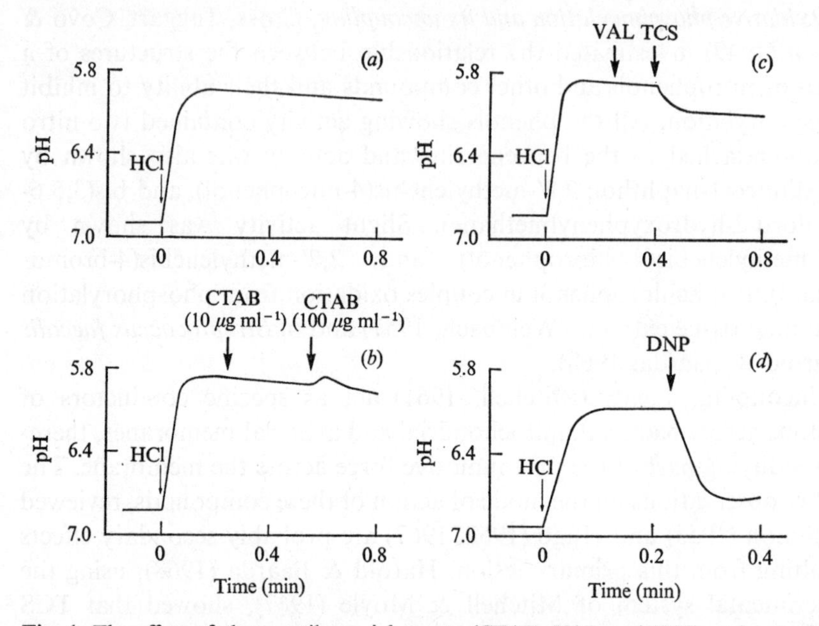

Fig. 1. The effect of three antibacterial agents (CTAB, VAL and TCS) and one phenol (DNP) on proton flux across the membrane of *Clostridium perfringens*. (*a*) Control suspension. (*b*) Additions were made as indicated. CTAB, cetyltrimethylammonium bromide. (*c*) Additions were made as indicated. VAL, valinomycin (0.5 $\mu g\ ml^{-1}$ final concentration); TCS, tetrachlorosalicylanilide (3×10^{-6} M concentration). (*d*) DNP was added as indicated to a final concentration of 5×10^{-5} M. In all cases HCl (indicated by small arrows) was added at 0 min. (Reproduced with permission from Daltrey & Hugo, 1974.)

discharged it (Fig. 1*c*) and DNP did so completely (Fig. 1*d*). Chlorhexidine and CTAB had no effect at 10 $\mu g\ ml^{-1}$, although with CTAB at 100 $\mu g\ ml^{-1}$ a further slight drop in pH was observed (Fig. 1*b*).

Inhibition of energy dependent transport. Another postulate of the chemiosmotic theory is that it provides an explanation of the source of energy for the transport of metabolites into the cell. Several workers have shown that chemicals which act as uncoupling agents, and are able to discharge the trans-membrane proton gradient, also inhibit energy-dependent transport processes. Thus, Gale & Folkes (1953) demonstrated that DNP (60 $\mu g\ ml^{-1}$) caused a 50 per cent inhibition of free glutamic acid accumulation in *Staphylococcus aureus*. Hamilton (1968) showed that glutamic acid uptake in *S. aureus* was completely inhibited by TCS and DNP, and Hugo & Bloomfield (1971*c*) and Bloomfield (1974) indicated that Fentichlor, at bacteriostatic concentrations, inhibited uptake of several amino acids.

Horecker, Thomas & Monod (1960) found that DNP inhibits the uptake of galactose in a galactokinase mutant of *E. coli*. Harold &

Baarda (1968) showed that a variety of uncoupling agents, including TCS, even at a concentration of 10^{-7} M, strongly inhibited the transport of phosphate, rubidium, potassium, glycylglycine, alanine and leucine in *Streptococcus faecalis*. They concluded that a significant relationship exists between the facilitation of proton conduction and the inhibition of membrane transport.

Later Gale and Llewellin (1972), Niven, Jeacocke & Hamilton (1973) and Niven & Hamilton (1974) made it clear that the modes of action of uncoupling disinfectants are linked with the mode of action of ionophoric antibiotics and transport phenomena in general. Niven & Hamilton (1974) were able to distinguish the components of the proton-motive force responsible for the transport of individual amino acids. Thus lysine was driven by the membrane potential, $\Delta\Psi$, isoleucine by the total protonmotive force, Δp, while glutamic acid was driven by the pH gradient, $-Z\Delta pH$. Chemical stress imposed by drugs capable of discharging the trans-membrane pH may thus effect amino acid transport in different ways according to the particular amino acid being studied.

A transport mechanism, believed to be limited to bacteria, was described by Kundig, Ghosh & Roseman (1964). It involves the combination of phosphoenolpyruvic acid with a sugar to give a sugar phosphate and pyruvic acid. Three enzymes and various factors are involved in the process and the sugar is transported across the membrane as a sugar phosphate. Amongst the compounds transported in this way are glucose, fructose, mannose, mannitol and sorbitol. Vinylglycollic acid, 2-hydroxy-3-butenoic acid, is a specific inhibitor of this process (Walsh & Kaback, 1973) and the fact that this transport mechanism is unique to bacteria poses the question of whether vinylglycollate could control human and animal infections. However, it would fail if alternative uninhibited energy sources, i.e. lactose or amino acids, were available to the organism as they would be in the animal body.

Interference with membrane enzymes. Harold, Baarda, Baron & Abrams (1969) examined chlorhexidine and found it to be an inhibitor of both membrane-bound and soluble adenosine triphosphatase (ATPase) and also of net K^+ uptake in *Streptococcus faecalis*.They deduced that ATPase is coupled with K^+ transport in membrane processes in this organism. Daltrey & Hugo (1974) have shown a similar inhibition of ATPase in *Clostridium perfringens* and produced evidence that the energy-dependent amino acid uptake in this organism is inhibited by chlorhexidine. Frederick *et al.* (1974) showed that hexachlorophene inhibits part of the membrane-bound electron transport chain in *Bacillus megaterium*.

Mercuric salts act at low concentrations (10^{-6} M), by combining with membrane enzymes containing thiol groups. At such concentrations this action can be reversed by addition of thiol compounds, e.g. thioglycollate or cysteine, to the reaction system (Passow, Rothstein & Clarkson, 1961). More recently Bowman & Stretton (1972) and Stretton & Manson (1972) have shown that 2-bromo-2-nitropropan1,3-diol (Bronopol) oxidizes thiol groups to disulphides in bacteria and that this action may be reversed by thiol compounds. These compounds also cause membrane damage as indicated by leakage of material which absorbs at 260 nm.

It is clear that the mode of action of the 'membrane-active' antimicrobial compounds is complex and subtle and that any investigation on their mode of action must include an evaluation of their effect on the energy processes associated with the membrane. For very comprehensive reviews on the subject the reader is referred to Hamilton (1971, 1975), Harold (1970) and Harold, Altendorf & Hirata (1974).

INTERACTIONS WITH THE CYTOPLASM

There are four targets for antibacterial drugs; the cytoplasm itself, cytoplasmic enzymes, the nucleic acids and the ribosomes. Hugo (1957, 1965*a*, 1967) has reviewed the first two of these aspects in considerable detail but a few of the main points are given below.

Irreversible coagulation of cytoplasmic constituents

This drastic lesion is usually seen at drug concentrations far higher than those causing general lysis or leakage. It was the first cytological effect to be reported and so most antiseptics were classified as general protoplasmic poisons or as protein precipitants. Indeed at concentrations used in many practical disinfection procedures this is undoubtedly the mechanism of rapid killing, the more subtle and more slowly fatal effects being completely masked.

The cytoplasmic components most likely to be coagulated or denatured are proteins and nucleic acids, most studies having been made on the former. Functionally, proteins are of two main kinds in living cells, structural and enzymic. The high specificity of enzymes is due to their unique surface contours and to the distribution of charges on this surface. The latter arise from residual charges, on the carboxylic acid or amino groups of the constituent amino acids, that are left after peptide bond formation. It is not difficult to imagine that a derangement of this uniquely contoured and electrically charged unit can upset its function.

As early as 1901, Meyer showed that the degree of antibacterial action of phenols was proportional to their distribution between water and protein, thus suggesting that protein was a prime target. Cooper (1912) came to a similar conclusion and decided that phenols destroy the protein, structure within the cell. Bancroft & Richter (1931) actually observed a coagulation of cell protein in *Bacillus megaterium* and *Klebsiella aerogenes* using ultraviolet microscopy.

Three main methods are available for studying protoplasmic coagulation – light and electron microscopy, light scattering and direct observation of cytoplasm obtained from smashed cells.

Examples of studies by electron microscopy include those of Salton, Horne & Coslett (1951) and Dawson, Lominski & Stern (1953) (with CTAB) on whole cells treated with antibacterial compounds, and those of Hugo & Longworth (1966) (with chlorhexidine), and Bringmann (1953) (with chlorine, bromine, iodine, Cu^{2+}, Ag^{+} and hydrogen peroxide) on thin sections. Hugo & Longworth found a correlation between the appearance of thin sections and cytoplasmic leakage. Thus at the concentration which caused maximum leakage, electron micrographs of thin sections showed a significant loss of electron dense material; at higher concentrations, which promoted no leakage due to a general coagulation and sealing in of labile protoplasmic constituents, the electron micrographs showed a dense granular cytoplasm differing markedly in appearance from that seen in untreated cells.

An interesting technique for studying protoplasmic coagulation is based upon protoplast formation and the fact that rod-shaped cells yield globular protoplasts in isotonic media. If this operation is carried out on cells which have been treated with a coagulant or fixative, then a rod-shaped protoplast is produced which, if coagulation is severe, does not undergo lysis or osmotic explosion on subsequent dilution of the medium (Tomcsik, 1955; Hugo & Longworth, 1964; Daltrey & Hugo, 1974). Coagulation of protein may affect the light scattering properties of cells and this is a sensitive method of assessing changes in treated bacteria.

Yet another method consists of disrupting the cells and investigating the action of antiseptic on cell-free extracts. Although this system is an artificial one and the relative concentrations of the protoplasmic constituents change and suffer enzymic degradation, it gives some indication of the order of concentration required to produce coagulation. Thus, Hugo & Longworth (1966) found that protein and nucleic acid were precipitated from cell-free preparations of *E. coli* at concentrations of chlorhexidine far higher than those causing leakage.

Effects on metabolism and enzymes

Despite the large volume of work stretching over many years and still being prosecuted, the general conclusion is that enzyme inactivation is only one of many events caused by chemical stress and is not likely to be a prime mechanism for death. In some circumstances, however, it may be a cause of bacteriostasis. The reviews of Rahn & Schroeder (1941) and Roberts & Rahn (1946) are worth reading for a discussion of this topic.

The nucleic acids

There are a number of antibiotics which affect the biosynthesis and functioning of nucleic acids. However, amongst the non-antibiotic antibacterial drugs, only three main compounds affecting these targets have been identified: these are the acridine dyes, formaldehyde and phenylethanol.

The acridine dyes. Acridine dyes, first introduced into medicine as trypanicidal agents, have been used extensively as antibacterial agents and a large amount of work has been carried out on their mode of action. As with the TPM (triphenylmethane) dyes it was realized early that the cation was the active ion. McIlwain (1941) showed that nucleic acids antagonized the antibacterial action of acridine dyes and Ferguson & Thorne (1946) after studying the effect of a series of acridine compounds on the growth and respiration of *E. coli* concluded that they inhibited reactions closely connected with synthetic processes. It emerged that if acridine dyes were only 33 per cent ionized as a cation, there was very little antibacterial action and for really high activity 98–100 per cent of the molecule should exist in the cationic form. Other studies showed that the flat area of the molecule should be from 0.38–0.48 nm^2. Albert (1966) has reviewed acridine chemistry comprehensively.

Studies of DNA binding with proflavine showed a first-order reaction which equilibrated with one molecule of proflavine binding to every four or five nucleotides. There was also a slower reaction of higher order in which a 1:1 binding occurred (Peacocke & Skerrett, 1956; Waring, 1965, 1966*a*; Blake & Peacocke, 1968). These interactions also involved marked spectral, viscosity and melting temperature changes.

From a review of available binding data, Lerman proposed (1961), and later elaborated (1964*a*, *b*), an intercalation model in which it was suggested that proflavine was bound by DNA, through the two primary amino groups, being held by ionic links to two phosphoric acid residues,

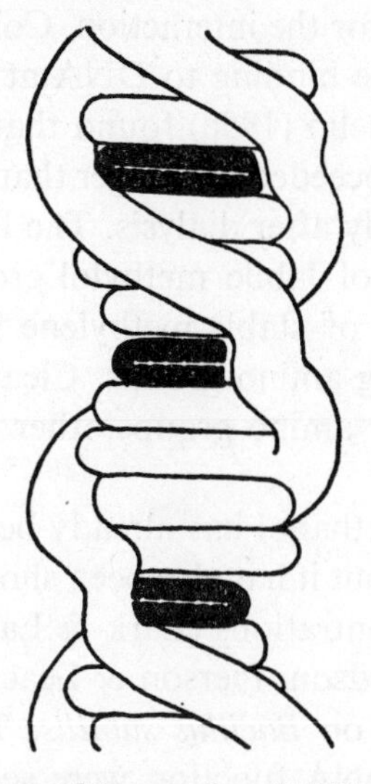

Fig. 2. A conjecture for the structure of the complex of DNA with proflavine (Lerman, 1964). The black area represents the proflavine molecule, the white area the DNA.

and the flat acridine ring system being linked on the purine and pyrimidine residues by Van der Waals forces (Fig. 2). Pritchard, Blake & Peacocke (1966) and Blake & Peacocke (1968) have suggested an alternative mechanism in which the acridine ring lies between two adjacent bases on the same polynucleotide with the primary amino groups now lying close to the phosphoric acid residues of the DNA. The intercalation theory is compatible with the observation of McIlwain (1941) that nucleic acids antagonize the action of acridines; with the observation, summarized in Albert (1966), that molecular size and shape are of importance in determining the relative potency of acridines; and with the observation that acridines inhibit synthetic processes (Ferguson & Thorne, 1946).

Ribonucleic acid polymerase has also been identified as a target for acridines in cell-free preparations from *E. coli*.

Nicholson & Peacocke (1965) suggested that inhibitory acridine molecules, which included proflavine and 9-aminoacridine, occupy sites on the polymerase which normally bind nucleoside triphosphates or the bases in the DNA molecule during copying. Waring (1965) discussed the ability of twenty-three drugs, including proflavine, to inhibit incorporation of adenosine monophosphate into RNA.

Formaldehyde. Grossman, Levine & Allison (1961) have shown a reasonably specific reaction to occur between nucleotides and formaldehyde. The amino groups of the purine and pyrimidine rings have been cited as likely sites for the interaction. Collins & Guild (1968) have shown that formaldehyde binding to DNA at 100 °C and pH 8 is irreversible. However, Stachelin (1958) found that binding with RNA was reversible if it had not proceeded for longer than 2 h; after 2 h the reaction was slowly reversible only after dialysis. The first reaction was thought to be due to formation of labile methylol groups, $-NH.CH_2OH$, the second to the formation of stable methylene bridges, $-NH.CH_2NH-$, between bases containing amino groups. Clearly reactions of this type could occur with cellular amino groups other than those in the nucleic acid molecule.

Phenylethanol. Phenylethanol has already been referred to as a membrane-active compound but it has also been shown to inhibit initiation of replication at high concentrations (Lark & Lark, 1966). Richardson & Leach (1969) and Richardson, Pierson & Leach (1969) re-examined the action of phenylethanol on *Bacillus subtilis*. They concluded that any effects of the drug on DNA function were secondary to its effects on membrane integrity and transport of RNA and DNA precursors into the cells.

Ribosomes

Ribosomes are associated with the formation of peptides from amino acids ordered by messenger RNA and assembled by transfer RNA. This process is a singular target for many antibiotics and there are some non-antibiotic antibacterial agents whose target lies here.

Treatment of cells with toluene releases ribosomes (Jackson & De Moss, 1965) but this may be taken as a manifestation of the extent of membrane damage by this compound such that a unit the size of a ribosome could be released.

Ethylenediaminetetraacetic acid (EDTA) is a specific chelator of certain metals, including Mg^{2+}, necessary for the integrity of the 50 S and 30 S ribosome units in prokaryotes. EDTA is not regarded as a primary antibacterial agent although it is used in conjunction with certain antiseptics to enhance their activity, especially against Gram-negative organisms (Russell, 1971; Leive, 1974). Its possible action on ribosome structure must be borne in mind when it is present in antibacterial systems.

Hydrogen peroxide dissociates the 30 S and 50 S subunits of the 70 S ribosome in *E. coli* (Nakamura & Tamaoki, 1968). This process is 80 per cent reversible upon adding 10 mM Mg^{2+}, providing the concentration of hydrogen peroxide has not exceeded 0.1 per cent. *p*-Chloromercuribenzoate (0.5 mM) dissociates the 100 S ribosomes of *E. coli* into 70 S monomers (Wang & Matheson, 1967). This reaction can be reversed by 2-mercaptoethanol but not by Mg^{2+}.

The ribosome cannot be considered a prime target for the specific selective action of any known non-antibiotic antibacterial agent although it may be destroyed by some chemical agents.

POSSIBLE MECHANISM OF SURVIVAL AND PERSISTENCE

Genotypic resistance to disinfectants

Resistance to certain metal ions has been discovered in the penicillinase plasmid of *Staphylococcus aureus* by Novick & Roth (1968) who found that the resistance markers conferred an increase in resistance ranging from 3- to 100-fold depending on the ion involved. Separate genetic loci for resistance to arsenate, arsenite, lead, cadmium, mercuric and bismuth ions were demonstrated. These workers did not attempt to identify the biochemical mechanisms associated with this resistance but it is interesting to note that a common target for all the resistant ions listed is the thiol group.

A most interesting mechanism for mercury-resistant, plasmid-bearing strains of *E. coli*, *S. aureus* and *Pseudomonas aeruginosa* involves the biochemical conversion of Hg^{2+} into a volatile organomercury compound (Summers & Lewis, 1973). These workers and Vaituzis *et al.* (1975) have summarized other examples of micro-organisms which can produce volatile mercury compounds from Hg^{2+}. Further investigation might reveal that other volatile organometallic compounds are produced by strains of microbes resistant to metal ions.

Thornley & Yudkin (1959*a*, *b*) and Sinai & Yudkin (1959*a*, *b*, *c*) studied the origin of bacteria resistance to proflavine. They concluded that no single factor determines drug resistance for a single species/single drug combination.

Russell (1972) showed that R^- and R^+ strains of *P. aeruginosa* were equally resistant to glutaraldehyde, chloroxylenol solution, lysol, chlorhexidine, CTAB and phenylmercuric nitrate, suggesting that transferable drug resistance was not occurring with these non-antibiotic antimicrobial agents. R^+ strains of *P. aeruginosa* have shown transferable drug resistance to antibiotics (Roe, Jones & Lowbury, 1971).

Biochemical basis for resistance to disinfectants

Conversion of a toxic substance to a non-toxic derviative. Many organisms are able to decompose aromatic compounds, many of which are used as disinfectants (Rogoff, 1961; Evans, 1963; Ribbons, 1965; Hugo, 1965*b*; Gibson, 1968). Beveridge & Hugo (1964) examined the ability of nine Gram-negative, non-sporing rods, mainly pseudomonads, to use aromatic compounds, many of them disinfectants, as a sole carbon source. *p*-Cresol, phenol, benzoic acid, *p*-hydroxybenzoic acid and salicyclic acid were readily attacked. The metabolic versatility of *Vibrio* O1, an organism now thought to be *Acinetobacter calcoaceticus*, in decomposing many compounds including some traditional disinfectants is amply demonstrated in the papers of Fewson (1967) and Beveridge & Tall (1969). Hugo & Foster (1964) found a strain of *P. aeruginosa* able to utilize methyl and propyl *p*-hydroxybenzoates as sole sources of carbon. Grant (1967) showed a similar capability with *Klebsiella aerogenes*. It is clear that this pattern of resistance can be of practical significance for the hydroxybenzoates were once used as preservatives in eye drops.

Changes in access to the vulnerable site: a possible involvement of lipid. The role of cellular lipid in resistance has attracted considerable interest in recent years.

The first hint that bacterial lipid was involved in determining the relative resistance of bacteria to detergents was provided by an experiment of Dyar & Ordal (1946). A strain of *S. aureus* with a high lipid content showed exceptional sensitivity to changes in electrophoretic mobility, induced by cetylpyridinium chloride, and acquired a greater apparent negative charge in the presence of sodium dodecylsulphate.

Following serial subculture, Chaplin (1951) obtained a 43-fold increase in the resistance of a strain of *E. coli* and a 200-fold increase in the resistance of *Serratia marcescens* to a series of quaternary ammonium compounds (QAC). He was unable to demonstrate resistance in *S. aureus*. Fischer & Larose (1952) found the adaptive process to be dependent on pH. Thus greater resistance was acquired more rapidly if the serial subcultures were performed at pH 6.8 rather than at pH 7.7. The effect of pH was thought to be on ionization of QAC and hence on drug uptake. In a further paper Chaplin (1952) stained cells with Sudan Black B and made electrophoretic mobility studies and showed an increase in the lipid content of the resistant cells of *S. marcescens*. Strong support for the existence of this lipid and its involvement in resistance came when lipase-treated cells lost their resistance.

Lowick & James (1957) trained *K. aerogenes* to grow in the presence of crystal violet and demonstrated by electrophoretic techniques that the surface of the resistant cells was predominantly lipid whereas in the case of untrained cells the surface was predominantly polysaccharide. Hugo & Stretton (1966) grew several micro-organisms in nutrient broth containing glycerol and increased their lipid content. These organisms showed an increased resistance to penicillins. Hugo & Franklin (1966) studied the effect of this lipid enhancement in the Oxford strain of *S. aureus* on its resistance to phenols. They used a series of 4-*n*-alkylphenols from phenol to hexylphenol. Enhanced cellular lipid protected the cells from the inhibitory action of only pentylphenol and *n*-hexylphenol. This protection is probably conferred by a non-specific blanketing mechanism in which the amyl- and hexyl-phenols are locked at the interface between the cellular lipid and the aqueous environment. In a similar study, Hamilton (1968) found no significant increase in resistance to tetrachlorosalicylanilide, tribromosalicylanilide, trichloroacetanilide, monochlorophenoxysalicylanilide or hexachlorophene in a glycerol grown culture of the same organism.

Lipid depletion in *S. aureus* and *E. coli* was achieved by Hugo, Bowen & Davidson (1971) by growing biotin-deficient organisms. Only a slight decrease in resistance to a series of phenols and alkyl trimethylammonium bromides was found in both organisms.

Vaczi has recently published a monograph on the biological role of bacterial lipids which includes a comprehensive review on lipid content and resistance to both antibiotic and non-antibiotic antimicrobial agents (Vaczi, 1973).

The special role of the Gram-negative cell wall in resistance. Gram-negative rod-shaped bacteria possess lipoprotein and lipopolysaccharide layers outside the peptidoglycan layer which afford protection to antimicrobial agents. The lipopolysaccharide may be partly removed by EDTA which results in an increased susceptibility to both antibiotic and non-antibiotic drugs (Russell, 1971; Leive, 1974; Haque & Russell, 1974*a*, *b*).

CONCLUSION

The foregoing review, which can only mention a small number of the thousands of papers which have appeared on the subject of chemical stress, its dynamics and mechanisms and the intricacies of resistance and survival, indicates that there are no ready rules or neat generalizations which can be made. Almost every drug–microbe interaction must be studied individually and by as many approaches as possible. There are very few examples where a single event, defined at the molecular level, can explain microbial death or survival following chemical stress.

REFERENCES

AINLEY-WALKER, E. W. & MURRAY, W. (1904). The effect of certain dyes upon the cultural characters. *British Medical Journal*, ii, 16–18.

ALBERT, A. (1966). *The Acridines*, 2nd edition. London: Arnold.

ALBERT, A., GIBSON, M. & RUBBO, S. (1953). The influence of chemical constitution on antibacterial activity. VI. The bactericidal action of 8-hydroxyquinoline (oxine). *British Journal of Experimental Pathology*, **34**, 119–30.

ARMSTRONG, W. McD. (1958). The effect of some synthetic dyestuffs on the metabolism of baker's yeast. *Archives of Biochemistry and Biophysics*, **73**, 153–160.

BAKER, Z., HARRISON, R. W. & MILLER, B. F. (1941*b*). The bactericidal action of synthetic detergents. *Journal of Experimental Medicine*, **74**, 611–620.

BANCROFT, W. D. & RICHTER, G. H. (1931). The chemistry of disinfection. *Journal of Physical Chemistry*, **35**, 511–530.

BEAN, H. S. & DAS, A. (1966). The adsorption by *Escherichia coli* of phenols and their bactericidal activity. *Journal of Pharmacy and Pharmacology*, **18**, 107S–113S.

BECKETT, A. H. DAS, R. N. & ROBINSON, A. E. (1959). Metallic cations and the antibacterial action of oxine. *Journal of Pharmacy and Pharmacology*, **11**, 195T–197T.

BECKETT, A. H., PATKI, S. J. & ROBINSON, A. (1958). Interaction of phenolic compounds with bacteria. *Nature, London*, **181**, 712.

BECKETT, A. H., PATKI, S. J. & ROBINSON, A. (1959). The interaction of phenolic compounds with bacteria. I. Hexylresorcinol and *Escherichia coli*. *Journal of Pharmacy and Pharmacology*, **11**, 360–366.

BECKETT, A. H., VAHORA, A. A. & ROBINSON, A. E. (1958). The interactions of chelating agents with bacteria. I. 8-Hydroxyquinoline (oxine) and *Staphylococcus aureus*. *Journal of Pharmacy and Pharmacology*, **10**, 160T–169T.

BENNETT, E. O. (1959). Factors affecting the antimicrobial activity of phenols. *Advances in Applied Microbiology*, **1**, 123–140.

BERRY, H. & MICHAELS, I. (1947). The evaluation of the bactericidal activity of ethylene glycol and some of its monoalkyl ethers against *Bacterium coli*. *Quarterly Journal of Pharmacy and Pharmacology*, **20**, 331–347.

BEVERIDGE, E. G. & HUGO, W. B. (1964). The resistance of gallic acid and its alkyl esters to attack by bacteria able to degrade aromatic ring structures. *Journal of Applied Bacteriology*, **27**, 304–311.

BEVERIDGE, E. G. & TALL, D. (1969). The metabolic availability of phenol analogues to bacterium NCIB 8250. *Journal of Applied Bacteriology*, **32**, 304–311.

BLAKE, A. & PEACOCKE, A. R. (1968). The interaction of amino acridines with nucleic acids. *Biopolymers*, **6**, 1225–1253.

BLOOMFIELD, S. F. (1974). The effect of the phenolic antibacterial agent Fentichlor on energy coupling in *Staphylococcus aureus*. *Journal of Applied Bacteriology*, **37**, 117–131.

BOLLE, A. & KELLENBERGER, E. (1958). The action of sodium lauryl sulphate on *E. coli*. *Schweizerische Zeitschrift für Pathologie und Bakteriologie*, **21**, 714–740.

BOWMAN, W. R. & STRETTON, R. J. (1972). Antimicrobial activity of a series of halo-nitro compounds. *Antimicrobial Agents and Chemotherapy*, **2**, 504–505.

BRAUN, H. & SOLOMON, R. (1918). Die Fleckfieber Proteusbazillen. *Zeitschrift für Bakteriologie*, **82**, 243–251.

BRINGMANN, G. (1953). Electron microscope findings on the action of chlorine, bromine, iodine, copper, silver and hydrogen peroxide on *Escherichia coli. Zeitschrift für Hygiene und Infectionskrankheiten*, **138**, 155–166.

BROCK, T. D. (1961). *Milestones in Microbiology*. Englewood-Cliffs: Prentice Hall.

BROWN, M. R. W. & RICHARDS, R. M. E. (1964). Effect of polysorbate (Tween 80) on the resistance of *Pseudomonas aeruginosa* to chemical inactivation. *Journal of Pharmacy and Pharmacology*, **15**, 51T–55T.

CHAPLIN, C. E. (1951). Observations on quaternary ammonium disinfectants. *Canadian Journal of Botany*, **29**, 373–382.

CHAPLIN, C. E. (1952). Bacterial resistance to quaternary ammonium disinfectants. *Journal of Bacteriology*, **63**, 453–458.

CHICK, H. (1908). An investigation of the laws of disinfection. *Journal of Hygiene, Cambridge*, **8**, 92–99.

COLLINS, C. A. & GUILD, W. R. (1968). Irreversible effects of formaldehyde on DNA. *Biochimica et Biophysica Acta*, **157**, 107–113.

COOPER, E. A. (1912). On the relationship of phenol and *m*-cresol to proteins; a contribution to our knowledge of the mechanism of disinfection. *Biochemical Journal*, **6**, 362–387.

COSTERTON, J. W., INGRAM, J. M. & CHENG, K. J. (1974). Structure and function of the cell envelope of Gram-negative bacteria. *Journal of Bacteriology*, **38**, 87–110.

COWAN, S. T. & ROWATT, E. (1958). *The Strategy of Chemotherapy. Symposia of the Society of General Microbiology*, **8**, London: Cambridge University Press.

CROSS, R. J., TAGGART, J. V., COVO, G. A. & GREEN, D. E. (1949). Studies on the cyclophorase system. VI. The coupling of oxidation and phosphorylation. *Journal of Biological Chemistry*, **177**, 655–678.

DALTREY, D. C. & HUGO, W. B. (1974). Studies on the mode of action of the antibacterial agent chlorhexidine on *Clostridium perfringens*. II. Effect of chlorhexidine on metabolism and on the cell membrane. *Microbios*, **11**, 131–146.

DAWSON, I. A., LOMINSKI, I. & STERN, H. (1953). An electron microscope study of the action of cetyltrimethylammonium bromide (CTAB) on *Staphylococcus aureus. Journal of Pathology and Bacteriology*, **66**, 513–526.

DELPY, P. L. & CHAMPSEY, H. M. (1949). Sur la stabilisation des suspensions sporulées de *B. anthracis* par l'action de certains antiseptiques. *Comptes rendus hebdomadaires des séances de l'Academie des Sciences, Paris*, **225**, 1071–1073.

DOBROGOSZ, W. J. & DE MOSS, R. D. (1963). Induction and repression of L-arabinose isomerase in *Pediococcus pentosaceus. Journal of Bacteriology*, **85**, 1350–1364.

DYAR, M. T. & ORDAL, E. J. (1946). Electrokinetic studies of bacterial surfaces. I. The effects of surface-active agents on the electrophoretic mobilities of bacteria. *Journal of Bacteriology*, **51**, 149–167.

EDDY, A. A. (1953). Death rate of populations of *Bact. lactis aerogenes*. III. Interpretation of survival curves. *Proceedings of the Royal Society of London*, Ser. B, **141**, 137–145.

EVANS, W. C. (1963). The microbiological degradation of aromatic compounds. *Journal of General Microbiology*, **32**, 177–184.

FERGUSON, T. B. & THORNE, S. (1946). The effects of some acridine compounds on the growth and respiration of *Escherichia coli. Journal of Pharmacology and Experimental Therapeutics*, **86**, 258–263.

FEWSON, C. A. (1967). The growth and metabolic versatility of the Gram-negative bacterium NC1B 8250 ('*Vibrio* O1'). *Journal of General Microbiology*, **46**, 255–266.

FISCHER, R. & LAROSE, P. (1952). Factors governing the adaption of bacteria against quaternaries. *Nature, London*, **170**, 715–716.

FREDERICK, J. J., CORNER, T. R. & GERHARDT, P. (1974). Antimicrobial actions of hexachlorophene: inhibition of respiration in *Bacillus megaterium*. *Antimicrobial Agents and Chemotherapy*, **6**, 712–721.

GALE, E. F. & FOLKES, J. P. (1953). The assimilation of amino acids by bacteria. 15. The actions of antibiotics on nucleic acid and protein synthesis in *Staphylococcus aureus*. *Biochemical Journal*, **53**, 493–498.

GALE, E. F. & LLEWELLIN, J. M. (1972). The role of hydrogen and potassium ions in the transport of acidic amino acids in *Staphylococcus aureus*. *Biochimica et Biophysica Acta*, **266**, 182–205.

GALE, E. F. & TAYLOR, E. S. (1947). The action of tyrocidin and some detergent substances in releasing amino acids from the internal environment of *Streptococcus faecalis*. *Journal of General Microbiology*, **1**, 77–84.

GHUYSEN, J. M. (1968). Use of bacteriolytic enzymes in determining of wall structure and their role in cell metabolism. *Bacteriological Reviews*, **32**, 425–464.

GIBSON, D. T. (1968). Microbial degradation of aromatic compounds. *Science, Washington*, **161**, 1093–1097.

GILBY, A. R. & FEW, A. V. (1960). Lysis of protoplasts of *Micrococcus lysodeikticus* by ionic detergents. *Journal of General Microbiology*, **23**, 19–26.

GILES, C. H., MACEWAN, T. H., NAKHWA, S. N. & SMITH, D. (1960). Studies in adsorption. XI. A system of classification of solution adsorption mechanisms and measurement of specific surface areas of solids. *Journal of the Chemical Society*, 3973–3993.

GILES, C. H. & MCKAY, R. B. (1965). The adsorption of cationic (basic) dyes by fixed yeast cells. *Journal of Bacteriology*, **89**, 390–397.

GRANT, D. J. W. (1967). Kinetic aspects of the growth of *Klebsiella aerogenes* with some benzenoid carbon sources. *Journal of General Microbiology*, **46**, 213–224.

GROSSMAN, L., LEVINE, S. S. & ALLISON, W. S. (1961). The reaction of formaldehyde with nucleotides and T2 bacteriophage DNA. *Journal of Molecular Biology*, **3**, 47–60.

HAMILTON, W. A. (1968). The mechanism of the bacteriostatic action of tetrachlorosalicylanilide. *Journal of General Microbiology*, **50**, 441–458.

HAMILTON, W. A. (1970). The mode of action of membrane-active antibacterials. *FEBS Symposium*, **20**, 71–79.

HAMILTON, W. A. (1971). Membrane active antibacterial compounds. In *Inhibition and destruction of the Microbial Cell*, ed. W. B. Hugo, p. 77. London & New York: Academic Press.

HAMILTON, W. A. (1975). Energy coupling in microbial transport. *Advances in Microbial Physiology*, **12**, 1–53.

HAQUE, H. & RUSSELL, A. D. (1974*a*). Effect of chelating agents on the susceptibility of some strains of Gram-negative bacteria to some antibacterial agents. *Antimicrobial Agents and Chemotherapy*, **6**, 200–206.

HAQUE, H. & RUSSELL, A. D. (1974*b*). Effect of ethylenediamine-tetraacetic acid and related chelating agents on whole cells of Gram-negative bacteria. *Antimicrobial Agents and Chemotherapy*, **6**, 447–452.

HAROLD, F. H. (1970). Antimicrobial agents and membrane function. *Advances in Microbial Physiology*, **4**, 45–104.

HAROLD, F. M. (1972). Conservation and transformation of energy by bacterial membranes. *Bacteriological Reviews*, **36**, 172–230.

HAROLD, F. M., ALTENDORF, K. H. & HIRATA, H. (1974). Probing membrane transport mechanisms with ionophores. *Annals of the New York Academy of Sciences*, **235**, 149–160.

HAROLD, F. M. & BAARDA, J. R. (1968). Inhibition of membrane transport in *Streptococcus faecalis* by uncouplers of oxidative phosphorylation and its relation to proton conduction. *Journal of Bacteriology*, **96**, 2025–2034.

HAROLD, F. M., BAARDA, J. R., BARON, C. & ABRAMS, A. (1969). DIO9 and chlorhexidine: inhibitors of membrane bound ATPase and of cation transport in *Streptococcus faecalis. Biochimica et Biophysica Acta*, **183**, 129–136.

HAROLD, F. M., PAVLASOVA, E. & BAARDA, J. R. (1970). A transmembrane pH gradient in *Streptococcus faecalis*; origin and dissipation by proton conductors and *N,N*[1]-dicyclohexylcarbodiimide. *Biochimica et Biophysica Acta*, **196**, 235–244.

HERZENBERG, L. A. (1959). Studies in the induction of β-galactosidase in a cryptic strain of *Escherichia coli. Biochimica et Biophysica Acta*, **31**, 525–538.

HERZOG, R. A. & BETZEL, R. (1911). Zur Theorie der Desinfektion. *Hoppe-Seyler's Zeitschrift für Physiologische Chemie* **74**, 221–226.

HORECKER, B. L., THOMAS, J. & MONOD, J. (1960). Galactose transport in *Escherichia coli.* I. General properties as studied in a galactokinase mutant. *Journal of Biological Chemistry*, **735**, 1580–1585.

HOTCHKISS, R. D. (1944). Gramicidin, tyrocidin and tyrothricin. *Advances in Enzymology*, **4**, 153–199.

HUGHES, W. H. (1956). The structure and development of the induced long forms of bacteria. *Symposia of the Society for General Microbiology*, **6**, ed. E. T. C. Spooner & B. A. D. Stocker, pp. 341–360. London: Cambridge University Press.

HUGO, W. B. (1957). The mode of action of antiseptics. *Journal of Pharmacy and Pharmacology*, **9**, 145–161.

HUGO, W. B. (1965*a*). Some aspects of the action of cationic surface active agents in microbial cells with special reference to their action on enzymes. In *Surface, Activity and the Microbial Cell: SCI Monograph* **19**, pp. 67–82. London: Society of Chemical Industry.

HUGO, W. B. (1965*b*). The degradation of preservatives by micro-organisms. In *Scientific and Technical Symposium. 112th Annual Meeting, American Pharmaceutical Association, Detroit*, CIII, pp. 1–7.

HUGO, W. B. (1967). The mode of action of antibacterial agents. *Journal of Applied Bacteriology*, **30**, 17–50.

HUGO, W. B. (1971). *Inhibition and Destruction of the Microbial Cell.* London & New York: Academic Press.

HUGO, W. B. & BLOOMFIELD, S. F. (1971*a*). Studies on the mode of action of the phenolic antibacterial agent Fentichlor against *Staphylococcus aureus* and *Escherichia coli.* I. The adsorption of Fentichlor by the bacterial cell and its antibacterial activity. *Journal of Applied Bacteriology*, **34**, 557–567.

HUGO, W. B. & BLOOMFIELD, S. F. (1971*b*). Studies in the mode of action of the phenolic antibacterial agent Fentichlor against *Staphylococcus aureus* and *Escherichia coli.* II. The effects of Fentichlor on the bacterial membrane and the cytoplasmic constituents of the cell. *Journal of Applied Bacteriology*, **34**, 569–578.

HUGO, W. B. & BLOOMFIELD, S. F. (1971*c*). Studies in the mode of action of the phenolic antibacterial agent Fentichlor against *Staphyloccocus aureus* and *Escherichia coli.* III. The effect of Fentichlor on the metabolic activities of *Staphylococcus aureus* and *Escherichia coli. Journal of Applied Bacteriology.* **34**, 579–591.

HUGO, W. B., BOWEN, J. G. & DAVIDSON, J. R. (1971). Lipid depletion in bacteria induced by biotin deficiency and its relation to resistance to antibacterial agents. *Journal of Pharmacy and Pharmacology*, **23**, 69–70.

HUGO, W. B. & DALTREY, D. C. (1974). Studies on the mode of action of the antibacterial agent chlorhexidine on *Clostridium perfringens*. I. Adsorption of chlorhexidine on the cell, its antibacterial activity and physical effects. *Microbios*, **11**, 119–129.

HUGO, W. B. & DAVIDSON, J. R. (1973). Effect of cell lipid depletion in *Staphylococcus aureus* upon its resistance to antimicrobial agents. II. A comparison of the response of normal and lipid depleted cells of *S. aureus* to antibacterial drugs. *Microbios*, **8**, 63–72.

HUGO, W. B. & FOSTER, J. H. S. (1964). Growth of *Pseudomonas aeruginosa* in solutions of esters of *p*-hydroxybenzoic acid. *Journal of Pharmacy and Pharmacology*, **16**, 209.

HUGO, W. B. & FRANKLIN, I. (1968). Cellular lipid and the antistaphylococcal action of phenols. *Journal of General Microbiology*, **52**, 365–373.

HUGO, W. B. & FRIER, M. (1699). Mode of action of the antibacterial compound dequalinium acetate. *Applied Microbiology*, **17**, 118–127.

HUGO, W. B. & LONGWORTH, A. R. (1964). Some aspects of the mode of action of chlorhexidine. *Journal of Pharmacy and Pharmacology*, **16**, 655–662.

HUGO, W. B. & LONGWORTH, A. R. (1965). Cytological aspects of the mode of action of chlorhexidine. *Journal of Pharmacy and Pharmacology*, **17**, 28–32.

HUGO, W. B. & LONGWORTH, A. R. (1966). The effect of chlorhexidine on the electrophoretic mobility, cytoplasmic contents, dehydrogenase activity and cell walls of *Escherichia coli* and *Staphylococcus aureus*. *Journal of Pharmacy and Pharmacology*, **18**, 569–578.

HUGO, W. B. & NEWTON, J. M. (1964). The adsorption of iodine from solution by micro-organisms and by serum. *Journal of Pharmacy and Pharmacology*, **16**, 49–55.

HUGO, W. B. & STRETTON, R. J. (1966). The role of cellular lipid in the resistance of Gram-positive bacteria to penicillins. *Journal of General Microbiology*, **42**, 133–138.

IVANOV, V., MARKOV, K. I., GOLOVINSKII, E. & KHARISANOVA, T. (1964). Importance of surface lipids for various biological properties of a *Pseudomonas aeruginosa* strain. *Zeitschrift für Naturforschung*, **196**, 604–606.

JACKSON, R. W. & DE MOSS, J. A. (1965). Effect of toluene on *Escherichia coli*. *Journal of Bacteriology*, **90**, 1420–1425.

JACOBS, S. E. (1960). Some aspects of the dynamics of disinfection. *Journal of Pharmacy and Pharmacology*, **12**, 9T–18T.

JAMES, A. M. (1965). The modification of the bacterial surface by chemical and enzymic treatment. In *Cell Electrophoresis*, ed. E. J. Ambrose, pp. 154–170. London: J. & A. Churchill.

JAMES, A. M. (1972). *The Electrochemistry of Bacterial Surfaces. Inaugural Lecture*. University of London: Bedford College.

JORDAN, R. C. & JACOBS, S. E. (1944). Studies on the dynamics of disinfection. I. New data on the reaction between phenol and *Bact. coli* using an improved technique, together with an analysis of the distribution of resistance amongst the cells of the bacterial population studied. *Journal of Hygiene, Cambridge*, **43**, 275–289.

JOSWICK, H. L., CORNER, T. R., SILVERNALE, J. N. & GERHARDT, P. (1971). Antimicrobial actions of hexachlorophene: release of cytoplasmic materials. *Journal of Bacteriology*, **108**, 492–500.

JUDIS, J. (1962). Studies on the mechanism of action of phenolic disinfectants. I. Release of radioactivity from carbon-14-labelled *Escherichia coli*. *Journal of Pharmaceutical Sciences*, **51**, 261–265.

JUDIS, J. (1966). Factors affecting binding of phenol derivatives to *Micrococcus lysodeikticus* cells. *Journal of Pharmaceutical Sciences*, **53**, 803–817.

KNAYSI, G. (1930). Disinfection. I. The development of our knowledge of disinfection. *Journal of Infectious Diseases*, **47**, 293–302.

KNAYSI, G. & MORRIS, G. (1930). The manner of death of certain bacteria and yeast when subjected to mild chemical and physical agents. *Journal of Infectious Diseases*, **47**, 303–317.

KNOX, W. E., STUMPH, P. K., GREEN, D. E. & AUERBACH, V. H. (1948). The inhibition of sulphydryl enzymes as the basis of the bactericidal action of chlorine. *Journal of Bacteriology*, **55**, 451–458.

KUHN, R. & BIELIG, H. J. (1940). Uber Invertseifen. I. Die Einwirkung von Invertseifen auf Eiweiss-Stoffe. *Berichte der Deutschen Chemischen Gesellschaft*, **73**, 1080–1091.

KUNDIG, W., GHOSH, S. & ROSEMAN, S. (1964). Phosphate bound to histidine in a protein as an intermediate in a novel phosphotransferase system. *Proceedings of the National Academy of Sciences, U.S.A.* **52**, 1067–1074.

LAMBERT, P. A. & HAMMOND, S. M. (1973). Potassium fluxes. First indications of membrane damage in micro-organisms. *Biochemical and Biophysical Research Communications*, **54**, 796–799.

LARK, K. G. & LARK, C. (1966). Regulation of chromosome replication in *Escherichia coli*: a comparison of the effects of phenylethyl alcohol treatment with those of amino acid starvation. *Journal of Molecular Biology*, **20**, 9–19.

LERCH, C. (1953). Electrophoresis of *Micrococcus pyogenes* var. *aureus*. *Acta Pathologica et Microbiologica Scandinavica*, **98**, *Supplement*, 1–194.

LEIVE, L. (1974). The barrier function of the Gram-negative envelope. *Annals of the New York Academy of Sciences*, **235**, 109–127.

LERMAN, L. S. (1961). Structural considerations in the interaction of DNA and acridines. *Journal of Molecular Biology*, **3**, 18–30.

LERMAN, L. S. (1964*a*). Acridine mutagens and DNA structure. *Journal of Cellular and Comparative Physiology*, **64**, *Supplement*, 1–18.

LERMAN, L. S. (1964*b*). Amino acid group reactivity in DNA–aminoacridine complexes. *Journal of Molecular Biology*, **10**, 367–380.

LEVINTHAL, C., SINGER, E. R. & FETHERHOL, K. (1962). Reactivation and hybridization of reduced alkaline phosphatase. *Proceedings of the National Academy of Sciences, U.S.A.* **48**, 1230–1237.

LOWICK, J. H. B. & JAMES, A. M. (1957). The electrokinetic properties of *Aerobacter aerogenes*. A comparison of the properties of normal and crystal violet-trained cells. *Biochemical Journal*, **65**, 431–438.

MADSEN, T. & NYMAN, M. (1907). Zur Theorie der Desinfektion. I. *Zeitschrift für Hygiene und Infectionskrankheten*, **57**, 388–395.

MCILWAIN, H. (1941). A nutritional investigation of the antibacterial action of acriflavine. *Biochemical Journal*, **35**, 1311–1319.

MEYER, H. (1901). Zur Theorie der Alkoholnarkose. III. Der Einfluss Wechselnder Temperatur auf Wirkungstark und Narcotics. *Archiv für Experimentelle Pathologie und Pharmakologie*, **46**, 338–342.

MICKELSON, M. N. (1974). Effect of uncoupling agents and respiratory inhibitors on the growth of *Streptococcus agalactiae*. *Journal of Bacteriology*, **120**, 733–740.

MITCHELL, P. (1961). Coupling of phosphorylation to electron and hydrogen transfer by a chemiosmotic type of mechanism. *Nature, London*, **191**, 144–148.

MITCHELL, P. (1968). *Chemiosmotic Coupling and Energy Transduction*. Bodmin: Glyn Research Ltd.

MITCHELL, P. (1970). Membranes of cells and organelles. In *Symposia of the Society for General Microbiology*, **20**, ed. H. P. Charles & B. C. J. G. Knight, pp. 121–166. London: Cambrdge University Press.

MITCHELL, P. (1972). Chemiosmotic coupling in energy transduction: a logical development of biochemical knowledge. *Journal of Bioenergetics*, **3**, 5–24.

MITCHELL, P. & MOYLE, J. (1967). Acid–base titration across the membrane system of rat-liver mitochondria: catalysis by uncouplers. *Biochemical Journal*, **104**, 588–600.

MOLTKE, O. (1927). *Contributions to the Characterisation of* Bact. proteus vulgaris (*Hauser*). Copenhagen: Levin & Munksgaard.

MOORE, C. D. & HARDWICK, R. B. (1956). Germicides based on surface-active agents. *Manufacturing Chemist*, **27**, 305–309.

NAKAMURA, K. & TAMAOKI, T. (1968). Reversible dissociation of *Escherichia coli* ribosomes by hydrogen peroxide. *Biochimica et Biophysica Acta*, **161**, 368–376.

NEWTON, B. A. & REYNOLDS, P. E. (1966). *Biochemical Studies of Antimicrobial Drugs. Symposia of the Society for General Microbiology*, **16**. London: Cambridge University Press.

NICHOLSON, B. H. & PEACOCKE, A. R. (1965). The inhibition of ribonucleic acid polymerase by acridines. *Biochemical Journal*, **100**, 50–58.

NIVEN, D. F. & HAMILTON, W. A. (1973). Valinomycin-induced amino acid uptake by *Staphylococcus aureus*. *FEBS Letters*, **37**, 244–248.

NIVEN, D. F. & HAMILTON, W. A. (1974). Mechanisms of energy coupling to the transport of amino acids in *Staphylococcus aureus*. *European Journal of Biochemistry*, **37**, 244–248.

NIVEN, J. F., JEACOCKE, R. E. & HAMILTON, W. A. (1973). The driving force for the transport of amino acids by bacteria. *FEBS Letters*, **29**, 248–252.

NORRIS, J. R. (1957). A bacteriolytic principle associated with cultures of *Bacillus cereus*. *Journal of General Microbiology*, **16**, 1–8.

NOVICK, R. P. & ROTH, C. (1968). Plasmid linked resistance to inorganic salts in *Staphylococcus aureus*. *Journal of Bacteriology*, **95**, 1335–1342.

PASSOW, H., ROTHSTEIN, A. & CLARKSON, T. W. (1961). The general pharmacology of the heavy metals. *Pharmacological Reviews*, **13**, 185–224.

PEACOCKE, A. R. & SKERRETT, J. N. H. (1956). The interaction of aminoacridines with nucleic acids. *Transactions of the Faraday Society*, **52**, 261–279.

PETHICA, B. A. (1958). Bacterial lysis. Lysis by physical and chemical methods. *Journal of General Microbiology*, **18**, 473–480.

PETHICA, B. A. & SCHULMAN, J. H. (1953). The physical chemistry of haemolysis by surface active agents. *Biochemical Journal*, **53**, 177–185.

PRITCHARD, N. J., BLAKE, A. & PEACOCKE, A. R. (1966). Modified intercalation model for the interaction of amino acridines and DNA. *Nature, London*, **272**, 1360–1361.

PROKOP, A. & HUMPHREY, A. E. (1970). Kinetics of disinfection. In *Disinfection*, ed. M. A. Benarde, pp. 61–83. New York: Marcel Dekker Inc.

PULLMAN, J. E. & REYNOLDS, B. L. (1965). Some observations on the mode of action of phenol on *Escherichia coli*. *Australian Journal of Pharmacy*, **46**, S80–S84.

PULVERTAFT, R. J. V. & LUMB, G. D. (1948). Bacterial lysis and antiseptics. *Journal of Hygiene, Cambridge*, **46**, 62–64.

RAHN, O. (1945). Factors affecting the rate of disinfection. *Bacteriological Reviews*, **9**, 1–47.

RAHN, O. & SCHROEDER, W. R. (1941). Inactivation of enzymes as the cause of death in bacteria. *Byodynamica*, **3**, 199–208.

RAZIN, S. & ARGAMAN, M. (1963). Lysis of mycoplasma, bacterial protoplasts, spheroplasts and L-forms by various agents. *Journal of General Microbiology*, **30**, 155–172.

RIBBONS, D. W. (1965). The microbial degradation of aromatic compounds. *Annual Reports on the Progress of Chemistry*, **62**, 445–468.

RICHARDSON, A. G. & LEACH, F. R. (1969). The effect of phenylethyl alcohol on *Bacillus subtilis* transformation. I. Characterisation of the effect. *Biochimica et Biophysica Acta*, **174**, 264–275.

RICHARDSON, A. G., PIERSON, D. L. & LEACH, F. R. (1969). The effect of phenylethanol on *Bacillus subtilis* transformation. II. Transport of DNA and precursors. *Biochimica et Biophysica Acta*, **174**, 276–281.

ROBERTS, M. H. & RAHN, O. (1946). The amount of enzyme inactivation at bacteriostatic and bactericidal concentrations of disinfectants. *Journal of Bacteriology*, **52**, 639–644.

ROE, E., JONES, R. J. & LOWBURY, E. J. L. (1971). Transfer of antibiotic resistance between *Pseudomonas aeruginosa* and other Gram-negative bacilli in burns. *Lancet*, **1**, 149–152.

ROGERS, H. J. (1967). Killing of staphylococci by penicillins. *Nature, London*, **213**, 31–33.

ROGERS, H. J. & FORSBERG, C. W. (1971). Role of autolysins in killing of bacteria by some bactericidal antibiotics. *Journal of Bacteriology*, **108**, 1235–1243.

ROGOFF, M. H. (1961). The oxidation of aromatic compounds by bacteria. *Advances in Applied Microbiology*, **3**, 193–221.

RUSSELL, A. D. (1971). Ethylenediaminetetraacetic acid. In *Inhibition and Destruction of the Microbial Cell*, ed. W. B. Hugo, pp. 209–224. London & New York: Academic Press.

RUSSELL, A. D. (1972). Comparative resistance of R^+ and other strains of *Pseudomonas aeruginosa* to non-antibiotic antibacterials. *Lancet*, **2**, 332.

RUSSELL, A. D. (1974). Factors influencing the activity of antimicrobial agents: an appraisal. *Microbios*, **10**, 151–174.

SALTON, M. R. J. (1950). The bactericidal properties of certain cationic detergents. *Australian Journal of Scientific Research*, B3, 45–60.

SALTON, M. R. J. (1951). The adsorption of cetyltrimethylammonium bromide by bacteria, its action in releasing cellular constituents and its bactericidal effect. *Journal of General Microbiology*, **5**, 391–404.

SALTON, M. R. J. (1957). The action of lytic agents on the surface structures of the bacterial cell. In *Proceedings of the Second International Congress on Surface Activity*, vol. II, ed. J. H. Schulman, pp. 245–253. London: Butterworth.

SALTON, M. R. J. (1963). The relationship between the nature of the cell wall and the Gram stain. *Journal of General Microbiology*, **30**, 223–235.

SALTON, M. R. J. (1968). Lytic agents, cell permeability and monolayer penetrability. *Journal of General Physiology*, **52**, 227s–252s.

SALTON, M. R. J., HORNE, R. W. & COSLETT, V. E. (1951). Electron microscopy of bacteria treated with cetyltrimethylammonium bromide. *Journal of General Microbiolgy*, **5**, 405–407.

SCHAECHTER, M. & SANTOMASSINO, K. A. (1962). The lysis of *Escherichia coli* by sulphydryl binding agents. *Journal of Bacteriology*, **84**, 318–325.

SCHULMAN, J. H. & RIDEAL, E. K. (1937). Molecular interactions in monolayers. I. Complexes between large molecules. *Proceedings of the Royal Society*, B**122**, 29–45.

SHAFA, F. & SALTON, M. R. J. (1960). Disaggregation of bacterial cell walls by anionic detergents. *Journal of General Microbiology*, **23**, 137–141.

SILVER, S. & WENDT, L. (1967). Mechanism of action of phenylethyl alcohol: breakdown of cellular permeability barrier. *Journal of Bacteriology*, **93**, 560–566.

SIMON, E. W. (1953). Mechanisms of dinitrophenol toxicity. *Biological Reviews*, **28**, 453–479.

SINAI, J. & YUDKIN, J. (1959*a*). The origin of the bacterial resistance to proflavine. III. The alleged rapid adaption to proflavine resistance in *Bacterium lactis aerogenes* (syn. *Aerobacter aerogenes, Klebsiella pneumoniae*). *Journal of General Microbiology*, **20**, 373–383.

SINAI, J. & YUDKIN, J. (1959*b*). The origin of bacterial resistance to proflavine. IV. Cycles of resistance in *Escherichia coli* and their bearing on variations in resistance in cultures. *Journal of General Microbiology*, **20**, 384–399.

SINAI, J. & YUDKIN, J. (1959*c*). The origin of bacterial resistance to proflavine. V. Transformation of proflavine resistance in *Escherichia coli*. *Journal of General Microbiology*, **20**, 400–413.

SPRAY, G. H. & LODGE, R. M. (1943). The effects of resorcinol and of *m*-cresol on the growth of *Bact. lactis aerogenes*. *Transactions of the Faraday Society*, **39**, 424–431.

SRIVASTAVA, R. B. & THOMPSON, R. E. M. (1966). Studies in the mechanism of action of phenol on *Escherichia coli* cells. *British Journal of Experimental Pathology*, **48**, 315–323.

STACHELIN, M. (1958). Reactions of tobacco mosaic virus nucleic acid with formaldehyde. *Biochimica et Biophysica Acta*, **29**, 410–417.

STRETTON, R. J. & MANSON, T. W. (1973). Some aspects of the mode of action of the antibacterial compound Bronopol (2-bromo-2-nitropropan-1,3-diol). *Journal of Applied Bacteriology*, **36**, 61–76.

STROMINGER, J. L. (1959). The accumulation of uridine and cytidine nucleotides in *Staph. aureus* inhibited by gentian violet. *Journal of Biological Chemistry*, **234**, 1520–1524.

SUMMERS, A. O. & LEWIS, E. (1973). Volatilization of mercuric chloride by mercury-resistant plasmid-bearing strains of *Escherichia coli, Staphylococcus aureus* and *Pseudomonas aeruginosa*. *Journal of Bacteriology*, **113**, 1070–1072.

THORNLEY, M. J. & YUDKIN, J. (1959*a*). The origin of bacterial resistance to proflavine. I. Training and reversion in *Escherichia coli*. *Journal of General Microbiology*, **20**, 355–394.

THORNLEY, M. J. & YUDKIN, J. (1959*b*). The origin of bacterial resistance to proflavine. II. Spontaneous mutation to proflavine resistance in *Escherichia coli*. *Journal of General Microbiology*, **20**, 365–372.

TOMASZ, A., ALBINO, A. & ZANATI, E. (1970). Multiple antibiotic resistance in a bacterium with suppressed autolytic system. *Nature*, *London*, **227**, 138–140.

TOMCSIK, J. (1955). Effects of disinfectants and of surface active agents on bacterial protoplasts. *Proceedings of the Society of Experimental Biology and Medicine*, **89**, 459–463.

VACZI, L. (1973). *The Biological Role of Bacterial Lipids*. Budapest: Akadémiai Kiadó.

VAITUZIS, Z., NELSON, J. D. JR, WAN, L. W. & COLWELL, R. R. (1975). Effects of mercuric chloride on growth and morphology of selected strains of mercury-resistant bacteria. *Applied Microbiology*, **29**, 275–286.

WAHLIN, J. G. & ALMADEN, P. H. (1939). Megalomorphic phase of bacteria. *Journal of Infectious Diseases*, **65**, 147–153.

WALSH, C. T. & KABACK, H. R. (1963). Vinylglycollic acid. An inactivator of the phosphoenolpyruvate–phosphate transferase system in *Escherichia coli*. *Journal of Biological Chemistry*, **248**, 5456–5462.

WANG, J. H. & MATHESON, A. T. (1967). The possible role of sulphydryl groups in the dimerization of 70 S ribosomes from *Escherichia coli. Biochemical and Biophysical Research Communications*, **23**, 740–744.

WARING, M. J. (1965). The effects of antimicrobial agents on ribonucleic acid polymerase. *Molecular Pharmacology*, **1**, 1–13.

WEINBACH, E. C. (1957). Biochemical basis for the toxicity of pentachlorophenol. *Proceedings of the National Academy of Sciences, U.S.A.* **43**, 393–397.

WILLIAMSON, R. L. & METCALF, R. L. (1967). Salicylanilides: a new group of active uncouplers of oxidative phosphorylation. *Science, Washington*, **158**, 1694–1695.

WOLDRINGH, C. L. (1973). Effects of toluene and phenylethyl alcohol on the ultrastructure of *Escherichia coli. Journal of Bacteriology*, **114**, 1359–1361.

WOODROFFE, R. C. S. & WILKINSON, B. E. (1966). The antibacterial activity of tetrachlorsalicylanilide. *Journal of General Microbiology*, **44**, 343–352.

INDEX